IMS Application Developer's Handbook
Creating and Deploying Innovative IMS Applications

Rogier Noldus

Ulf Olsson

Catherine Mulligan

Ioannis Fikouras

Anders Ryde

Mats Stille

ELSEVIER

AMSTERDAM • BOSTON • HEIDELBERG • LONDON • NEW YORK • OXFORD
PARIS • SAN DIEGO • SAN FRANCISCO • SINGAPORE • SYDNEY • TOKYO
Academic Press is an imprint of Elsevier

Academic Press is an imprint of Elsevier
The Boulevard, Langford Lane, Kidlington, Oxford, OX5 1GB
225 Wyman Street, Waltham, MA 02451, USA

First published 2011

Notices
Knowledge and best practice in this field are constantly changing. As new research and experience broaden
our understanding, changes in research methods, professional practices, or medical treatment may become
necessary.

Practitioners and researchers must always rely on their own experience and knowledge in evaluating and
using any information, methods, compounds, or experiments described herein. In using such information or
methods they should be mindful of their own safety and the safety of others, including parties for whom they
have a professional responsibility.

To the fullest extent of the law, neither the Publisher nor the authors, contributors, or editors, assume any
liability for any injury and/or damage to persons or property as a matter of products liability, negligence or
otherwise, or from any use or operation of any methods, products, instructions, or ideas contained in the
material herein.

British Library Cataloguing-in-Publication Data
A catalogue record for this book is available from the British Library

Library of Congress Number: 2011927093

ISBN: 978-0-08-101601-5

For information on all Academic Press publications
visit our website at www.elsevierdirect.com

Printed and bound in the United Kingdom

11 12 13 14   10 9 8 7 6 5 4 3 2 1

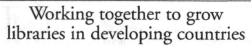

Contents

Foreword xi
Preface...xiii
Acknowledgements xvi
About the Authors xvii

CHAPTER 1   Introduction 1
   1.1 Why Was IMS Developed?..1
   1.2 Observations 2
   1.3 Network Vision: Enable and Simplify 2
      1.3.1 Billions of Mobile Handsets...4
      1.3.2 The Multi-Talented Mobile Handset...5
      1.3.3 Extending Existing Behavior 6
      1.3.4 Voice-Over IP Over Broadband 6
      1.3.5 The Mobile Phone, Boosted...8
   1.4 IMS Architecture for Those That Don't Need to Know 9
      1.4.1 Services...12
      1.4.2 The Home Network Concept 12
      1.4.3 The Residential Opportunity 13
      1.4.4 The Enterprise Opportunity 13
   1.5 Setting the Scene: The Story So Far 14
      1.5.1 IMS VoIP on Existing IP Networks 14
      1.5.2 Rich Communication Suite (RCS)..14
      1.5.3 Push-to-Talk...15
   1.6 Doing Useful Work: The Service Story 15
      1.6.1 The Communication Service Layer 17
      1.6.2 IMS and Web 2.0 20
   1.7 The Concept Applied 21
   1.8 Multimedia Telephony 21
      1.8.1 Multimedia Telephony: What Is It? 22
      1.8.2 Why MMTel – What are the Driving Requirements?.....................23
      1.8.3 Multimedia Telephony: The Origins..25
   1.9 Summary..26

CHAPTER 2   Business Modeling for a Digital Planet 27
   2.1 Introduction...27
   2.2 Basic Economic Concepts for Developers...27
      2.2.1 Economies of Scale..27
      2.2.2 Transaction Costs 28
      2.2.3 Open APIs and Transaction Costs...28
      2.2.4 Factors of Production...32

2.2.5 Capital Goods Software 32

2.2.6 Consumer Goods Software 33

2.3 Value Creation and Capture in Modern Communications Industries 33

2.3.1 The Role of the Individual in a Digital World 35

2.3.2 The Mobile Broadband Platform 37

2.4 The Business Case for IMS 38

2.4.1 Global Interoperable Standards – a Developer's View 39

2.4.2 Regulation and the Right to Private Communications.........................41

2.5 Business Models for a Digital Planet...42

2.6 Toward a Diagramming Technique 44

2.7 Practical Examples – Application to IMS 47

2.8 Conclusions...48

CHAPTER 3 Service Deployment Patterns..49

3.1 Introduction...49

3.2 Back to Basics...50

3.3 Client-Side Application 51

3.4 Server-Side End-Point Application 51

3.5 Web Server-Side End-Point Application 52

3.6 Web Client-Side End-Point Application 53

3.7 Mid-Point Application 55

3.8 Client-Side Application, Building on a Standardized Service.............................56

3.9 To-Do List...57

3.10 Summary..58

CHAPTER 4 Applications in the IP Multimedia Subsystem 59

4.1 Introduction...59

4.2 IMS Service Creation 60

4.2.1 Service Composition 60

4.2.2 Composition Through Chaining 61

4.2.3 IMS Service Chaining Architecture..62

4.3 IMS Service Composition...64

4.3.1 Initial Filter Criteria..64

4.3.2 Two-Tier Composition and the Service Capability Interaction Manager.........65

4.3.3 Unified Web Services and IMS Composition......................................67

4.3.4 Next-Generation Intelligent Networks and Migration to IMS 68

4.4 IMS Application Servers...69

4.4.1 The Converged SIP Servlet Container...69

4.4.2 SIP Application Types 75

4.4.3 SIP Application Composition in JSR116 77

4.5 Conclusions...80

CHAPTER 5   Service Development **81**
  5.1 Virtual Call Center Use-Case 82
      5.1.1 Use-Case Architecture 83
      5.1.2 Use-Case Business Logic 83
      5.1.3 Constituent SIP Applications 87
  5.2 Web-Based Do-Not-Disturb Use-Case 93
      5.2.1 Use-Case Architecture 93
      5.2.2 Constituent Components 95
      5.2.3 Use-Case Business Logic 98
      5.2.4 AJAX/SIP Interaction 102
  5.3 Conclusions 104

CHAPTER 6   Introduction to IP-Based Real-Time Communications **105**
  6.1 Introduction 105
  6.2 Basics of Voice Over IP 105
      6.2.1 Digital Speech Transmission 105
      6.2.2 OSI Reference Model 109
      6.2.3 Data Transmission Using the Real-time Transport Protocol 111
      6.2.4 Real-time Transport Control Protocol 118
      6.2.5 Control Plane Versus User Plane 118
      6.2.6 Multi-Party Communication Session 129
  6.3 Registration 130
      6.3.1 Initial Registration and Call Establishment 133
      6.3.2 De-registration 136
      6.3.3 Re-registration 136
      6.3.4 Mobility Versus Nomadicity 137
  6.4 Locating the Registrar 137
  6.5 Registration Relationships 141
      6.5.1 Subscriber Administered in VoIP Network, but Currently not Registered 141
      6.5.2 Subscriber Administered in VoIP Network and Currently Registered 142
  6.6 Network Domains 142

CHAPTER 7   Introduction to Session Initiation Protocol **145**
  7.1 Introduction 145
  7.2 The SIP Standard 145
  7.3 SIP Session Versus Media Session 145
  7.4 SIP Transaction Model 147
      7.4.1 Command Sequence 152
  7.5 SIP Transaction State Models 154
  7.6 Proxy Roles 157
      7.6.1 Stateless Proxy 158

7.6.2 Stateful Proxy 158
7.6.3 Back-to-Back User Agent 160
7.7 SIP Session Establishment 161
7.7.1 Request Message 162
7.7.2 Response Message 163
7.7.3 Initial Request Message Routing 163
7.7.4 Response Message Routing 168
7.7.5 Building an SIP Routing Path for Subsequent SIP Requests 173
7.7.6 Exchanging Contact Addresses for Subsequent SIP Requests 179
7.7.7 Subsequent Request Message Routing 181
7.8 SIP Transport Considerations 183
7.8.1 Internal DNS Versus External DNS 185
7.8.2 Reliability of SIP Requests and SIP Responses 185
7.9 Canceling a SIP Transaction Request 194
7.10 SIP Dialogs 197
7.10.1 Multiple Early Dialogs 201
7.10.2 Target Set 205
7.10.3 Early Media 206
7.11 Media Transmission: Offer–Answer Model 209
7.11.1 A Closer Look at the SDP Structure 215
7.11.2 Some SDP Examples 219

CHAPTER 8 Introduction to the IMS Network **223**
8.1 Introduction..223
8.2 Overview of IMS Standards and Releases..223
8.3 IMS Network Architecture – A Global View......................................224
8.3.1 IMS Core Network 227
8.3.2 IMS Access Network 229
8.4 IMS Network Architecture – A Closer Look......................................232
8.4.1 Core Network Entities 232
8.4.2 Network Border Gateway Nodes..242
8.5 Registration...249
8.5.1 Registration Relationships..259
8.5.2 Periodic Re-Registration and De-Registration 260
8.5.3 Implicit Registration Set...262
8.5.4 Third-party Registration 266
8.5.5 Application-initiated Registration...268
8.6 Session Establishment 270
8.6.1 Media Gating...284
8.7 Using Phone Numbers 285
8.7.1 Number Normalization 286
8.7.2 ENUM Query 288

         8.7.3 Public ENUM versus Carrier ENUM..290

         8.7.4 Phone Number Representation Through SIP URI...............................291

     8.8 Application Servers in IMS 292

         8.8.1 Introduction and Concept 292

         8.8.2 The ISC Reference Point 294

         8.8.3 Service Chaining 298

         8.8.4 SIP-AS as Proxy, B2BUA, UAC, or UAS...................................300

         8.8.5 Public Services 304

         8.8.6 Service-initiated Session Establishment.......................................312

         8.8.7 User Interaction 316

         8.8.8 Unregistered Service Invocation...320

     8.9 Messaging in IMS 324

         8.9.1 Instant Message 325

         8.9.2 Messaging Session..328

CHAPTER 9   **MMTel and Other IMS Enablers** **329**

     9.1 Introduction...329

     9.2 A More In-Depth Look into MMTel...329

     9.3 Basic MMTel Architecture...330

     9.4 Going Deeper and Wider 331

     9.5 Adding to MMTel 334

         9.5.1 ISC Chaining 334

         9.5.2 Northbound Interface...335

         9.5.3 Forwarding to Extension Logic 335

         9.5.4 Web Interfaces on the Client Side...336

     9.6 Use-Case: Calendar-Based Routing..336

     9.7 IMS Presence 337

         9.7.1 Presence as Defined by OMA..338

         9.7.2 Interacting with the Presence System..340

         9.7.3 The Presentity Data Model...343

         9.7.4 XDM Data Management 345

     9.8 Finding the right devices...346

     9.9 Conclusion 349

CHAPTER 10 Charging **351**

     10.1 Introduction...351

     10.2 Obvious and Not So Obvious Ways of Getting Paid 352

     10.3 Money Makes the App Go Around..352

         10.3.1 Selling to the End-user Through a Store 352

         10.3.2 Selling Over and Over Again..353

         10.3.3 Pay-per-use 354

         10.3.4 Advertising..354

10.3.5 Letting Someone Else do the Heavy Lifting 355
10.3.6 Sell Something Else...356
10.3.7 Count on your Fellow Man..356
10.3.8 Benefit in an Entirely Different Dimension................................356
10.4 The Mechanics of Charging...357
10.4.1 Offline Charging 358
10.4.2 Online Charging 359
10.5 Summary..362

CHAPTER 11 Interworking with Legacy Networks **363**
11.1 Introduction..363
11.2 The Bigger Picture – Connecting IMS to the Outside World 363
11.3 Interworking Through MGCF and IM-MGW 365
11.3.1 General 365
11.3.2 Protocol Mapping 367
11.3.3 MGCF SIP Signaling Capability 371
11.3.4 User-plane Interworking...376
11.4 Video Interworking...378
11.5 Supplementary Service Interworking 380
11.5.1 Calling Line Presentation and Calling Line Presentation Restriction 382
11.5.2 Connected Line Presentation and Connected Line Presentation
        Restriction...383
11.5.3 Call Hold and Resume...386
11.5.4 Call Forwarding..388
11.6 Applying Legacy VAS in the IMS Network 389
11.6.1 The Starting Point: VAS in the CS Network and VAS in the
        IMS Network 389
11.6.2 The Challenge: Safeguarding Legacy VAS Investment..................393
11.6.3 Service Capability Interaction Manager 399

CHAPTER 12 Rich Communication Suite **401**
12.1 Introduction..401
12.2 The Basics of RCS..402
12.2.1 What is RCS?...402
12.2.2 Why RCS? 402
12.3 Overview of RCS Release Functionality 404
12.4 RCS Release 1 405
12.4.1 Enriched Call 406
12.4.2 Enhanced Messaging 414
12.4.3 Enriched Phone Book 417
12.5 RCS Release 2 418
12.5.1 Broadband Access 418
12.5.2 Multi-Device Environment 419

12.5.3 Enriched Call – Multi-Device...419

12.5.4 Network Address Book...420

12.5.5 RCS Provisioning 420

12.6 RCS Release 3 421

12.7 RCS Release 4 422

12.8 RCS-e...423

12.8.1 Capability Discovery in RCS-e 424

12.9 Using RCS Applications to Capture Value 425

12.10 Conclusions...430

CHAPTER 13 Evolved IP Multimedia Architecture and Services **431**

13.1 Introduction...431

13.2 Overview of the Evolved IMS Architecture 431

13.3 GSMA VoLTE – IMS Profile for Voice and SMS...........................432

13.4 VoLTE Considerations for Service Designers 436

13.5 Single Radio Voice Call Continuity (SRVCC) 436

13.5.1 SRVCC Architecture in 3GPP Release 9.................................437

13.5.2 SRVCC High-Level Use-case Explained 438

13.5.3 SRVCC Architecture in 3GPP Release 10................................440

13.6 IMS Centralized Services (ICS) 441

13.6.1 ICS Solution with Evolved MSC...443

13.6.2 ICS Solution Using Existing ISUP/Mg and CAMEL 444

13.6.3 Terminating Access Domain Selection (T-ADS).........................445

13.7 SRVCC and ICS Considerations for Service Designers...................445

CHAPTER 14 Future Outlook: Market and Technology **449**

14.1 What is Next in Store for IMS?...449

14.2 TV...449

14.3 Smart Pipes 449

14.4 Home Networks 450

14.5 Web Clients...450

14.6 Machine to Machine (M2M) 450

14.7 Vehicle Automation 450

14.8 WAC and Other App Stores 450

14.9 Secure, Non-Anonymous Comms: The Alternative Network 451

14.10 Conclusion 451

References..453

Abbreviations..455

Index 463

Foreword

IMS – the IP Multimedia Subsystem of the 3GPP family of telecommunication standards – may very well be at the same time the worst and the best kept secret of the telecom world. "Secret" because it is essentially designed to be invisible – the modern version of the infrastructure that delivers communication to the world. "Worst kept" because it has dominated the strategies for communication evolution in the past years, and has thus been very visible, at least to those in the industry. And finally "best kept", as it is right now sneaking up to become the key technology it was built to be, with the advent of new personal communication concepts like RCS (Rich Communication Suite) and the way IMS will provide telephony services to the emerging LTE radio standard (the VoLTE initiative). Thus, telecommunication is now fully transforming itself into IP-based technology. In order to do this, some basic capabilities from the classical technology space needed to be provided, such as interoperability between peer communications providers. IMS was designed to ensure such interoperability, hinting that the "M" in IMS could just as well be interpreted as multi-operator. It doesn't stop there: IMS is also multi-access and multi-device. Interestingly, as this book shows, it is also multi-service: IMS provides the infrastructure to build and deliver all kinds of interesting, useful user features, with the added benefit of potential worldwide interoperability.

Traditionally, the focus of technology vendors has been on how to build the actual networks and the standard services that run in and on them. The view from the outside, as seen by a software developer or service provider, has been harder to find. The time has come to change this, as the industry – and essentially not just the telecom industry, but the whole converging information and communication technology space – is going through a game-changing phase. With the advent of open APIs and the new way of creating software that we see emerging, where it is natural to build new capabilities by creatively using and combining existing assets, a new way of approaching IMS is becoming apparent. With this developer-oriented mindset, the important issues are not so much how you go about building an efficient IMS network, but rather how you can use what it provides with minimum effort and maximum efficiency; i.e. the things a developer should have to know in order to build something useful and profitable should be exactly what he or she *needs* to know – hiding the details and providing the right abstractions is the key property here, in addition to all the classical attributes like performance, availability, and robustness.

Therefore, the book you have before you right now is a very timely contribution to the IMS community, aiming to give the developer an outside-in view: how applications interact with the IMS network, which of the inner workings you need to know about, how IMS can support your business model, and also – unusually for this kind of text, but very interesting reading for those of us who do not spend much of our time thinking about what we do in the language of economics – how IMS and the application ecosystem around it can be described in terms that business school graduates might want to use.

To summarize: in the following pages you will hopefully find information that will help you design your services and bring them to market. I look forward to being amazed and amused by your creative new IMS-based applications!

Håkan Eriksson, CTO Ericsson Group
San Jose, February 2011

Telephony has been with us for over a century and we have been awaiting the dawn of a new age of multimedia communications for many years. That wait is finally over. IMS, the IP Multimedia Subsystem defined by 3GPP, is set to revolutionize the communications world. Originally defined almost a decade ago, we are finally seeing a broader deployment from fixed and cable operators and of course mobile operators, spurred on by the commercial launch of LTE and by initiatives such as the GSMA's own Rich Communication Suite (RCS). Not only will RCS provide a wealth of interoperable multimedia capabilities for person-to-person communication across device and network boundaries, but it will also provide a range of new APIs to developers, to embed those capabilities into their own applications.

This is a multifaceted and complex topic, covering protocols, devices, and of course the all-important applications. Getting to grips with IMS is not for the faint hearted and that is why a book such as this one is essential. Written by seasoned industry professionals, it serves as an accessible introduction to the subject for beginners, as well as a reference work, for those already engaged in the development of multimedia services and applications.

Alex Sinclair, CTO GSM Association
London, UK, February 2011

Preface

THE REASONING BEHIND THIS BOOK

Many books have been written about IMS, so why do we think another is needed? Most of the existing books are written from the perspective of those who implement the technology, either network vendors or operators. There is no such focus for developers. The standards that form the basis of IMS are complex – as they are designed to solve complex problems – and require specialized knowledge to understand. Developing services and applications on IMS requires a different set of skills and knowledge, however, and these are generally overlooked in existing books. This book covers these aspects, from creating small applications to utilizing the full features of IMS Communication Services and RCS.

This is a unique IMS book, therefore, written not from the perspective of *building* an IMS system, but from *using* it to create new and interesting services. We base this on many years of practical engineering experience, pointing out the important bits as we go along so you can avoid getting lost in the detail. This includes a walk-through of the IMS infrastructure, but in a novel way: starting from first principles, then gradually introducing the core concepts. We also provide examples of how services are built: general service composition principles as well as standard services like Multimedia Telephony, and industry standard service profiles like Rich Communication Suite (RCS).

READER'S GUIDE

In order to help you focus on your particular interests, this list of chapters describes what subjects are respectively covered:

Block 1: The Context

1. **Introduction.** Some of the background and the basic concepts. Includes a brief introduction to what is potentially the most commercially important IMS service: Multimedia Telephony.
2. **IMS and business modeling for a digital planet.** The business context. This is a rather unorthodox chapter for a book like this, but the value it brings is that it provides some economic theory and practice. This is very useful in building more understanding of IMS as a way of supporting – and sometimes driving – current changes in the business landscape.

Block 2: The Service Developer View

3. **Service delivery deployment patterns.** Describes a number of answers to the question, "Where does an application connect to the IMS infrastructure?" Applications attach to the IMS at different points; APIs and platforms depend heavily on where your app is.

4. **Applications in the multimedia subsystem.** Basic principles for server-side application creation. This section gives an overview of modern service composition as applied to IMS/SIP-based applications.
5. **Service development.** Some concrete examples of how applications can be structured, applying the principles from Chapter 4 in practice.

Block 3: How IMS Works

6. **Introduction to IP-based real-time communications.** Building the architecture from the ground up. The chapters in this block are a bit more technical; in Chapter 6 we discuss the general technologies needed to deliver media streams over digital networks.
7. **Introduction to Session Initiation Protocol.** What you need to know about SIP. Building on Chapter 6, it discusses how SIP provides the necessary control capabilities.
8. **Introduction to IP Multimedia Subsystem.** How IMS puts SIP and other protocols into an architecture. This is where it all comes together: the logical entities in an IMS network, why they are there, and what they do.
9. **Multimedia Telephony and other IMS enablers.** A brief description of some of the key services that IMS supports. Part of the chapter describes how the IMS service architecture is applied to produce standardized services; another part shows how those standardized services can be extended.
10. **Charging.** How to make money out of IMS-based services. The basic scenarios are laid out, and an overview is given of how IMS charging mechanisms work.

Block 4: IMS Deployment and Evolution

11. **Interworking with legacy networks and services.** How does IMS interconnect with the existing telecom world? This is one of the key differentiating properties of IMS; it is designed from the ground up to work with existing networks.
12. **Rich Communication Suite.** RCS packages a set of IMS-based services to provide a rich user experience. RCS terminals and systems are being deployed as this is written; thus, it provides a good starting point for the introduction of new services building on the same enablers as for RCS.
13. **Evolved IP multimedia architecture and services.** This chapter is aimed at explaining the main new concepts and evolution of IMS supporting mobile telephony evolution. The intent is that it should provide background and create awareness of how value-added service developers need to understand this evolution.
14. **Future outlook: Market and technology.** Some final notes on where IMS might be going.

Depending on your viewpoint and needs, you may want to approach this book from different angles. Feel free to read it as you please, but we would like to suggest a couple of selections from the menu. If you are interested in:

• A general overview of the technology, see Chapters 1, 3, 4, and 12.
• Mainly the business aspects, see Chapters 1, 2, 3, and 10.

- Service design, see Chapters 1, 3–5, 12, 13, and Appendix A.
- How IMS works in more detail, see Chapters 1 and 3–13.

And then to round off, Chapter 14 may give some more food for thought regarding where this technology is going, and what your place in it might be.

But now, let's get down to the business at hand: introducing you to what IMS is, why it was designed, and what you can do with it. Happy reading!

Acknowledgements

It is an amazing experience to be involved in the development of a system within the telecommunications industry due to the sheer number of collaborations that are necessary to get things working together. So, while only six authors are involved in this book, the ideas outlined here are the result of many thousands of person-hours of discussions and engineering development. This book would not have seen the light of day had it not been for the assistance received from various colleagues within Ericsson and colleagues within the telecommunications industry at large. Both our reviewers and sparring partners have shared their knowledge and insight with us, for which we owe them a big thank you. Whilst it is impossible for us to name everyone individually, we would like to acknowledge and thank the following individuals for their contributions to the ideas outlined in this work or for their reviewing activity: Bo Åström, Christer Boberg, Gregory Bond, Martin Börjesson, Gonzalo Camarillo, Eric Cheung, Ross Demirel, Sjaak Derksen, Hans-Erik van Elburg, Göran Eriksson, Jonas Falkenå, Eugen Freiter, Carsten Garburg, Kristoffer Gronowski, Magnus Hallenstål, Henk van den Heuvel, Martien Huysmans, Roman Levenshteyn, Salvatore Loreto, Jörg Niemöller, Lennart Norell, Håkan Österlund, Marcello Pantaleo, Kari-Pekka Perttula, Per Roos, Konstantinos Vandikas, Henk van der Velden, and Patrik Wiss.

Most importantly, we owe thanks to our respective families for granting us the (evening and weekend) time to work on this project. May this book give further insight into what their loved ones are working on!

About the Authors

Rogier Noldus is an expert at Ericsson Telecommunicatie B.V. in Rijen, the Netherlands. He has been involved in Intelligent Networks (IN) standardization and has driven the development of CAMEL within Ericsson. He has subsequently made a switch to the IP Multimedia System (IMS) and is now focusing on the integration of GSM and IMS networks, covering areas such as next-generation IN, fixed mobile convergence, media transmission, multi-access, value-added services (e.g. enterprise services such as IP Centrex), and next-generation networks. He holds a B.Sc. degree (electronics) from the Institute of Technology in Utrecht (the Netherlands) and an M.Sc. degree (telecommunications) from the University of the Witwatersrand (Johannesburg, South Africa). He joined Ericsson in 1996. Rogier's telecommunications roots lie in South Africa, where he worked for Siemens, Telkor, and Telecommunications Manufacturers of South Africa (TMSA). Rogier is the author of the book *CAMEL, Intelligent Networks for the GSM, GPRS and UMTS Network* (Wiley, 2006) and is the author of various patents/patent applications in the area of Intelligent Networks, IMS, and fixed mobile convergence.

Ulf Olsson is Senior Expert at Ericsson's Business Unit Multimedia, with a main interest in application architecture. He entered the world of programming 40 years ago, and has been working on large-scale software system architectures for the last 30 years. Initially, these efforts were in the field of distributed high-performance systems for shipborne command and control, but as it turned out the principles and experiences from that field were surprisingly applicable also to the design of mobile packet data systems. He has been with Ericsson since 1996, being involved with systems like GPRS, PDC, UMTS, cdma2000, and – of course – IMS. His professional focus has recently shifted to the next level of abstraction: how to support and automate the business processes of a communications service provider. He holds an M.Sc. in engineering physics from Stockholm's Royal Institute of Technology, having also spent a scholarship year at Dartmouth College, New Hampshire. He is the co-holder of a number of patents in the mobile communications area, and is a frequent contributor to *Ericsson Review*.

Catherine Mulligan is the Transitional Research Fellow in Innovation Studies at Horizon Digital Economy Research at the University of Nottingham. She holds a Ph.D. in economics from the University of Cambridge, and an M.Phil. in engineering, also from the University of Cambridge. She received first-class honours for her B.Sc. (business information technology) from the University of New South Wales, Australia. Prior to her current post, Catherine worked for 15 years in the IT and telecommunications industries, including 10 years at Ericsson contributing extensively to IMS – in particular representing Ericsson within several standardization forums. She holds various patents in core network areas. Catherine is also the co-author of several books, including *SAE and the Evolved Packet Core: Driving the Mobile Broadband Revolution* (Elsevier, 2009), and the sole author of *The Communications Industries in the Era of Convergence* (Routledge, 2011), which investigates the economic and technical factors driving the communication industries.

Ioannis Fikouras is currently Chief Architect for Services and Software at Ericsson Research. He joined Ericsson in 2005 to pioneer the application of service composition for IN, IMS, and Internet services within Ericsson. His work produced the Ericsson Composition Engine (ECE) and other technologies. Ioannis then made the switch to the real world to work as Strategic Solution Manager for Ericsson Global Services in the area of IMS and Service Delivery Platform (SDP). He has been active as a technology strategy consultant for the European Commission Directorate General for the Information Society and other national European research organizations since 2001. Ioannis holds a degree in computer science from the University of Bremen, Germany, where he also earned a doctorate degree on service composition. He is the author of numerous papers and book contributions on service composition as well as various patents on service-oriented technologies in the telecommunications domain.

Anders Ryde is an expert in network and service architecture within Ericsson AB in Sweden. He joined Ericsson in 1982 and has a background in network and service architecture development for multimedia-enabled telecommunication, targeting both enterprise and residential users. He has been working on the evolution of IMS and IMS-based services for more than a decade, and is currently engaged in the ongoing evolution of mobile telephony service and networks to all IP and IMS. He holds an M.Sc. in electrical engineering from the Royal Institute of Technology, Stockholm.

Mats Stille currently holds an Expert position with Ericsson in Stockholm. He has a background in mobile telephony core network system management related work around standardization, network, and service architecture development including terminal aspects, and also acts as technical leader in teams.

He joined Ericsson in 1985 and started working with core network functions of 1G analog mobile telephony systems such as TACS and AMPS, but was soon pioneering 2G GSM standards and its development in the late 1980s and early 90s. He has also worked with the Japanese 2G PDC system, 3G UMTS, and 4G systems.

Mats has been representing Ericsson for four years in the GSMA/RCS committee where he was focusing on IMS core, video, and voice related services, and has been the GSMA official editor of the committee's specification on MMTel packet switched voice.

He has studied mathematics at the University of Stockholm, Sweden.

Introduction

1.1 WHY WAS IMS DEVELOPED?

The communications industry has undergone profound changes over the past decades, driven by similar economic forces across mobile, fixed, and IT/computing networks. Economies of scale and scope have stimulated equipment vendors and operators alike to pursue lower cost technologies, most often based on IP technology. It was within this atmosphere that the IP Multimedia Subsystem (IMS) was initially designed.

At its simplest, the IMS is a set of IP-based technologies that allow for ubiquitous access to multimedia services from any terminal, be it a mobile, landline phone, or PC. It is designed from the same conceptual basis as any mobile network technology, to provide global interoperability between all handsets and all operators worldwide. In addition, however, the IMS is designed to handle universal service access – wherever you are roaming in the world, your handset should provide you with the same set of services. With the IMS, this happens automatically, without the developer – or, more importantly, the user – needing to do anything; the standards handle this complexity without the developer needing to worry about anything except his or her own service logic.

Like any technology, the IMS grew out of the existing political landscape of the industry in which it was developed. This book will help explain the reasoning that went into the development of this architecture, with particular focus on how it enables developers to create innovative services. In addition, this book will explain how developers can create business models within the value chain of the telecommunications world.

The IMS was designed to provide a wide range of possibilities for service creation. This means that it covers two fundamentally different market needs: firstly, what we all expect in terms of standardized services, where we can reach anyone without having to worry what operator the person at the other end has chosen, what device he or she prefers. Secondly, the operator – possibly together with content and service partners – can choose to provide innovative, differentiating services that make that operator the most attractive choice for end-users in the market. From an operator perspective, the IMS is also about reducing transaction costs: delivering innovative services while reducing time to market. Throughout this book, we will show how we can pull these rabbits out of the IMS hat. If you wish, you can find out about some of the detailed magic that the standardization people have designed to make this happen, but if not we will guide you to what you need to know – and only what you need to know – to build applications on top of the IMS platform.

1.2 OBSERVATIONS

In a very real sense, the IMS is the response by the telecommunications community to the emergence of a number of key technologies. So, let's take a look at the key properties of any modern communication infrastructure:

- IP technology is the basis for all present and future mass market communication.
- Communications are becoming multimodal, including voice, video, presence, messaging, etc, sometimes on appliances that might not even support voice.
- End-users expect quality of service to increase (HD video, high-quality audio); this requires negotiation capabilities and a network that actively assures that this is delivered.
- The market expects rapid creation and introduction of multiple services, fully utilizing the capabilities of access network technology.

There is only one comprehensive, multi-vendor, multi-operator, internationally standardized architecture that fulfills all of the above requirements: the IMS.

A number of initiatives have appeared over the years that address one or more of these requirements. However, they tend to miss various parts of the puzzle: for instance, voice-over-IP (VoIP) solutions that assume pure IP connectivity have no means to assure the proper bearer handling in resource-constrained networks such as cellular systems. Other examples are systems that work well as long as you need to reach only the ones that have made the same technology selections as you, e.g. with a Skype client, you can talk to another Skype user but not one using Windows Live. In a sense, therefore, the IMS is about evolutionary necessity – once you reach a certain number of end-users on a technology, a global standard becomes more cost effective and more reliable.

However, as history has repeatedly shown, it is not always the perfect technology that prevails, but rather the alternative that is capable of attracting real market support. In the case of the IMS, this means being an attractive development platform. We will show that IMS clearly has the right qualities to be the tool of choice for application developers. In particular, we explore how the advanced and varied features of the IMS can be packaged and exposed simply and effectively to the developer and user communities.

1.3 NETWORK VISION: ENABLE AND SIMPLIFY

How does the IMS network support the vision described above? In the preceding section, we talked about future communications networks in terms of the need for them to allow everyone to choose their operator freely, and still achieve global reachability, i.e. multi-operator support. Figure 1.1 illustrates two more such "multi" aspects. Firstly, **multi-access**: all

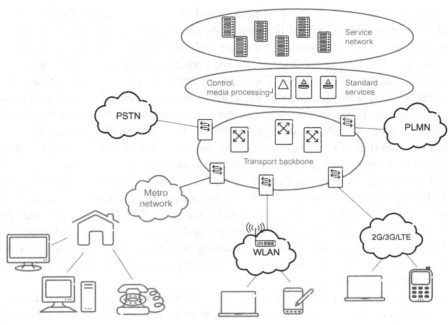

FIGURE 1.1

Reference architecture.

services can be delivered over all access forms: fixed, wireless, even legacy devices. Note that this does not mean that we need to bring all services down to the level of the lowest common denominator; if a certain access has capabilities that the service can utilize, it is allowed to do so. The negotiation capabilities and ability to announce intentions that allow this are a fundamental part of IMS.

Secondly, Figure 1.1 illustrates the **multi-device** aspect. Users are allowed to choose whatever device they want, and at the same time expect the communications system to adapt to whatever device – or indeed devices – that they have chosen to be reachable on at any given time. Again, negotiation and the ability to signal capabilities are essential to support this level of end-user choice. It should be noted that we do not link the term 'devices' to the connectivity mechanism that they are using to access the network. A mobile phone nowadays can be attached via WiFi, making it essentially part of the fixed access world. To blur things even more, it could actually still provide telephony services that the user perceives as plain old cellular services, using bridging architectures like UMA (Unlicensed Mobile Access; the ability to use a WLAN to connect a phone to the GSM/WCDMA core network). Conversely, a laptop can be connected over mobile broadband, which means that even though its services are defined by PC expectations, it belongs to the wireless side as far as connectivity is concerned.

You might argue that the property of being used by a wide variety of devices is not that unusual: after all, a basic property of IP technology allows for the delivery of anything to anyone over any access. An IP connection, however, only identifies a network connection point – an IP address. The IMS, however, focuses on the end-user or the individual that you want to reach, irrespective of their IP address. The network finds the correct *person* that you wish to establish a connection with and selects the correct device, even when multiple devices are active. As an example, it will not attempt to establish a video session to a mobile handset that does not support video. The IMS is designed to handle this complexity seamlessly.

The IMS ensures that the level of complexity associated with handling all the differences and combinations of operator, access, and handset manufacturers is not something an application developer needs to concern themselves with – it is handled automatically on his or her behalf. The application designer is left to concentrate on the actual logic of the service. As will be seen later, this happens on two levels: the core IMS network provides functions concerning the finding, authenticating, charging, and managing of the end-user. "Finding" across network boundaries using a uniform naming scheme is the key concept, as the main role of the network is to provide reachability, "find-and-connect" being the basic service as experienced by the end-user.

The second level – the service layer, as it were – adapts, personalizes, and delivers content. The interplay between user, subscription, devices, bearers, and media flows is extremely interesting, because this is where person-to-content and person-to-person services overlap. This illustrates rather vividly the aspect that IMS is not just another voice replacement technology (although that certainly is a very viable use), but enables a much richer set of opportunities. New approaches are needed to fulfill the promise of new access technology without creating an unmanageable tangle of options and restrictions for the end-user. Therefore, the two key value propositions for the IP multimedia subsystem (IMS) are essentially what was stated in the heading of this section, enable and simplify: "enable" as it supports such a wide range of access, service and interconnectivity options; "simplify" as it provides one and the same infrastructure across all these technology variants.

Before we get into more technical detail, at this point it might be useful to examine a few of the relevant current technical trends.

1.3.1 Billions of Mobile Handsets

While a handset is the most personal device imaginable – it is always with you, it is the way you are reached as a person, it is becoming the preferred way of keeping in touch with your social network(s) – it is also important to realize that for the majority of people, a mobile device will define how they use the Internet. This level of personalization provides opportunities for tailoring services and service behavior to the needs and wishes of the user, as well as the means to provide context and user profile adapted content. Mobile device identifiers do not identify a place (as is the case with classical phone numbers) but rather a person. We can thus associate several attributes to a person by way of the identity of the device.

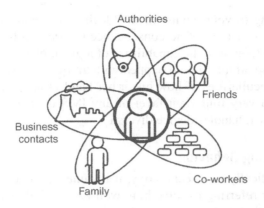

FIGURE 1.2

Relationship circles.

Extending the argument above, it is also very clear that communication occurs in many different contexts. You want to keep in touch with your inner circle – family and close friends – as well as with acquaintances, colleagues, business contacts, shops, companies they deal with as private customers, health services, officials of various kinds, banks, and so on. Applications that are created must be able to manage these different contexts in a coherent and understandable fashion without trying to squeeze them all into one mold. Additionally, the mechanisms that keep track of these partially overlapping circles cannot be application specific: each new application should not need to be told who your friends are. Conversely, a set of contacts you want to keep track of – say, the parents of your child's classmates – is not defined by what applications they prefer, but rather from the contact network that you share. Coupling this idea of multiple spheres of influence (Figure 1.2) with the desire to keep information from leaking from one circle to another, you realize that the communications infrastructure should also allow you – when you so desire – to use different identities for different purposes. The IMS provides such mechanisms in a standardized and consistent fashion.

1.3.2 The Multi-Talented Mobile Handset

These days, a mobile handset is so much more than merely a phone. It is a camera, a calendar, a music player, a radio, a note-taker, a voice memo recorder, a game console, and a credit card. And on top of that, mobile Internet devices are also increasingly general-purpose computing platforms. Indeed, it is just about anything you might carry in your pocket except possibly a handkerchief and a comb (but of course a mobile with a front-facing camera works quite well as a mirror!).

However, as mobile application developers have discovered – often painfully – over the years, the diversity of platforms and device capabilities is also a major problem: either you produce literally hundreds of build variants of your application, or you figure out some way of detecting and adapting to your environment. Ideally, your code should not have to consider platform differences that are of no consequence to your application; at the same time, you also want to be able to exploit the capabilities of a given platform that can give your app that extra edge. Of course all this should happen without the end-user ever needing to understand or deal with the peculiarities of how his or her handset happens to be built in terms of low-level detail. Thus, a very important facet of any IMS device-side application development environment is how it handles device diversity.

1.3.3 Extending Existing Behavior

When implementing radically new technology, it is quite important that users can relate to the new functionality by referring to something well known. Telephony may be an (almost) ancient concept, but the find-and-connect and human-centric way of thinking about communication is the basis for both the SIP protocol and the IMS. Of course, the IMS provides all the relevant support for well-known addressing schemes. Telephone numbers (known in the trade as "E.164"), and mail-style addressing must be supported. Additionally, it helps tremendously if there is someone to contact from day one; thus, interoperability with existing networks is a must. It is, however, most certainly not restricted to the "I-call-you" model; the IMS also encompasses many other forms of communication popularized by the ubiquitous social networks: publish-and-subscribe for presence-style posting of what I want the world to know about me; instant messaging – multimedia enabled, of course – for chatting with friends and people with shared interests, etc. More importantly, however, it can be used in conjunction with the existing social networks – it enhances web technologies by providing a global standard for multimedia connectivity.

1.3.4 Voice-Over IP Over Broadband

Voice-over IP (VoIP) has clearly moved from being a technical possibility to a market reality. New entrants and incumbents like Skype, Vonage, and TeliaSonera have implemented first-, second-, and third-line telephony services with considerable market impact.

Why has this happened now? Obviously, the main enabler is the availability of ubiquitous broadband access, over whatever medium that happens to be convenient (xDSL, cable, Metro Ethernet, fiber, mobile broadband). Also, the fact that the average (even the below-average) PC of today has the capability to handle the media plane without resorting to special hardware has helped move this area from the hardware to the software domain.

Assume, then, that you are signing up for a VoIP service. Then you try to phone a friend, and find out that he or she hasn't signed up for that vendor, but s/he has an ordinary phone – can't you try that? With a bit of luck, your provider will allow you to dial out – for a fee – to the public telephony system (Public Switched Telephony Network: PSTN, if you are into alphabet soup, and as you will see we are – but we will try to explain everything as we go along). But should your friend want to call you, you want to be reachable, so – for a higher fee – you can also typically get a phone number to post on your Facebook page.

But can't you just get your friend to download the same software? Of course, but she might have already selected another, and you don't really want to give up your choice, do you? And cluttering up your entire desktop with clients seems a bit unnecessary (and even more cumbersome if you try to do it on a small screen device). So you end up connecting, but via a PSTN link connecting the two IP networks, delivering 3.1 kHz audio. And that seems such a shame, when both of you were going digital. Again, IMS comes to the rescue, having figured out all the details of how you connect not just within a network, but also between them, including also how you interconnect with PSTN when needed.

At this point, we should also point out that a key value of IMS – one that we will return to – is the fact that it is deliberately built on trusted authentication mechanisms. Between the network and the user, authentication mechanisms (based on SIM cards for the mobile side) are used to ensure that the person connecting is who they say they are; between peering networks it is based on IPSec and contractual trust relations. This is actually one of the ways to combat unwanted and/or malicious communications; it is not nearly as attractive to misbehave if you can't be anonymous.

In a sense, what we have described above is an interesting illustration of the fact that communications solutions are not created by protocols alone, but by careful development of architectures in which the protocols operate. In addition to providing the framework for multi-access, multimedia, multi-operator, multi-vendor (important for the operators!) and multi-device operations, it is also designed from the ground up for massive scalability (billions of connected devices), with the possibility to charge for services. The latter is still an important factor: one way or another, infrastructure must be paid for. However, this does not necessarily mean the traditional "subscriber pays" model: it could be sponsored, ad-financed, revenue shared, or any other interesting new business model. Whatever the case, the network still needs the ways and means to keep track of what is delivered to whom and what is consumed in the process. Again, this is something the standardization groups have considered.

A word of caution while we are still on the subject of VoIP: a fixed operator can typically consider implementing voice over IMS as a way of delivering telephony at lower cost, phasing out obsolete legacy equipment and replacing it with IP-based infrastructure. One way of achieving this is to connect legacy end-user equipment through various kinds of gateways converting from IP connectivity to classical two-wire analog technology. The end-user should not really notice anything: the same services and the same devices are being used (see Figure 1.3). Recently, this strategy has been mimicked by the creation of the VoLTE way of delivering voice over LTE (which does not have any circuit-switched bearers in its arsenal). Interestingly, the "invisible change" property implies a very interesting set of requirements; of course, reliability, scalability, and performance of the new way of doing business must match or preferably exceed the classical methods. This means that IMS had to be designed to allow instant very-large-scale implementation, something not easily achieved when creating new technology.

Now, if voice were all we wanted to do, the problem would be much simpler. However, voice-over IMS is not called VoI (or whatever); it is known as MMTel (Multimedia

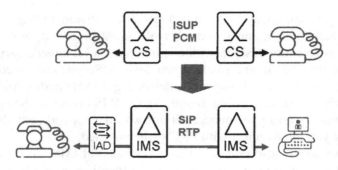

FIGURE 1.3

Replacing the switch.

Telephony). Of course, it can deliver voice only, but as the name implies the whole control machinery is available for you to add other media streams to the call. Thus, you can start with something fairly simple, but the machinery is ready for bigger and better things, such as the applications we expect you to create, once you have finished reading this book!

1.3.5 The Mobile Phone, Boosted

Having considered fixed access above, what about mobile multimedia? In a sense, there are no fundamental differences between the access forms (disregarding speed, latency, and error rates for the time being). However, the mobile device has evolved over the past few years, from a simple voice device through messaging (SMS) and simple email to a veritable pocketful of entertainment and information options. So, it does not come as a surprise that you might want to use those capabilities to add to the communication experience. Extending the concept even further, you may well compare the resulting device with something coming from a directly opposite direction: any kind of consumer electronics is nowadays likely to be upgraded with the ability to talk to its surroundings, turning it into a CCE (Connected Consumer Electronics) device. At some point, it will be tricky (and maybe irrelevant!) to decide if a device is a phone with a camera or if it is a camera that you happen to be able to talk into.

All this obviously couldn't happen if it wasn't for the almost unbelievable improvements in processing power, battery life, display capacity, user interface flexibility, bit rates, etc. Downloadable applications – and the app stores to find them in – have appeared to exploit the opportunity to treat the handset more like a general-purpose computing platform than a phone. However, note that in order to manage essential capabilities like, for example, secure authentication and audio/video streams, you still need platform support in the device, as improper use of such functions may jeopardize the stability and integrity of the device, or in extreme cases the network itself. And as the network is in a very real sense – particularly in the mobile case – a shared resource, it needs to be protected.

FIGURE 1.4

The multi-service mobile.

Developers do not want to worry about all of this, in the same way as the user expects the device platform to handle other mundane chores. Again (pardon the repetition, but we really like this part), IMS helps with this.

Above, we have discussed the mobile scenario in terms of its multi-service aspects. The key to being able to create, implement, and deploy a new service is to get as much for free as possible: this means that the threshold is low; thus, it is cheap to try many things. This is good news for service providers, but also for long tail developers: you can concentrate on the logic of your own application and let the infrastructure take care of the heavy lifting. In particular, this means that the great opportunity is most likely not in inventing yet another VoIP client, but rather the ability to add communication capabilities to *your* application: a space exploration game, a fly-fishing site, a training tool for auto mechanics, etc.

1.4 IMS ARCHITECTURE FOR THOSE THAT DON'T NEED TO KNOW

In this section, we will consider a few of the key things underpinning IMS. Don't worry, the details you need to know we cover in Chapters 6 and 7. Your application – client side, network side, or both – will, however, essentially only interact with an IMS network through end-point APIs. Here, we will provide a high-level overview so you have some idea of what makes the machine tick. So, down the rabbit hole we go, using Figure 1.5 as a map!

IMS is built using SIP as the core signalling protocol. SIP is used on the UNI (user-to-network interface) when the device communicates with the infrastructure, and in most interfaces between the involved network elements. Basically, SIP supports routing of signalling messages between the end-points, allowing the devices to negotiate things like codec choices. The IMS architecture builds on this by harnessing SIP in a framework of network elements that employ various SIP routing tricks to fulfill different roles regarding service triggering, authentication and authorization, media plane resource invocation, network inter-connect, topology hiding, scalability, resilience, and many other essential network properties.

- The basic call/session control function (CSCF) node, which in essence is the purest SIP server in the IMS architecture, can be configured for three different roles: serving (the registrar and SIP router), proxy (managing the access), and interrogating (single point of entry to the network). One of its most interesting functions is to ensure the correct application servers handle a session (or, indeed, combination of ASs).
- Application server(s) providing the capability to initiate, terminate, and/or modify sessions. This is where network services are actually executed (or at least the control plane part of the services).
- Media plane resources (MRFP), providing services like media translation, conferencing, and voice/video mail.
- Control of the media plane resources (MRFC).
- Session border gateway, providing protection of network resources and subscribers, letting through traffic from authenticated sources only, and ensuring that service-level agreements are honored. Also, the SBG supports resolving issues like Network Address Translation (NAT) in the signalling path.

A feature of IMS that turns it into the "missing link" between existing networks and a fully IP- and SIP-based environment is that it has been designed to integrate into the existing cellular infrastructure. It does so by relying on:

- The HSS (including the HLR) to provide subscriber information. One of the essential pieces of information is the IFC (initial filter criteria) – this is what the serving CSCF uses to match requests to the correct AS.
- The SIM card to assist in verifying the end-user, device, charging, media gateways, and signalling gateways.

Figure 1.5 can be extended in two very interesting ways. First, in Figure 1.6 we add the capability of locating the application server on the other side of a business-to-business (trust border) interface.

Second, if you take two half-networks like the one in Figure 1.5, you get the result shown in Figure 1.7. This shows how that the same infrastructure can support person-to-person communication (from one user device to another) as well as person-to-service/content – where the network side of the session is terminated on an application server inside the network, or delivered through a gateway to a partner service provider. Of course, the initiative can come from the network side; IMS faithfully sets up the session in either case.

A key feature of IMS is that it was designed to fit into evolutionary scenarios. It is very hard to motivate oneself (or one's boss, as the case may be) to invest in a shiny new phone, if a call is interrupted or completely lost when you move out of the downtown area where

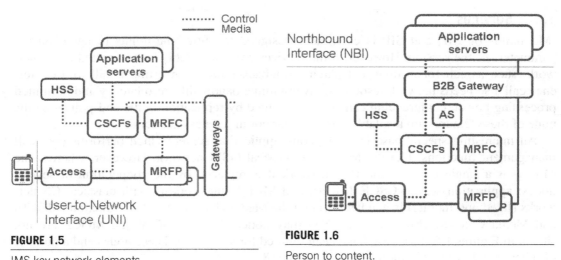

FIGURE 1.5

IMS key network elements.

FIGURE 1.6

Person to content.

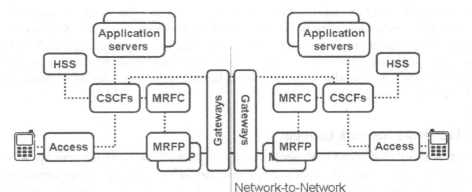

FIGURE 1.7

Across operator borders.

the equally shiny new service was available. Therefore, IMS was designed to manage the interaction with legacy cellular networks, handing over calls in both directions as appropriate. This also includes supporting SMS: launching something that looks and behaves like a mobile without supporting SMS would be a very unlikely proposition indeed.

One of the interworking variants of interest to a network-side application developer is the case where a user is "anchored" in the IMS domain, regardless of what technology and what device is used. That is, call control is always handled in the IMS domain. This means that applications are always invoked from the IMS side, and implementation is that much simpler.

1.4.1 Services

IMS builds on SIP, and SIP is essentially designed to allow end-points to agree on the parameters of the session. However, some services are unavoidably best executed in the network, such as call forwarding and playing ringback tones. Converting a call to a conference call also needs network resources: given infinite bandwidth, zero latency and unlimited processing power, conferencing can be implemented by replication in the end-points. Sadly, none of these features are available, in particular not in wireless networks.

An interesting special case is a telephony application server, which performs core call management functions, like the MSC in a classical GSM network. However, whereas the MSC was a single node, in the IMS network it is replaced by the cooperation of several nodes: application server, CSCF, MRFC, and MRFP. Interworking with classical CS networks (fixed and mobile) is handled through the Media Gateway Control Function (MGCF) and Media Gateway (MGW) nodes. You may notice that the MGW and MGCF are not shown in Figures 1.5–1.7; these have been omitted for simplicity. These nodes and a number of others are described in more detail in Chapter 8.

It must be stressed that the 3GPP IMS architecture is expressed in terms of logical network elements. Actual products can come in many shapes and forms: full-scale deployments can involve multiple racks per node (or node cluster); small startup configurations may combine most nodes into a single preconfigured cabinet. Either way, starting small or large, an operator can easily scale up from market trials to mass-market scenarios. As an application developer, you will essentially never see the difference, except in terms of traffic volumes and – hopefully – revenues.

1.4.2 The Home Network Concept

If you have travelled with a GSM phone, you will be well aware of the concept of roaming (i.e. being served by a network other than the one owned by the operator you signed a contract with). Typically, you become aware of this at the end of the month, when you realize that the charges levied can differ drastically depending on where you happen to be. Another effect of more interest to us here is that the so-called supplementary services in GSM (and, unless noted specifically, when we say GSM here it applies equally to WCDMA) are executed in the visited network. This has caused major interworking issues; different networks have different capabilities, different equipment, etc. Attempts have been made to fix this (e.g. the CAMEL protocol versions), but in IMS a radical approach was taken: all signalling is always brought to the home network, and all services are executed there. For the application developer, this means that the interfaces do not change, regardless of where the user is. And the customer – the IMS user – gets his or her services in the same way wherever they happen to be. Note that this applies to control signalling: media typically takes the shortest path available in order to avoid latency. It might have been prudent to start this section to say that "Home" here is not where you hang your hat (as the Bard described it), but rather the network directly associated with the operator that you signed a contract with. "Home" in the sense we normally use it is of course called something a bit more complicated in this context, as described in the next section.

1.4.3 **The Residential Opportunity**

A typical home these days hosts a number of communication-enabled devices: mobile phones, TV sets, PCs, Internet radios, WiFi-enabled picture frames, connected power meters, etc. According to many prediction scenarios we have only seen the beginning of this evolution. Conceivably, you don't want to buy a new communications subscription every time you add a device to your home WLAN, so what is needed is some kind of demarcation point between the controlled external network (provided by your ISP) and the more ad-hoc structure of your home devices. Even so, you would want an incoming video call to be transferable from your mobile – where you picked it up – to the HD TV + camera combo in the home. And you certainly want routing to be smart enough to avoid delivering the call to the voice-only phone in the study. Different strategies exist to achieve this: termination in or near the end-user device (set-top boxes), or termination in a residential gateway (RGW), which then acts like a bridge between IMS and – typically – a DLNA[1] home network. Note that the RGW can be a physical box, or a simple L2 gateway managed by an RGW control plane part running in a cloud somewhere.

An access and service anchor point like the RGW may seem an unnecessary complication in a home network. However, it is in an ideal position to be the point where the operator's responsibility ends, clearly defining who can control what parts. If the home owner wishes to hand over some control (agreeing that the operator is allowed to see the internal home network configuration), an operator can actually sell home network management as a service, based on the presence of the RGW. Many mass market households may actually prefer to hand over the responsibility of ensuring that their home network is safe and functioning well, as it typically requires a large and growing skill set. This space is actually another business opportunity for application developers: the trend for gateway and set-top devices is to incorporate frameworks like OSGi[2], making it possible to deploy additional functionality as desired.

1.4.4 **The Enterprise Opportunity**

What about enterprises? IMS provides a large number of features that are very attractive for an enterprise user. To mention a few: strong authentication, same services over multiple device types (including standard cell phones, fixed phones, SIP phones, etc., within the limits of their respective capabilities), interworking with classical networks, advanced routing, multimedia and multi-service capabilities. Also, being IP based it is comparatively easy for an operator to provide hosted PBX functionality ("IP centrex"), providing excellent economies of scale. For the moment, that is about as much as we are going to say about enterprise applications, knowing full well that this market is likely to be one of the more important segments.

---

[1] www.dlna.org
[2] www.osgi.org

1.5 SETTING THE SCENE: THE STORY SO FAR

In the next few sections, we will give some examples of services that were and are being built on IMS.

1.5.1 IMS VoIP on Existing IP Networks

This is at present one of the areas with the most commercial activity. A number of fixed-line telephony operators are faced with the need to rejuvenate and improve their networks, partly in order to be able to deliver new services, but also to replace aging equipment. Moving to an IP-based transport infrastructure can mean radical improvements in CAPEX and OPEX spend; moving to an IP (SIP)-based control infrastructure is just carrying this through to its logical conclusion.

A number of entrepreneurial operators have trialled this for a number of years, in order to gain experience and innovator credentials. It actually works, and for a number of years has provided essential feedback. Now, with the advent of LTE (Long-Term Evolution, the latest step in the continuously evolving 3GPP architecture) and later versions of HSPA (3G terminology for High Speed Packet Access, sometimes known as Turbo3G), the bearers and control mechanisms are in place to do this for real.

Various experiments were also done early with VoIP over WLAN. Again, in overprovisioned networks (i.e. networks that are idle for most of the time, with only a small risk of congestion), this works well, but as experience has shown, good quality directly implies the need to categorize and prioritize traffic.

The "real" way of delivering voice and many other media streams has been standardized by ITU TISPAN and 3GPP in cooperation: Multimedia Telephony. However, this is important enough to warrant its own section (section 1.8).

1.5.2 Rich Communication Suite (RCS)

This initiative originally arose from a group of operators who realized that person-to-person services needed to be ubiquitous in order to succeed, at least in regions (also a recurring theme in this book). They then proceeded to prioritize and package up a number of basic functions that would constitute a compelling set of end-user features. Originally, these features were presence, messaging, and file/image/video sharing, and the concept of a network-based, active address book. And, of course, voice. In the first release, the voice part for cellular use is based on normal CS (circuit-switched) bearers, but in RCS 2.0 this is augmented by MMTel (as it is already in 1.0 for fixed access). This concept of mixing CS voice with packet-based enrichments is sometimes called Combinational Services, and was a cornerstone in working out how to get IMS to the market even before the radio IP bearers were really up to full-duplex, low-latency speech.

RCS is actually an interesting example of how different organizations cooperate and build on each other's result to create the end user experience. IETF provided the basic protocols, 3GPP did the core architecture (and, with a detour through ETSI, TISPAN, MMTel), OMA did the actual services such as presence, group management, and messaging.

1.5.3 Push-to-Talk

A number of years ago, push-to-talk was generating considerable interest, particularly in the USA. Nextel was generating ARPU (average revenue per user, for obvious reasons, one of the most interesting performance indicators for a service provider) way above the average, and with very low churn rates to boot (churn is the rather odd name for when users decide for various reasons to switch operators). A PoC (push-to-talk over cellular) service was standardized in OMA (Open Mobile Alliance), based on common IMS service enablers and infrastructure, such as group, list and presence management, multi-party conferencing, security, charging, and O&M. For a number of reasons the market uptake has so far been very low. It seems like PTT is not necessarily as popular outside the USA; also, some players decided to launch proprietary implementations in order to get market share. In a sense, this in itself is a proof point for the theory that interoperability and an open business model is a prerequisite to reach mass market status: fragmentation may be appropriate in experimental phases, but is very detrimental to the creation of technology with true global reach.

1.6 DOING USEFUL WORK: THE SERVICE STORY

So far, we have been discussing the fundamental core and access basis for IMS. However, now the time has come to start approaching the service development side of the house: the main focus of this book. What we need to do then is – a standard trick in the software trade – to raise the level of abstraction, hiding unnecessary detail and focusing on what is actually of use to an outside observer, or in this case an application developer. Thus, the way we would like you to see IMS as a developer is not the daunting expanse of logical network elements you might find in the 3GPP specs, but rather a black box view (Figure 1.8).

In this way, a compact and easily understandable API is all developers need to worry about, allowing them to concentrate on the actual application. As with all simplifications, this needs to be elaborated further, but as a fundamental frame of reference it is important to carry this simple notion with you as we go along. Please note that the figure shows only half the truth (we try to spare your sensibilities at this early stage), i.e., half of the network (the originating half, if on the caller side). Connecting two such halves provides a whole connection, but more of that in Chapter 3.

Now, the first elaboration is on the API itself. Is it a homogeneous, one-level entity? No, of course not. Simply put, we have so far discussed issues up to the session level in the stack in Figure 1.9.

At this point, we need to consider where application developers want to be in this scheme of things. The access level is typically too detailed, and somewhat defeats the purpose of being able to write widely deployable applications. Getting in on the SIP level – issuing and

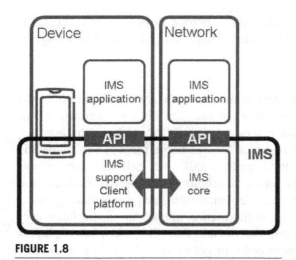

FIGURE 1.8

IMS as a black box.

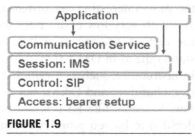

FIGURE 1.9

The abstraction stack.

responding to individual SIP messages – gives full control, but at the cost of having to read (and understand!) an amazing amount of 3GPP and IETF standardization. SIP and its entourage of associated RFCs is very comprehensive and capable; one may be tempted to wonder why this is all necessary and whether there is an easier way to do it. There is, but it has nothing to do with the protocol syntax itself. Note that syntactically, SIP is a direct descendent of HTTP; the differences derive from solving a different problem – symmetric, asynchronous for SIP vs. asymmetric, request-response for HTTP. The purist might note that SIP manages sessions as well as IMS (the next step up the ladder, discussed in the next paragraph – and, of course, in almost all of this book); the rationale behind the figure is essentially that the sessions on SIP level can be seen as technology tools that support the more user-oriented sessions on the IMS level.

Instead, the solution must be found in raising the level of abstraction. Using Java as an example, JSR180 is the Java API for SIP in Java ME (the device-oriented profile of Java). However, if you don't really want to tweak individual control messages (which may have interesting side-effects in the core network, if you are doing things not catered for in the standards) but instead want to invoke basic IMS procedures like registering, initiating a call, and subscribing to a set of events, then you use JSR281 (IMS Services API).

Figure 1.10 illustrates how a simple game could be built on JSR281-style abstractions. Note that the details of setting up the session are completely hidden from the application; all it sees is a notification like "OK, session is up" (for those of you who wonder what an IARI is, all will be revealed in the next section).

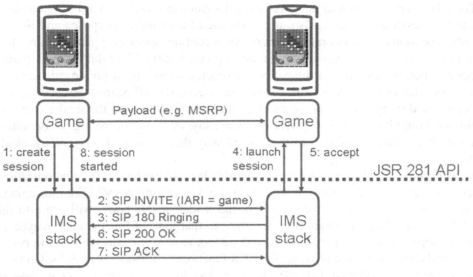

FIGURE 1.10

An application using JSR281.

1.6.1 The Communication Service Layer

Going up the stack in Figure 1.9, what do we find at the Communication Service level? An even higher level of abstraction, of course. We are now at a level where things happen that are of direct use to a human user: making a video call, posting a presence update, starting a game, etc. The "only" thing that needs to be added is a user interface (and of course some application-specific logic!); these days, this is typically something very graphical or touch-based. Now, once you start thinking in terms of multi-service devices, it becomes obvious that some sort of concept on top of the basic session is necessary. If an IMS terminal receives an incoming INVITE message (relax, in later chapters we will give you all the details about how this works), how can the base platform figure out what to do with it; i.e. what other software component should be invoked? Note that this is where the asynchronous aspects of person-to-person communications really have an effect on architecture: in a classical client–server setup, the user is always the initiator (even though this may be technically hidden behind the scenes, resulting in polling applications). But in telephony or messaging, it is the person at the other end that initiates; your device must be ready to respond whenever something comes in.

In order to fix this issue, 3GPP has defined a way for the participating devices to do the right thing. Essentially, it means that session signals are tagged with an identifier (ICSI: IMS Communication Service Identifier). Not only does it make it possible to route signals in the end device; it also means that the right application servers can be linked in along the signal path. Thus, both end-points and the core networks can agree on "the kind of behavior we all expect as this session evolves". Also, it can be used for charging purposes and SLA monitoring when sessions cross operator borders: an accepting operator typically wants to know what resources an incoming session might end up consuming. Note that in most places (the USA being a notable exception) the caller is expected to pay for a phone call; it is reasonable to expect the same behavior for a multimedia call. But IP bearers are usually paid for by the connected party, so unless we can flag that this session (and the media flows associated with it) should be accounted differently from, say, everyday browsing the operators will have a very hard time explaining to their users why their bills look very different all of a sudden.

Now, as Figure 1.9 indicates, we have yet another layer to explore. The communication service layer was useful in and of itself, but once this kind of capability (MMTel, messaging, presence, etc.) is available, smart people like you, the reader of this book, will begin to think in terms of "hmm, if I had a programmatic interface to that kind of functionality, maybe I could use it in my app?". And of course this can be done. For example, contributions have been made to the Android community that provides the RCS functional components as APIs. On the server side, the SIP Servlet standard in JSR289 allows services to be built that can act as end-points in MMTel sessions, for example, making it possible to build interesting applications such as highly interactive audiovisual customer service portals. The interested reader has now already deduced the following dilemma: if we can now install an application that can use MMTel as a service, how do we make an incoming INVITE with an MMTel ICSI end up in my virtual curling game rather than in the MMTel dialler client? Again, the answer lies in identifying the session signals properly, this time by something called IARI (IMS Application Reference Identifier). The only thing you need to know at this stage, however, is that there are ways to keep your signals coming to you, and not someone else's app installed on the same device (or server). The namespaces have been designed so that they are derived from domain names; if you have control over at least a part of a DNS namespace (which you do if you own a domain), you are home free.

The detailed story is described in 3GPP TS 24.229. It identifies a developer's application regardless of whether it operates on top of a standard communication service as "bearer" (i.e. in conjunction with an ICSI) or directly on top of the IMS core (no ICSI). This identifier is used both by the network and by the device for different purposes.

An example of an application identity would be a service-urn like this:

```
Urn:urn-7:3gpp-application.ims.iari.worldsbestappcompany.poker
```

It is carried by the IMS infrastructure by a feature tag called +g.3gpp.iari-ref (3GPP TS 24.229, based on IETF RFC 3840/41). This feature tag in turn is included in the SIP header Accept-Contact. For example, when the inviter activates the application, his or her device will translate that into an SIP signal like this:

```
INVITE tel :+12129876543
Accept-Contact:*; +g.3gpp.iari-ref= "urn%3Aurn-7%3A.3gpp-application.ims.iari.
worldsbestdeveloper.poker".
```

Note: The "%3A" replaces ":" when put in the context of a feature tag according to IETF rules.

The network uses this identifier to find a matching invitee device to deliver the application to. A matching device is an invitee device that has installed the same application as the inviter and registered that to the network.

If the application offer includes a standard service, identified by the IMS communication service identifier (ICSI), that the application uses as "bearer", the network links in the SIP-AS that is associated with that ICSI, otherwise not.

Although the network has the ability to link in a SIP-AS also for an application invitation operating on the IMS core only (no ICSI included in the offer but only an IARI), it is strongly recommended that the developer group assigns (and register over the API, e.g. worldsbestdeveloper-applications) their own proprietary ICSI instead, used in the 3GPP ICSI feature tag, e.g.:

```
+g.3gpp.icsi-ref="urn:urn-7:3gpp-service.ims.icsi.
worldsbestdeveloper-applications"
```

Then the network operator can define an IFC (Initial Filter Criteria) for this proprietary ICSI linking in an SIP-AS, in case server-side support is needed.

Figure 1.11 shows how incoming SIP signals can be routed to the right service handler, based on the ICSI tag (providing the necessary matching of device protocol behavior to server protocol behavior). If no IARI tag is present, the default application behavior is executed (e.g. the call is delivered to a telephony GUI, if the incoming signal was an MMTel

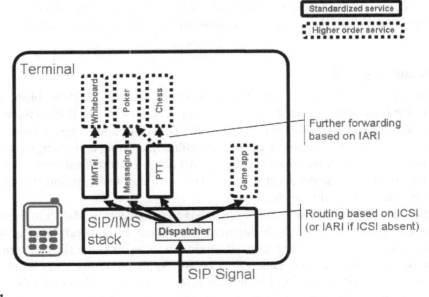

FIGURE 1.11

Delivering a signal inside the terminal.

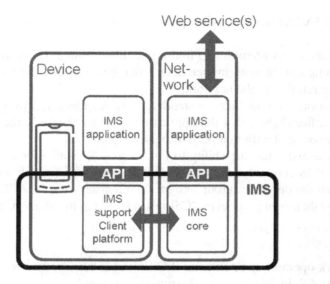

FIGURE 1.12

Exposing network capabilities as web services.

INVITE). The IARI tag is used to ensure that applications using standardized services as components actually get control over the session. As an example, an application using IMS messaging to convey in-game state changes would like to avoid having those messages turn up in the user's mailbox! For comparison, think of the tagged SMSs used in over-the-air activation.

1.6.2 IMS and Web 2.0

We have already mentioned REST style APIs as a way of accessing IMS enablers. This leads us directly to the interesting – indeed, essential – subject of how to link IMS to the web world. A few years ago, some people saw this as a conflict between parallel approaches; we would much rather see it as another example of the web utilizing existing and emerging capabilities as simply and efficiently as possible, leaving the details to be sorted out behind simple and straightforward web APIs the details still need to be sorted out, which means that there is still enough for all of us to do in this world.

So, another addition to the very simple black box view appears as shown in Figure 1.12.

In some cases, this may mean transparent translation of network-level abstractions. However, it is much more likely that the web services will be on much simpler semantic levels, corresponding to the Communication Service and Application levels in Figure 1.9. Such efforts are under way within the GSMA OneAPI project, in cooperation with OMA.

At this point, it may be useful to consider how this makes IMS link in nicely to another modern system design concept: the Service-Oriented Architecture approach (SOA).

Exposing IMS capabilities in web service form makes them available as components in service composition machinery; this idea will be explored in Chapters 4 and 5. These chapters cover in detail how a developer can use the IMS to creative innovative, breakthrough applications for the emerging digital economy, as is discussed in Chapter 2.

1.7 THE CONCEPT APPLIED

A typical example of the total concept may go like this (as implemented in a project involving one of the authors). The basis was an existing multi-player poker game, where the developers wanted to enhance the gaming experience with a real-time voice. They quickly realized that this is a non-trivial exercise if you want to do it from the ground up. Instead, it was built using the push-to-talk capabilities in a pre-standard implementation on JSR281. The result was a much more compelling gaming experience, with limited developer effort (a few weeks from idea to deployment).

1.8 MULTIMEDIA TELEPHONY

One of the most important services in IMS – probably, most would consider it the defining use – is multimedia telephony, in most parts of this book abbreviated to MMTel. As has been explained earlier, the IMS core itself does not provide telephony; it is essentially a find-and-connect machine (as you should know by now, it does do a lot more, but we will keep things simple for the time being). Multimedia telephony, as the name implies, is the IMS equivalent of telephony as we know and love it, from GSM and PSTN (ISDN), for example, but set within the context of an IP transport network.

MMTel in its purest form is basically a SIP application server (SIP-AS) connected using the ISC interface from the S-CSCF. The standard procedures apply (of course), i.e. an incoming session (the initial INVITE) is matched against the IFC rules for this user; a properly tagged INVITE is then routed to the right AS instance. This can be statically assigned per subscriber, or – in more advanced implementations – based on allocation decisions made when this user registered. As MMTel fills the role of classical telephony, it must also behave like a customer would expect telephony to do: the basic services as well as what is known as supplementary services, e.g. call diversion on various criteria, or more advanced services like VPNs and black-list/white-list. The standard service set in MMTel is actually richer than in any previous telephony system, building on many decades of experience as to what appeals to private and enterprise customers. Many features that are realized through value-added services (VAS) in legacy circuit-switched networks are standard features in MMTel; thus, there will be considerably fewer vendor-specific variants. Also, given that MMTel is essentially modeled upon the internationally standardized GSM/3G network capabilities defined by 3GPP, this level of commonality and common behavior is now also brought to the fixed environment. As fixed-line telephony evolved for the most part in circumstances of monopoly PTT, each country tended to evolve its own specific services, protocol variants, etc. With the advent of GSM and roaming, this was obviously no longer practical (a mobile must work on all networks); also, this level of standardization is what created a global market for handsets with such volumes that prices could come down to levels that enabled the current situation where essentially the entire globe is reachable. Now, with MMTel the corresponding level of standardization and commonality is also possible for fixed-line devices. This is particularly important for advanced new features like multi-terminal support, multiple user identities, and parallel alerting.

As in essentially any IMS service, MMTel is, by design, access network independent. Hence it may be used over wireline access, wireless LAN, and even packet data cellular mobile access. The latter case will be particularly prominent with the rollout of LTE (3GPP Long-Term Evolution), as LTE does not implement any circuit-switched bearers. Thus, voice for a LTE device can be implemented two ways: either by MMTel and SIP, or by various ways of leveraging existing infrastructure. But maintaining multiple infrastructures is not terribly smart for an operator, and multiple radios in a mobile device drive cost and power consumption. Hence, LTE does really mean MMTel. And as we have already pointed out many times, in an IMS context, many different kinds of media are supported. Therefore, even though we refer to voice telephony as the model for how MMTel works conceptually, we need to reiterate that audio is a special case here: any media flow can be negotiated using SIP.

1.8.1 Multimedia Telephony: What is it?

We now consider some more details about what defines MMTel. We have hopefully established the main premise that MMTel is a new global standard based on the IP Multimedia Subsystem. But what can it really do? As it turns out, it can perform a number of interesting tricks. Among other things it can be used for:

- **Voice, video, and chat in real time one at a time or combined.** The session concept is essentially independent of what kind of media is transported.
- **Sharing media (video, images, and files).** In addition to the continuous media streams that are the main reason for having a session established, associated media can be transferred between end-points, "piggybacking" on the established session. The benefit of this concept is somewhat more profound than may be thought initially; the result is a user experience where addressing is implicit rather than explicit. For example, consider making a phone call, then realizing that you want to send a picture of what you see to whomever you are talking to. Technically, it is of course possible to fire up a camera app, take a picture, start the mail client, figure out the email address of the person you talked to, dig up the correct picture ("was it DSC0347 or DCS0374?"), mail it, and hope that your friend hasn't gotten bored and hung up. Instead, the ability to associate file, image, and video clip transfer to an ongoing voice session means that the user experience is simply a "take and transfer picture" button in the GUI.
- **Converged fixed and mobile multimedia services.** This is really one of the key issues. MMTel is designed to work across as many devices as possible: mobile phones, fixed IP

phones, PC clients, even legacy phones (through TAs – terminal adapters). Additionally, IMS was not defined in a vacuum: the ability to call out to – and be called from – legacy networks (note the subtle difference between connecting a legacy *network* and a legacy *device*) is absolutely essential to make the whole thing a viable business proposition.

- **Users can add and drop media streams during an ongoing session.** It was mentioned above that a session can have any combination of media in it. But it doesn't stop there. The media handling is fully dynamic: if the devices have the right capabilities, media streams can be added and removed as appropriate. The negotiation functionality built into IMS/SIP means that devices can inform each other about what options are available; therefore, the end-user experience will be the best possible given the circumstances, not a safe lowest-common-denominator approach.
- **Users can add and drop other parties during an ongoing session.** The definition of MMTel also includes how to set up, extend, and shrink conference calls. And of course all this includes the multimedia dimension, and multi-device, multi-access, multi-vendor, and multi-operator. By now, it should be clear that we should all be grateful that MMTel knows how to deal with all this; all a developer needs to know is how to attach to it.

1.8.2 Why MMTel – What are the Driving Requirements?

In the previous section we provided an outline of what MMTel is. But wait: aren't you supposed to state the requirements before you design it? Absolutely. But as we are all engineers, the previous discussion has given us the context that we need to understand the requirements. So, very briefly, there follows a short list of properties that IMS-based multimedia telephony needs to have:

- **Provide a framework for many services.** As mentioned above, MMTel is defined to have a number of compelling services built in from the beginning. However, innovation does not stop there; it is essential that new multimedia telephony services can be added simply and efficiently. Note that this is in addition to the inherent multi-service properties of IMS, where many different application server types can be provided to all users, sharing the same core.
- **3GPP Long-Term Evolution (LTE) radio access only supports packet-switched access.** LTE was designed from the ground up to be fully optimized for high-bandwidth, low-latency, high-efficiency packet data traffic. Thus, it is essential to be able to provide an efficient and interoperable voice service on top of IP.
- **Provide a service on top of the IMS core.** It may be self-evident, but of course it is important for IMS to have compelling applications. An infrastructure with nothing using it is not very useful.
- **Phase out circuit-switched technologies, replacing them with an all-IP multimedia solution.** Many operators are faced with costly maintenance and/or replacement scenarios for their existing telephony switches and transport backbone. If investments have to be made, they should of course be towards technology that brings additional benefits.

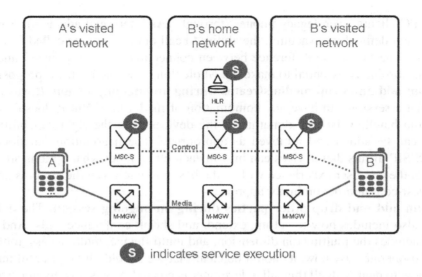

FIGURE 1.13

Mobile circuit-switched telephony service topology.

- **Cost reduction.** The cost of deploying and operating an IMS-based telephony network must be significantly lower than that of the legacy technology it replaces. Much of this is realized simply by the conversion to IP-based technologies like SIP and RTP (allowing sharing of IP-based infrastructure for many uses, and capitalizing on the rapid evolution in this space), but in addition the ability to leverage media plane handling for many different purposes is essential.

Before we go any further, we need to make a brief comparison between circuit-switched and IMS-based telephony (specifically, the mobile version). Figure 1.13 shows how a call is delivered from one roaming phone to another: neither are in their "home network", i.e. the network belonging to the operator that they have signed contracts with. The call signalling – indicated by the thin arrow – is first routed to the home network (based on analyzing the called party's number). Then, the home network figures out – based on location information in the HLR – where the target is, and forwards the signalling there. Once the connection is established, the involved servers determine which media resources need to be told to set up the required media paths.

Any services executed for this session are taken care of by originating and terminating Mobile Switching Centres (MSCs), i.e. in the visited networks. This means that in order to have consistent service behavior, all networks essentially have to implement the same things. With the advent of CAMEL, the home network could handle some service logic, but even then compatibility and implementation level led to slow progress in terms of service diversity.

A design feature of IMS is that service invocation is always in the home network; visited networks are always restricted to providing access only. As Figure 1.14 shows, signals

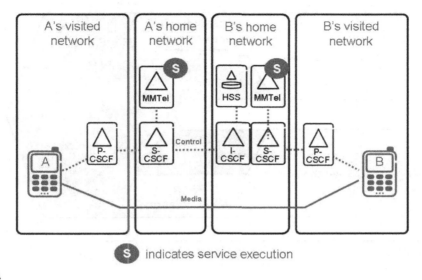

S indicates service execution

FIGURE 1.14

IMS telephony service topology.

are routed through both home networks (calling party home network as well as called); this ensures that a user always gets the same service logic he or she signed up for, even when roaming.

Note that in this case media do not pass any intermediaries, as long as no extra processing (such as conference bridging) is required. That is, media routing can be optimized. Also, note that although this example has the MMTel application server as the invoked service, the same pattern applies to any IMS service.

1.8.3 Multimedia Telephony: The Origins

The example in the preceding section talked about mobile IMS telephony. However, MMTel is designed to be useful both in mobile and fixed scenarios, with a very active cooperation having taken place between the ITU's TISPAN and 3GPP. As Figure 1.15 indicates, MMTel is the common target, but the path to get there – and the drivers for that migration – is different. Let us consider that for a moment:

- Fixed:
  - Growth of broadband and the decline of the traditional PSTN drive the evolution of the wireline narrowband (NB) networks.
  - The opportunity to have a converged service offering for the migration of POTS to new access types such as Fiber/xDSL and fixed wireless broadband access drives an IMS telephony solution.

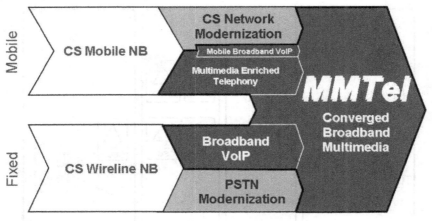

FIGURE 1.15

Converging to MMTel.

- In the longer term, the close-down of the PSTN network to reduce operational cost will emerge as a driver.
- Mobile:
  - The 3G/GSM telephony service will remain as the voice service for narrowband CS access for the foreseeable future.
  - Mobile CS networks are generally being modernized by introducing MSC servers and media gateways to replace monolithic switches, and by replacing PCM-based transmission technology by IP transport. This in itself makes introduction of fully IP-based systems easier, but also makes it obvious that if the media plane has gone IP, the control plane should also be modernized to embrace IETF protocols.
  - By introducing RCS, this will be extended into enriched telephony by adding operator-provided presence, active address book, and various IMS-based multimedia enrichments.
  - Finally, the introduction of LTE, which is a PS-only access, drives SMS over IP and IMS telephony to replace SMS and CS telephony.

1.9 SUMMARY

This chapter covered the purpose of the IMS – to merge the best properties of the extremely successful classical telephony model with the wide new range of possibilities that IP-related technologies make available. The following chapters cover how a developer can capture value from the IMS within the emerging economy.

Business Modeling for a Digital Planet[1]

2.1 INTRODUCTION

A common statement regarding IMS is the apparent lack of business models associated with it, that it might be a "solution in search of a problem to solve". This chapter is dedicated to dispelling those myths and aims to highlight the "evolution of value" within the communications industries and shed light on the emerging global economy. A further motivation for the present chapter is to extend the discussion of IMS beyond the widespread view of it being merely a replacement for SS7. Rather, we view IMS from the perspective of the role it has to play in the world economy, both in the capital goods markets and as a key enabler of breakthrough consumer products and services. From this point, we then discuss the business case for IMS and how to develop business models that leverage IMS within the context of a world economy that is rapidly becoming dominated by digital technologies.

2.2 BASIC ECONOMIC CONCEPTS FOR DEVELOPERS

As a developer, it is important to understand a few basic aspects of the economics driving different parts of the industry. These concepts will help to sell applications and hopefully help to clarify what sort of vocabulary venture capitalists and similar are looking for when you are talking with them. This chapter is based on several different veins of economic theory, bringing together several complex concepts. We only have space for a brief review of the relevant aspects of theory in this section. We have provided suggested further reading in the footnotes for anyone interested in more detail.

2.2.1 Economies of Scale

"Economies of scale" is a term used regularly within the communications industries. It refers to the idea that as you produce more of a particular item, the average cost of producing each one decreases. This is because the costs of production can be spread over a greater

---

[1] This chapter is based on research into business models for the digital economy by Dr Catherine E. A. Mulligan and Dr Robert Houghton, funded by RCUK through the Horizon Digital Economy Research grant EP/G065802/1.

IMS Application Developer's Handbook: Creating and Deploying Innovative IMS Applications.

number of goods. For example, Ericsson is able to split the research and development costs for a Long Term Evolution (LTE) radio base station across a large number of operators worldwide. Reducing the cost of each individual base station means that more operators can buy a system, thus increasing the spread of the technology worldwide.

Economies of scale are therefore often associated with larger companies that have a regional or global reach and are able to access a greater number of potential markets for their products. Also, economies of scale often lead to the creation of oligopolies, whereby several large companies form the dominant "core" of an industrial structure. This could arguably be considered the case within the mobile network industry, with only three major equipment manufacturers (Ericsson, Alcatel-Lucent, and Huawei) and typically only a small number of main mobile operators in each country (e.g. Vodafone, T-Mobile, and O2 in the UK).

Leveraging products from companies with economies of scale means that a developer can produce applications more cheaply than they would otherwise be able to do. By developing applications on platforms that impose lower costs upon operators and others to buy means that those applications will have the greatest reach possible.

2.2.2 Transaction Costs

Transaction costs, first defined by Coase in 1937, are essentially the costs of doing business within the economy. They may be viewed as the costs of exchange of goods and services – for example, the costs of establishing contracts between two companies, or the costs of obtaining information about goods or services.

Consideration of transaction costs is particularly important within the mobile communications industries, especially with regard to roaming agreements between operators. Roaming agreements allow end-users to use another operator's radio network + core network when they have moved out of their own coverage area, and are based on complex legal documents that define service level agreements between operators. The cost of establishing such contracts (identifying which operator to create roaming agreements with, and the legal cost of creating the contract) may be viewed as a form of transaction costs.

Such arrangements also reduce the transaction costs for the end-user; with roaming agreements already in place, a user can travel with his or her mobile device and not need to establish a new contract with a mobile service provider when they arrive in a new country. They are able to use their existing contract with their service operator, albeit at a higher cost.

2.2.3 Open APIs and Transaction Costs

Transaction costs are probably the most important for developers to understand, in particular as they relate to the Open API phenomenon experienced within the communications industries during the past few years. Every startup has an Open API strategy, as have many well-established companies.[2] A key question to ask, therefore, is why has this become so important in the last few years in comparison to the previous decades.

As any software developer knows, an API functions as a technical boundary – an interface that allows the developer to connect different modules together in order to more effectively develop software, without the need to create everything from scratch. An API allows them to reuse software that other people have developed and (hopefully) tested before. An Open API, however, serves a much wider purpose within the digital economy than any normal API used internally within a product. Open APIs act to reduce transaction costs: they act as both a technical and an *economic* boundary.

Let's investigate a little exactly what this means – let's say you wish to develop an application for providing book recommendations to someone who is entering a bookstore. In order to provide relevant recommendations, you would need to understand the user's age, their tastes in books, and what books they have already read. You may potentially wish to provide book recommendations based on their tastes in music, films, and also what books their social contacts have read and left positive reviews for. You may also wish to provide reviews from public websites and perhaps select ones that are from the same age group as the user in question. In addition, it might be nice to direct the end-user to the correct location in the bookstore for the books that they have reviewed online or have created a reading list for.

In order to implement such a service, you would require access to social network information, the end-user's location, the layout of the particular bookstore they are in, the RFID information for the books in question, and some personal information about the individual end-user. In the days before Open APIs, a developer faced quite a problem and significant costs, as illustrated in Figure 2.1:

- Information gathering in terms of which bookstores, mobile devices, publishing houses, and social networks would be interested in such an application.
- Development of business relations with these companies and the time associated with explaining the concept of the application and why it would benefit the companies (as well as the end-users).
- Contract negotiation and the associated legal costs and Service Level Agreements (SLAs).
- The establishment of Non-Disclosure Agreements (NDAs).

These are all transaction costs – i.e. the cost of doing business. Before Open APIs, they were necessary evils for development of applications that brought together many different data sources. This, however, was extremely expensive and remained the domain of large companies that could afford to develop relationships in this manner.

As the communications industries have moved towards a participatory value chain, however, end-users have gained significantly more control over the selection of services that they can access from their mobile devices. As increasing amounts of data are made available

---

[2] See, for example, labs.ericsson.net, or http://openapiservice.com/

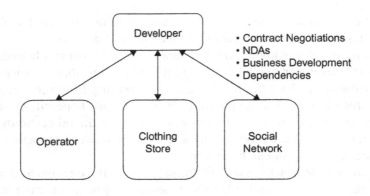

FIGURE 2.1

Open APIs between several actors.

through smartphones, sensors, RFID, and social networks, it is extremely difficult to know what combination of these will be successful or liked by end-users. This leads to uncertainty in the development of the market around these data sets. Open APIs are the market's response to this uncertainty. An Open API acts to reduce the costs of establishing a new market around the emerging technologies and platforms of the digital economy.

This is due to rapidly changing technology and platforms – for example, Google Android recently outstripped Nokia's Symbian platform in terms of market share in February 2011.[3] These gales of "creative destruction" (Christensen, 2001) create a large amount of risk for those companies and developers involved in these industries. Which platform should they support? Which social network will provide longevity for their application? Which mobile internet devices will be around the longest? In the current industrial structure, many of these are hard to predict. There is therefore great risk associated with such decisions.

Open APIs, however, remove the cost of establishing these sorts of relationships and contracts. An Open API allows a connection to be established without ever needing to meet the company in question.[4] Open APIs therefore reduce the transaction costs associated with establishing a market. In addition, Open APIs mean that developers can move relatively easily between companies – for example, the application in the above example could be extended to include other social networks, or another smartphone model. In addition, the book stores in the example would benefit from having many developers create a number of different applications. Open APIs therefore create a market for innovative capacity as well, as illustrated in Figure 2.2.

---

[3] http://www.nytimes.com/2011/02/01/technology/01android.html?ref=technology

[4] It should be noted that the use of Open APIs does not remove the necessity for contracts altogether; contracts are still required in many other aspects of business, but they are not required for the development of a loosely coupled relationship based on the exchange of data. Developers may essentially "date" many companies through their Open APIs before settling down to establish a longer-term relationship with one.

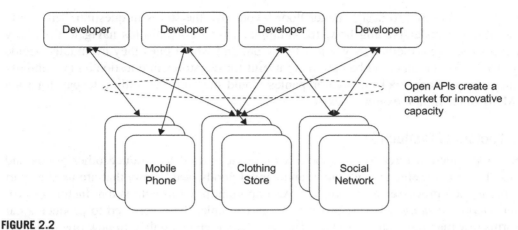

FIGURE 2.2

APIs and Market Creation.

From a developer's perspective, therefore, Open APIs work to reduce transaction costs within communications industries through reducing the number of contracts that a developer needs to implement in order to create an innovative application (Mulligan, 2011a, b). These Open APIs allow developers to take advantage of the economies of scale of both the platforms developed by companies such as Google, Ericsson and IBM, and the scale of network operators (Mulligan, 2011a).

So far, so good – but how about funding it? You have all this great information and, thanks to Open APIs, you can create the service from the data that companies have kindly agreed to share with you – but how are you going to make money? You can always go with the advertising model, but perhaps there is another method to create some revenue. Applications such as the one described above open up several different revenue opportunities, as there are many different economic actors involved in the process. If we take the example of a retail scenario – one where a developer is guiding a potential customer through a store – there are several different methods to charge for this service.

The most direct method is of course to charge the end-user for the application and/or use of the application. Alternatively, using app stores such as the ones created for handsets running the Android Operating System, you are able instead to use an advertising revenue stream if you wish to provide your app free to end-users. Another alternative exists, however, in the reuse of the data from a number of end-users. As a developer, your application will (hopefully) be very successful and downloaded and used by as many people as possible. This aggregated data[5] could be used to improve the efficiency of a store by several means.

---

[5]There are several issues with the use of personal data from end-users. While it is beyond the scope of this chapter to discuss these in detail, it is recommended that a developer investigate methods of protecting an end-user's personal data through enabling them to opt out or through other technical means.

Firstly, it could be used to design better floor layouts for the stores in question. This could be done through understanding how, for example, 18- to 24-year-olds navigate their way through a store, what clothes they select to try on, and which ones they eventually decide to buy. This information could be extremely useful for the stores in question and essentially help them move more stock; many companies would therefore be willing to pay for such data. Many other similar examples exist.

2.2.4 Factors of Production

Factors of production refer to commodities or services used to produce other goods and services. These may refer to land, labor, and capital goods (see below) that are used as part of the production processes of a company. An important point to note is that factors of production *remain unchanged* by the process. As an example, the labor used to produce a car transforms raw materials into a product. The labor can then be applied to new raw materials to create a new car. The raw materials, in this case aluminum or steel, are fundamentally changed by the process and cannot be reapplied.

The changing nature of the factors of production, specifically with regard to capital goods, within the modern global economy is fundamental to understanding how to create business models in the digital era. This will be discussed in greater detail in the subsequent sections.

2.2.5 Capital Goods Software

Capital goods are those goods used in the intermediate processes of creating other products for consumption – for example, manufacturing equipment. Within the communications industries, capital goods were one of the original types of software produced, e.g. IBM's mainframe technology (Campbell-Kelly, 2003). Capital goods software is software that is used in the production of other goods or even the creation of other software. As the industry has progressed over several decades, capital goods software has become one of the most important aspects of any company's efficiency. For example, Oracle, SAP, and IBM provide software platforms that are used within corporations to increase their efficiency and effectiveness. Since the introduction of such systems during the 1970s, these platforms have evolved to form the basis of much of the global economy as it functions today. The platforms that each of these companies develops have also slowly become open platforms, allowing for rapid integration between different systems. The large number of mergers and acquisitions between companies seen during the 1980s, 1990s, and 2000s were, to some extent, enabled through the development of open platforms that allow companies to rapidly and simply integrate their systems with one another. These platforms now form the basis of a large part of the world's supply chains and business operations. Most companies identify these technical systems as "critical" to their operations (McKinsey, 2011).

From a developer's perspective, there is therefore a significant financial opportunity in building systems that are able to increase the efficiency and effectiveness of organizations.

In the following section, we discuss the evolution of value within the communications industries and how a developer can use this evolution to create significant value in relation to the evolution of capital goods software. This is in preparation for Chapter 5, which discusses how to use IMS to develop applications within a capital goods software context.

2.2.6 Consumer Goods Software

At its simplest level, consumer goods are those purchased by end-users that they then "consume". There are many different examples, including mobile phones, TVs, or any range of different devices.

Consumer goods software became important within the communications industry with the development of the PC industry and the development of small packages of software that an end-user could buy off the shelf. The best-known examples are products like Microsoft Office, and Lotus Notes. Consumer goods software has now made its way to mobile devices in the form of the many different apps available for operating systems such as the iPhone and Android. This is just the start for consumer goods software on mobile devices, however.

Consumer goods software, in contrast to capital goods software, can be viewed as a very personal thing – today, smartphones are able to capture a great deal of information about an end-user that previous generations of mobile devices simply could not. They are able to transmit this data to cloud computing environments for processing due to the reduction in the cost of bandwidth (back to those economies of scale again).

This brings up the inherent role of the end-user within the value chain of modern communications industries. As end-users gain more choice and ability to influence the nature of applications developed on mobile devices, the manner in which value is created and captured by companies and developers working in these industries changes.

The following section provides a brief overview of these changes in order to place the discussions that follow in subsequent sections and chapters in context. IMS fulfills a certain role within the "spaces" created by the changes at work within the global economy today.

2.3 VALUE CREATION AND CAPTURE IN MODERN COMMUNICATIONS INDUSTRIES

Communications industries have undergone significant industrial change since 2005, in particular with the advent of mobile Internet devices that allow end-users to connect directly to Internet and web services, rather than having to rely on operator-based services. More importantly, however, the introduction of the mobile broadband platform has also triggered changes within the wider global economy. Understanding these changes is likely to be important in constructing a business model capable of remaining viable within this changed economic environment. This section therefore provides a high-level background of the evolution of value as a precursor to the discussions about business modeling that follow in section 2.5.

Until now, interest in "business models" for IMS has in fact tended to focus largely upon the *business case* for operator procurement of IMS system. However, relatively little attention has been placed upon *business models*, i.e. understanding how profit can actually be generated through the use of IMS. One explanation of this apparent conflation of business case and business model is that both were so strongly implicit within the industrial structure of the industry as to lack clear points of differentiation. Since the 1960s an operator would purchase a system, implement it, and recoup its investment through providing voice services to end-users. This evolved very little over the decades, apart from the fact that the lion's share of research and development and interoperability testing came to be handled by equipment manufacturers such as Ericsson and Alcatel-Lucent (Ljungberg, 2010). The model worked extremely well for development and rollout of both GSM and WCDMA.

With the development and rollout of LTE and Enhanced Packet Core (EPC),[6] however, the move to an all-IP network is unleashing another series of technical and market pressures that are now fundamentally redefining the industrial structure of the communications industries. Operator networks are being challenged to become open platforms for mobile application development. This was first seen with the popularity of the iPhone and the end-user's ability to directly access Internet services in an easy-to-use manner. Initiatives such as WAC[7] or GSMA One API[8] are also indicators of this trend. Essentially, in the same way as computing platforms form the nexus of contracts within the ICT industries (Economides and Katsamakas, 2006), mobile network platforms are becoming the nexus of contracts within the telecommunications industry (Mulligan, 2011b).

In conjunction with the development of smartphones, end-users now have significantly more control over their selection of services and service providers. They can use Skype on their mobile device, or they can use Google Maps for location-based services. The mobile broadband platform created by LTE and EPC is therefore best viewed as a *participatory value chain*. Understanding the role of the end-user and their interaction with the mobile broadband platform is therefore increasing in importance for actors in the telecommunications industry; it is no longer enough just to provide an excellent voice service on a mobile handset.

This is only one part of the story, however, as it is not just handsets that are now connected via mobile radio technologies. Cheap radio technologies and ICs delivered by companies with significant economies of scale have led to a broad-scale consumerization of technology.[9] Everything from cars to the apples you buy in a supermarket are now in some way reliant on semiconductor, radio, and computing technologies. The mobile broadband platform therefore actually represents a sea change in the nature of value creation and capture not just within the communications industries, but the wider economic system. The mobile broadband platform is not just the nexus of contracts for the communications industry, but is rapidly becoming *the nexus of contracts for*

---

[6] It is beyond the scope of this book to cover EPC or LTE in any detail; two excellent references in this area are Olsson et al. (2009) and Dahlman et al. (2008).

[7] http://www.wholesaleappcommunity.com/default.aspx

[8] http://oneapi.aepona.com/

[9] Otellini, CEO of Intel, 2008.

the whole world's economy (Mulligan, 2011b). This is an unprecedented, and somewhat unexpected, change in the nature of the boundaries of the firm (Penrose, 1995) and transaction costs within the global economy (Coase, 1937). Our planet is now truly digital and so is our economy.

Even if IMS is therefore viewed as merely voice replacement technology on a low-cost IP infrastructure, the nature of communications industries' value chain has changed so fundamentally that it seems unlikely the same business model – one operator selling voice services to end-users – will remain viable in itself outside a wider set of offerings. While voice and SMS are still the main revenue streams for the operators,[10] this situation is rapidly changing as data traffic is starting to outstrip voice traffic in operator networks (Eriksson, 2010).

At first sight this may present a somewhat depressing scenario for the future of the telecoms industry. Market share is being hollowed out by technologies readily available on the Internet, which are ostensibly "free" to the end-user. Operators are reduced to "bit pipes", or at best "smart pipes", where they form partnerships with content and other service providers in order to try to capture some nominal value from being platform providers. Equipment manufacturers, meanwhile, face a volatile market for their service layer products (Vestberg, 2010) and appear to be relegated to providers of radio and core network technology with rapidly declining operating margins in comparison to pre-2005 (Verwaayen, 2010).

However, within the challenges shifting industrial structure have created for the telecoms industry also lie a set of opportunities in terms of a series of new "spaces" for the telecommunications and communication industries to enter; through convergence, neither of these industries functions within a vacuum any longer; its role as a fundamental part of the world economy is now unavoidable for operators, and network vendors and developers alike. With industrial turbulence comes risk, but also great opportunity. Success in the mobile broadband platform now lies with understanding the flows of information in and around the world economy. Network vendors and operators are rapidly becoming a core part of global commodity chains in the same way as IBM and Oracle did in the mid-1990s. This means that profit now comes from understanding the flows of information that form the basis of the global economy, rather than just providing a voice system or service to end-users.

As will be discussed below, the core technologies of IMS provide one of the most powerful bases available for this evolution of the global economy. Quite inadvertently, the telecommunications industry has created an enormously powerful technology for business and consumer enablement in the modern digital economy.

Within the emerging industrial structure, there are two main areas of value that need to be understood: the new role of the individual and the role of the mobile broadband platform. The following two sections provide high-level overviews of these nascent areas.

2.3.1 The Role of the Individual in a Digital World

One of the most challenging economic developments to understand and come to terms with is the enabling effect technology has upon the individual and how this might conceivably

---

[10] Alcatel-Lucent (2010). Why IMS? White Paper.

change their relationship with market structures. Users are now able to interact with one another in a manner previously unimagined not just in the digital age since the development of the semiconductor in the 1960s, but in the 250 years that have been defined by the capitalist market economy.[11] Examples of this are starting to emerge in concepts like crowd sourcing, where a group of people contribute their knowledge or labor to a common task. The most commonly cited example is Wikipedia, where many people contribute to create a living encyclopedia. The transition of the individual from a mere consumer to a participant and in some sense a peer alongside existing business in value creation constitutes a potential challenge for established economic literature. This section provides a brief overview of these changes and their impact on business modeling.

In general, consideration of the individual has had a limited role within economic literature. Typically, individuals have been viewed as either "faceless" workers in a factory or a company, or as the equally faceless "rational man" who acts as the consumer of the products those companies produce (Galbraith, 1958). Workers were generally viewed as factors of production in the form of labor and, later, knowledge. The labor or knowledge of many workers was pooled together in order to develop the necessary products for corporations. Each worker was viewed as differentiated primarily in terms of the replacement cost of their labor. It was difficult for an individual to work alone if their skills were best combined with another's. For example, an innovation or patent in radio technology is of little use to an individual who is outside a company that can combine it together with the other patents that are required to build a mobile phone.[12]

Consumers, meanwhile, were conceived only in the role of purchasers of the output of companies (Galbraith, 1958). While there has been some attempt by a company to "insert themselves in customers processes" (Vargo and Lusch, 2008), defined as "co-creation of value", it is generally designed to create a unique, personalized experience for an end-user. It also often relates to obtaining feedback about product usage in order to understand how to better develop products in the future. This dominant focus of this literature is still the company, rather than the customer or individual, however.

Obviously, the Internet enabled new interactions within the global economy. As an example, an artisan can create ceramics in Cornwall and sell them via the Internet to consumers in Japan. To some extent, it is possible for certain skill sets to live where they choose and still participate in the economy.

The mobile broadband platform brings a new aspect to this dynamic: mobility. New technologies now even allow individuals to take localized payment by credit card on their mobile devices.[13] Individuals are now able to directly establish connections between one another in order to create flows of work between them, rather than merely as a nameless

---

[11] Naturally, this is rather a simplistic overview of economic theory. For an excellent and more detailed overview of the history of economic thought, see Galbraith (1958) *A History of Economics: The Past as the Present*.

[12] While it may be possible to register a patent, it is nearly impossible for a single individual to protect intellectual property against infringement from larger companies that have economies of scale in terms of legal teams.

[13] http://www.adelante.co.uk/mobilepos.html

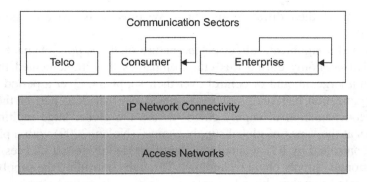

FIGURE 2.3

Evolving role of the consumer.

entity within a corporate environment. This is a significant change in the nature of value creation and value capture in the digital world, as illustrated in Figure 2.3.

The IMS was initially developed as a "person-to-person" communication technology, aimed at allowing end-users to interact and communicate with one another. It is perfectly suited to developing and implementing community-based workflows for groups of individuals within the mobile broadband era. This is an area of huge potential for developers to earn significant money, and the technical implementation of the building blocks is covered in detail in Chapter 12.

2.3.2 The Mobile Broadband Platform

As discussed previously, the mobile broadband platform is in the process of becoming the nexus of contracts within the global economy, rather than just for the mobile industry. This section analyzes and explores the implications of this claim in terms of its economic effects.

The increasing level of network connectivity has driven the most recent era of globalization between different companies across the globe. Corporations are able to run 24 hours a day, with different teams of people located in different time zones. This global interconnectivity has also enabled a significant level of mergers and acquisitions, as companies are able to integrate core technologies together at a speed that was not possible previously.

A key enabler of this process was the development of enterprise platforms such as those from SAP and Oracle. During the late 1980s, computing platform providers such as Oracle realized that it was no longer enough for them to provide just a database for their customers, but instead it was needed to develop applications and system integration units for particular industry verticals, e.g. pharmaceuticals, oil and gas, and financial services.[14] By the mid-1990s, these industry verticals were provided with open interfaces that allowed companies

---

[14] To the extent that Larry Ellison, CEO of Oracle, regularly describes Oracle's role within the global economy as a "business enabler".

to quickly interconnect their different enterprise systems. This drove costs of merger and acquisition down.

This, in turn, had a significant effect on the nature of the supply chains that formed the basis of the global economy during this era. Companies that implemented this technology were able to gain a large amount of control over their suppliers. Over a period of time, every multinational organization had such systems in place as a standard part of their operations. This led to the concatenation of supply chains across the globe: large multinationals positioned themselves at the "apex of global supply chains" (Nolan, 2009) with a plethora of suppliers that were controlled by ICT technology. Suppliers rapidly found it necessary to comply with the multinationals' supply chain technology. Through controlling the supply chains, these companies were able to exploit this advantage to significantly reduce costs: it is this level of control that has allowed Apple, for example, to produce the iPhone for a relatively low price.

Nearly every product for sale across the globe has "IT inside" as these global supply chains powered by digital technology work to manage supply and demand dynamically. For example, WalMart makes extensive use of digital technology in order to dynamically manage stock in their stores: as soon as a particular product is seen to been selling well, the system automatically orders more of that product and organizes delivery.

As discussed, the mobile broadband platform is now becoming a critical part of the infrastructure for the global economy. As technologies such as RFID and M2M communications proliferate, the mobile broadband platform and the companies that provide the technology that forms its basis become critical parts of these global supply chains. In effect, the telecommunications industry finds itself in a similar position to Oracle and SAP during the late 1980s – needing to move from being merely platform providers to business enablers (Oracle Archives, 1998).

The mobile broadband platform, in conjunction with cloud computing capabilities, therefore allows for the further concatenation of global supply chains as companies move to establish even greater efficiency and control of stock levels and other supplies. Companies are now able to monitor the prices of raw materials and renegotiate contracts on a monthly basis, something that had previously been done at most on a quarterly basis.[15] Essentially, this is about the development of capital goods software that takes into account the mobile broadband platform as a part of the digital economy.

As a developer, the key to understanding how to develop applications for such an environment is *scale and abstraction*. This is discussed in detail in section 2.5, which goes into more detail about how to think about creating business models.

2.4 THE BUSINESS CASE FOR IMS

The traditional business case for IMS is often built around the need for globally interoperable standards. This section describes what a globally interoperable standard is and, more

---

[15] For example, semiconductor companies have made savings based on monthly reviews of their input materials, e.g. silicon and precious metals.

importantly, what it provides developers. This section also highlights some other points with regard to the IMS standard that are useful for developers to understand.

2.4.1 Global Interoperable Standards – a Developer's View

IMS is developed by 3GPP, a global standards organization dedicated to the development of different mobile technologies, from radio to core network. This forum is formed from representatives of hundreds of different actors across the telecommunications and computing industries,[16] ranging from semiconductor manufacturers to the mobile operators that own the final networks. Many of the companies participating in these forums are in direct competition with one another, which raises the question: why do they all work together to form the standard that they are then going to compete on?

The platforms that are developed within these forums are actually unique in comparison to traditional platforms from the computing industry. When IBM or Oracle develops their platforms, they focus on delivery of their platform only. If it is required for an Oracle platform to talk to an IBM one, this is taken as a system integration project, where developers work to build the necessary connecting infrastructure. In the mobile industry, however, this is not possible as operators wish to provide end-user consumers with as much choice as possible, in particular with regard to the sort of handsets they are able to purchase and use on a network. Operators also wish their networks to be as cheap to run as possible and so they push their suppliers to interwork with one another from the very beginning of the process.

What does this mean in reality though? Essentially, it means that any platform developed for the telecommunications industry needs to provide the following core principles:

1. **Multi-operator.** The platform must allow operators to interconnect with one another without a major systems integration project.
2. **Multi-vendor.** The platform must allow for different components from different network vendors to connect together seamlessly, without a major system integration project – it should just work. This allows operators to ensure that they are able to execute price pressure on their suppliers in order to keep network costs down. In turn, network vendors execute on their economies of scale in order to ensure the platform is produced as cheaply as possible.
3. **Multi-access.** The platform should support as many forms of access technology (fixed and mobile) as possible, including legacy systems, in order to ensure that those on older terminals are able to participate as fully as possible in the communication systems of the world. This also allows operators to run several different types of radio network at the same time, e.g. GSM, WCDMA, and LTE, providing full support for legacy interworking.
4. **Multi-device.** The platform should support as many different types of device as possible, in order to ensure that consumers have as much choice as possible. This requirement also implies that any handset should be able to connect to any network equipment – for

---

[16] And indeed, other interested parties such as car manufacturers, etc.

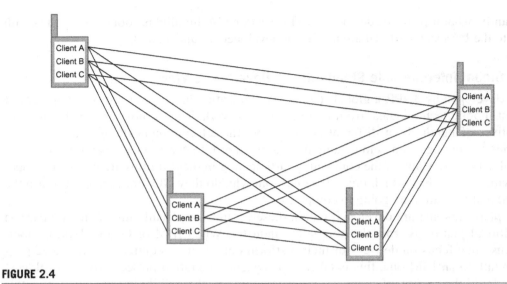

FIGURE 2.4

The client view.

example, any Nokia terminal should be able to connect to any Ericsson radio equipment, etc.

Some of this functionality might seem a bit boring: why should a service developer worry about all of these rather mundane features of a platform? Using the correct underlying technology will ensure as wide a reach as possible for your applications. Let's take a look at how a globally interoperable standard works for IMS.

Without IMS and the global interoperability it provides between handsets, equipment manufacturers and operators, implementing an application on a mobile device is quite limited, even with a smartphone. Applications are currently siloed, i.e. they need to run on a particular device, with a particular operating system. Most applications need a particular level of radio access to operate fully. Each time a consumer wants to use an application, e.g. Skype or another application, they first need to download the client to their device. This is not the end, however, as they also have to ensure that the person they wish to communicate with has also installed the same client software so that they are able to establish communication with one another. There is no guarantee that the operator that different friends and colleagues are using supports the application in question. In addition, if anyone has a legacy device, the communication format is more than likely not possible. This is illustrated in Figure 2.4 – for each client they download, the end-user must ensure that it is interoperable with all the other people that they wish to communicate with.

Using IMS as one part of the application development process allows developers to exploit the fact that IMS is designed to be globally interoperable. This means that an application developed using some IMS functionality will work across operators, across different

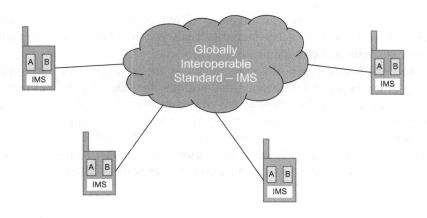

FIGURE 2.5

The globally interoperable view.

access networks, and across different service delivery platforms. Moreover, IMS provides for interworking towards legacy terminals, meaning that while a client may be downloaded on to a smartphone, for example, the users of that device will still be able to communicate with end-users on legacy terminals. This is illustrated in Figure 2.5.

2.4.2 Regulation and the Right to Private Communications

Voice has been the fundamental method of communication since humanity's first dinner party nearly 10,000 years ago. We are quite simply a species driven to communicate our thoughts, our ideas, and our business plans with one another. To communicate is to express our humanity. The right to discuss freely and openly is the basis of most democratic nations.

The majority of governments worldwide have established regulation ensuring that it is illegal to tap a person's phone calls without first obtaining a warrant.[17] IMS was developed within this regulatory framework and therefore had privacy and security designed in at an early stage of development, rather than added as an afterthought, as is the case in many Internet services such as HTML-5.

Digital technologies provide huge opportunities, but may also leave end-users exposed to persecution or discrimination based on the content of their phone calls or chat sessions.[18] This leaves end-users vulnerable to violation of the fundamental human right to freedom of speech[19] if unauthorized parties can intercept their communications relatively easily

---

[17] Generally called lawful intercept.

[18] http://www.theatlantic.com/technology/archive/2010/11/chinese-woman-sentenced-to-one-year-of-hard-labor-for-tweet/66765/

[19] As outlined in Article 19 of the Human Rights Convention (Everyone has the right to freedom of opinion and expression; this right includes freedom to hold opinions without interference and to seek, receive and impart information and ideas through any media and regardless of frontiers).

and without their knowledge. The evolution to HTML-5, despite the huge benefits that it provides, also creates a series of new security risks with regard to gathering information about end-user behavior and communication.[20]

Issues with regards to privacy of communications are set to increase, as end-users gain understanding of the types of interception that are possible. Developers will therefore need to take these issues into consideration as the communication industries continue to evolve. Several governments are currently investigating the role of "privacy by design" within their regulatory frameworks,[21] i.e. that all systems will need to be built to ensure the privacy of the end-user in question. Using IMS as a base allows a developer to make use of telecommunications grade privacy and security mechanisms without needing to implement them themselves.

2.5 BUSINESS MODELS FOR A DIGITAL PLANET

Business modeling has been defined as a managerial competency that is growing in recognition (Chesborough and Rosenbloom, 2002; Johnson et al., 2008; Chesborough, 2010). A growing body of literature focuses on the need for "design" thinking with regard to business models (Boland and Collopy, 2004; Dunne and Martin, 2006; Martin, 2010; Pandza and Thorpe, 2010). The idea behind such methods is to attempt to create breakthrough understanding and response to changes based on so-called wicked problems (Churchman, 1967).

While these approaches address the need for different thinking patterns when attempting to understand business models, they do not provide a structured approach to dealing with the uncertainty associated with technological, market, and societal unknowns (Pandza and Thorpe, 2010). Traditionally, companies have tried to handle the problem of complexity within their industries through a process of simplification – "acquiring competitors, locking in customers, producing standardized products and services, etc." (Naughton, 2010). As discussed in the previous section, the telecommunications industry is currently undergoing a lot of turbulence; as a result, these tactics are unlikely to be of significant use when developing business strategies or models. Design thinking methodologies provide recognition of this fact, but no real methodology for how to create a business model; instead they often focus on applying creativity to finding a solution.

This "think creatively" approach is actually an extremely difficult process to apply to the communications industries, especially in companies where value chains are traditionally modeled as a set of modular elements occurring in a set sequence of events, as illustrated in Figure 2.6.

This approach to understanding a company's role within the value chain (and hence the business model they implement and run in the industry) is deeply seated in the industrial

---

[20] http://www.nytimes.com/2010/10/11/business/media/11privacy.html?_r=1&pagewanted=all
[21] For example, the UK and several European countries.

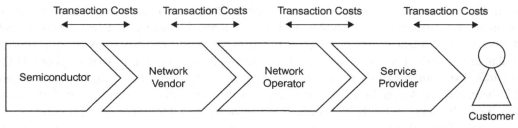

FIGURE 2.6

Simplified traditional value chain diagram.

process model view of business activity, which originated in the early decades of the previous century. A defining characteristic of this style of business thinking was that the business landscape was relatively stable (Rasmussen, 1990). As a result, many of the components of a company's business processes and interaction with the wider economic environment could be represented as modular elements that occurred in a sequence. Value creation and capture related to a set of transactions through a relatively linear process. Business modeling as a result has often tended to focus on optimizing models of production, e.g. supply chain improvement, rather than understanding where imbalance within the economic environment creates threats and new opportunities. Many examples of this exist within business literature, where companies do not change their models of doing business until forced to do so by threat of bankruptcy or a series of disastrous results for their shareholders. Perhaps the best-known example of this is IBM during the 1990s, which took the decision to completely restructure its entire business practices (and hence business model) in order to recover its position as a world-leading IT company. Companies such as IBM are success stories, however. Those companies that do not have large cash reserves or the scale to pull themselves out of such situations are generally relegated to business history books.

Within the digital economy, however, as discussed in the previous section, it is unlikely that the established roles of corporations during the industrial era remain relevant. Given that the roles and relationships between individuals and companies (and indeed parts of companies) are in flux, we might therefore turn our focus away from making existing flows and relationships more efficient toward mapping the possible space in which new relationships and structures could emerge. Indeed, one can think of the numerous elements within the value chain diagram in Figure 2.6 as a complex industrial subsystem, defined by its own economic forces, technological regimes, and set of competing companies. The interactions between these complex industrial systems are often where the most significant opportunities for value capture occur. Thus, by guiding our attention toward how new relationships could be formed, we can find new "spaces for value capture" that are perhaps otherwise easy to overlook without at least a semi-structured form of enquiry.

The way we propose thinking about this space is then from the perspective of *scale* and *abstraction*, i.e. by defining the scope and boundary of understanding required about a particular system. The relationships that are of interest in the digital economy are not just between mono-lithic industrial structures but relationships that leap *across* boundaries between levels of scale and abstraction. A clear example of this is found in the music industry. Previously it was the case that artists under contract to record labels released their music on pressed vinyl or CD that was purchased from record shops. Today it is possible for bands to record their own music and sell it themselves as MP3s direct to consumers over the web. This is not merely an "efficiency" within a supply chain, as might have been represented by an independent label pressing its own records and offering them directly to record shops (thus cutting out the record company and dis-tributor as middlemen), but rather meeting a high-level end (the sale of music) from the deploy-ment of disparate means in terms of both scale and abstraction (i.e. a computer and a web page). Certainly we might imagine that computers featured in many aspects of the "old" record indus-try model, but here they were the means to ends that in themselves only contributed indirectly via a chain of abstract purposeful activity to realizing the overall ultimate purpose of selling music (e.g. stock control, which enables efficient logistics to ship records out to shops in order for them to be sold). Now the computer, via the Internet, is in itself sufficient to realize the ulti-mate goal, and therefore has a direct and unmediated relationship to it and in doing so effects radical change upon the business models within the music industry.

2.6 TOWARD A DIAGRAMMING TECHNIQUE

In order to visualize these relationships, we have adopted an approach to diagramming based very loosely on the "abstraction hierarchy" work of Rasmussen (1990) to relate abstraction to scale within a grid. We give the example of a media production company.

Scale is represented in the diagrams below (Figure 2.7) from left to right (e.g. we repre-sent the company, a component part of that company – the media production department – and a sub-component within that department) while the layers of abstraction are represented from top to bottom. We represent the industrial structure as a set of interacting boxes. Each box represents a purpose located at a given level of abstraction and scale. Abstraction is plotted on the Y-axis with the highest-level, most abstract purpose of the system at the top (e.g. sell music). Each successive level downward represents a slightly less abstract, slightly more concrete purpose. The relationship of these boxes is thus that the box above represents the "why" and the box below the "how". If we extend the grid down we create a self-documenting explanation for why and how different purposes are fulfilled within the industrial design. Thus, for example, a high-level purpose might be to "generate revenue"; this is achieved (how?) through "selling media" and "selling advertising", and thus reading upward we have documented the reason for selling media and advertising (why?): to gener-ate revenue.

In terms of how scale works, note that the media production department's top-level pur-pose is to produce media. While this certainly constitutes an important part of the company's business model, profit is only generated from this activity at the company level. Similarly,

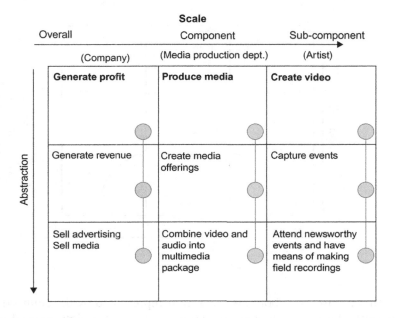

FIGURE 2.7

A practical example.

while the artists themselves contribute to the production of media, their highest level function is to create video; it is the production department that creates the media offering.

Having built up the diagram in Figure 2.7, we can use it to ask different types of questions about the nature of business in this space. The first is based on noting that the descriptions we have used in each of the boxes is of a purpose; while we identify how and why in each case (reading up and down the hierarchy of abstraction; we could extend the diagram downward infinitely but have not done so here for brevity) we do not identify the actual means by which these things are physically achieved. Thus, the first kind of question we can ask is about how these purposes can be met. There may be many different ways a media company can fulfill these purposes and in itself changing them may or may not affect the underlying business model. For example, in terms of "making field recordings" we could choose to use a broadcast video camera. An alternative way of doing this may be to use a camera-phone. If this is merely a straight substitution, the relationship between means and ends remains the same. However, by identifying this change we might be caused to consider whether a citizen-journalist or event participant equipped with a camera-phone could in fact fulfill the purposes of the artist within this business model. We might alternatively reflect that there might be ways of generating revenue that go beyond selling media and advertising; perhaps we could sell the production department's expertise as a service to others.

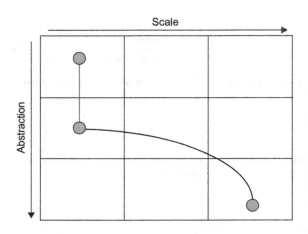

FIGURE 2.8

Radical model.

A second form of reasoning about the diagram is represented by the connections between purposes represented by the circle-and-line connectors. The "straight up and down" structure in Figure 2.7 is representative of a hierarchical structure where sub-components all fulfill purposes culminating in the top-level company purpose of generating profit. Is it possible to achieve a high-level end at the company-level scale through a lower-level means? In the above example we might consider that there now exists the means where, for example via the web, we can generate revenues through directly uploading captured footage without having to compile a finished video, prepare a multimedia offering, or even sell the advertising or the media itself ourselves, a leap across scale and abstraction as per Figure 2.8, with labeling removed for clarity. In general we believe this "bottom right to top left" pattern of connectivity is stereotypical of highly disruptive digital economy business models. Viewed from this perspective, it is perhaps clear why such changes within industries are so surprising to participants; the affairs they generally concern themselves with occur at a given level of scale where the relationship between means and ends are, if not traditional, at least time-served. By leaping across these demarcations, radical new business models do not merely disrupt flows within a value chain, but cut deeply into the core underlying assumptions about "what companies do and how they do it".

An additional way of using this diagramming technique (and thus using the scale and abstraction lens to view business models and structures) is to attempt to plot the linkages of means and ends within an industrial sector. In Figure 2.8 we compare two companies, each with different ultimate high-level purposes. Company A is perhaps a broadcasting company (using the labeling above) whereas Company B is perhaps more of a content producer. Note that because stacked boxes and links indicate means and ends, we see that both companies are pursuing the same end for at least one point in their business models. Recognizing this commonality should cue us to ask questions about whether there is some sort of relationship they could have. Perhaps they are competitors in this area, but perhaps from Company A's perspective they could source footage from Company B or perhaps consider spinning off part of their existing "monolithic" operation to compete directly with Company B in its market niche.

2.7 PRACTICAL EXAMPLES – APPLICATION TO IMS

Here we take a look at some different business models for IMS, specifically non-traditional business models that may not be widely discussed. Many reasons for developers to use IMS exist, in particular access to a global standard, as discussed in section 2.4. These reasons are covered many times within this book. Here we will instead discuss a few examples of how to create a business model for IMS as a developer, not as a network vendor or a mobile network operator.[22]

IMS enables a few different things that developers may make use of to develop business models that both create and capture value for end-users in the emerging structure of the communications industries. We consider *reachability* first.

Reachability is a key factor within the emerging industrial structure of the global economy. As smartphones gain significantly more functionality in regard to gathering information about individual end-users, including their location and other personal data, the ability to successfully locate the exact individual increases in value. In comparison to other platforms, IMS delivers something extra to developers – remember, it is a globally interoperable platform. IMS does not depend on Google Android, or iPhone, or Symbian – it functions across any mobile operating system (including Android, iPhone, and Symbian).

So, how does reachability allow you to create new applications? One example is to take IMS's ability to connect you to a person, irrespective of which device they are using. As an example, we consider an application that creates radical innovation within the delivery system for couriers. Let's take a company such as UPS, responsible for the delivery of packages, documents, and other communications worldwide.

Anyone who has ever been waiting for a delivery from a courier knows that it can be rather irritating with the necessity to reorganize your life just to be in the right place at the right time to receive the delivery. From the company's perspective, the cost of organizing redeliveries for those occasions when people are not at home or have not received the notification of a delivery are reasonably high.

Using IMS, however, a package could theoretically be routed to an individual, rather than a specific address.

Say, for example, that I register my address with the courier company as my SIP address. Instead of routing to my address, the courier is able to take a series of much smarter decisions with regard to the delivery of my package. For example, through my location, they

---

[22] They already know why they need it.

would know that I was on semester and would not be in the country to receive a package and could schedule the delivery on my return. More importantly, through knowing my SIP address, they could route the package directly to me, rather than to my address.

How would this work? Firstly, within IMS, my SIP address refers to me and I am able to indicate the device that I am currently using. For example, using IMS, you are able to know where I currently am – at home, at my office, or perhaps at the local coffee store, where I am reading the papers for an hour and have my mobile device with me. Using IMS's forking procedures, the "call" from the courier about my availability to receive a package would be received on all devices at the same time. When I receive the communication and answer, the other devices would stop ringing. Using IMS's communication capabilities, the courier company would be able to either establish a call or a messaging session with me in order to organize the most convenient time and place for me to receive the package.

This is a radical innovation for the courier companies – but would provide a significant competition (or complementary good) to the services that couriers already deliver.

2.8 CONCLUSIONS

This chapter has covered a series of topics that are relevant to application developers wanting to understand the evolution of the global economy and the role that IMS may play in developing applications. The terminology used in this chapter will recur throughout the book as we illustrate in more detail how the concepts outlined here fit together in the wider context of IMS. In particular, Chapters 5 and 12 illustrate how to create applications related to consumer and capital software goods.

Service Deployment Patterns

3.1 INTRODUCTION

As you may have hopefully realized from the discussions in Chapter 1, IMS applications can appear in different parts of the network. Why does an application provider need to worry about this? Actually, it will affect you quite a bit, as the nature of your application will govern where it makes most sense to put it (and sometimes parts of it will have to go in different places). By "nature" in this context we don't mean whether it is a game, a recipe catalog, an office productivity tool, or anything else along those lines, but rather where it turns up in the general scheme of things: which side of the IMS black box it connects to. The real issue at hand for application developers is that the interfaces look different on different sides of the box; we will go into more detail on this later in the book. But as an overview, these are essentially the options:

- **Client-side applications.** These are deployed in user devices, possibly building on capabilities provided by standard services like 3GPP Multimedia Telephony (MMTel) and the Open Mobile Alliance (OMA) standard services such as chat, file transfer, presence, messaging, and GSMA streaming video and image transfer during a phone call.
- **End-point services in the network.** These are addressable by the end-user through a client application – for example, a timetable service responding by voice. Note that deployment scenarios vary: classically, they would be deployed inside operator networks. Increasingly, operator business partners can be responsible for hosting services, across B2B interfaces. The extreme version of this type of scenario is exposure of capabilities to long-tail developers and small-scale service providers; in such cases, REST-style APIs are becoming prevalent.
- **Mid-point services in the network.** These tend to be more about modifying the behavior of sessions (e.g. smart call forwarding based on callee calendar status).

Tooling and deployment scenarios will be different, as will most definitely the business models. Also, it will probably be very common that an application will need both client side elements and network side elements. Another emerging trend is – as mentioned above – that the actual APIs may be provided using web technologies like REST; note that in semantic terms nothing (much) has changed, and it is mostly a question of protocol technology choice. There is a lesson in that; as experienced designers will know, it is sometimes easy to confuse low-level mechanisms with the really important issues: the semantics of what support you are planning to invoke.

We will now explore these variants further.

3.2 BACK TO BASICS

Remember the "black box" figure we introduced in Chapter 1? If not, let's revisit it, as it will be the basis for a number of scenarios in what follows.

Figure 3.1 illustrates two of the major interface categories:

- **The client-side API.** This will be either a native API if you are coding directly to a specific platform, or the JavaME JSR281 interface. JSR281 allows you to think in terms of setting up and using sessions, without worrying about the detailed exchange of signals that has to happen below the surface. Additionally, client-side applications can invoke IMS services indirectly through REST interfaces delivered through network-side gateways. Standards for this are evolving, in particular from GSMA and OMA. Using HTTP-based technologies like REST makes it easy to also include IMS functionality in PC-based client applications; thus, the phone icon in the figures in this section could of course equally well have been PCs, laptops, tablets, fixed SIP phones, or anything else with an IMS stack inside. This will be discussed further in a later section.
- **The server-side API.** Here, telco services are very likely to be written in Java. In Java terms, the level of support typically found in commercial servers is at least JSR116 (SIP application server support), now increasingly the evolved version: JSR289. The latter is capable of more intelligent composition of SIP application components, moving in the direction of component-based application development. This concept will be described

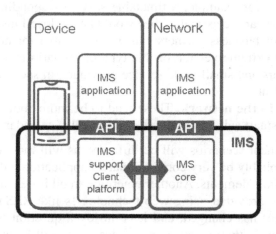

FIGURE 3.1

IMS as a black box, revisited.

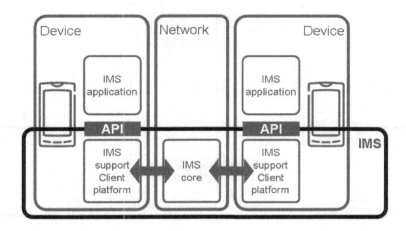

FIGURE 3.2

Client-side application.

in detail in Chapters 4 and 5. The level of abstraction here is typically individual SIP signals, so in this case you may have to know a bit more about IMS protocol details. Alternatively, you can use the higher level web service and REST-style APIs defined in GSMA, OMA, and elsewhere with semantics comparable to JSR281.

3.3 CLIENT-SIDE APPLICATION

If all you need to support your application is to be able to find and connect a session to another terminal equipped with your application, then you really don't need to know anything except the client-side API (Figure 3.2). The IMS core will handle all the details around client registration, SIP signal routing, etc.

Now, let's make it a bit more exciting. Assume that the two participants in the session above have made – for a number of very valid reasons, no doubt – different decisions regarding which operator to choose. Do you care? Should you care? Of course not! IMS routing will ensure that your two application instances can find and connect to each other (see Figure 3.3). And as far as the application code goes, nothing has changed from the same-operator case.

3.4 SERVER-SIDE END-POINT APPLICATION

Many interesting applications such as two-party gaming can be made on top of client-to-client communication. However, in many cases (probably most) you need some kind of server-side assistance. For example, anything requiring three parties or more to share a conversation (voice, video, text, document updates, game moves, etc.) benefits from a central conference server merging information streams and deciding who has precedence. In some

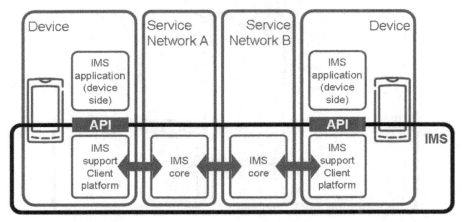

FIGURE 3.3

Client-side application, across operator boundary.

cases, you can get away with doing multiparty apps client to client, either by using a fully meshed set of communication links or by designating one of the parties as a controller. If you need to scale to more than a (small) handful of participants, bandwidth and latency limitations will soon be a problem; there is no such thing as a free lunch (read "infinite bandwidth and zero latency" for free lunch in this context). This applies particularly if your end-points are connected on a mobile network.

Figure 3.4 illustrates a case where the connectivity provider is also the service provider. Of course, this case can also be extended in the same way as the client-to-client scenario, with services being delivered across network boundaries.

If you look carefully, you can see that in this figure we are indicating a number of clients connecting to the same server: it is implied here that, in this case, each client has its own unique session with the central server. In typical applications like this, another mechanism is normally responsible for correlating the sessions that together form the user experience (a conference focus for the video conference, the team name in the multiplayer game, etc.).

3.5 WEB SERVER-SIDE END-POINT APPLICATION

Now, let's add the web dimension (see Figure 3.5). Apologies for using the term "web" here; essentially, we mean any service delivery scenario where the server-side end-point is outside the trust domain (or at least the classical network technology domain) of the telco operator. But you know what we mean. The interesting issue is, however, that IMS in this case is – again – the infrastructure that delivers find and connect, interoperability, global roaming, and controlled QoS – packaged and delivered in protocols that are taken from the web world, and hence known to a wider range of software creators. Classically, the telco industry has tried

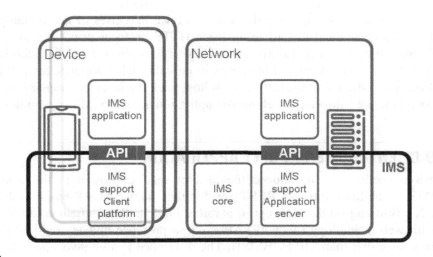

FIGURE 3.4

Server-side end-point application.

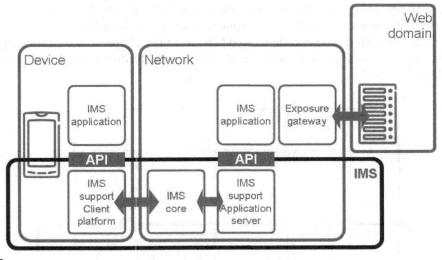

FIGURE 3.5

Web server-side end-point application.

to find ways of achieving this, but previously the semantics of the operations provided have been designed more to allow complete and detailed control over the full range of service capabilities. You might be forgiven for suspecting that the mantra was "if it is in there, I must be able to control it". With the advent of Parlay X, the focus changed to "what does the developer actually need, and how can we make it easier to deliver that?" Parlay X was designed using web service tools well aligned with the web service initiatives of late (notably WS-I[1]). Recently, the very clear trend toward using basic HTTP protocol primitives to deliver the semantically interesting protocol primitives (REST, Representational State Transfer) has meant that GSMA and OMA have launched initiatives to provide such mappings. Semantically, the job to be done is still the same; the difference is how thick the layer of wrapping is until you get to the goodies inside: allowing a web-based application access to IMS capabilities.

3.6 WEB CLIENT-SIDE END-POINT APPLICATION

In the previous section, we introduced the idea of services hosted in the web domain reaching IMS capabilities through REST (or for that matter SOAP-based web services like Parlay X). Building on that concept, it is rather obvious that a client could likewise be provided with web technology (network) APIs; e.g. a presence update could be done with HTTP operations rather than SIP PUBLISH. The difference is how invocations are authenticated and authorized: typically, client calls would be accompanied by a security token; server side applications interacting over a business-to-business interface may be implicitly authenticated by being delivered through a secure tunnel. Note that the semantics are still that of a client interacting with the network, even though the protocol has changed.

As Figure 3.6 shows, the API is now conceptually more from the client directly to the network, rather than from client code to middleware in the client device (drawn in this figure as a laptop; the emergence of web client APIs is mostly driven by the desire to create PC-based IMS clients).

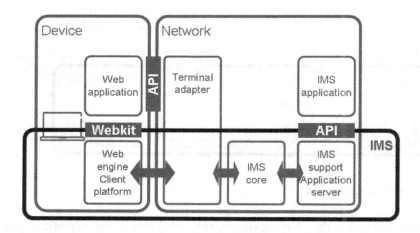

FIGURE 3.6

Web client-side end-point application.

---

[1] www.ws-i.org

It should be noted that although REST is HTTP based and therefore in some sense associated with the browser environment, such APIs are by no means restricted to usage from inside web-page scripts. Any kind of application (client or server side) can use them, provided that it can present the proper credentials to the IMS core.

3.7 MID-POINT APPLICATION

If your job description says that you are more of a classical telephony service developer, you may be more into what is called "supplementary services". The name implies what this scenario is all about: managing and modifying sessions behind the scenes, rather than being the end-point that the user is directly aware of. Services in this area concern things like virtual private networks (number translation, call policies, etc.), black-lists/white-lists, call diversion on various conditions, etc. In legacy telephony, these services are typically implemented using Intelligent Network (IN) technology: calls being processed by a switch trigger service requests to other nodes specializing in supporting these supplementary services.

Figure 3.7 is applicable to both standard and operator-specific services, as long as the network-to-network interoperability interface rules are upheld. Thus, one operator could offer call diversion dependent on a user's Google Calendar status (using a link to a web service as indicated in the previous section); this would not in any way affect interoperability, as it is not visible on the other side of the interconnect between A and B.

In scenarios like this, services are typically specified as originating (triggered in the caller's network) or terminating (triggered in the callee's network). If you bundle them together you get the canonical service in the IMS space: this is, of course, Multimedia Telephony (MMTel; more about this in later chapters). It is interesting to note that a number of features

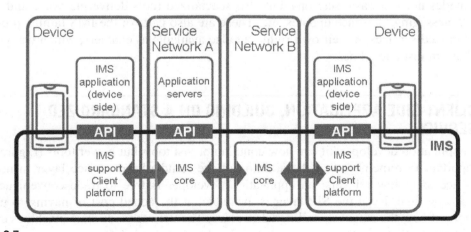

FIGURE 3.7

Mid-point application.

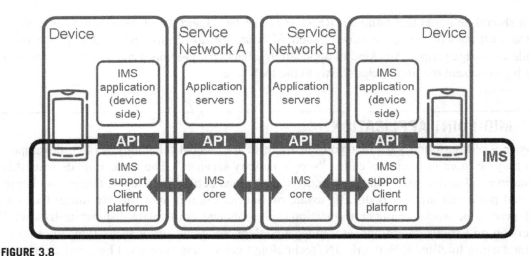

FIGURE 3.8

Mid-point application across operator boundaries – a standardized service.

that were typically implemented outside the switch in GSM have been made mandatory in MMTel. Thus, the baseline level of support that can be expected in an IMS network is higher. The complete picture would look like that shown in Figure 3.8.

Essentially, this is the blueprint for MMTel and other interoperable IMS real-time person-to-person services. But if such services are standardized, interoperable, and available, why shouldn't application developers be able to make use of them as components to build upon? But of course they should. And then we have raised the level of abstraction one more step: IMS provides not just basic sessions, but also specialized tools delivering voice and video sessions, messaging, presence updates, etc. These are also parts of the IMS family, provided as standardized services in their own right, but also available as enablers, which we will put to good use in just a few lines.

3.8 CLIENT-SIDE APPLICATION, BUILDING ON A STANDARDIZED SERVICE

So, the application developer kit can now contain not just tools but power tools (Figure 3.9). It is important to remember that the addition of the standardized service layer in no way restricts access to lower layers: an application developer is free to trade convenience for control at any time. Even the SIP layer is available, at the added cost of having to understand more of how the actual signalling takes place. The network-side state machines expect certain behavior that must be respected; this is already encoded for your convenience in JSR281.

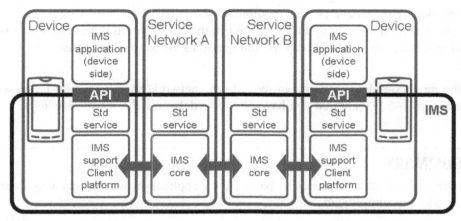

FIGURE 3.9

Client-side application using standardized service.

3.9 TO-DO LIST

At this point, it is hopefully clear that in order to get an application running, you may have to interact with the system in several places. This section provides a short list of what you need to do to deploy a service.

Let us assume the following scenario. A retail store chain has built an app where you can browse the offerings and mark what you are interested in. It also contains a quick-dial button that allows you to talk directly to an agent. The mid-point server intercepts the call, picks up the current content of this user's shopping cart, uses that info to select the right specialist, and pushes customer and cart info to the specialist's smartphone where it is picked up by the salesperson's version of the app (because, obviously, he or she is not tied down to a desk).

Preconditions are that the client app and server app have been designed, compiled, and built using your favorite development environment. The example indicates that we need a client-to-client scenario with a server-side app interacting with the session, i.e. a service with two client-side components and a mid-point component. Further, let's assume that you are running this in a test scenario (which means that you need to do some of the basic plumbing yourself; always a good learning experience).

The steps to get an app running on a test network are:

1. Install and start HSS, CSCF, DNS, AS.
2. Populate DNS with domain names of HSS, CSCF, AS.
3. Deploy server app to AS, initialize app.
4. Define IFC in HSS (remember, the initial filter criteria are what the CSCF will use to ensure that your AS is linked in to the session).
5. Define users in HSS.

6. Deploy client apps in devices (real or simulated).
7. Start client apps and check that they register in the AS. Note: in the example this is a terminating service, so normally only the salesperson's client would register.
8. Run!

This is the most complete scenario; in real life the network operator would be responsible for at least steps 1–5 (except that you have of course delivered the server-side app).

3.10 SUMMARY

If this listing of the various places to put an IMS application was in any way confusing, please consider this simplification:

- If you don't need a network side component to your app, life is really simple: use your favorite client API.
- For the server side, use the JSR289 SIP servlet framework if your project requires SIP level control (this is the main subject of Chapters 4 and 5). If not, check out GSMA's OneAPI (http://www.gsmworld.com/oneapi/) and/or OMA NGSI (Next Generation Service Interfaces, http://www.openmobilealliance.org/).

Applications in the IP Multimedia Subsystem

4.1 INTRODUCTION

Competition from new players, convergence of legacy circuit-switched (fixed and mobile), IMS and Internet services, and the resulting disruptive innovation drive operators to improve their efficiency and agility. Service-Oriented Architecture (SOA) is a paradigm shift that leads to software architectures driven by business objectives and thereby enables agile systems capable of flexibly and rapidly executing strategic decisions.

SOA requires building and managing software systems focusing on the services they provide, rather than the technology they use. Such systems typically achieve efficiency by rapidly assembling new services by creating their functionality through reuse of existing services. This allows the quick creation and deployment of services to respond to rapidly changing business needs. Moreover, adoption of the SOA way of working also leads to new disruptive ways of doing business, a prime example of which is Software-as-a-Service (SaaS) business models.

In traditional telecommunications terms, such as functionality may be compared to value-added services. The term *value-added services* describes a technique that encompasses a set of *protocols* and supporting *architecture* that telecommunication network operators employ to offer customized end-user services. To end-users, value-added services represent a user experience that goes beyond standard network services.

The business relevance of SOA-based systems requires a way of working that is analogous to traditional engineering, involving different types of reusable services and processes that reflect varying requirements for operation, provision, assembly, and management. This is enabled by means of service composition technology.

Value-added services provide operators with a means of differentiating themselves from each other. A rich portfolio of value-added services helps them attract new subscribers and to reduce churn. In addition, the time that operators need to develop, test, and deploy value-added services is typically much shorter than for standardized network services.

The significance of this area for telecommunications is clearly demonstrated by the rapid rise of a whole host of technologies and their incarnations as different nodes, such as SCIM and IM-SSF, that help operators to start an efficient migration of legacy circuit-switched services to IMS and at the same time open up their networks to the Internet.

This chapter will introduce the basic principles of IMS application development, focusing on service chaining as the main means to create new services from existing constituent components. Then, a more in-depth analysis will present a number of technologies and architectural variants that may be used in this context to help the operator create valuable and differentiating IMS applications.

Section 4.2 of this chapter gives a high-level overview of the IMS architecture from the point of view of the application developer. This section intends to familiarize the developer with functions critical for deploying applications in IMS.

Section 4.3 gives an overview of possibilities for composing SIP applications in the larger context of IMS and their advantages and disadvantages. Furthermore, the composition of functionality external to IMS, i.e. implemented using HTTP rather than SIP and web service technology, is discussed and unified composition architecture is introduced.

Section 4.4 provides details of IMS AS architecture based on the cross-industry Java Enterprise Edition standards and the converged SIP servlet container in both its JSR116 and JSR289 incarnations. The section details the types of SIP applications that may be developed and how they can be linked into chains using the mechanisms provided by the SIP container.

4.2 IMS SERVICE CREATION

This section introduces a number of fundamental concepts that will be used extensively in this and the following chapter to explain IMS service creation through chaining and composition.

4.2.1 Service Composition

Service composition is most efficient when services are designed according to guidelines that allow them to be reused efficiently and effectively. This section will introduce some basic SOA design principles and discuss how they apply to the design of SIP applications and service composition in IMS.

A fundamental decision with respect to designing reusable, composable services is the question of the correct size, or so-called granularity. SOA introduces the concept of granularity to describe the functional scope for which the service is responsible. The larger the scope, the coarser the granularity; conversely, a focused scope also means finer-grained granularity. Note that the actual amount of physical implementation behind the interface does not reflect greater or less granularity. A service may offer great functional scope behind its interface (i.e. full charging functionality), but may contain only a minimal implementation. On the contrary, a focused interface (i.e. a logging function) may hide a large and complex implementation.

The ideal granularity of a service to achieve maximum reuse is determined by the business context. If the business logic of the application is understood, the proper level of granularity for the components will become apparent. Unfortunately, many developers do not perceive the connection between the overall business process and its application. This is critical in successful service design. On the other hand, the need for reuse is contrasted with the need for efficiency. Making things more general may increase reuse, but may also increase overheads. The rule of thumb is that services should be usable as tools that fulfill a specific function that has a distinct role and value in the business logic of the overall application.

A further concept critical to successful service design as regards reuse and composition is loose coupling. Coupling typically refers to a measure of dependency between two functions. Loose coupling advocates the reduction or elimination of dependencies between the service and its consumers. Consequently, loose coupling between services may be translated into a lack of dependencies during the implementation of a service while still guaranteeing interoperability with present and future consumers.

Simply put, a set of loosely coupled services can be replaced without the need for changes to the rest of the architecture. This means that improvements to the overall business logic may be implemented as needed in an agile and efficient manner. Moreover, this flexibility also extends to the set of tools and technologies available to developers. In a loosely coupled architecture, developers are free to use the best technology for the task at hand without being restricted by technical dependencies.

Finally, the concept of separation of concerns introduces a tool for reducing the overall application design complexity by reducing the amount of interaction between different features inside the application. In essence, a larger problem is easier to solve when broken down into a set of smaller problems (i.e. concerns). This principle allows developers to design and implement individual logical blocks of the overall business logic one at a time without having to comprehend or concern themselves with the overall complexity of the total problem. Moreover, such blocks of functionality are inherently reusable since they are expressly designed to address a well-defined partition of the overall problem and as such may be applied to other similar problems in the same area.

4.2.2 Composition Through Chaining

The ability to easily enhance applications and create differentiating, value-added versions thereof is a recurring requirement related to invoking telecommunications services.

Composition of applications based on the sequential invocation of components, also referred to as chaining or Distributed Feature Composition (DFC), is used to create distributed software systems that establish and control sessions (Jackson and Zave, 1998). Sessions are set up in terms of components or features in order to ensure flexibility.

A feature is defined to be an "increment of functionality with a coherent purpose", a definition which closely corresponds to the definition of a service in a Service-Oriented Architecture. Components communicate via internal calls through a common protocol. The description of a corresponding component type and the rules for including instances of the component into compositions specifies a new feature.

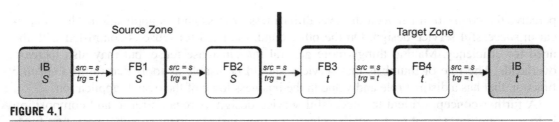

FIGURE 4.1

Telecommunications feature composition.

Source: IEEE Transactions on Software Engineering, *24(10).*

A composition in the DFC is mapped to a dynamically assembled configuration of instances of components, according to the features to be applied to that particular call (see Figure 4.1). The resulting configuration is analogous to an assembly of pipes and filters where feature components are independent, they do not share state, they do not know or depend on which other feature components are at the other ends of their calls (pipes), they behave compositionally, and the whole set is easily enhanced (Garlan and Shaw, 1993).

For the IMS network, multiple service invocation is solved through IFC chaining. Multiple SIP-AS may be invoked within a SIP session, depending on the contents of IFC. The IFC for a particular user, for example, might stipulate that, for an originating call, the S-CSCF must first invoke a service in SIP-AS #1, followed by a service in SIP-AS #2. The sequential invocation of SIP-AS within a single session is called IFC chaining.

However, IFC chaining does not manage service interaction; consequently, different services may provide different, possibly conflicting, instructions to the network. Therefore, IFC chaining offers limited possibilities for invoking multiple constituent services to create differentiating IMS applications. This chapter will discuss ways to implement enhanced feature interaction logic inside an IMS AS based on service composition principles.

4.2.3 IMS Service Chaining Architecture

IMS uses the Session Initiation Protocol (SIP) to control the establishment of voice and multimedia sessions. SIP runs end to end, i.e. it originates from and terminates in end-users' terminals. Accordingly, it is carried over the access network (for IMS, this may be 3G or 4G, WLAN, or any type of fixed access). In IMS applications are implemented on a SIP application server (SIP-AS), which may be augmented by logic on the terminal equipment. In the context of this book we will focus on the network side of IMS services.

The SIP-AS is a service execution platform on which one or more services are deployed. It may also be connected to the MRF to control media-related functionality, such as playing call progress announcements. Similarly, its connection with the HSS is used for storing and retrieving static or dynamic subscriber data. Figure 4.2 shows a generic architecture for value-added services in IMS.

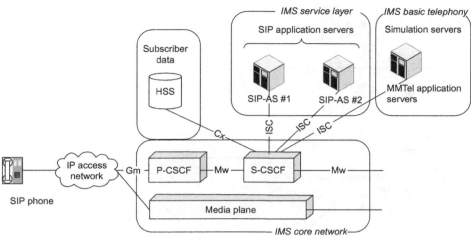

FIGURE 4.2

SIP-AS in the IMS network. Key: AS, application server; Cx, reference point between CSCF and HSS; Gm, reference point between SIP user agent (terminal) and P-CSCF; HSS, home subscriber system; ISC, IMS service control interface; Mw, reference point between CSCF and CSCF; SIP-AS, SIP application server.

The SIP-AS is logically connected to the:

- Serving CSCF (S-CSCF)
- Interrogating CSCF (I-CSCF)
- Multimedia resource control function (MRF)
- Home Subscriber System (HSS).

Whenever a terminal registers with the network, the S-CSCF fetches the service profile of the subscriber from the HSS. A core element of this data is called initial filter criteria (IFC). The S-CSCF uses information stored in the IFC to decide which IMS applications or constituent components thereof to invoke and how to reach the SIP-AS that host these applications.

The interface between the S-CSCF and SIP-AS is called the IMS service control interface (ISC). The SIP signaling carried over ISC is enhanced compared to vanilla SIP, e.g. to ensure that SIP messages sent from S-CSCF to SIP-AS can be returned to the S-CSCF.

Subsequent to the invocation of a SIP-AS, the CSCF routes call-related SIP signaling through it for the particular session. The linking of the SIP-AS to the SIP session flow gives the SIP-AS full control over the SIP session (Figure 4.3). The Session Description Protocol (SDP), which describes the media for a SIP session, is included in the transparent SIP signaling through the SIP-AS.

By pushing a route on to the SIP route header (in much the same way as the CSCF routes to an IMS application server), the application router may also instruct the container to route the request to a SIP application deployed on another server. A detailed description of this mechanism is provided in the following section of this chapter.

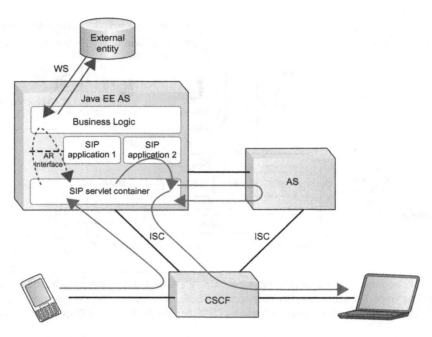

FIGURE 4.3

Influencing AS chaining.

Source: Ericsson Review, *No. 3, 2007. Service composition in IMS using Java EE SIP servlet containers.*

Consequently, IMS applications may be invoked by and subsequently control any type of session- and non-session-related communication service in the IMS network, such as but not limited to voice calls, video calls, messaging, and chat sessions. A SIP-AS may thus influence how media is used and routed for a SIP communication session. Applications may be invoked for originating SIP sessions (*initiated by* a served IMS subscriber) and for terminating SIP sessions (*destined for* a served IMS subscriber).

4.3 IMS SERVICE COMPOSITION

This section gives an overview of the different possibilities available for composing SIP applications in IMS and their peculiarities.

4.3.1 Initial Filter Criteria

Upon receipt of an INVITE, the S-CSCF forwards it to the appropriate AS based on the service profile of the user and the information stored therein. This profile is retrieved in advance by the S-CSCF upon registration of the terminal from the HSS. For the purpose of selecting an AS, the main content of the service profile is the initial filter criteria (IFC) list.

IFCs trigger on service point triggers (SPTs) in order to send the SIP request to the correct AS. Filter criteria contain the following information:

- Address of the AS
- Sequence of the filter criteria
- Trigger points composed of service point triggers (SPTs)
- SPTs chained through Boolean operators
- Default handling
- Optional info to be added to the message body.

The following SPTs are possible:

- Initial SIP methods (e.g. INVITE)
- Registration type (initial/re-registration/de-registration)
- Existence of specific headers
- Content of specific headers or Request-URI
- ID of the user – mobile originated (MO) or mobile terminated
- Session description information.

The IFC may only address an individual AS; since an end-user will very likely be subscribed to multiple services deployed on different AS, multiple IFCs are needed, one per AS. Moreover, IFCs may not address individual services on the AS; consequently, a workaround is needed in order to accommodate for multiple applications hosted on the same AS. Typically this issue creates the need for significant additional logic inside the AS. This logic may be seen as a meta application that attempts to infer which service is appropriate based on the properties of the signaling. This application would receive all SIP signaling and then decide which application on the AS will handle it. This typically will be part of the implementation of the application router on the JSR289 SIP container. Unfortunately, this implies that the AR implementation needs to be extended with every new application deployed on the AS. A programmable implementation of the AR that may be programmed to execute different orchestrations solves this problem.

Another way to deal with this issue is to deploy multiple AS that cater for different user groups with different service configurations.

An in-depth description of the IFC mechanism is provided in the following chapters.

4.3.2 Two-Tier Composition and the Service Capability Interaction Manager

IMS foresees rudimentary and manual service interaction coordination through the IFC mechanism. However, for cases with larger numbers of services or frequently changing services, the IFC mechanism is time-consuming, complex, and expensive.

Due to the limitations of the IFC mechanism and the strong requirements for a composition mechanism that may efficiently support the creation of enhanced value-added IMS applications, an orchestrating entity, the Service Capability Interaction Manager (SCIM), was defined by the 3GPP envisioned to act between the S-CSCF and the various other application servers over ISC (Figure 4.4).

The SCIM is an entity in the application layer of IMS that was introduced by 3GPP as part of the REL-5 version of the 3G network. Capabilities interaction management refers to co-ordinated execution of potentially conflicting services. Its original purpose was the co-ordination of service interaction. However, the introduction of the SCIM allows for increased flexibility through its compositional capacity. The SCIM may be triggered multiple times in the course of a session, before or after any AS, in order to execute additional services, perform some manipulation, consult external databases, or perform any other task as part of the composition. Effectively the SCIM introduces an additional point of in direction that enables a second level of composition above that of the S-CSCF.

It should be emphasized that the strict application of this design pattern may also introduce a potential bottleneck, since it requires all signaling to traverse the SCIM, even for sessions that have no value-added functionality.

In order to avoid this bottleneck, IMS architecture application servers (AS) that offer high-performance basic services need to be directly accessible from the S-CSCF, bypassing the SCIM in case no composition is required. Such an AS should also be able to seamlessly integrate itself in a composition executed by the SCIM in order to make its basic services part of a larger value-added use-case.

MMTel (Multimedia Telephony) – supporting basic telephony and supplemental features in IMS – is such a basic high-volume service that needs to be offered directly to millions of

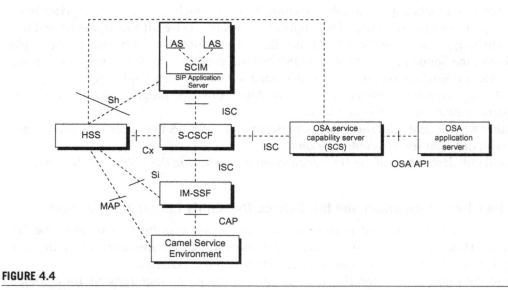

FIGURE 4.4

SCIM according to 3GPP.

Source: 3GPP TS 23.002 v7.1.0, figure 6a. © 2010 3GPP™ TSs and TRs are the property of ARIB, ATIS, CCSA, ETSI, TTA and TTC who jointly own the copyright in them. They are subject to further modifications and are therefore provided to you "as is" for information purposes only. Further use is strictly prohibited.

subscribers, while in other cases it will certainly be useful as part of a larger composition. An MMTel AS can be deployed as an AS in parallel to the SCIM so that, based on mechanisms such as IFCs and service identifiers, the S-CSCF may route traffic directly to the AS when the service does not require composition.

4.3.3 Unified Web Services and IMS Composition

The design principles discussed in the previous section allow developers to reduce the complexity of component design and to build reusable application building blocks. The ability to rapidly develop a rich portfolio of value-added IMS applications by focusing on implementing a business process, rather than SIP signaling flows, analogous to the SOA way of working increasingly implemented by Enterprise systems hinges on the existence of a platform to facilitate rapid application composition out of constituent components. The aggregation of such composable constituent components into coherent applications is a task that may be achieved through a variety of means.

A number of mechanisms have been standardized for telecom service composition. Previous sections have introduced the use of the Service Capability Interaction Manager (SCIM) as defined by the 3GPP to enable the creation of enhanced value-added IMS applications, an orchestrating entity for IMS over the ISC interface. This chapter will further discuss in great detail the SIP servlet API based on Java Enterprise Edition for the implementation of SIP and IMS applications.

Moreover, given the proliferation of SOA-inspired systems both in the Enterprise and on the Internet, there exists a need to efficiently and effectively closely integrate IMS applications with web services of different types (SOAP, REST, etc.). Invariably, IMS applications need to be composed out of web services and SIP applications to form value-added services that synergistically combine telecom functionality with the wealth of data accessible via web service APIs. This composition of web services with telecom services is called "unified composition" (Bond et al., 2009). Refer to Figure 4.5. One example of unified composition would

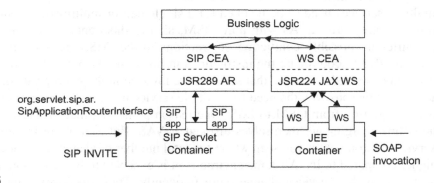

FIGURE 4.5

Unified web services and IMS composition.

be a customized service that displays the social network personal profile of a caller (Web 2.0 service) for an incoming call (SIP application).

A composition function that fulfills the requirements toward an SCIM may be implemented based on the foundation provided by the converged SIP servlet container as standardized in JSR289. Two examples of such unified composite applications created out of SIP and web services constituent components using the application router (AR) interface and the Java API provided by JSR289 are presented in detail in Chapter 5.

4.3.4 Next-Generation Intelligent Networks and Migration to IMS

Value-added services in IMS and the evolution of legacy value-added services toward all-IP and IMS are generally termed the *Next-Generation Intelligent Network* (NG-IN). NG-IN embodies the service environment in next-generation networks. It is this service environment that makes it possible to control (a) calls in circuit-switched networks, such as GSM and PSTN/ISDN, and (b) calls and sessions in packet-switched networks, such as IMS. NG-IN helps operators to safeguard investments made in legacy value-added services.

The Intelligent Network standard adheres to the principle of *single point of control*. This implies that not more than one IN service may, at a specific moment, control a call in the GSM network. Nevertheless, operators do wish to deploy multiple IN services in their network, with these multiple IN services working on a call. Vendors often apply proprietary multiple service invocation techniques for this method, usually a "glue" service that combines the original services or for the simplest cases different services triggered in sequence (service chaining). Simple rules for invocation priorities allow sufficient control with reasonable effort, when using such glue logic. Using proprietary methodology has, by its nature, the limitation that it can be used in one network only and often requires extensive system integration and testing.

When migrating from GSM to IMS and deploying legacy IN in the IMS network, the above-described issue of multiple IN service invocation needs to be addressed as well. An operator may want to combine legacy IN service invocation with IMS VAS. An architecture for multiple service invocation is given in Figure 4.6.

When multiple services need to be invoked for a single call or multimedia session, *service* interaction becomes crucial. As with pure GSM, also in this context different services may provide different, possibly conflicting, instructions to the IMS core network. Service composition is needed to coordinate the instructions in this case. Also, careful provisioning of subscriber data is needed, to ensure that services are invoked in the appropriate order.

There are several reasons why the need for multiple service invocation becomes even more profound when migrating from GSM to IMS. One reason is the fact that MMTel simulation services and online charging are themselves invoked as VAS. But even more importantly, as in IMS, a service environment is foreseen where the total number of available services is considerably higher compared to IN. Also, the environment is expected to become more dynamic in the sense that available services change more frequently. Thus, operators want to benefit from VAS capability in IMS to build flexible, short-time-to-market services. So, multiple services may need to be invoked in IMS. When a legacy IN service is applied in the IMS

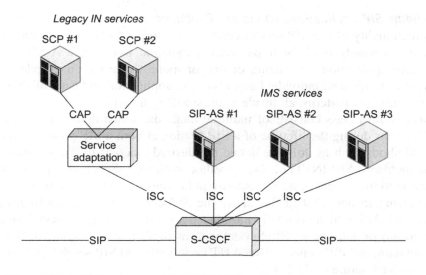

FIGURE 4.6

Multiple service invocations, GSM services and IMS services combined.

network, this legacy IN service will have to be combined with IMS service(s). In addition, a (mutual) dependency between the legacy IN service and the services in IMS may exist.

4.4 IMS APPLICATION SERVERS

This section will give an overview of technologies related to IMS SIP-AS with respect to IMS application development starting from the triggering of an AS via the initial filter criteria, continuing with the industry standard converged SIP servlet container technology used on Java Enterprise Edition application server platforms to develop and execute SIP applications and the chaining mechanisms used thereupon.

4.4.1 The Converged SIP Servlet Container

From the multitude of mechanisms available for telecom service composition, a few have been standardized for the implementation of telecommunications applications over IP: at the architecture level, the 3GPP IMS standard; within application servers the JAIN SLEE, a Java implementation of the Service Logic Execution Environment, a telecommunications-specific standard; and the SIP servlet API based on Java Enterprise Edition, the standard enterprise middleware platform across many industries.

This section will introduce developers to the concepts of SIP application composition based on the mechanisms of JSR289 and its predecessor JSR116. These mechanisms may be used to compose services into more complex composite applications, both within a single SIP container, as well as across multiple AS within IMS. Examples of specific IMS applications built using this technology are detailed in the following chapter.

4.4.1.1 Building SIP Applications on the SIP Container

The basic functionality of the SIP servlet container is similar to that of conventional HTTP servlet containers widely used for implementing dynamic web pages. A SIP servlet container hosts SIP applications consisting of one or more servlets each. Servlets within an application may encapsulate individual aspects of the application and can collaborate to create complex interaction patterns, all inside a single SIP application.

A SIP servlet Java class consists of multiple methods, which define reactions to events that typically occur during the lifetime of a SIP session at both a SIP protocol and SIP Java stack level. Methods such as doInvite define the desired reaction to the receipt of an initiating signal such as a SIP INVITE, and a method such as doSuccessResponse defines the desired reaction to the receipt of a SIP response indicating success (e.g. 200 OK).

It is interesting to note that in this sense the SIP container is similar to Java Servlets Specification JSR 315, which originally introduced a similar concept for handling the HTTP protocol by means of defining an HttpServlet with methods like doPost and doGet. There is, however, a fundamental difference between HTTP servlets and SIP servlets, which relates to the session-oriented nature of the SIP protocol.

The SIP container may receive a multitude of SIP signals in run time in no particular order and in an asynchronous manner. This occurs due to the asynchronous and event-driven nature of session management, which is fundamentally different, more complex, yet also much more powerful than the simple request–response operation of a protocol such as HTTP. This complexity creates the need for a mechanism that may handle all possible permutations when it comes to SIP events processing.

The complexity of session management is further aggravated by the fact that the same doSipEvent methods (e.g. doSucessResponse) used by a SIP servlet are triggered even if the events occur on different call legs. As a result, servlet code should also take care of properly identifying which leg the current signal belongs to before taking any action. All these subtleties generally greatly increase the complexity of correct state handling in SIP application development.

This has brought forth a number of commercial and OpenSource tools geared toward reducing the complexity that the IMS/SIP developer has to deal with. Such tools are the Ericsson Composition Engine[1] as a commercial composition creation and execution platform for unified Internet/IN/IMS services, as well as eCharts. ECharts is a state machine-based programming language for event-driven systems derived from the standardized UML Statecharts language.[2]

4.4.1.2 Application Routing on JSR289

A Java EE AS provides the network services over which SIP requests and responses are sent and received. The SIP container then implements an application selection and chaining mechanism much like the one used by IMS on the level of the S-CSCF and on a higher level

---

[1] http://www.ericsson.com/ourportfolio/consumer-and-business-applications/composition-engine
[2] http://echarts.org/

by the SCIM (Service Capability Interaction Manager). This mechanism is the application router (AR) standardized in the SIP servlet API JSR289.

When the S-CSCF passes on an initial SIP request (i.e. a request that is not part of an existing SIP dialog or call) to the AS, the SIP servlet container uses the application router (AR) interface to consult an external composition logic with respect to what services to execute in order to establish a SIP chain that corresponds to the needs of the particular session and subscriber. Given that the service being executed passes on the initial request to continue the call, the container queries the AR again to select the next service to invoke. In this way, a chain of services is formed until all the desired services are invoked. Any subsequent requests and responses within established SIP dialogs are routed along the SIP chain. The AR is not triggered by such subsequent requests.

Each invoked SIP application on the container operates as an independent SIP entity. It can send and receive SIP requests and responses to both external and internal entities. In the context of chaining within the container, such messages are passed between adjacent applications (upstream and downstream) in the chain executing in the same container.

It is important to realize that to be able to take part in a SIP session and manipulate the call state, a SIP servlet needs to be on the communication path between the caller and the original callee right from the session establishment phase and may not be easily added to the chain later on.

This peculiarity of one-dimensional application logic introduces many problems when designing more complex applications that may only be successfully managed through the enforcement of a clear separation between the time when the SIP component is put on the communication path (SIP chain in SIP jargon) corresponding to a SIP session, and the time when this component may act at the SIP level. It should be pointed out that this mechanism bears great similarity to the way IMS AS are chained by the S-CSCF based on the initial filter criteria as described in previous sections of this chapter.

With a few rare exceptions, SIP components can be put on the SIP chain only during SIP session establishment. Once the session is established, its SIP chain can be modified only with difficulty. Therefore, any SIP component that may be needed at a later stage should be put on the SIP chain during its establishment (e.g. during initial SIP INVITE processing), even if later due to some dynamic conditions it turns out that this component is not going to be used after all.

It should be noted that in theory the application chain may also branch and merge, and can change dynamically at any point in the lifetime of a communication session. However, this not only increases the complexity of managing the SIP chain; it also introduces additional complexity related to handling the internal application state.

We have now clarified when to put SIP components on SIP chains. But when do these components start acting? Sometimes, SIP components (usually SIP proxies) act directly on the initial SIP signaling when they are put on the SIP chain. In other situations, SIP components may also act when they received certain subsequent signaling, e.g. specific SIP responses. Sometimes, components do not act at all, e.g. because none of the required conditions was met, and just let SIP signaling flow through them transparently.

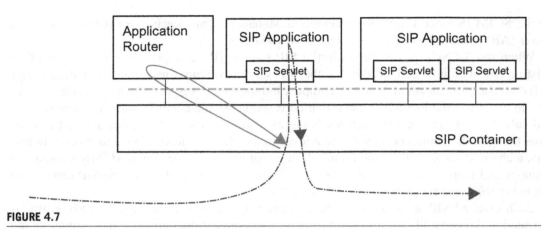

FIGURE 4.7

Service composition inside the SIP container.

Application routing in the SIP servlet standard is loosely based on the concepts defined by the Distributed Feature Composition (DFC) architecture. Figure 4.7 illustrates service composition in a SIP servlet container. The container orchestrates a linear sequence of services for a given originating or terminating subscriber. The services party A subscribes to are executed prior to services party B subscribes to.

The following section provides a detailed introduction to the development of an application router and the possibilities offered to developers interested in creating a flexible router able to create composite IMS services.

4.4.1.3 Implementing a JSR289 Application Router

JSR289 defines an API for implementing the application router functionality described in previous sections. This API is used by the SIP container to interact with the application router and as such it focuses on:

- Notification of the application router regarding the deployment and undeployment of JSR289-compliant SIP applications on the container.
- Consulting the application router with regard to which SIP application should process the current SIP request. This is a critical function for implementing composition functionality.

In addition to this API, JSR289 also allows the definition of custom application routers and outlines the rules for their packaging and deployment, thus allowing the creation of tailor-made application routers for specific needs.

It is important to realize that this API only defines a contract between a SIP container and an application router. It does not prescribe how the selection process for the next SIP application should look and how its logic should be implemented. Different SIP container vendors or third parties may provide their own versions of application routers.

The JSR289 standard mandates only that at least one application router with a default standard behavior should be provided by any JSR289-compliant SIP container, namely the default application router (DAR) as defined in Appendix C of the JSR289 specification. DAR is a very simple application router configured by means of a text configuration file containing routing rules that describe what should be done upon receipt of specific SIP requests, e.g.:

```
INVITE: ("CallForwarding", "DAR:To", "TERMINATING", "","NO_ROUTE", "0")
```

In this example, the DAR is configured to invoke an application on initial INVITE request, for the terminating half. The application is identified by its name as it was defined in the SIP application deployment descriptors.

To define a custom application router, a special Java class implementing the javax.servlet.sip. ar.SipApplicationRouter interface should be implemented. This interface is defined as follows:

```
public abstract interface SipApplicationRouter
{
  public abstract void init();
  public abstract void init(Properties props);
  public abstract void destroy();
  public abstract void applicationDeployed(List<String> sipApps);
  public abstract void applicationUndeployed(List<String> sipApps);
  public abstract SipApplicationRouterInfo
  getNextApplication(SipServletRequest initialRequest,
  SipApplicationRoutingRegion region, SipApplicationRoutingDirective directive,
  SipTargetedRequestInfo targetRequestInfo, Serializable previousState);
}
```

Init and destroy methods are invoked once during the deployment and initialization of the application router or during its undeployment.

The applicationDeployed and applicationUndeployed methods are invoked by the container when SIP applications are being deployed or undeployed. These methods receive lists of applications as parameters. This provides the application router with exact information regarding the currently deployed SIP applications.

getNextApplication is the most important method of the SipApplicationRouter API. The chain of services to be executed is selected based on the following four parameters:

- The subscriber (party A or B) URI from the initial request message.
- The current routing region (e.g. originating or terminating). An application router must distinguish between regions because a subscriber's services can differ between its originating (user is party A) and terminating (user is party B) regions.
- The routing directive (NEW, CONTINUE, or REVERSE). A service is free to specify the party A or B URI in an outbound request as well as specify an associated routing directive. The routing directive enables a service to give a hint to the composition logic in the container with respect to how to handle a subsequent service invocation based on the addresses specified in the outbound request.
- The application selection state associated with the request. This data structure contains a reference to previously invoked services. Where multiple services may be executed in a region, this hint may be used by the composition logic to choose the next service to invoke.

The task of the logic inside this method is to decide based on the information at hand which SIP application to select for further processing of the current initial SIP request.

In addition to the aforementioned sources of information, the AR is free to use any other local or remote source of information to decide what services to execute in order to complete the chain. Examples of internal information may include the time of day, network conditions, and state of the AS. Examples of external sources of information may be subscriber or service profiles in external databases, remote systems, i.e. Enterprise back-ends, or even Internet services.

In addition to defining an application router class, one also needs a deployment descriptor file called:

```
META-INF\services\javax.servlet.sip.ar.spi.SipApplicationRouterProvider
```

This is a simple configuration text file that contains only the name of the class implementing the application router. This class, together with the deployment descriptor and any further required classes, is archived in the form of a JAR file and is to be deployed on the SIP container. Upon deployment, the SIP container registers the presence of the deployment descriptor for a custom application router and starts using it.

4.4.1.4 Signaling between SIP Applications

The SIP servlet container provides the means for controlling the order and parameters of SIP servlet invocations based on the standard SIP mechanism described in RFC3261. Dynamic manipulation of the dispatching chain and consequently dynamic composition execution can therefore be achieved based on routing control of SIP requests inside the container. Such operations are supported by the SipServletRequest interface that enables the modification of proxying parameters.

This API provides the functionality to intercept, re-route, and even change the actual content of SIP messages and should therefore be used with caution. IMS application servers (AS) operate without dependencies on other SIP entities, as such changes to SIP signaling that assume additional functionality in other AS downstream are impossible to implement reliably. Functionality based on such extensions cannot be guaranteed to be executed correctly, or even be executed at all, as this may depend on the correct interpretation of the SIP message by the AS.

According to the SIP RFC, a SIP proxy is able to mandate that a request visit a specific set of proxies prior to delivery (section 16.6 of the RFC). This set of proxies is represented by a set of SIP URIs. This set must be pushed into the Route header field ahead of any existing values, if present. If the Route header field is absent, it must be added, containing that list of URIs. Based on this specification the route taken by a SIP request can be determined through modification of header fields such as Route. Correspondingly the pushRoute method of the SipServeletRequest interface defined in the API of JSR289 allows the manipulation of this header field, as illustrated below:

```
public void pushRoute(SipAddress proxyAddr);
public void pushRoute(SipURI proxyAddr);
```

In order to ensure that an application stays on the return signaling path, a servlet may be added to the SIP record route using the setRecordRoute method. This method creates a reverse dispatching chain that contains all servlets that need to receive return signaling. Responses will route through the set of Record-Route proxies traversed by the request in the reverse order.

The following example illustrates the approach:

```
Proxy p = req.getProxy();
p.setRecordRoute(true);
p.proxyTo(destination);
```

The first two lines of code in the previous fragment illustrate how to set the Record-Route flag. The proxyTo method in the third line sets the Record-Route flag that causes the request to be proxied to the required destination.

In addition to the RFC3261 compliant methods for SIP signaling redirection, the SIP container offers internal redirection functionality as part of its javax.servlet.sip.Proxy interface. The proxyTo method can redirect a request to a specific proxy without defining a record route. However, it should be noted that using this method introduces additional uncertainty since any downstream proxy may also invoke proxyTo once more and thereby redirect the request. The proxyTo argument identifies a proxy (or list of proxies) that should be visited before the request reaches it final destination. If an application generates its own final response it cannot proxy to more destinations. The proxyTo methods of the proxy class javax.servlet.sip.Proxy are defined in the following way:

```
//Proxies a SIP request to the specified destination
public void proxyTo(URI uri);
//Proxies a SIP request to the specified destinations
public void proxyTo(java.util.List uris);
```

A SIP servlet using this proxyTo API acts as a forking proxy either parallel or sequential, depending on the requested mode of operation. These container mechanisms can be used to implement complex interaction between SIP applications on the same container, as well as across IMS AS, as will be discussed in alter chapters.

4.4.2 SIP Application Types

This section provides an overview on the development of SIP applications on JSR289. The two basic types of SIP entity described are SIP proxy and SIP back-to-back user agent.

4.4.2.1 SIP Proxies in SIP Servlet API

A SIP proxy application does not terminate SIP signaling it receives. As such, a proxy application provides a relatively simple way to implement operations on SIP signaling without heavy impact on the signaling flow itself. According to RFC3261 a proxy may enforce policy by interpreting or even rewriting parts of the SIP signaling before forwarding it to the next entity in the chain.

The SIP application "redirect_sip" described in detail in later sections of this chapter is a typical example of a SIP proxy app. This section will focus on the different ways to activate proxying functionality in a SIP application.

Basic proxy functionality according to RFC3261 is implemented by means of the Record-Route header field. This mechanism can be used by proxies to insert a new route that will force future requests in the dialog to be routed through the proxy. Recording such a route may be implemented by simply including the following code fragment in a SIP application deployed on JSR289:

```
Proxy p = req.getProxy();
p.setRecordRoute(true);
```

There is, however, an alternative to the RFC3261-compliant mechanism for proxying of SIP signaling. The JSR289 javax.servlet.sip.Proxy interface and its proxyTo method provide developers with a simpler mechanism that requires no manipulation of SIP signaling to implement a proxy.

The argument for the proxyTo method identifies one or multiple proxies that will be chained before the request reaches it final destination. ProxyTo may also be used multiple times in sequence. This chain will be broken if one of the applications within it generates a final response and thereby terminates the chain.

An explanation of the proxyTo API was provided in the previous section. JSR289 provides further mechanisms for customizing proxy functionality based on the proxyTo API. Such customization enables a proxy to implement additional operations before passing on the SIP request to the next entity in the chain. A significant caveat is that such customization applies to all proxies defined by proxyTo, i.e. in the case where proxyTo defines three SIP URIs to be included in the SIP chain, all of these three proxies will receive the signaling with the prescribed modifications.

A way to avoid such inflexibility is the use of the ProxyBranches mehod. ProxyBranches acts in a way that resembles proxyTo, but enables the developer to define branches on the chain, so that each SIP URI may have its own set of customizations. The following code fragment illustrates the creation and customization of such branches:

```
public void doInvite(SipServletRequest local_request) {
  Proxy local_proxy = local_request.getProxy();
  local_proxy.createProxyBranches(
  sip:voicemail@sipservlet_example.com);
  local_proxy.startProxy(); }
```

The significant distinction between the proxyTo example and the previous ProxyBranches example is that the ProxyBranches may be used repeatedly to add new branches until one of the entities generates a final response; each of these branches may be created with a different SIP URI and different properties. Such properties may be the following:

- Push branch-specific SIP route headers
- Issue a branch-specific SIP CANCEL request

- Different values for SIP header fields like Record Route and Path
- Different recursive properties of 300-class SIP responses
- Different branch time-out values.

4.4.2.2 SIP B2BUA in SIP Servlet API

According to the SIP RFCs, a back-to-back user agent (B2BUA) is a SIP entity that receives a request and processes it as a user agent server (UAS). In order to determine how the request should be answered, it acts as a user agent client (UAC) and generates requests. A B2BUA differs from a SIP proxy application as it is required to maintain dialog state and typically performs additional operations before relaying signaling between the dialogs it has established.

A B2BUA may be particularly difficult to implement compared to a proxy, since it may need to manipulate all SIP requests (both session setup and termination) that are relayed to it. Clearly, this puts the B2BUA into a powerful position for implementing advanced manipulations that would not be within the scope of a simple proxy. As such, B2BUAs are an increasingly common design pattern in the architecture of SIP applications. Indeed, as SIP and IMS applications evolve and grow in complexity and functionality, it is recognized that B2BUAs are the only way of implementing applications of significant complexity. This point is clearly illustrated by our example use-case that is largely composed of B2BUA, and contains but a single proxy application.

A JSR116 B2BUA developer was required to implement functionality necessary for relaying between the two (or more) SIP dialogs connected by the application. One of the improvements introduced by JSR289 toward making the life of the application developer easier was the addition of the B2buaHelper class. B2buaHelper contains functionality that automatically reproduces SIP requests between the different dialogs, while always correctly translating the various header fields such as "To" and "From" and preserving unknown fields to avoid breaking application-specific communication implemented on the basis of proprietary header fields.

The use of B2buaHelper is illustrated in the implementation of the CallTransferB2BFrontend class of the CallTransfer component in later sections of this chapter.

4.4.3 SIP Application Composition in JSR116

While JSR116 may not provide the application router (AR) mechanisms referred to in previous sections, it nonetheless provides sufficient functionality to implement composition of SIP applications, albeit with a little more development effort compared to JSR289. The functionality of the missing application router may be reproduced through a SIP servlet deployed on the container for the purpose of controlling the dispatching chain. Such control can be implemented in SIP signaling and related JSR116 APIs (Figure 4.8).

The deployment descriptor format as used in deployment configuration files (e.g. sip.xml) is used to describe the set of SIP servlets provided by a given SIP application.

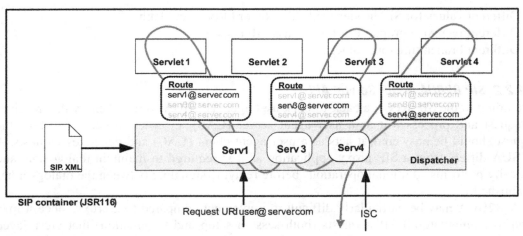

FIGURE 4.8

JSR116 SIP servlet chaining overview.

```
<?xml version="1.0" encoding="UTF-8"?>
<!DOCTYPE sip-app
  PUBLIC "-//Java Community Process//DTD SIP Application 1.0//EN"
  "http://www.jcp.org/dtd/sip-app_1_0.dtd">
<sip-app>
        <display-name>SIP Servlet Sample</display-name>
        <servlet>
                <servlet-name>echo</servlet-name>
                <servlet-class>com.sample.EchoServlet</servlet-class>
        </servlet>
        <servlet-mapping>
                <servlet-name>echo</servlet-name>
                <pattern>
                        <or>
                                <equal>
                                        <var>request.method</var>
                                        <value>REGISTER</value>
                                </equal>
                                <equal>
                                        <var>request.method</var>
                                        <value>INVITE</value>
                                </equal>
                        </or>
                </pattern>
        </servlet-mapping>
</sip-app>
```

FIGURE 4.9

JSR116 sip.xml configuration file.

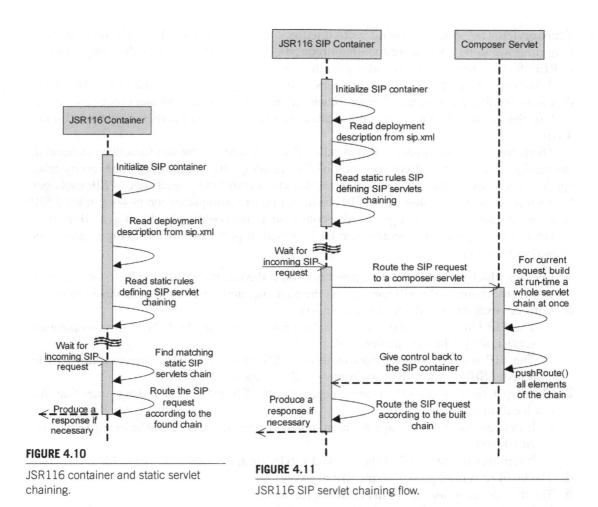

FIGURE 4.10

JSR116 container and static servlet chaining.

FIGURE 4.11

JSR116 SIP servlet chaining flow.

The deployment descriptor defines rules that specify the conditions (type of SIP request, headers of incoming requests, etc.) under which SIP applications and servlets are invoked, as well as the order of their invocation.

Multiple servlets may be invoked as a result of the same incoming SIP request. In this case, the container first composes the chain of execution of SIP servlets inside a SIP application based on the static rules defined in the deployment descriptor of the application and then starts composition execution according to this chain. The conditions formulated in the deployment descriptor format therefore implement limited composition functionality, allowing for the static definition of SIP servlets chains that cannot be altered without a change in the deployment descriptor.

The example in Figure 4.9 represents a sip.xml file for a servlet "echo". Its servlet section defines that the name of the servlet is "echo" and it is defined by a class com.sample.

EchoServlet. The servlet-mapping section defines, for this servlet, which SIP request mapping rules are to be applied for incoming requests. Specifically, it defines that only INVITE or REGISTER SIP requests should trigger this servlet.

A SIP request may traverse several proxies on its way to a SIP server. Each will make routing decisions, modifying the request before forwarding it to the next element and eventually putting itself on the so-called Record-Route, if it wants to be visited when a response is sent back (Figure 4.10).

Therefore, any servlet may use the so-called Record-Route mechanism (described in detail in the previous section), to enforce reception of SIP signaling after it has been processed by other applications on the container. This consequently allows control of the exchange of SIP messages between servlets participating in the SIP chain. In order to manipulate composition-related SIP signaling, the SIP application implementing composition functionality should act as a SIP proxy.

Once the composition execution servlet is started, it performs the following sequence of steps (Figure 4.11):

1. Decides which servlet is to be executed next or, if there is no further servlet to be executed, stops processing of the SIP message on the container and sends it out to its destination.
2. Enforces execution of the next selected servlet:
   - The SIP URI of the next servlet on the route, as well as the URI of the composition servlet, are pushed via pushRoute().
   - The SIP request object send() method or SIP proxy proxyTo() method are used to transfer SIP message delivery control to the container dispatcher.
3. The dispatcher examines the routing headers of the SIP message and determines that this is a local route:
   - It delivers the SIP message to the required servlet, i.e. the servlet selected to be executed next.
   - It removes the SIP URI of the invoked servlet from the Route header field.
4. The invoked servlet processes the SIP message.
5. The dispatcher is once more in control of the process:
   - It determines that the next destination point is a local route and the composition servlet is invoked.
   - The process of selecting the next servlet starts again at step 1.

4.5 CONCLUSIONS

This chapter introduced in detail the mechanisms and concepts involved in creating IMS applications based on common off-the-shelf software technologies such as Java Enterprise Edition and reuse of existing services according to Service-Oriented Architecture principles. The next chapter will present in detail the step-by step implementation of two IMS applications, making extensive use of these mechanisms.

Service Development

In Chapter 2, we introduced the concept of capital goods software; this chapter provides an explanation of how to implement such IMS applications by means of two use-cases implemented using IMS service composition, which we described in Chapter 4. The first use-case is a virtual call center application. Its implementation relies extensively on JSR289-compliant SIP applications and demonstrates some of the finer points of SIP signaling manipulation using a SIP container as part of the development of an IMS application. The second use-case is web oriented; it implements a web-based do-not-disturb feature using a third-party calendar service. This application blocks calls based on the availability of the callee in the third-party calendar service.

The description of the implementation of both use-cases is broken down starting with the overall composition of the business logic, its constituent components (both SIP and web/RESTful services), and finally a detailed description on how to implement those using JSR289 API mechanisms.

In both of the use-cases that follow, implementation of the IMS application will be described based on the concept of service composition introduced in Chapter 4. The application will therefore focus on the implementation of the core business logic, whereas most functionality related to signaling handling will be delegated to external services that are orchestrated by the main application. This applies to both SIP servlets and web/RESTful services.

The development of the core business logic may be achieved using traditional Java-based development, i.e. on JSR289 itself. Such a composite application may indeed be an implementation of a custom application router as specified by JSR289 (see Chapter 4). However, modern tooling based on interactive graphical design is also available today[1] that allows the orchestration of functionality on the converged SIP container based on a visual way of working. Such tooling allows for overall rapid development and greater reuse of the constituent components. Examples of such tools include the Ericsson Composition Engine (Niemöller et al., 2009) as a commercial composition creation and execution platform for unified Web/IN/IMS services, as well as eCharts. ECharts is a state machine-based programming language for event-driven systems derived from the standardized UML Statecharts language (Smith and Bond, 2007).

The following sections will make use of such a graphical notation to illustrate the business logic implemented inside such a custom Application Router (AR) implementation.

---

[1] Refer to www.elsevierdirect.com/companions/9780123821928 For freely available web based service development environment and introduction.

IMS Application Developer's Handbook: Creating and Deploying Innovative IMS Applications.

81

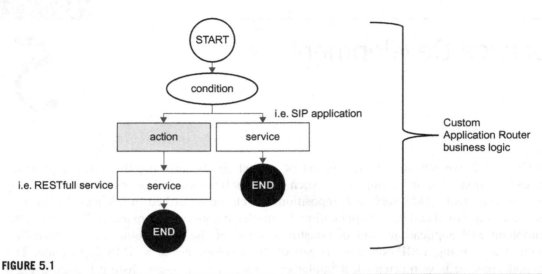

FIGURE 5.1

Simple graphical notation for application logic.

In the notation as illustrated by Figure 5.1 the following naming conventions are used with regard to JSR289 related components:

- AR will be referred to as application and its internal workings as business or composition logic (overall diagram)
- SIP applications, SOAP and RESTful services used as constituent components will be referred to as services (white boxes)
- start (white circle) and end (black circle) of the logic,
- conditional elements to signify a choice in the flow (oval)
- elements to signify an action by the custom AR itself (gray box).

5.1 VIRTUAL CALL CENTER USE-CASE

In this section we show how IMS service composition can be used in order to implement routing control for a virtual call center (VCC) application. This virtual call center employs groups of human agents with specific competencies. A customer shall be routed to an agent who has the right competence matching the customer's request. When calling a service number the customer is first routed to an interactive voice recognition system (IVR). The leg is to be subsequently terminated and a new leg is to be set up to the actual call-center agent.

The IVR presents the available competencies in a voice menu and requests the customer to input their selection. The call is then routed to an agent with matching competence. The composite service that controls this call center will also ensure that customers never call agents directly.

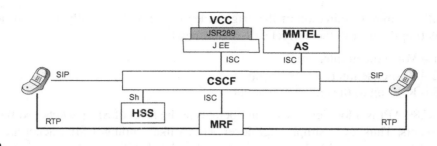

FIGURE 5.2

VCC use-case architecture.

5.1.1 Use-Case Architecture

SIP functionality in this application is implemented by means of the Java SIP servlet APIs as standardized in JSR289, a widely accepted platform for Java-based development of SIP/IMS applications. As such, a JSR289 SIP container on a Java EE AS is the core of the architecture we describe.

All XML web services components are implemented in Java by means of the JAX-RS API, which is a part of the Java EE specification. Since web services as such are not the focus of this book, we omit the details of their implementation and simply refer to their functionality. These services could be developed by means of a number of technologies including Java EE, .NET, or even scripting languages.

Figure 5.2 shows the network nodes used in this use case. A JSR289 Java EE application server (i.e. SailFin) hosts the IMS application that executes the VCC composite service. Upon receipt of a call to the service number, the VCC composite service is invoked through the ISC interface. The VCC composite service contains the overall logic to utilize and coordinate the constituent components. This includes interaction with the MRF/IVR in order to identify the customer's requirements, as well as the manipulation of SIP signaling in order to route the call accordingly. This requires the possibility to react to SIP responses and detailed manipulation of SIP signaling.

IVR functionality (streaming of the audio announcement and user interaction) is used extensively in this case. The Media Resource Function (MRF) typically provides such functionality in the IMS. An MRF supports streaming, broadcasting, and multiplexing of audio and video from one or multiple sources to one or multiple destinations. In our case, to receive an audio stream, one needs to establish a SIP session to the MRF and provide as input the name of the VoiceXML file to play. When a session is successfully established, the MRF would stream audio to the caller according to the required VoiceXML description.

5.1.2 Use-Case Business Logic

The business log of the virtual call center application is broken down into three phases. The first of these deals with IVR interaction and blocking of direct calls to a specific agent. The second one deals with the configuration of response interception, essential for triggering the new call leg upon termination of the original leg to the MRF. Finally, the third leg deals with connecting the user to the correct call-center agent.

Phase 1 – Voice menu interaction
Phase 2 – Response interception set-up
Phase 3 – Routing to the call-center agent.

The JSR289 API is a low-level API that operates at the level of SIP packets and their headers and contents. Thus, even simple high-level actions like sending a SIP packet are actually implemented by a comparably long sequence of low-level API calls that first create a SIP packet with a required payload and then send it. The resulting code is usually rather long and obscures the actual high-level logic behind a forest of low-level details. Going through the full Java code implementing the use-case would be too tedious and time-consuming for the reader. Therefore, this chapter describes the logic of SIP components by means of message flow/interaction diagrams, concentrating mostly on the high-level logic and message flows. Code fragments are embedded in the text to illustrate the API calls needed to achieve critical steps in the logic. Finally, for the sake of completeness, the complete Java source code for the described components is available for study and review in Appendix A[2].

5.1.2.1 Phase 1: Voice Menu Interaction

The first phase of the service VCC service is the routing of the user to the voice menu. The business logic sections for this phase are shown in Figure 5.3. The application first checks the dialled phone number. If the customer has dialled something other than the special service number, an announcement is played. This prevents the customer from calling an agent directly. The right branch in the business logic implements this simple announcement to the user and call termination. In order to redirect the call to the announcement, the services replace_system_headers and redirect_sip are used. Subsequently the composite service is completed, i.e. the SIP session setup is continued by the CSCF. As a result of the change in destination, the call is terminated with an announcement.

The left branch of the business logic is taken if the customer has dialled the correct service number. The first service used in the composition is call_transfer. This step only prepares a later breaking and re-establishment of the SIP session by embedding the call transfer service in the SIP chain. The next step is routing the call to the IVR in order to acquire the customer's requirements. This redirection is achieved by invoking the replace_system_headers and redirect_sip services.

As a result of this phase the AR releases control of the signaling and the user is routed to the voice menu. The flow diagram in Figure 5.4 represents SIP flows corresponding to the first phase of the use-case: the user is routed to the voice menu.

The following phase will deal with the application logic needed to receive the responses from the IVR/MRF.

5.1.2.2 Phase 2: Response Interception

At this stage of the business logic, the application needs to intercept the subsequent BYE request coming as a consequence of the termination of the call leg to the IVR/MRF after its

---

[2] see www.elsevierdirect.com/companions/9780123821928

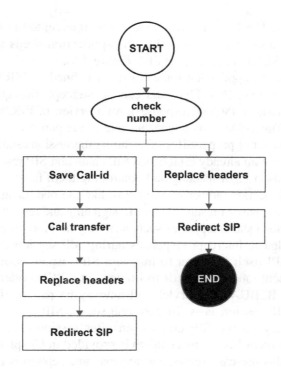

FIGURE 5.3

VCC Phase 1 business logic.

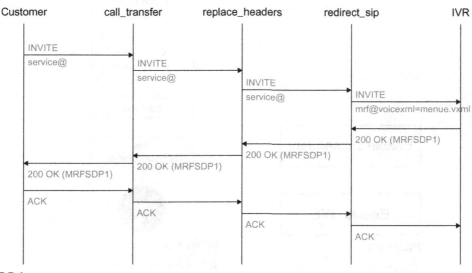

FIGURE 5.4

VCC Phase 1 flow diagram.

interaction with the user. The BYE request needs to be intercepted in order to stop the call from being torn down. In order to achieve this, the application needs to be set up so that it will receive subsequent SIP requests (e.g. BYE) (Figure 5.5).

It is worth noting that an application router (AR) as defined in JSR 289 is only triggered on the initial SIP requests (e.g. INVITEs), and cannot intercept subsequent SIP requests and responses (e.g. BYE requests or 486 OK responses). An overview of JSR289 and the AR is given in the previous chapter. This behavior is understandable as the purpose of the application router is to control the establishment of proper SIP chains during the session establishment phase. Since subsequent SIP requests use an already established SIP chain and SIP responses cannot alter the SIP chain, there is no reason for inspecting and controlling them from a session establishment point of view. Nevertheless, more advanced scenarios, like the one being described here, often require additional triggering based on subsequent SIP signaling and responses.

This shortcoming is not a showstopper as such, as it may be overcome in a number of ways.

The basic tools needed to intercept responses during SIP session establishment are provided by the JSR289 API itself. In order to intercept SIP responses or subsequent requests after session establishment, one only needs to implement a SIP servlet that will stay on the chain. In the case of a B2BUA, the servlet will anyway be part of the SIP chain as long as the corresponding SIP session lasts. In case one sues a SIP proxy, the servlet needs to record-route itself to stay on the SIP chain even after session establishment. A detailed explanation of using the record-route mechanism is provided in Chapter 4.

For the purposes of this use-case subsequent requests interception is required which can be implemented with the help of the corresponding handler method for the SIP method of interest,

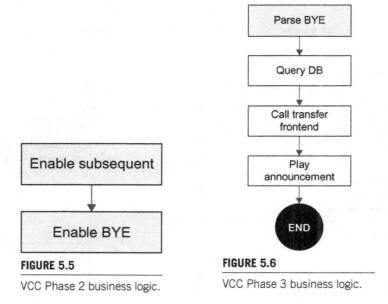

FIGURE 5.5

VCC Phase 2 business logic.

FIGURE 5.6

VCC Phase 3 business logic.

e.g. doBye (). A SIP stack will trigger this handler every time it detects a matching subsequent SIP request. Chapter 4 also provides a detailed explanation of SIP servlets triggering mechanisms.

Alternatively, a framework may be used to hide the more arcane details of working with SIP signaling interception on JSR289 and instead expose a more abstract view on the underlying SIP stack. This abstraction may be implemented by automatically inserting help apps on demand, which in turn may intercept any subsequent messages and notify the custom AR/business logic. The tooling associated with such a framework typically will provide mechanisms for the user-friendly definition of interception rules for any type of SIP signal.

The following phase will be initiated by the receipt of the responses from the IVR/MRF.

5.1.2.3 Phase 3: Routing to the Agent

The third phase of the use case implements the assignment of an agent to the user. The user is put on the waiting queue of the agent and hears an announcement, while waiting for the agent. Finally, the agent answers the call and the announcement leg is terminated.

The composition logic in Figure 5.6 details the final part of the VCC composite service. This part is invoked when finally the subsequent SIP request BYE is received from the IVR. Consequently, application execution is on hold while the IVR is active and it resumes at reception of the BYE issued by the finished IVR. As a result the outcome of the interaction with the user, i.e. the required agent skill, is added to the BYE response as a payload. Parsing the message to retrieve this information from the BYE is therefore the first action when application execution resumes.

The RESTful service querydb is used in order to find a suitable agent based on the already identified required skill. The agent data returned from the database contains the SIP URI of the agent. Subsequently, the call shall be re-directed to this URI.

Given that the establishment of the original SIP session was already completed, the established outgoing leg needs to be released and a new leg needs to be set-up toward the new URI. This function is implemented by the service call_transfer. The call transfer step has been prepared in advance through the addition of call_transfer to the already established SIP chain.

The composite application will now use the HTTP interface call_transfer_frontend in order to activate the call_transfer converged application. Subsequently, call_transfer starts setting up a new call leg towards the selected agent. While this is ongoing the composite application adds the service play_announcement to the new leg. The play_announcement service creates a further leg towards IVR. This way an announcement is played to the customer while the call transfer is still proceeding. Finally, when the new call leg toward the agent is set up, the SIP leg to the IVR is released (Figure 5.7).

5.1.3 Constituent SIP Applications

This section focuses on explaining the inner workings of the more complex SIP components of our use-case. There are two such components. The first is the "play announcement" component and the other is the "call transfer" component. As is usually the case with SIP, most

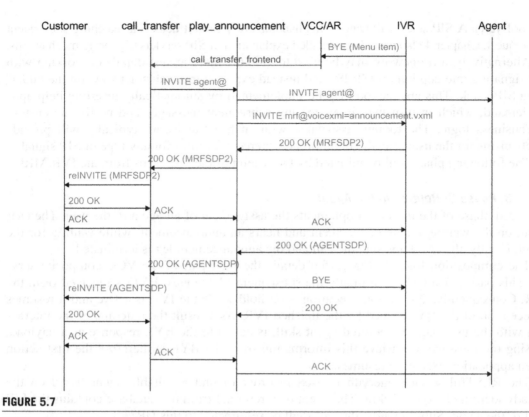

FIGURE 5.7

SIP flow Phase 3 (intermediate announcement and call to agent).

of the difficult components are B2BUAs (back-to-back user agents). This is for a good rea-son: such components typically control and manipulate multiple incoming and outgoing call legs simultaneously and have a nontrivial logic for deciding when and how these legs are created, connected to each other, or terminated.

Besides explaining the logic of implementation, we will also see how such important and useful features of the JSR289, such as SIP-related annotations, resource injection, and sup-port for converged applications, are used in the implementation.

Covering all constituent components in depth in this chapter would result in considerable repetition. To avoid this, simple constituent components, with self-explaining functionality, are omitted. Nevertheless, full source code is included in Appendix A[3].

The following constituent components are used and integrated by the composite service.

redirect_sip
This service takes the role of a SIP proxy. It redirects SIP signaling to a new destination. This is reached by changing the SIP request URI. This redirection can be used during ses-sion establishment.

---

[3] see www.elsevierdirect.com/companions/9780123821928

replace_system_headers
This service implements a back-to-back user agent (B2BUA). It flexibly allows replacing SIP headers. This includes system headers like "to", "from", "contact", or "via". Together with redirect_sip, this service is used to redirect the call.

play_announcement
This service implements a back-to-back user agent. It creates a new SIP leg to the MRF. Furthermore, this service releases the additional leg on receipt of 200 OK on the main SIP leg.

call_transfer/call_transfer_frontend
This service is a converged service. It exposes two interfaces and therefore appears as two separate services. Call_transfer is a back-to-back user agent and thus a SIP service. Call_transfer_frontend is an HTTP interface towards the same service. The purpose of the overall service is the redirection of an already established SIP session. This is reached by setting up a new leg, while the existing outgoing leg is released. In order to do so, the service needs to be inserted into the SIP service chain at that particular pivot location from where the outgoing leg will be replaced. The SIP part call_transfer of the service is used in order to insert it at that position during session establishment. The HTTP interface call_transfer_frontend can be used later in order to initiate and control the replacement of the outgoing leg.

querydb
This service allows search of a database. Here this generic component is used to search the agent database in order to find an agent with the required competence. querydb is a RESTful service.

These components are generic. They implement generic actions that are not specific to the described use-case and can thus be reused easily in other contexts.

5.1.3.1 Play Announcement Component
The "play announcement" service used in the VCC business logic is implemented by the class announcementB2B based on the JSR 289 API.

A SIP servlet class announcementB2B consists of multiple methods, which define reactions to various events that may occur at the SIP protocol and SIP stack level. A couple of basic examples are the doInvite method, which defines the appropriate reaction to the receipt of an INVITE, and the doSuccessResponse method, which specifies the appropriate reaction to the receipt of a success SIP response (e.g. 200 OK). The full source code used to implement this functionality is provided in Appendix A[4]. A detailed explanation of the full set of methods defined by the API is not within the scope of this book. Both the JSR289 standard itself and other literature provide detailed explanations of the API.

In addition to defining reactions to SIP events according to the SIP servlets API, announcementB2B also makes use of the following JSR289 features:

- SIP-specific annotations for defining the name of the SIP servlet and SIP application:
  ```
  @javax.servlet.sip.annotation.SipServlet(applicationName =
  "AnnouncementB2B", name = "AnnouncementB2B", loadOnStartup = 1)
  ```

---

[4] see www.elsevierdirect.com/companions/9780123821928

- Resource injection for obtaining a SipFactory object from the SIP container. According to JSR289, the SipFactory object can later be used for creation of SIP requests and SIP addresses:
  ```
  @Resource
  private SipFactory sipFactory;
  ```

The high-level logic of this SIP application implements the following functionality used in the overall business logic illustrated in previous sections.

When an incoming call leg is received by this component, it would create two outgoing legs. One of the outgoing legs goes to the required callee and expects him or her to accept (or reject) the call, while the other one is supposed to be used to play an announcement to a caller in the meantime. As soon as the callee reacts to the incoming call leg and responds (i.e. accepts the call), the announcement leg is terminated and the callee is connected to the caller by connecting the corresponding incoming and outgoing legs of the announcementB2B servlet.

There are of course some interesting aspects about both outgoing legs. They are being established in parallel. Signaling on each of the legs may happen asynchronously and in any relative order; SIP signaling, therefore, should be handled with care.

Let's see what can happen:

- IF the callee responds before the announcement leg can be established, then the announcement leg should be cancelled (by sending CANCEL) and the callee leg should be connected to the incoming leg in the usual way, by simply propagating the response to the caller and then transparently propagating subsequent SIP signaling in both directions.
- IF the callee responds after the announcement leg has been established, but before this leg was connected to the incoming leg, then the announcement leg should be terminated (by sending BYE) and the callee leg should be connected to the incoming leg in the usual way, by simply propagating the response to the caller.
- ELSE, if the callee responds after the announcement leg was established and it was connected to the incoming leg, then the announcement leg should be terminated (by sending BYE) and the callee leg should be connected to the incoming leg by sending REINVITE to the incoming leg (this corresponds to Phase 3 of the overall business logic).

This logic is implemented mostly by the doSuccessResponse method of the annoucementB2B class. As SIP responses could be received on any of the outgoing legs, but reactions should be dependent on a specific leg, the implementation of the servlet marks different legs (more precisely, SipSession objects representing call legs) using custom attributes and setting them by means of the SipSession.setAttribute (attributeName, attributeValue) API. For example, an attribute called "role" indicates if a given leg is an incoming, outgoing, reinvite, or announcement leg. Another attribute called "status" indicates if a given leg is in an active or inactive state. The following is an example from the code of the doRequestForwarding method:

```
new_request.getSession().setAttribute("role", "announcement");
new_request.getSession().setAttribute("status", "active");
```

With this approach, when a response is received, the servlet checks the value of the "role" attribute to identify which leg generated this response and then acts accordingly. For example, this piece of code from the doSuccessResponse method checks if an incoming response was received on the announcement leg:

```
String invitePurpose = (String)
aServerResponse.getSession().getAttribute("role");
String status = (String)
aServerResponse.getSession().getAttribute("status");
if (invitePurpose != null&&
invitePurpose.equalsIgnoreCase("announcement")) {
// 200 OK for the announcement leg
}
```

Usage of such custom attributes is a typical trick used in a SIP servlet programming model for saving call leg-specific information for later usage during subsequent SIP events processing.

Implementations of SIP B2BUAs often make use of SipSession attributes for maintaining correspondence and identifying relationships between related SIP signals on different logically connected B2BUA legs. For example, a SIP session corresponding to an outgoing SIP signal of a B2BUA may have an attribute (like "linkedrequest" in our example) that keeps the related SIP signal from the incoming leg. For example, the doRequestForwarding helper method contains this piece of code:

```
// Associate outgoing request with the incoming request
serverRequest.getSession().setAttribute("linkedrequest", req);
```

A SIP session corresponding to the SIP signal from the incoming leg may in turn have a "linkedrequest" attribute that keeps the outgoing SIP signal. By doing it this way, B2BUA crosslinks certain SIP signals and/or SIP sessions. This correspondence is not always one-to-one, as complex B2BUAs may have, for example, one-to-many relationships between incoming and outgoing SIP signaling.

The establishment of the announcement leg and its connection to the incoming leg before receiving a response from the callee is, in essence, a specialized case of the "call transfer" scenario. In this specific case the original call is established between the caller and the IVR playing the announcement, while at the same time a call transfer command is initiated instantly to transfer the call from the IVR to the callee.

5.1.3.2 Call Transfer Component
The "call transfer" functionality is implemented as a converged application, combining SIP and HTTP servlets. The HTTP servlet part, which is implemented by the class

CallTransferB2Bfrontend (see Appendix A[5]), is responsible for accepting call transfer requests over HTTP and triggering the corresponding machinery at the SIP level. The SIP servlet CallTransferB2B (see Appendix A[6]) in turn takes care of the initial call setup and proper handling of the call transfer SIP signaling.

The internal implementation of the call transfer functionality is similar to the "Play Announcement" logic described in the previous subsection. The main difference lies in the triggering of transfer functionality. Whereas, in the case of the announcement function, a call transfer was initiated almost simultaneously to the establishment of the announcement leg, here we want to have more freedom when it comes to selecting a time for performing a call transfer. Therefore, a dedicated mechanism for triggering the call transfer is developed. As mentioned previously, this mechanism is implemented by means of an HTTP servlet called CallTransferB2Bfrontend. CallTransferB2Bfrontend receives two input parameters that are retrieved in the doGet method handling HTTP GET requests: the unique identifier of the original call (we use the SIP Call-ID for this) and the new SIP address, which should serve as the destination of the transferred call. Having these parameters in place, the HTTP servlet first looks up the SIP session corresponding to the original call and then initiates the transfer to the new desired SIP destination. Call transfer SIP message flow is illustrated in Phases 1 and 2 of the business logic.

It is important to realize that to be able to look up a SIP call by its unique identifier and eventually manipulate the call state, the CallTransferB2B SIP servlet doing this should be on the communication path between the caller and the original callee right from the beginning of the session. For more insights on when to correctly include SIP applications in the SIP chain, see the discussion on SIP application routing in the previous chapter.

As a consequence of the fact that CallTransferB2B is placed on each SIP chain that potentially wants to use the call transfer functionality, the servlet should be able to keep track of all those SIP sessions and remember enough information about them so that it can look up any concrete SIP call based on the unique identifier.

Using Java platform mechanisms, this may be easily achieved by means of using collections containing SIP session objects and doing a lookup inside these collections. In some cases, a JSR289-based Java SIP stack may provide specialized functionality to achieve this more comfortably as part of a framework. The interested reader may want to investigate the JSR289 specification descriptions of the SipApplicationSession class and SipSessionsUtil.getApplicationSessionByKey API that are used in the doGet method of the CallTransferB2Bfrontend to obtain a list of ongoing SIP sessions controlled by this converged application.

Having seen all the peculiarities of developing SIP servlets, one may ask why it is necessary to implement the triggering functionality by means of an HTTP servlet, i.e. yet another protocol, and not in SIP directly. There are different reasons for this, outlined below.

Consider that in many scenarios the decision to perform a call transfer is not taken in the SIP/IMS domain, but elsewhere, i.e. in a call-center web GUI or an attendant application.

---

[5] see www.elsevierdirect.com/companions/9780123821928
[6] Ibid.

For this reason, it may be problematic to directly send a proper SIP signal from the application, as most web frameworks used for implementation of such applications do not even support SIP as a protocol. On the other hand, most frameworks support the HTTP protocol and have the ability to fetch information from a given HTTP URL.

It should be pointed out that JSR289 provides the possibility to access SIP sessions from HTTP servlets and HTTP sessions from SIP servlets. This capability is complemented by the mechanism of annotations. Annotation may be added to the code of SIP applications to implement access to SipFactory objects inside an HTTP servlet (this is called resource injection). This allows the rapid development of SIP signaling-related functionality. HTTP servlets may, however, only generate initial SIP signaling and may not receive subsequent SIP signaling. This limitation enforces an architecture where a converged application will most likely consist of at least two servlets, an HTTP servlet to interface with external systems via HTTP and a SIP servlet that implements proper session management functionality. Together they act as a team and represent a converged application.

5.2 WEB-BASED DO-NOT-DISTURB USE-CASE

The case described in this section is implemented by means of a converged IMS application that integrates SIP applications and RESTful web services to enrich user experience for a basic telecommunication service such as do-not-disturb (DND). This improvement is achieved by seamlessly integrating functionality from supplementary web-based services with IMS call control.

The user experience will be that success of call establishment depends on the availability of the callee as determined by his or her appointments in a third-party web-based calendar application (such as Google Calendar or any other web-based groupware). Subsequently, a follow-up appointment may be introduced in the calendar as a reminder for a later call. Moreover, an instant message may be sent to the caller to explain why the call was blocked.

In addition, this section concludes with a description of how IMS service composition can help to further enrich user experience by reusing SIP components to implement an AJAX communications monitor that can be viewed in a browser on the caller's smartphone (Figure 5.8).

Such a mashup application can provide access to the entire call establishment process (call is in progress, call is being rescheduled, call is successfully rescheduled) through a common browser. Additional features include a call log enriched through calendar information (i.e. showing missed calls together with scheduled follow-ups), along with contact information derived from Internet social networking services.

5.2.1 Use-Case Architecture

The deployment architecture for this case is similar to the setup described in previous sections; the IMS AS is used as a converged SIP/HTTP AS that allows for hosting and execution of composite applications that make use of both SIP- and HTTP-driven constituent components respectively implemented as SIP and HTTP servlets. Such an application is executed in a single converged context, thereby allowing for resource and state sharing.

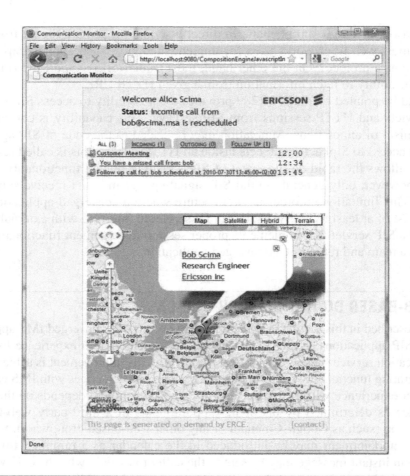

FIGURE 5.8

DND application AJAX GUI.

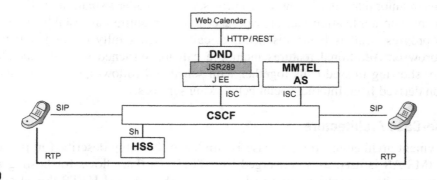

FIGURE 5.9

DND use-case architecture.

As shown in Figure 5.9, the DND application has access to a third-party calendar application that offers the functionality of hosting a user's agenda, authenticating the access to that agenda, and managing its contents (creating new events, deleting events or editing).

5.2.2 Constituent Components

The following constituent components are used and integrated in this converged application. Source code for these constituent components is available in Appendix A[7].

profile_info

This is a RESTful service implemented according to JSR311. It is used to query a database containing user information. In particular, the information it retrieves is as follows: the user's first/last name, the user's location (home/office), and authentication credentials that would be used later on as input to the agenda_service, in order to access the user's agenda. Such a service can be implemented using a relational persistence framework such as Hibernate. The implementation of this component goes very much in the direction of implementing a Java wrapper to a relational database, the details of which are outside the scope of this chapter. Therefore, for the sake of simplicity, the implementation details of this component are omitted.

send_sip_message

This SIP servlet is implemented according to the JSR289 API, and creates a SIP message and then transmits it. The input parameters for this service are the From, To addresses for the SIP message, the text message to be sent, and the content type for this text (e.g. "text/html"). Using this input, a SIP MESSAGE is generated and sent. This SIP servlet is implemented by overriding the doInvite method:

```
@Override
protected void doInvite(SipServletRequest req) throws ServletExecution {
}
```

A simple way to pass additional data (i.e. the content of the SIP MESSAGE) over the SIP signaling sent to this servlet is by adding additional custom SIP headers and/or payload. For this SIP servlet, we have used additional headers, namely "text", "msgfrom", "msgto", and "content-type". Using these custom headers, it is possible to retrieve this information inside the body of this function with the req.getHeader("HEADER_NAME") method and extend functionality of the SIP servlet. An example is shown in the following snippet:

```
String content_type=req.getHeader(PARAM_CONTENT_TYPE);
if ((content_type==null) ||
(content_type.length()==0))
content_type="text/plain";
```

---

[6] see www.elsevierdirect.com/companions/9780123821928

send_response

This SIP servlet is implemented according to JSR289 API. Its purpose is to generate a specific SIP response to the SIP request it has received as input. The type of response can be configured as an input parameter to this service. In this particular example it generates a SIP Busy response to a SIP invite, as a means of showing that the user called is currently busy. This SIP servlet is implemented by overriding the doRequest method:

```
@Override
Protected void doRequest(SipServletRequest req) throws ServletException {
}
```

In order to make this servlet more parameter driven, the same technique for custom header fields as used in the send_sip_message servlet is used here as well:

```
String response_phrase=req.getHeader("response_phrase");
SipServletResponse response;
if ((response_phrase!=null) && (response_phrase.length()>0)) {
response = req.createResponse(code,response_phrase);
} else {
response = req.createResponse(code);
}
```

agenda

This is a RESTful service, provided by a third party that allows for querying a user's agenda, the appointments that a user has made available there. Third-party vendors such as Google (Google Calendar) or Yahoo (Yahoo Calendar) usually provide services of this type and implementation details may vary, therefore. The API for accessing such a service typically requires as input in the HTTP GET request's query string, the user name whose agenda will be queried, and a public/private token key. For security reasons, the user provides the latter and it is a form of permission for accessing one's agenda.

As an example the end-point for an agenda may look like the following URL: http://3rd. party.vendor/user_name/token/calendar_name.

The reader can access this by sending an HTTP GET request. The response can either be an XML or JSON payload with a list of events stored in this calendar:

```
<event_list>
 <event>
  <title>Client Meeting: Rob Thomas account</title>
  <when start="20101111103000" end="2010111111113000"/>
 </event>
 <event>
  <title>Lunch</title>
  <when start="201011120000" end="20101111130000"/>
 </event>
</event_list>
```

By parsing the XML snippet, it is possible to isolate and identify interesting events, for example meetings that occur at a certain time period. To decrease the amount of information produced by this request, some third-party vendors make it possible to provide a timeframe as input parameter in the HTTP request, which signifies the time period of events of interest.

calendar_login

This RESTful service is used as means of creating an authentication token that can later be used by the create_reminder service, as a means of authenticating the transaction to the external third-party service that stores a user's agenda. The authentication takes place via the third-party vendor that offers the calendar service and therefore the mechanism may vary from a simple ClientLogin to the more sophisticated OAuth API. For more implementation details on such mechanisms, the reader is referred to the definition of JSR 311. In most cases, the input is a username and password. In our example, the username and password are retrieved by the profile_info service:

```
POST /login HTTP/1.1
Host: 3rd.party.vendor
User-Agent: Java/1.5.0_06
Accept: text/html,
Connection: keep-alive
Username; <enter_username_here>
Password: <enter_password_here>
```

If the authentication is successful, an authentication token will be returned within the 200 OK HTTP response message.

create_reminder

This RESTful service submits the reminder to the user's agenda. Similarly to the previous two services, a third-party vendor provides this service. It is normally implemented as an HTTP PUT request to an end-point provided by the third-party vendor that requires an authentication token (generated by the calendar_login service) and a payload written in the expected format for the targeted agenda. As a reference the following format can be used:

```
<agenda_entry>
 <title>New reminder</title>
 <when startTime="20100809102000" endTime="20100809104000"/>
<agenda_entry>
```

As an example, this can be implemented as an HTTP PUT request to the user's calendar. More specifically, the HTTP request could look like:

```
PUT /user_name/token/calendar_name HTTP/1.1
Host: 3rd.party.vendor
Authorization: AuthSub token="yourAuthToken" {Note: this is produced by the
calendar_login service }
User-Agent: Java/1.5.0_06
```

```
Accept: text/html,
Connection: keep-alive

<agenda_entry>
 <title>New reminder</title>
 <when startTime="20100809102000" endTime="20100809104000"/>
<agenda_entry>
```

If the request is successful, a 201 Created HTTP response will be returned, denoting that the newly posted resource was created.

5.2.3 Use-Case Business Logic

The business logic description for this use-case is split into five phases:

- Phase 1: Determining the location of party B
- Phase 2: "Out of office"
- Phase 3: An appointment
- Phase 4: "Party B is busy" message when the user is in the office
- Phase 5: Normal call establishment when the user is available.

For the sake of simplicity, in the following description we assume that party A is Bob's SIP phone and that party B is Alice's SIP phone. An additional assumption is that invocations of RESTful services are always successful (respond with a 200 OK).

5.2.3.1 Phase 1: Determining the Location of Party B

The first phase of this use-case aims at determining the called user's location. The business logic for this phase is depicted in Figure 5.10.

The composite service is triggered on the reception of a SIP INVITE message generated by Bob's IMS phone. The converged application initially retrieves Alice's current location (i.e. home/office) from her profile. In order to retrieve this information, a service called

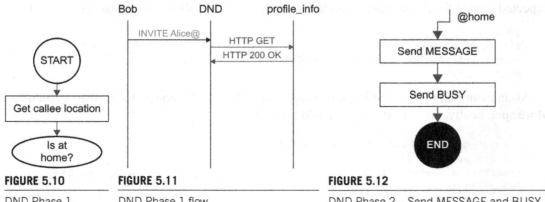

FIGURE 5.10

DND Phase 1.

FIGURE 5.11

DND Phase 1 flow.

FIGURE 5.12

DND Phase 2 – Send MESSAGE and BUSY.

"profile_info" is invoked. This is a RESTful service and it is triggered with an HTTP GET request. If the result of this invocation is "home", the left branch is executed.

The converged SIP/HTTP signaling flow for this phase is shown in Figure 5.11. As shown in the SIP flow, once the converged application, DND, intercepts a SIP INVITE it sends an HTTP request to the profile_info service in order to find where the user is, at home or in the office.

5.2.3.2 Phase 2: "Out of Office Message"

In this alternative, a message is sent to Bob's SIP Phone, with a text stating that Alice is currently unavailable. The segment of the business logic used for this phase is shown in Figure 5.12.

The first service in this branch generates and sends the SIP MESSAGE with the content "The user you are calling is currently out of the office". The next service will then send a SIP BUSY response to the caller, to Bob's SIP phone in this case. The converged SIP/HTTP flow for this phase is shown in Figure 5.13.

The converged SIP/HTTP flow here extends the one described in Phase 1 with the addition of the send_sip_message and the send_sip_response services. In this case the send_sip_message SIP servlet will be placed on the SIP chain. This servlet receives a set of parameters that allow it to generate a message originating from Alice to Bob with a message saying, "The user is currently out of the office". As a final step in this branch the send_sip_response servlet will be placed on the SIP chain. This servlet would generate a SIP BUSY response to the SIP INVITE that initiated this process in the first place.

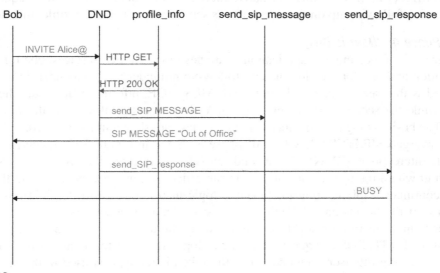

FIGURE 5.13

DND Phase 2 flow – Send MESSAGE and BUSY.

FIGURE 5.14

DND Phase 3 – Get agenda and parse.

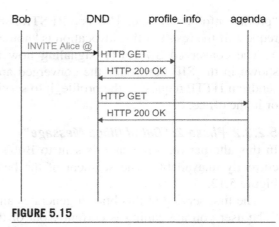

FIGURE 5.15

DND Phase 3 flow – Get agenda and parse.

5.2.3.3 Phase 3: An Appointment

When the service "Get called user's location" responds with "office", the right branch of the business logic is executed. The first service in this branch performs a lookup on Alice's calendar in order to find out if she is currently in a meeting or not. The adjacent conditional element evaluates whether there are any meetings. The segment of the business logic used for this phase is shown in Figure 5.14.

The converged SIP/HTTP flow for this phase is shown in Figure 5.15. As shown in the converged SIP/HTTP flow, in this alternative there is a subsequent HTTP GET request to the agenda service following up on the one that was sent previously to the profile_info service.

5.2.3.4 Phase 4: "User is Busy"

When Alice is in a meeting, the application continues to invoke two services, one for calendar authentication and one for creating the call follow-up reminder in the calendar. Subsequently, we proceed with a service that will send a SIP MESSAGE, informing Bob that Alice is in a meeting, while the second service sends a BUSY. The execution flow is finalized at the end element. The business logic segments used in this phase are shown in Figure 5.16.

The converged SIP/HTTP flow for this phase is shown in Figure 5.17. It can be seen that once DND intercepts a SIP INVITE, it sends an HTTP request to the profile_info service in order to find where the user is, at home or in the office. In this case the user is in the office, so DND continues implementing the business logic instructions from the rightmost branch. In this case, it first sends an HTTP GET request to the agenda service in order to find out if the user is in a meeting or not. In this alternative, the user is in a meeting. The next step is to send an HTTP POST request to calendar_login in order to create an authentication token for accessing the user's agenda. Once the token is ready, it is used in the HTTP PUT request, along with the content of the reminder, and it is sent to Alice's agenda. Afterwards, DND places the send_sip_message SIP servlet in the SIP chain. This servlet receives a set of

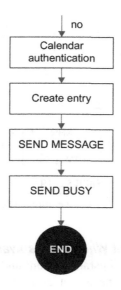

FIGURE 5.16

Send BUSY and remember.

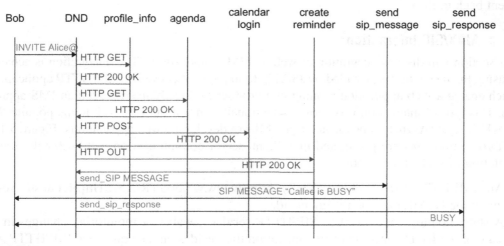

FIGURE 5.17

Phase 4 – Converged SIP/HTTP flow.

parameters that allow it to generate a message originating from Alice to Bob with a message saying, "The user is currently busy in a meeting". The final step in this branch is for DND to place the send_sip_response servlet in the chain. This servlet would generate a BUSY response to the SIP INVITE that initiated this process in the first place.

FIGURE 5.18

Phase 5 flow.

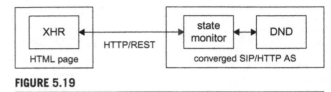

FIGURE 5.19

Logical component architecture for AJAX/SIP interaction.

5.2.3.5 Phase 5: Call Establishment When User is Available

The fifth and final phase is that of call establishment and it occurs if two conditions are met: the called user is in the office and she is not currently in a meeting. The converged SIP/ HTTP flow for this phase is shown in Figure 5.18. In this case DND application allows call establishment by proxying the SIP INVITE to Alice. If Alice picks up the phone, a 200 OK is sent back to Bob.

5.2.4 **AJAX/SIP Interaction**

This section will discuss integration of web and IMS applications. This integration is achieved by using the mechanisms provided by JSR289 to implement converged SIP/HTTP applications, which enable a Web application running in a browser to directly interact with an IMS application. This type of interaction was chosen intentionally since it fits very well to the popular Web 2.0 AJAX style (Asynchronous JavaScript XML) of developing web applications (Figure 5.19).

Two components are positioned in a client–server communication pattern for the implementation of such an interaction:

- An XMLHTTPRequest object (XHR) (http://www.w3.org/TR/XMLHttpRequest/), residing in the HTML page on the client side.
- A state monitor – a converged SIP/HTTP servlet, capable of monitoring changes in the state of the DND converged SIP application that resides on the converged SIP/HTTP AS.

Since the state monitor is implemented as a converged SIP servlet, it is already registered to an HTTP listener, which is provided by the underlying application server.

The execution flow for this interaction initiates on the client-side HTML page. This is done with the execution of a script written in JavaScript that specifies which update of the converged application the client side is interested in. In the context of the DND application, possible candidates may be that call establishment is in progress, the call is being rescheduled, the call has been successfully rescheduled, and so on.

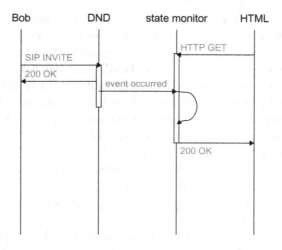

FIGURE 5.20

Flow for a converged SIP/AJAX application.

More specifically, this translates to sending an HTTP GET request to requestUrl, which is the listening end-point for the state monitor. The following snippet describes how this functionality is implemented by means of the XHR object:

```
xhr.onreadystatechange = subscriptionHandler;
xhr.open("GET", requestUrl, true);
xhr.setRequestHeader("event", event);
xhr.setRequestHeader("Content-Type", "text/xml; charset=utf-8");
xhr.setRequestHeader("Accept", "text/xml; charset=utf-8");
xhr.send("");
```

The event header in this HTTP request is used in order to specify the desired event to which a subscription is being made. Using this snippet, once the specified update is available a response will be sent back to the XHR object and the subscriptionHandler callback will be invoked to handle it. The following snippet describes a possible implementation of the subscriptionHandler:

```
function subscriptionHandler() {
if (xhr.readyState >= 3) {
 if (xhr.status >= 200 && xhr.status < 300) {
  if (xhr.readyState == 4) {
  alert("Call establishment is taking place");
  }
 }
}
}
```

More precisely, the code first checks the state of the XHR request and if a response of 200 OK has been received, an alert message pops up on the browser window.

On the server side, the state monitor, once it receives a request from an HTML page, can store it in a list for future reference. In DND, once an event occurs, then the state monitor can be statically invoked with some information regarding the event that has occurred. The state monitor then checks its list to see if an HTML page has asked for such an update. If this is the case, it responds back to that HTML page with an HTTP 200 OK response.

For example, if DND receives a SIP INVITE, it invokes the state monitor indicating that a SIP INVITE has been received. The state monitor would then check if an HTML page has requested such information and, if such an HTML page has, then it will receive an HTTP 200 OK. This functionality is illustrated in Figure 5.20.

The proposed implementation for the state monitor is only mentioned here as an example. The reader is encouraged to implement this functionality in a more sophisticated approach, better suited semantically and also performance-wise to the context of the application one is applying this approach to.

To further assist web developers, the aforementioned HTTP-based functionality can be encapsulated in a Javascript API, which can be referenced by an HTML page. Such an API acts as an adapter to the various HTTP requests (i.e. GET, POST, PUT, DELETE), toward end-points and the content-type of messages that need to be sent to the converged SIP application. Moreover, it deals with how to handle the responses and raising them as Javascript callbacks back to the Document Object Model they have originated from. Commercial tools, such as the ECE, already provide such APIs for the web developer's convenience.

For the purposes of scalability, the converged HTTP/SIP application server's HTTP listener can be extended in order to support HTTP long-polling requests. Using a framework such as Comet, you can easily make such an extension.

5.3 CONCLUSIONS

This chapter has presented the steps involved in implementing constituent service components in Java and how they are integrated into an overall application by means of service composition. The development of business logic is high level and easy to understand. The development of the necessary constituent component services is at least one order of magnitude more complex due to the use of low-level APIs for detailed protocol handling.

This approach provides an efficient way to integrate web services with IMS applications into converged applications that can offer a far richer user experience than with traditional telecommunications applications.

The methodology bears great potential for rapid and cost-efficient development. Constituent components implemented in a generic way lead to the creation of a library of components that can be broadly reused for various composite applications. The actual business logic of a composite application is then implemented as a high-level application that relies on constituent components to provide the implementation of protocol-level functionality.

Introduction to IP-Based Real-Time Communications

6.1 INTRODUCTION

This chapter introduces the reader to voice over IP (VoIP) and multimedia over IP. Although IP communications networks support a wider spectrum of services than voice calls, the technique for transporting voice via the Internet is used as a reference model. This chapter serves as a preamble to the subsequent chapters, where the principles of VoIP are placed into network context, such as the Session Initiation Protocol (SIP) and IP Multimedia Subsystem (IMS). Understanding VoIP may be considered a prerequisite for understanding IMS. Throughout the coming chapters (SIP, IMS), we will utilize and refer to the basics of VoIP.

6.2 BASICS OF VOICE OVER IP

When considering "voice over IP", we firstly need to gain an understanding of the "voice over digital transmission system". The "IP" in VoIP relates to the communication network through which the "V" (voice) is transported. Digital transmission of voice occurs in networks such as ISDN and GSM/UMTS, where the voice is transported, in digitized form, over *digital circuits*. A circuit is a virtual connection between two end-points. When a circuit is established, data (media) can be transported to the remote end of the circuit by sending the data "into the circuit". The circuit then ensures that the data is transported to the circuit end-point. The use of "circuits" for transporting voice and other forms of real-time media has led to the term "circuit-switched networks" (CS networks). IP networks, meanwhile, fall in the category of packet-switched (PS) networks. Each message (containing media, for example) that is to be transported to a remote end-point has to be addressed explicitly with that remote end-point's address.

Essentially, transporting digitized speech through an IP infrastructure has much in common with transporting digitized speech through a CS network. To appreciate the intricacies of digitized speech transmission, we shall study that aspect first.

6.2.1 Digital Speech Transmission

Figure 6.1 shows a schematic example model of speech transmission over a digital communication network. In the example in Figure 6.1, human voice is sampled (by an analog-to-digital, AD, converter), resulting in a constant stream of *codes* (samples). The sampling

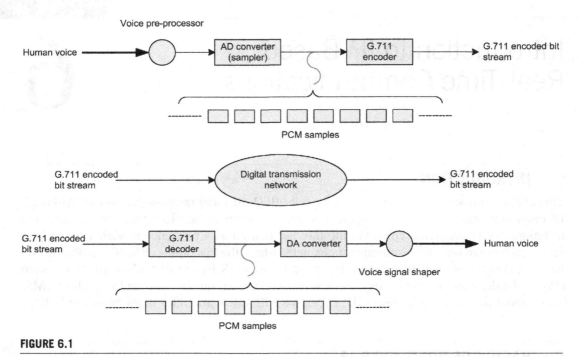

FIGURE 6.1

Digital speech transmission: a schematic representation.

frequency is determined by the maximum frequency of the analog signal for which transmission is required. According to the Nyquist theorem, the sampling frequency is twice the maximum frequency of the analog signal. For human voice, where transmission is traditionally applied in the range 300–3400 Hz, the sampling frequency is set to 8 kHz. A code size of 8 bits/code yields an effective bit stream of 64 kb/s.

Each code represents a sample of the speech. This process of speech digitization is known by the generic term pulse code modulation (PCM). Different forms of PCM exist. Each PCM code may, for example, be a linear or logarithmic representation of the absolute signal amplitude, or may represent a difference in signal amplitude. The latter is commonly known as differential PCM (DPCM). A further enhancement is the use of adaptive code books, as in adaptive differential PCM (ADPCM).

A coder-decoder (codec) subsequently processes the constant stream of PCM samples. The codec, as its name implies, is a combination of coder and decoder. For transmission of digitized speech, a coder is required. For the reception of digitized speech, a decoder is required. The task of the coder is to transform the constant stream of PCM samples into a bit stream that is suitable for transmission over a digital transmission system. For the resulting bit stream, i.e. the output of the coder, exact encoding of the PCM code points is specified, ensuring global compatibility between digital transmission systems.

Various speech coding standards are in use. The standards vary in terms of sampling frequency, code size, and encoding of the codes into the bit stream. Table 6.1 provides an overview of commonly used codecs.

The choice of codec is dependent on the application (high quality versus low quality) and on the transmission network (transmission capacity). The codecs listed in Table 6.1 are defined by the ITU[1] and by 3GPP,[2] except for iLBC, which is defined by the IETF.[3] Codecs are also differentiated through the manner in which they are adapted to their specific application. A codec may, for example, be optimized for human voice. Such codecs are typical for GSM/UMTS.

The bit stream resulting from the codec is suitable for transmission over a digital transmission network. Examples of digital transmission systems include:

- Circuit-switched (CS) connections, as used in ISDN, for example. A 64 kb/s PCM-encoded bit stream may be transported through one time slot of an E1 communication link. One E1 communication link may be divided into 32 slots for data transmission (of which 30 are used for media transfer). This form of synchronous multiplexing of time slots is known as time division multiplexing (TDM).
- Packet-switched (PS) networks, e.g. the Internet. The PCM-encoded speech is encapsulated in IP packets and transported through the IP network to the receiver. The encapsulation of the speech samples in IP packets is described in more detail later.
- GSM Radio Access Network (RAN). When a GSM voice call is established, the GSM phone transmits the encoded speech over a circuit through the RAN. This circuit has sufficient bandwidth to transport the gross bit rate of the GSM-encoded speech. Within the *core network* of the GSM network (including 2G network and 3G network), speech data is traditionally converted to 64 kb/s PCM for further transmission, typically through a time slot of an E1 link.
- Currently, speech may also be transported through the core network without conversion to 64 kb/s, leading to more optimized usage of transmission capacity in the network and leading to less speech processing. For interconnection to other networks, it is often still necessary to convert to 64 kb/s, though.

The speech data, e.g. G.711-encoded data, which is received at the receiving side of the communication link, is decoded by the codec, yielding a constant stream of PCM samples. This constant stream of PCM samples should be equal to the constant stream of PCM samples that was generated by the AD converter at the sending side. The only difference should be (minimal and constant) latency. The stream of PCM samples is then fed into the DA converter, which produces an analog signal. The analog signal, in turn, is applied to analog signal processing circuitry to produce audible sound (human voice).

Generally, when the PCM bit stream is transported over a CS connection, the PCM bit stream does not need further encapsulation. The CS connection provides synchronous,

---

[1] Refer to www.itu.int for the respective recommendations.
[2] Refer to www.3gpp.org for the respective technical specifications.
[3] Refer to www.ietf.org for the standard.

Table 6.1 Overview of Voice Codecs (not exhaustive)

Codec	Description
G.711	8 kHz sampling, 64 kb/s PCM. Commonly used in ISDN. Two variants exist: • A-law (ITU-T standard) • μ-law (used in North America) A-law and μ-law differ in the manner in which samples are represented by code values.
G.722	64 kb/s ADPCM standard suitable for 50–7000 Hz bandwidth speech at superior quality. It applies 16k samples/s. The codec is designed specifically for speech encoding.
G.723.1	Coder for multimedia communications at 5.3 and 6.3 kb/s. May be used in combination with video telephony, over a 64 kb/s data channel.
G.726	ADPCM codec for 40, 32, 24, 16 kbit/s signal output.
G.729	Conjugate-structure algebraic-code-excited linear prediction (CS-ACELP); 8 kb/s speech.
iLBC	Internet Low Bitrate Codec; developed specifically for VoIP applications. Defined by the IETF; refer to IETF RFC 3951.
GSM_FR	GSM Full Rate. Refer to GSM TS 06.10. Standard GSM codec methodology, in accordance with the Regular Pulse Excitation–Long-Term Prediction linear predictive coder (RPE-LTP). Average effective bit rate of 13 kb/s and gross bit rate (including bits for error correction, etc.) of 22.8 kb/s. Used for GSM radio access.
GSM_HR	GSM Half Rate. Refer to GSM TS 06.20. Codec based on Vector-Sum Excited Linear Prediction (VSELP) algorithm. Average effective bit rate of 5.6 kb/s and gross bit rate of 11.4 kb/s. Used for GSM radio access.
GSM_EFR	GSM Enhanced Full Rate. Refer to GSM TS 06.60. Codec based on Algebraic Code Excited Linear Prediction coder (ACELP) algorithm. Provides improved speech quality compared to Full Rate. Average effective bit rate of 12.2 kb/s and gross bit rate of 22.8 kb/s. Used for GSM radio access.
AMR	Adaptive Multi Rate. Refer to 3GPP TS 26.090. Set of codec standards developed for the third-generation mobile network, based on the multi-rate ACELP (MR-ACELP). Provides effective bit rate between 4.75 and 12.2 kb/s.
AMR_WB	Adaptive Multi Rate – Wide Band. Refer to 3GPP TS 26.190. Introduced in 3GPP Release 5 (2002). Applies a sampling rate of 16k samples/s, leading to effective bit rate between 6.6 and 23.85 kb/s. Superior speech quality compared to AMR and other narrow band codecs.

jitter-free transfer of the speech between two end-points (the CS connection offers a 64 kb/s data channel). Hence, the required transmission bandwidth of the CS connection may be equal to the PCM-encoded bit stream.

When the PCM bit stream is transported over a PS network, the PCM bit stream needs additional encapsulation in order to make it suitable for transmission. The details of this encapsulation are described in a later section. While a CS connection provides synchronous, jitter-free, guaranteed transmission of the speech data, speech data transmission over a PS network may suffer from the following problems:

- Latency
- Jitter
- Data packet loss
- Change in packet order at the receiver, compared to transmission source.

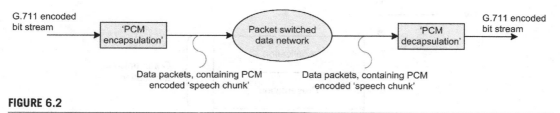

FIGURE 6.2

PCM encapsulation for transfer over a PS network (example).

To compensate for these characteristics of PS network-based speech transmission, the encapsulation method has to apply a designated mechanism to detect these anomalies and to recover from them (see Figure 6.2).

Encapsulation of the PCM-encoded speech serves not only to overcome the characteristics of the PS network, such as jitter, packet order change, etc., but is also required to adapt the speech data to the transmission method applicable in the PS network. One example is the Real-time Transport Protocol (RTP). RTP encapsulates chunks of PCM-encoded speech data into RTP messages. RTP messages may be transported over UDP/IP, to the intended receiver of the speech data (the remote party).

In the examples given above, the sender and receiver are using the same codec. Speech data encoded in accordance with the G.711 standard, for example, will be decoded in accordance with that same G.711 standard. A subsequent section explains how the sender and receiver know (negotiate) which codec to use for a communication session.

6.2.2 OSI Reference Model

The OSI (Open System Interconnection) reference model, developed by the International Organization for Standardization (ISO), is a model that is used in communication systems to divide a communication channel into logical tasks. A "seven-layer stack", as depicted in Figure 6.3, forms the OSI model. The OSI model is also used in the context of VoIP. Throughout this book, protocols are described for communication between two or more entities. Each of these protocols is positioned on one of the seven OSI layers. Each layer provides designated and well-specified services to the adjacent upper layer, using services from the adjacent lower layer. We will frequently refer to this reference model when describing protocols.

Above layer 5, the distinction becomes a little blurred. For example, in a SIP phone application, layers 6 and 7 are combined into the application that uses SIP (and RTP, RTCP) for a phone call.

An entity in a communications network does not need to support all seven layers of the OSI stack. Two entities may communicate on a particular communication level, whilst various nodes in the network provide lower-layer service. This is depicted in Figure 6.4 with a rudimentary example. The two SIP phones in the example in Figure 6.4 establish a communication session for setting up a phone call. The phone applications communicate with one another on layers 6 and 7. A SIP session, residing on layer 5, is established for that purpose

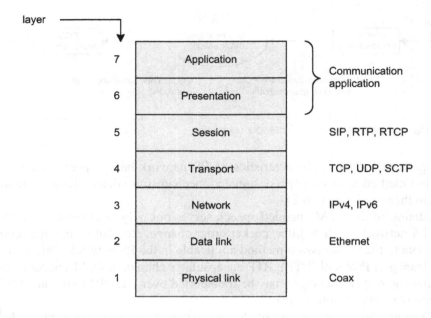

FIGURE 6.3

OSI reference model (with example layer implementation).

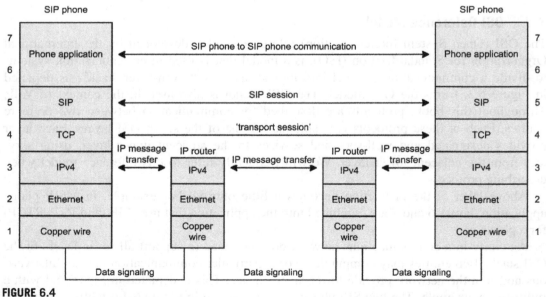

FIGURE 6.4

Example OSI layer representation for communication session between two SIP phones.

between the two phone applications, whereby the SIP session runs end to end between the two phones. A transport connection (layer 4) is established between the two SIP phones. A "TCP socket" offers an application on a PC (or host in general) a connection to a remote address (IP address and port). On layer 3, IP communication takes place. IP communication is both end-to-end between the phone applications, and hop-by-hop between the IP routers along the path followed by an IP message. The IP message has a source address and a destination address relating to the respective end-points. However, when the IP message is sent from one phone application to the other phone application, the IP interface determines, based on the destination address, the next hop for this IP message, whereupon this next hop, in turn, also determines the next hop for the message. On layer 2 in the example, MAC addresses and the Address Resolution Protocol (ARP) are used to send the IP message over a local area network (LAN), from one node to the next node. Data signaling techniques are used on layer 1 for the actual data (symbols) transmission between communicating nodes.

When two entities establish an SIP session, there may be a mix of lower layers. Some examples include:

- An SIP session that is carried over a combination (chain) of TCP and UDP transport layers
- A TCP connection that is carried over a combination of wireless LAN (WLAN) and Ethernet.

6.2.3 Data Transmission Using the Real-time Transport Protocol

The Real-time Transport Protocol (RTP) is an essential component in real-time communication systems such as VoIP. RTP, an OSI layer 5 protocol, provides the capability to transport the digitized speech data over the Internet. As described in an earlier section, VoIP and other (multimedia) communications over IP require that a data stream, resulting from speech sampling (PCM) and encoding, and hence representing speech in digitized form, is transported over IP infrastructure to a remote party, where this data stream can be reverse processed, resulting in audible speech. RTP provides the framework for real-time, reliable transmission of data streams over IP, whereby a data stream may be encoded in accordance with one of a large number of encoding standards. RTP does not define an encoding standard as such, comparable with G.711 or G.722. Rather, RTP defines a mechanism for transporting a data stream of G.711-encoded speech, or another encoding standard, over IP. This is depicted in Figure 6.5, which shows speech transmission in one direction. The same speech-processing steps apply in the reverse direction.

The "RTP encapsulator" in Figure 6.5 represents the process of packing a defined number of PCM code words into an RTP message. For real-time voice communication, RTP messages are typically constructed from digitized voice spanning a period between 10 and 80 ms. Twenty milliseconds is the commonly used packetization time for real-time voice. Long packetization time leads to less RTP message overhead, and thus more efficient RTP transmission, but also leads to longer transmission delay. In addition, long packetization

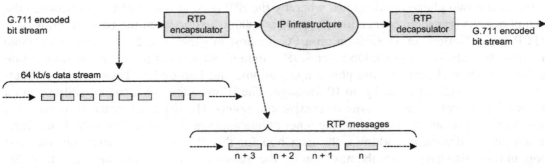

FIGURE 6.5

RTP encapsulation and decapsulation.

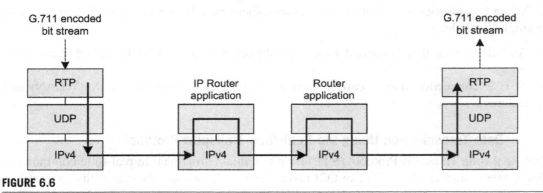

FIGURE 6.6

RTP in relation to lower layer protocols.

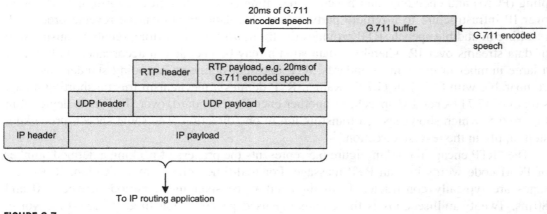

FIGURE 6.7

G.711 encapsulation in RTP message, UDP message, and IP packet.

time requires a larger transmit and receive buffer, and makes the transmission less resilient to packet loss (a lost RTP message disrupts the received voice for a long(er) duration). The numbering in Figure 6.5 (n, $n + 1$, $n + 2$, etc.) depicts the numbering of RTP messages. This numbering is needed since RTP uses UDP as transport protocol and RTP message transmission over an IP network is not guaranteed; RTP messages may arrive at the destination in a different order than they were sent or RTP messages may get lost. The RTP numbering helps the receiver to re-order the messages and to detect whether a message has been lost.

Figure 6.6 shows the relation between RTP and the underlying communication protocols. Layers lower than OSI layer 3 are not depicted in Figure 6.6, as that is very specific to the network that is used. The solid arrow depicts the flow of the G.711-encoded bit stream through the IP network. The G.711-encoded data stream, as in this example, is subject to a number of encapsulation steps. This process of encapsulating the encoded real-time voice, or other communication, is depicted in Figure 6.7.

At the receiving end of the communication session, the reverse process takes place, i.e. the G.711-encoded speech is retrieved from the IP packet payload, retrieved from the UDP message payload, and retrieved from the RTP message payload. At the sender's side, the encoded speech may be encapsulated into RTP messages at a constant speed. The RTP encapsulation process will consistently obtain 20 ms of G.711-encoded speech, construct an RTP message, and initiate transmission of the RTP message over UDP/IP. At the receiver's side, however, a jitter buffer is required. IP packets containing G.711-encoded speech destined for a particular application (e.g. SIP phone) on a particular host may arrive "out of order". An IP packet could also be lost. Hence, to compensate for jitter and for possible packet loss, the RTP receiver will place the received G.711 code words in a jitter buffer, in a position that corresponds to the RTP message number in which these G.711 code words were received. The jitter buffer outputs a G.711 data stream in correct order. A jitter buffer introduces additional delay. A large jitter buffer increases the resilience of the speech transmission against jitter. The downside of a large jitter buffer is the long(er) delay it introduces and the computing resource usage.

RTP is specified by the IETF. This began in 1996, in IETF RFC 1889. IETF RFC 3550 later superseded IETF RFC 1889 in 2003. RTP specifies the structure of RTP messages and describes how RTP messages are transported over the underlying network protocols, such as UDP and IPv4. RTP is typically transported over UDP, not TCP. One reason for not using TCP for RTP is the fact that TCP provides "reliable transport", including packet retransmission when a packet is determined to be lost. While retransmission will increase the success of the data transfer, i.e. the original data stream has a higher chance of arriving in good shape at the remote end, it will also increase transmission delay. As a result, the use of TCP may decrease the effective speech quality, leading to UDP being the preferred transport protocol for RTP.

RTP itself does not specify how encoded speech or encoded video is encapsulated in an RTP message. RTP only specifies the structure of the RTP message. More specifically, RTP specifies the *RTP header* (see Figure 6.8).

An RTP message is constructed as a sequence of groups of four octets. The length of an RTP message is derived from the length of the transport message, e.g. UDP message. One

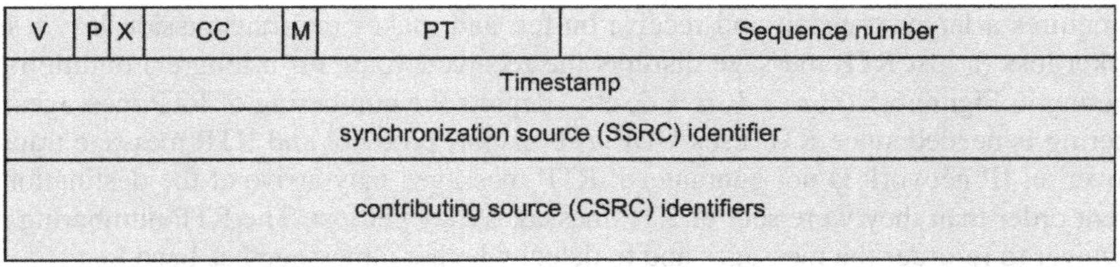

V	P	X	CC	M	PT	Sequence number
					Timestamp	
					synchronization source (SSRC) identifier	
					contributing source (CSRC) identifiers	

FIGURE 6.8

RTP header structure.

UDP message cannot contain more than one RTP message, so the length of an RTP message equals the UDP message length minus the UDP header length. Table 6.2 describes the RTP header fields.

The structure of the RTP payload, following the RTP header, is defined in separate IETF standards. Hence, IETF RFC 3550 does not specify the values of the Payload Type field in the RTP header. IETF RFC 3551, for example, specifies a set of RTP payload types for a variety of commonly used audio and video coding standards. Table 6.3 provides an overview of the audio coding standards supported in IETF RFC 3551. IETF RFC 3551 replaces the older IETF RFC 1890. It should be emphasized that IETF RFC 3551 does not specify the encoding standards as such; it merely specifies how the data stream resulting from a particular coding standard will be encapsulated in RTP.

The "RED" payload type (not listed in Table 6.3), as also specified in IETF RFC 3551, is not really a payload type by itself. Instead, it specifies a means to transport a *redundant* set of RTP payload information, in addition to the *regular* RTP payload information. The redundant RTP payload information constitutes a compressed version of the related regular RTP payload information. The use of redundant payload transfer aims to increase the resilience of real-time media transfer against packet loss.

Table 6.4 lists the video coding standards supported in IETF RFC 3551. For each of the audio coding standard supported in IETF RFC 3551, there is a reference to the standard where the encoding is defined, e.g. ITU-T recommendation G.711. For some of the audio coding standards, the encapsulation in RTP messages is defined in the same standard document as where the actual encoding standard is defined. This may originate from the fact that those coding standards were defined targeting specifically RTP as application protocol for transporting the information over an IP network.

H.264, not listed in Table 6.4, is currently preferred for Multimedia Telephony (MMTel) and Rich Communication Suite (RCS).

For the video coding standards supported in IETF RFC 3551, IETF RFC 3551 refers to the respective coding standards for the encapsulation of those real-time media in RTP messages.

The combination of RTP header and RTP payload would result in an RTP message as shown in Figure 6.9. The RTP message in Figure 6.9 does not include contributing source identifiers, indicating that the RTP data stream relates to a single media source.

Table 6.2 RTP Header Description

Header Field	Description
Version (V)	Identifies the version of the RTP protocol applicable for this message. For both RTP according to IETF RFC 1889 and RTP according to IETF RFC 3550, the version has value 2.[a]
Padding (P)	Padding indicator used to indicate that the RTP payload contains one or more padding octets. Padding is needed when the RTP encapsulation of a chunk of encoded media does not yield a multiple of four octets.
Extension (X)	Extension marker used to indicate that the RTP header includes a *header extension*. Header extension allows for applications to transport application-specific information in a media session. The contents of RTP header extension and the use thereof are proprietary for an application. SIP and SDP (Session Description Protocol; see Chapter 7 for description of SIP and SDP) do not specify means for negotiating, between applications using RTP, the use of RTP header extension.
Contributing source (CC) count	This field indicates how many CSRC identifiers are included in the RTP header. See below.
Marker (M)	This field is used to indicate that the RTP payload contains a designated event, specific for the encoding standard for this RTP message. This designated event conveys information other than encoded real-time data; e.g. it may signal a frame boundary.
Payload type (PT)	This field indicates the type of media that is included in the RTP message body, such as G.711 data, G.722 data. Although the PT field may change during a media session, it is not intended for multiplexing multiple media streams in a single RTP session. A video call, for example, comprises both an audio stream and a video stream. These streams will be transported through separate, dedicated RTP sessions.
Sequence number	Incremented by 1 for each subsequent RTP message. The receiver may use the sequence number for re-ordering packets and for determining that a packet was lost. According to RTP standard, the initial packet number should be random (for security). This has the effect that the receiver of an RTP stream cannot know which packet constitutes the start of the RTP stream and should be placed in the front position of the de-jitter buffer.[b]
Timestamp	Represents the absolute time applicable to the PCM sampling of the first octet of the RTP payload. It is used by the receiver to calculate jitter and to apply media synchronization. Different encoding standards may use different clock frequency for the timestamp calculation.
Synchronization source (SSRC)	Identifies the source of the encoded media, the entity where the sampling and encoding took place. When a media session consists of audio stream and video stream, then the respective RTP sessions will have equal SSRC, indicating that these media streams are related to each other.
Contributing sources (CSRC) identifiers	List of different media sources from which output media are mixed to form an aggregate media stream. When a media stream consists of media of a single media source only, then this field is absent from the RTP header. A maximum of 15 CSRC identifiers can be contained in an RTP header.

[a]*RTP version 0 and 1 are not commonly used. Applications that intend to establish a media session normally do not negotiate the RTP version. RTP version 2 is assumed.*
[b]*Although this is not standardized for protocols like SIP and SDP, it might be useful to signal the initial sequence number to the intended receiver of the media stream, allowing that receiver to render correct speech from the first packet onward.*

Table 6.3 Audio Coding Standards Supported in IETF RFC 3551

Encoding Standard	Code Size (bits/sample)	Sampling Rate	Description
DVI4	4	Variable	Adaptive differential pulse code modulation (ADPCM). Uses compressed samples. The header block of DVI4 RTP payload contains information needed to decompress the code words.
G.722	8	16,000	7 kHz audio coding within 64 kb/s channel.
G.723.1	N/A[a]	8000	Dual-rate speech coder for multimedia communications transmitting at 5.3 and 6.3 kb/s. Is used, among others, as audio channel for video telephony over CS networks.
G.726-40	5	8000	This set of coding standards specifies how PCM A-law or PCM µ-law encoded media may be transcoded from 64 kb/s into lower bit rates of 40, 32, 24, and 16 kb/s. The resulting bit stream is encoded according to an ADPCM method.
G.726-32	4	8000	
G.726-24	3	8000	
G.726-16	2	8000	
G.728	N/A	8000	Coding of speech at 16 kb/s using low-delay code excited linear prediction.
G.729	N/A	8000	Coding of speech at 8 kb/s using conjugate structure-algebraic code excited linear prediction (CS-ACELP). Efficient coding technique for low bandwidth channels, e.g. cellular radio links.
G.729D	N/A	8000	
G.729E	N/A	8000	
GSM	N/A	8000	The speech encoding standard used in GSM networks. The encoding standard results in an effective data rate of 13.2 kb/s. When GSM-encoded data is conveyed over the radio access network, additional framing is added yielding a gross bit rate of approximately 22 kb/s.
GSM-EFR	N/A	8000	GSM enhanced full rate. Variant of the GSM codec.
L8	8	Variable	PCM encoding with designated offset for reflecting a sample in a code word.
L16	16	Variable	PCM encoding with two's complement representation of the PCM sample values.
LPC	N/A	8000	Linear predictive encoding – experimental coding standard, yielding 5.6 kb/s data stream.
MPA	N/A	Variable	MPEG audio, as specified by ISO/IEC.
PCMA	8	Variable	PCMA and PCMU are specified in ITU-T recommendation G.711. PCMA and PCMU differ in the logarithmic scaling that is applied for the sampling process, to convert from analog sample to PCM sample code. Although PCMA and PCMU support 64, 56, and 48 kb/s modes, only the 64 kb/s mode can be conveyed over RTP.
PCMU	8	Variable	

(Continued)

Table 6.3 (Continued)

Encoding Standard	Code Size (bits/sample)	Sampling Rate	Description
QCELP	N/A	8000	Qualcomm Code Excited Linear Prediction. Used for speech transmission in CDMA wireless networks. Generates output of 8 or 13 kb/s.
VDVI	Variable	Variable	Variable-rate version of DVI4. It generates a data stream between 10 and 25 kb/s.

[a]N/A, not applicable.

Table 6.4 Video Coding Standards Supported in IETF RFC 3551

CelB	Proprietary video encoding defined by SUN Microsystems
JPEG	Defined by the Joint Photographic Expert Group (JPEG) and adopted by ISO and ITU as video coding standard
H.261	Video coding standard for video transmission at multiples of 64 kb/s
H.263	Video coding for low bit rate communication. Successor of H.261
H.263-1998	Improved version of H.263 standard
MPV	MPEG video streams
MP2T	MPEG-2 transport streams, for either audio or video. Hence, MP2T may be used to transport audio, using a video-designated Payload Type value

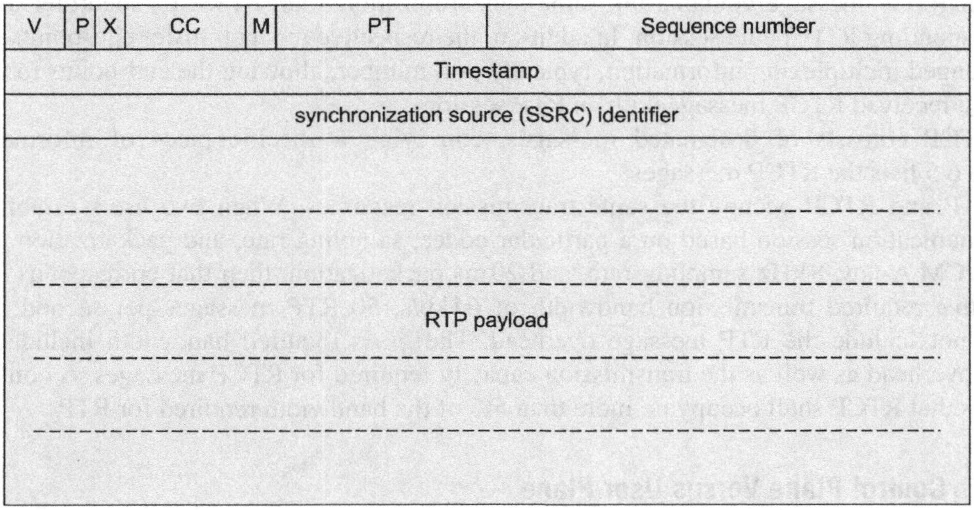

FIGURE 6.9

RTP message, including RTP header and RTP payload.

6.2.4 Real-time Transport Control Protocol

The Real-time Transport Protocol (RTP) is accompanied by a control protocol, the Real-time Transport Control Protocol (RTCP). RTCP is also specified in IETF RFC 3550. When an RTP media session is established between two hosts, an RTCP control session will be established at the same time between these hosts. An RTCP control session is carried through the user plane IP infrastructure, as is the associated RTP media session. Multiplexing RTP and RTCP sessions on the same host is typically done through the use of distinctive port numbers.

RTP media streams are not *always* accompanied by RTCP. Some SIP phones do not support RTCP. In addition, for RTP-based media transfer over the Nb reference point (connection between two media gateways in the CS network), the use of RTCP is optional.

RTCP allows the respective hosts to exchange information related to the RTP session, about the quality of the media transfer. This feedback on the quality of media transfer may, in turn, be used to dynamically adapt encoding and transmission parameters or to maintain a log of the transmission characteristics of the network. If one of the parties decided to adapt any encoding or transmission parameter for the RTP session, then this may first have to be negotiated with that remote party. Such negotiation may be done with a control plane protocol such as the Session Initiation Protocol (SIP).

The quality of the media transfer parameters that may be provided through RTCP include: jitter, packet count (from sender), octet count (from sender), percentage lost packets, total lost packets, and transmission latency.

Figure 6.10 depicts the use of RTP versus RTCP. If a multimedia session between hosts comprises two RTP media sessions, then each RTP session will have an RTCP session associated with it (Figure 6.11).

An RTCP message contains the same synchronization source (SSRC) identifier as the corresponding RTP media session. In addition, the respective media transfer end-points have exchanged multiplexing information, typically port number, allowing the end-points to associate a received RTCP message with an RTP session.

RTCP consists of designated messages, conveying a specific piece of information. Table 6.5 lists the RTCP messages.

RTP and RTCP occupy the same transmission resources. When two users establish a communication session based on a particular codec, sampling rate, and packetization time, e.g. PCM A-law, 8 kHz sampling rate, and 20 ms packetization, then that corresponds to an *effective* required transmission bandwidth of 64 kb/s, 50 RTP messages per second. That does not include the RTP message overhead. The *gross* required bandwidth includes the RTP overhead as well as the transmission capacity required for RTCP messages. A common rule is that RTCP shall occupy no more than 5% of the bandwidth required for RTP.

6.2.5 Control Plane Versus User Plane

A concept found throughout contemporary communication networks is the differentiation between control plane and user plane. It has been described so far that hosts on the Internet

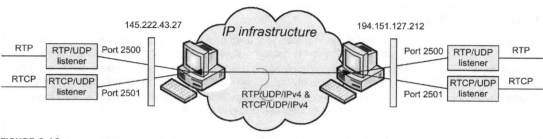

FIGURE 6.10

RTP session and RTCP session, for single media call.

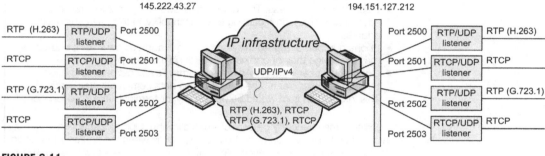

FIGURE 6.11

RTP session and RTCP session, for multimedia call.

may establish a data session, e.g. voice communication or voice + video communication, by setting up an RTP session and an associated RTCP session (or multiple RTP + RTCP sessions). The RTCP session, as described in the previous section, acts as a control channel for an associated RTP session. The RTCP control channel assists the respective end-points in the data transmission, such as controlling the quality of the received data, synchronizing the various media streams (e.g. voice + video, or multiple voice components), and adapting the transmission rate.

For establishing the RTP communication session, the hosts that will be engaged in the RTP session would need to have knowledge of the IP address and port numbers used by the respective peer for receiving RTP and RTCP packets. In addition, the hosts need to be prepared to receive RTP and RTCP packets, by installing RTP and RTCP listeners on the indicated IP address and port number. The hosts also need to be prepared for the use of specific codec(s) for the media transfer. In other words, before the hosts can establish a communication session, they need to communicate with each other the intention to establish said communication session. In addition, they need to exchange the details characterizing the intended communication session.

To address the above (and for other reasons), the concept of control plane versus user plane is applied. This concept is widely used within the telecommunications industry. Figure 6.12 gives a conceptual representation, postulating on previous examples.

Table 6.5 RTCP Messages

Sender report	A *sender report* message is used by an RTP entity to convey quality of media transfer, if that entity has sent any media during the last reporting interval.
Receiver report	A *receiver report* message is used by an RTP entity to convey quality of media transfer, if that entity has not sent any media during the last reporting interval.
Source description	The *source description* message allows for exchanging various descriptive information items between end-points, including, but not limited to: • Canonical end-point identifier. This item should be a consistent and unique identifier of the source of media transfer, e.g. john.smith@145.222.43.27. • User name. This item may be set by the end-user and provides an informative indication of the person on whose behalf the media is generated, e.g. "John Smith at work". • Electronic mail address. This data item allows end-users to exchange additional information regarding their reachability, in this case an email address. • Phone number. Similar to the electronic mail address data item, end-users may exchange their phone number through the user plane. • Geographic user location. An end-user may reveal his/her information through the user plane. The format of this data item is not standardized. This information item would be useful only for terminals in a fixed location. Operators may generally be careful when allowing source description information to be passed to end-users, especially when the source information contains IP address or location information. When topology hiding is applied in a VoIP network, these information elements may be removed, depending on the topology hiding implementation.
End of participation	An end-user may signal the end of participation in the media session. This would be useful for a conference call, with many participants. For a two-person session, the end of a media session may rather be signaled through the control plane, indicating that the entire RTP session will be terminated.
Application-specific function	The application-specific function is a powerful mechanism found in many communication protocols. It allows, on the one hand, for experimental extension to RTP and RTCP. On the other hand, vendors may develop proprietary extensions to RTP and RTCP, to benefit their specific application.

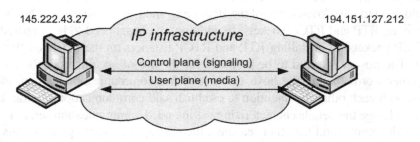

FIGURE 6.12

Control plane versus user plane.

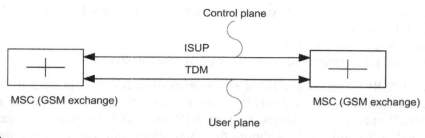

FIGURE 6.13

Control plane and user plane in ISDN, monolithic MSC.

The concept of control plane is also used in the Integrated Services Digital Network (ISDN). This is illustrated, by means of an example, in Figure 6.13. An ISUP connection is established between two monolithic mobile switching services centers (MSC), as well as a TDM media transmission session. The depicted MSCs are "monolithic MSCs", handling both media control plane and user plane.

The ISDN user part (ISUP; ITU-T Q.763) or bearer independent call control (BICC; ITU-T Q.1902) is the protocol used in the control plane of the ISDN-based network. ISUP comprises a set of call control messages and parameters for establishing and releasing a call (voice, video, etc.). In addition, ISUP contains messages and designated parameters for supplementary services; ISDN supplementary services augment the user experience of basic calls.

The user plane is formed by a time division multiplexed (TDM) data stream or, increasingly, by RTP data session. The control plane and user plane may be carried through the same transmission network or through different transmission networks. An example of the former is common channel signaling (CCS). CCS is a method whereby signaling (ISUP) and media (TDM) are multiplexed on a synchronous communication channel ("E1 channel") consisting of 32 time slots of 64 kb/s each. Of these 32 time slots, 30 are used for media transport and two are used for signaling. Increasingly, the control plane and user plane in ISDN-based networks are separate and use designated transmission networks, with different network topologies. These designated networks may be dimensioned separately.

For multimedia over Internet (e.g. VoIP), the control plane carries a specific protocol, just as the user plane carries a certain protocol (namely RTP and RTCP or another media transport protocol). When an application on an IP host intends to initiate a data session, e.g. RTP for voice, with an application on a peer IP host, the initiating application establishes a communication session on the control plane. It would therefore be required that the peer host has a listener installed for receiving messages related to the control plane. By initiating a communication session on the control plane, the respective applications on the IP hosts can exchange the following information:

- Intention to establish a communication session (from initiating application to responding application)

- Indication of acceptance of the proposed communication session (from responding application to initiating application)
- Codec(s) to be used for the session
- IP address and port numbers to be used for the media transfer.

The use of SDP for the above-listed information exchange is explained in Chapter 7. Based on the information exchange on the control plane, the applications in the respective hosts can install appropriate listeners, such as RTP and RTCP listeners, on specific local IP address and port numbers. In addition, the applications will, based on the information exchanged, seize software and hardware resources in the host, for encoding captured media, such as voice, using the agreed codec, and streaming the media to a process for encapsulating the data stream in RTP packets and transmitting these to the peer. The application will also seize resources in the host, for decoding media received in incoming RTP packets and sending these to a DA converter, such as a sound card.

When the communication session on the user plane is established, the communication session on the control plane may be used to tear down the communication session, i.e. indicate from one host to the other host the intention to stop communicating. As a result of this, the applications may release the software and hardware resources that are used for the media handling and media transmission and may de-install the aforementioned RTP and RTCP listeners.

One can visualize that, for the purpose of initiating communication between two hosts, a third, virtual, communication plane is needed. That is, two hosts on the Internet need to have exchanged a limited set of information, enabling them to establish a control relationship, i.e. to set up a communication session on the control plane. This is visualized as a stepwise approach in Figure 6.14. The control plane in this figure is formed by the Session Initiation Protocol (SIP), which will be elaborated upon in later sections.

When the phone applications in the two hosts are in idle mode, there will be at least a user relationship between these two applications. This user relationship entails that the applications are aware of one another's existence and have the ability to reach one another. More specifically, each application has the information needed to establish a control relationship with the other application. This information may be available explicitly or implicitly:

- **Explicit user relationship.** The users of the two applications have exchanged their IP addresses, the port number on which the control plane (SIP) listener is installed, and the transport protocol that will be used to address the listener, i.e. UDP or TCP.
- **Implicit user relationship.** The users have exchanged a public identity, such as sip:alice.jones@ericsson.se and sip:bob.smith@etsi.fr. In this case, a control relation may be established by Alice toward Bob by addressing sip:bob.smith@etsi.fr, instead of addressing an IP address and port number. The use of a public identity requires network support and will be described later.

When in idle mode, the hosts have a control plane listener installed, allowing other users to establish a control relationship with that host.

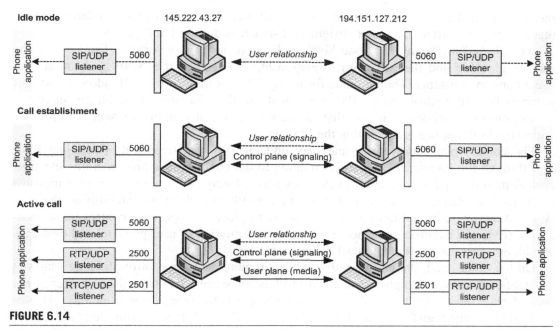

FIGURE 6.14

Subscriber relationships.

When call establishment is initiated, a control communication session is established (e.g. SIP session) between the two phone applications, using the aforementioned user relationship. Details relating to the user plane communication session (to be established) are exchanged through the control plane.

During the active phase of the call, a user plane exists (e.g. RTP and RTCP for voice communication). The control relationship remains in place during the active phase of the call, allowing the end-users to signal to the peer the intention to release the call. When the call is established, the control relationship could be shifted to another port number. This may be established during the initial message exchange on the control plane, namely to inform the peer that further control plane messages relating to this control relationship will be sent to another port number. In this manner, there will effectively be distinctively different data communication sessions for control plane signaling related to call establishment and for control plane signaling related to call continuation (including call release).

Figure 6.14 shows that the user relationship is retained when call establishment is initiated and when the call is active. The rationale is that the end-users may, when being engaged in a call, receive or establish another call or receive or establish a message exchange session. Figure 6.14 illustrates that, for the active call, the user relationship and the control plane use the same listener, i.e. are connected to the same IP address and port number. A VoIP application may, however, reserve a default listener for the user relationship and for call control messages related to call establishment. Once a call is established, the control relationship for that call may be shifted to another designated listener, as depicted in Figure 6.15.

As reflected in Figure 6.15, one VoIP application may apply separate listeners for the user relationship and the call control relationship, while the other VoIP application may use a combined listener. The rationale for using different transport (i.e. IP address and port number) for call control and for the established call than for session establishment is that control messages related to an established call may be handled in this manner with priority within the VoIP application or within the IP network.

The control plane and user plane may undergo different treatments (including routing) through the Internet. Data packets relating to the control plane are generally less time critical than data packets relating to the user plane. Delay or retransmission of a message or IP packet relating to the control plane may result in a delay in establishing the call. Applications may, in addition, have procedures in place to cope with control plane messages arriving in incorrect sequence. Delay or retransmission of user-plane related messages would, however, have a direct impact on voice quality.

On the other hand, the end-to-end reliability requirement for control-plane messages at the application level is higher than for user-plane messages. Incorrect data transfer at the control-plane level results in failure to establish a call (or other error situations). Hence, incorrect message transfer on the control plane generally leads to retransmission (with a limited number of retries). A dropped or corrupted packet on the user plane may, on the other hand, be experienced as a "click" in the voice stream.

Operators may structure the IP network such that priority (with respect to handling queued IP packets) is given to data sessions relating to real-time transfer (voice, video), above data sessions relating to the control plane. More specifically, shorter transmission latency would apply for the packets relating to real-time sessions. In the case of network overload, packets relating to real-time sessions may be dropped, whilst packets relating to the control plane should not be dropped. It will be shown in later sections that this is a very important aspect of multimedia communications network design, namely to minimize the network resource usage from the user plane and to minimize transmission delay for user-plane messages (especially for the real-time user plane).

The advantage of strict separation between control plane and user plane is best explained by a brief comparison with network architecture in circuit-switched (CS) technology. Figure 6.16 shows the architecture of a GSM network containing monolithic MSCs (mobile service switching centers). Monolithic MSCs handle both control plane (ISUP) and user plane (TDM).

The MSCs shown in Figure 6.16 are connected through E1 "trunks". E1 trunks are the atomic data connections between signaling points in a synchronous digital hierarchy (SDH) network. Control plane (ISUP messages) and user plane (TDM messages, carrying PCM-encoded speech) are multiplexed into a common set of E1 links. When a call is established between two GSM subscribers, as in Figure 6.16, the user-plane message will follow the same data path as for the control plane. As the call establishment request message is

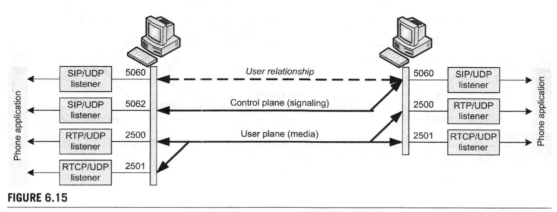

FIGURE 6.15

Separated listener for user relationship and call control relationship.

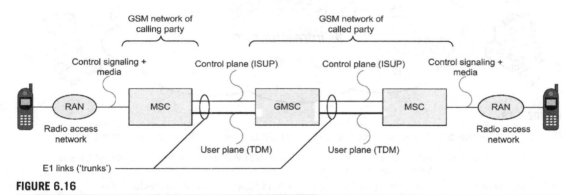

FIGURE 6.16

Combined control plane and user plane for GSM network with monolithic MSCs.

forwarded from one switch to the next, a media path ("circuit") is reserved from one switch to the next. This circuit is reserved even though the call establishment may fail, e.g. the called party may be busy or may not answer the call. As the call establishment failure is signaled in the backward direction, to the initiator of the call, the circuit is released again and the transmission resources they occupied may be used for other call (establishment).

Around 2000, 3GPP introduced the "layered architecture", whereby the control plane and the user plane, for a call in the GSM network, are handled by separated entities: the MSC server for the control plane and the media gateway (MGW) for the user plane. This allows different transmission networks to be applied for control plane and user plane (see Figure 6.17).

Bearer independent call control (BICC) is the functional successor of ISUP. BICC messages may be transported through the IP network infrastructure. The user plane in the layered architecture may consist of RTP and RTCP sessions. The user-plane messages may follow a different path than the control-plane messages. H.248 is a generic reference to the

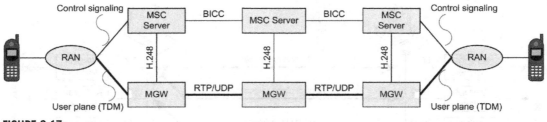

FIGURE 6.17

Separated control plane and user plane in CS network.

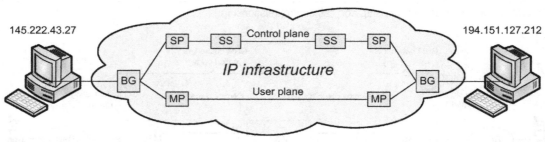

FIGURE 6.18

Differentiated routing of control-plane messages and user-plane messages through the Internet. Key:

BG	Border gateway	Media and signaling between user and VoIP network traverses border gateway (protection, among others)
MP	Media proxy	User plane (media) may be routed through a number of media proxies
SP	Session proxy	Control plane (signaling; call control) is routed through signaling proxy
SS	Session server	VoIP session is controlled by Session server

media gateway control protocol; an MSC server controls an MGW through an H.248 control session (H.248 message exchange). H.248 messages may be transported over IP.

The importance of separation of control plane and user plane, allowing for optimized user-plane data transmission, is also seen from the following initiatives in the mobile network development by 3GPP:

- Control plane and user plane use separate network topology, both being optimized and structured for their respective uses.
- Optimized MGW selection. When a call is established in the CS network, the various switches involved in the call establishment chain assign the same MGW for transmitting the media between the connected parties.
- Transit switches not controlling an MGW. When a transit switch does not need to control media, it does not need to seize an MGW and steer the media transmission through that MGW.

Figure 6.18 shows how the separation of user plane and media plane can be exploited in the VoIP network. The control plane is routed through various signaling proxies and servers in the IP network. The user plane, however, is routed directly between the end-users, potentially traversing a media proxy at the edge of the IP network.

The signaling end-points, in this case the phone applications in the respective hosts, exchange IP addresses and port numbers to be used for the user plane for this call. It is only when the call is actually answered that IP infrastructure capacity is required for the media transfer.

An additional and very powerful mechanism resulting from the strict separation of user plane and control plane is the media negotiation during call establishment. User A may initiate a call to user B and "offer" a set of codecs. User A indicates, in this way, which codecs he or she is prepared to use for this call. User B may accept a subset of the offered set of codecs. Users A and B have now "negotiated" a set of codecs that may be used for the call, and may now apply "resource reservation" over the access network, based on the results from the codec negotiation. In this manner, the resource reservation may be adapted to the media that will be used in this call.

Compare the above with video call establishment in a circuit-switched network. When establishing a video call in a 3G mobile network, for example, media transmission resources are reserved, allowing for video transfer; this requires 64 kb/s data transmission capability over the radio access network and in the core network. Should the called party not be in a position to answer the video call, e.g. because (s)he has a non-video capable terminal or (s)he is currently connected to a radio access network that does not have video capability, then the call will fail. One network option in such a case where the called party cannot answer the video call, is that a designated media entity in the network downgrades the video call to a voice-only call, so the call can be connected. The calling party has, however, reserved a 64 kb/s channel over the radio access network, but a maximum of 6.3 kb/s will be used for the voice transfer. This 6.3 kb/s is the bandwidth used for G.723.1 speech transmission in a CS video call.

A further mechanism resulting from the strict separation of user plane and control plane is the change of media capability *during* a call (see Figure 6.19).

Designated control-plane messages are used between the two peers in the call, by means of which they may request the respective peer to accept the addition of an extra media component, such as video. If the parties agree on this additional capability, an additional user plane is established and video may be transferred. Before a party initiates the addition of video capability, the phone terminal may have to reserve resources in the access network, so as to ensure that sufficient resources are available for the additional video component. Likewise, that party's peer may also apply resource reservation, before accepting the additional media component.

In-call media update is also specified for the ISDN-based network. 3GPP TS 23.172 specifies Service Change and UDI Fallback (SCUDIF). SCUDIF allows for toggling between voice call and video call during establishment of the call and during the active phase of a call. Due to the complexity of this feature, it is not generally available in 3G networks.

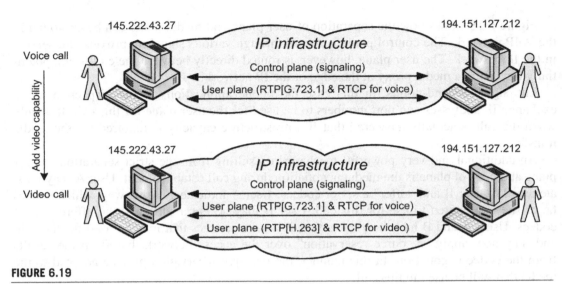

FIGURE 6.19

Adding media component(s) during a call.

Besides call establishment and call release, the control plane is also used for additional services, such as call forwarding, call hold, and call transfer. When a phone application on a host on the Internet receives a request to establish a voice session, this application may signal over the control plane that the call shall be forwarded to an alternative destination, e.g. a voicemail box. When a voice session is established, one of the parties connected in the call may signal to the peer that the media session is temporarily placed on hold, i.e. voice communication with that peer is temporarily suspended, and then subsequently signal that the media session be resumed.

ISDN-based networks, such as the PSTN and the GSM network, have an extensive set of supplementary services (Appendix C[4] for an overview of these services). Supplementary services augment the basic user experience for subscribers of the network. Supplementary services are controlled by the core network. PSTN or GSM subscribers need a subscription to a supplementary service in order to use this service. A supplementary service may be *requested* by the subscriber, e.g. a request to transfer an incoming call to an alternative destination. The core network will execute the service, provided that the PSTN/GSM subscriber has subscribed to that service. The PSTN/GSM network specifies designated control-plane messages to be used over the access network to request the execution of a supplementary service as well for administrative control (activation, deactivation, interrogation, etc.) of these services.

Voice over IP does not specify any network-controlled services, such as the aforementioned call forwarding, call hold, and call transfer. The control protocol used for VoIP, the Session Initiation Protocol (SIP), assumes that services are executed by the end-points (we are not considering application servers here). This form of service assertion is also known

---

[4] see www.elsevierdirect.com/companions/9780123821928

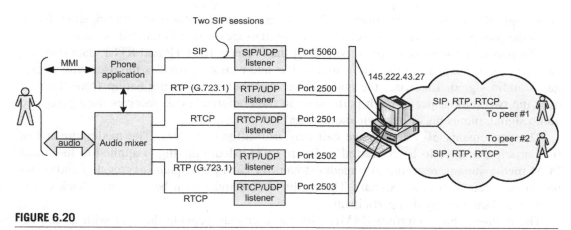

FIGURE 6.20

Terminal-based multi-party call.

as peer-to-peer communication. SIP contains various messages and parameters for that purpose. The subscriber does not need to subscribe to receive this service; the availability of a service to a VoIP user is generally dependent on the capability of the terminal (e.g. phone application). For the uptake of VoIP in public networks, it is, however, desirable that end-users have a set of services available that is at least of the same level as that available on PSTN and GSM. To address this requirement, 3GPP has introduced a set of network-based services for the IMS network. This set of services forms part of Multimedia Telephony (MMTel). MMTel is the network architecture and service definition for IP-based voice and multimedia communication for public networks. The inclusion of supplementary services in MMTel is an essential component in a telephony network. MMTel and its service definition are explained in Chapter 9.

6.2.6 Multi-Party Communication Session

The RTP and RTCP, as specified for Internet communication, support the establishment of a voice communication session between two parties (peer to peer), as well as a voice communication session between three or more parties (multi-party call). This section describes two architectures for realizing these types of call.

6.2.6.1 Terminal-Based Multi-Party Call

Refer to Figure 6.20. The phone application in the computer has established two peer-to-peer VoIP sessions. These two sessions are established and maintained independently of one another. There are two RTP sessions with the user, each RTP session accompanied by an RTCP session. The RTP sessions and the RTCP sessions, i.e. the data streams and the media flow control channels respectively, are not multiplexed on to one IP address + port number, but have their own IP address + port number. Unlike RTP and RTCP, SIP sessions can be multiplexed on a single IP address + port number. SIP messages contain SIP session identification, facilitating the phone application to associate incoming messages with different SIP sessions.

The phone application, being in control of the listeners for RTP and RTCP data streams, connects these listeners to the audio mixer. The audio mixer provides the user with aggregated audio signal. Likewise, the audio mixer distributes audio originating from the user over the two remote peers. Each participant in the conference call receives the aggregated audio signal, minus his or her own audio signal.

The participants of a conference call may use different codecs. The media streams of the participants have to be converted to *linear PCM*. Media mixing is applied to the linear PCM media stream, resulting in a media stream for each participant (aggregated audio signal minus their own audio signal). The media stream must then be converted back to the codec standard used by that participant.

The man–machine interface (MMI) with the user may provide the user with an indication about which peers are currently connected in the call. The audio mixer may, in addition, provide the phone application with information about the media transfer in the call, e.g. use the RTCP messages, received from the respective peers, to provide the status of the speech quality of the conference participants.

A terminal-based conference call requires no network support. However, it leads, typically, to double transmission bandwidth requirements (for the phone where the conference application runs), since media is transferred to and from two peers. This situation will be even worse for conference calls with more than three participants. Audio mixing is a resource-intensive task (let alone media transcoding), so terminal-based conference calls are not common practice for VoIP.

6.2.6.2 *Network-Based Multi-Party Call*
Figure 6.21 depicts the architecture for a network-based conference call. The depicted conference server includes the audio mixing function. With a network-based conference server, each phone application has a single media stream to/from the Internet. A network-based conference server will even allow for a variety of codecs to be used, based on the fact that a network-based conference server will typically have more computing power than the end-user applications, especially in the case of mobile terminals. RTP permits indicating in the media stream those participants (devices) that contribute to the combined media stream. In this manner, even with a network-based conference server, end-users can get an indication of a conference call.

The application depicted in Figure 6.21 represents a case where the involved parties call into a conference server. For the individual phone applications, this does not constitute a conference call as such, since each phone application is involved in a single call only.

6.3 REGISTRATION
This section introduces an essential component in voice-over-IP networks, namely the concept of user registration. In earlier sections, examples were given where one user established

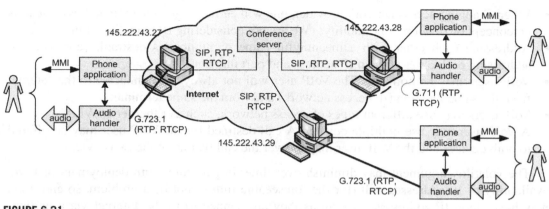

FIGURE 6.21

Network-based conference call.

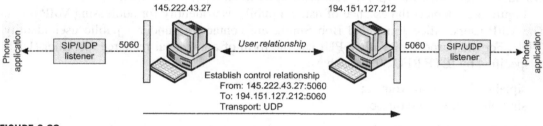

FIGURE 6.22

Establishing a control relationship using IP addresses.

a control-plane relationship with another user by addressing that other user on their IP address + port number, using a particular transport protocol (UDP, TCP).

The situation sketched in Figure 6.22 is based on the assumption that a *user* has a standing relationship with an *IP address* and that that user, if contactable, is always using a particular *port number* and *transport protocol* for his/her phone application.

Obviously, this is not a desirable situation. VoIP users do not want to address each other through IP addresses. One theoretical reason for this undesirability is the fact that IP address and transport protocol represent layer 3 and 4 protocols in the OSI data communication protocol stack. Phone applications represent layer 7 on the said protocol stack. VoIP architecture will not bestow the obligation on the end-user to communicate on levels 3 and 4. Further reasons include:

* IP addresses are inherently volatile. An IP address is obtained through DHCP of the access network. DHCP often applies dynamic IP address allocation. A DHCP client, e.g. a PC connecting to the Internet, may receive a different IP address with DHCP than the previous time it connected to the Internet.

- When changing access network provider, one will certainly get a new IP address. There is no concept of IP address portability! (We are not considering IP tunneling at this stage.)
- Addressing each other with a meaningful name or number is desirable, compared to addressing each other with an IP address and port number.
- VoIP users may be nomadic. The VoIP user will not always be connected to the Internet from the same location (IP access network) and from the same terminal.
- VoIP users may use different types of access network, such as Ethernet or WLAN.
- A VoIP user may use multiple devices. A call destined for the VoIP user may be offered to both devices and the VoIP user may answer the call on one of the two devices.

The volatility argument may diminish over time. In particular with deployment of IPv6, availability of IP addresses will (for the foreseeable future) not be a problem, so end-users may have static IP addresses, as long as they are connected to the Internet via a particular access network. However, the other drawbacks listed above remain. Hence, the need for *public addresses*, which may be used to address each other, independent of access network provider, access network type, IP subnetwork, terminal type, location, etc.

Figure 6.23 depicts the essence of using a public user identity for addressing VoIP users. The VoIP users, Alice Jones and Bob Smith, are contactable under a public user identity (PUI). For the purpose of VoIP, a PUI has the form of a universal resource identifier (URI) as specified by IETF RFC 3986, such as:

sip:alice.jones@my-voip.se;
sip:bob.smith@my-voip.se.

The PUI used for VoIP is built up as shown in Figure 6.24.

- **Schema.** The schema identifies the application for which the PUI is applicable. The indication sip: in the example PUIs above means therefore that Alice and Bob are contactable by means of SIP, i.e. are capable of receiving SIP session establishment requests. The schema is also used to indicate the format of the PUI. A sip: schema indicates that the PUI conforms to the URI definition (IETF RFC 3986). A tel: schema indicates that the PUI conforms to the telephony URI (Tel URI) definition, IETF RFC 3966, e.g. +31161249400. Such "overloading" of the schema is unfortunate. To establish a communication session with tel: +31161249400, for example, it must be determined from the context that the protocol to be used for this session establishment will be SIP.
  A further overloading of the meaning of the schema is the sips: schema ("secure SIP"). When sips: is used as the schema, this indicates that: (a) the PUI conforms to the URI definition; (b) the protocol to be used for establishing a communication session with that PUI is SIP; and (c) transport layer security (TLS; IETF RFC 5246) should be used for this communication session.
- **User.** The user identifies the user within the domain identified in the PUI. The user part is unique within this domain. If the PUI is defined as a Tel URI, then the PUI does not follow the URI structure, but consists (implicitly or explicitly) of a global E.164 number only.

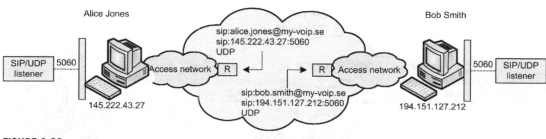

FIGURE 6.23

Public user identities. R, registrar.

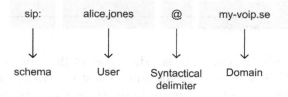

FIGURE 6.24

Public user identity for VoIP.

- **Domain.** The domain constitutes the Internet domain within which this VoIP subscriber is a user. The domain for the user does not have to be equal to the domain of the VoIP operator, as will be described below. There exists an SIP server within this domain capable of receiving SIP session establishment requests for the identified user.

The PUI of the VoIP user is used for identifying the VoIP subscriber over the Internet, in order to establish a VoIP session with that subscriber. Hence, a person initiating a VoIP session does not need to know the address of that subscriber, i.e. IP address, port number, and transport protocol. The VoIP subscriber has access to the Internet and is accessible from the Internet through an IP address, not a URI. Therefore, a designated SIP server acting on behalf of the VoIP subscriber will take care of translating between the URI and IP address. This designated SIP server is the registrar.

6.3.1 Initial Registration and Call Establishment

Registration entails that a VoIP subscriber provides his or her VoIP PUI and contact address to a registrar in the VoIP network. The registrar is able to receive VoIP session establishment requests and to forward these session establishment requests to the registered VoIP subscriber, using that person's registered contact address. This is depicted in Figure 6.25.

The VoIP session establishment request is routed, over (public) Internet, to the VoIP network my-voip.se. Regular Internet-based routing is used for this purpose; the DNS service

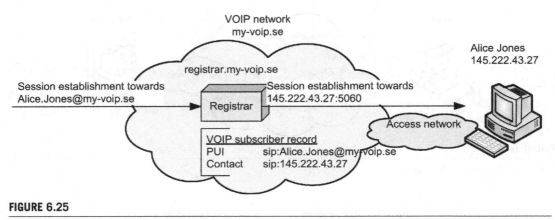

FIGURE 6.25

Using a registrar for VoIP session establishment.

is used for determining the IP address of an SIP server of the domain my-voip.se. Within this VoIP domain, the VoIP session establishment request is routed to the registrar holding the contact address of user Alice.Jones@my-voip.se. The registrar holds the contact address of the addressed user and is capable of forwarding the VoIP session establishment request to this user. The contact address through which the user is known in the registrar may be a public IP address or local IP address – that is, local from the viewpoint of the VoIP network.

The role of the registrar is restricted to call establishment, including the forwarding of the VoIP session establishment request message to the VoIP subscriber and the returning of the response message to the session initiator. Subsequent signaling related to the VoIP session does not need to traverse the registrar. This is reflected in Figure 6.26.

The contact address of the VoIP user is provided to the registrar though *registration*. The VoIP subscriber has to send a registration request message to the registrar (see Figure 6.27).

The registration message is a designated message, forming part of the VoIP protocol. A prerequisite for registration as a VoIP user at the registrar is having access to the Internet, i.e. having an IP address, access to DNS services, access to DHCP services, etc. Access to the Internet may be gained through various kinds of access network; in principle, VoIP places no restriction on the access network used for accessing VoIP services.

Other data that are configured in VoIP applications include (not exhaustive):

- **Home realm.** This is the domain name of the VoIP operator serving the VoIP user; this domain name is used for routing the registration request. When the VoIP application routes the registration request, through the (public) Internet, to the Home realm, the VoIP user uses DNS to obtain the IP address of an SIP server of the VoIP network.
- **PUI.** The VoIP user's public user identity (PUI) will be configured in the application; this PUI is provided to the registrar in the VoIP Home realm.
- **Transport protocol.** The VoIP application will be configured to use a particular transport protocol (UDP, TCP, other) for sending the registration request.

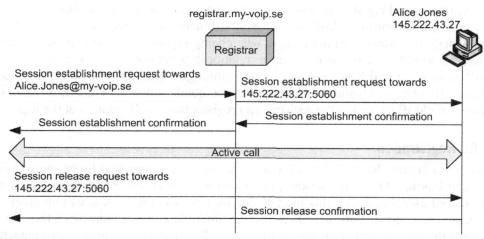

FIGURE 6.26

VoIP session establishment and subsequent signaling.

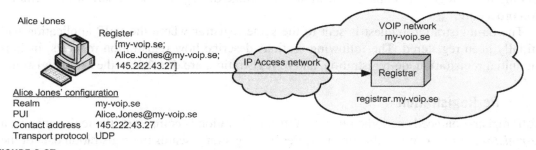

FIGURE 6.27

VoIP user registration.

The contact address provided by the VoIP subscriber may include a port number, informing the registrar that it will use that particular port number when establishing a VoIP session toward this user. In the absence of a port number, the registrar uses a (standardized) default port number for this purpose. A transport protocol indicator may accompany the contact address, informing the registrar that it should use that particular transport protocol when establishing a VoIP session toward this user. Without such an explicit indication in the contact address, UDP will be used.

The registrar typically has a subscriber database, containing the subscription record per subscriber that is allowed to register in this registrar. Hence, prior to registration, there is already an *administrative relationship* between a VoIP user and the registrar. When the registrar has accepted the registration, it confirms the registration to the VoIP user and stores the subscriber's PUI and contact address. The aforementioned administrative relationship between VoIP user and registrar is now changed into a *registration relationship*. The registrar is now capable of forwarding a VoIP session establishment request message to the VoIP user. The VoIP user's IP address will not change when being registered in the registrar; a change in IP address would destroy the registration relationship between VoIP user and registrar. The VoIP application will hence regularly renew (through DHCP query) the lease period of the obtained IP address. Should the VoIP application obtain a new IP address, it has to then de-register the old IP address at the registrar and register the new IP address at the registrar.

6.3.2 De-Registration

De-registration entails that the VoIP user indicates to the registrar that he or she wishes to remove their binding. On de-registration, the registrar discards the IP address from its internal user record associated with this user. If this user does not have at least one other contact address registered (a VoIP user may register multiple contact addresses), the registrar will discard the user record from internal memory. The user is now no longer contactable under this public user identity. De-registration may be associated with termination of the VoIP application. When the VoIP application is terminated, there is no need to maintain the binding in the registrar, so the application will initiate de-registration as part of the termination procedure.

The de-registration request is sent to the same registrar where the VoIP application had initially been registered. The following sections describe how registration requests, including initial registration, de-registration, and re-registration, are directed to the same registrar.

6.3.3 Re-Registration

VoIP registration needs periodic renewal. When registration is completed, the registrar starts a *re-registration timer*. When this timer expires, the registrar discards the registration data of the VoIP subscriber. More specifically, the registrar discards the contact address of the VoIP subscriber (a VoIP user may register multiple contact addresses). Hence, in order to remain registered at the registrar (with this contact address), the VoIP user has to perform re-registration within a defined time-out period. This registrar informs the VoIP user, when confirming the registration, about the re-registration time.

The rationale of periodic re-registration is that VoIP applications may switch off or be otherwise disconnected from Internet access without performing a regular de-registration. Examples include:

- Ethernet LAN-based SIP phone, being unplugged from the network (LAN).
- Cellular mobile VoIP application losing high-speed packet-switched connectivity (e.g. UMTS or HSPA) due to a change of location.
- WLAN application leaving WLAN coverage, thereby losing Internet access.

The registration interval, agreed between registrar and VoIP user, will be set with care. Too short an interval leads to unnecessary network load (Tuffin and Jestin, 2008). Too long an interval increases the chance of lost synchronization between terminal and registrar, e.g. when the VoIP application terminates without de-registration, there may be a long delay before the registrar discards the binding associated with that terminal. Other applications in the network may also remain under the impression that the terminal, i.e. the user, is still contactable.

Note

Minimum and maximum registration interval are normally generic system parameters for a registrar. This implies that machine applications registering as VoIP users, or other VoIP applications that may remain reliably registered for a long duration, often have to apply more frequent re-registration than would be necessary for their nature. This is one area where optimization may have to be applied over time – for example, allowing for adaptive maximum registration time, depending on the (asserted) type of application and depending on the access network in use.

6.3.4 **Mobility Versus Nomadicity**

A VoIP application (terminal) registers its contact address, typically an IP address, during registration. As long as the VoIP terminal is contactable through that contact (IP) address, the registration in the registrar remains valid. Fixed terminals, e.g. terminals connected to the Internet via Ethernet and DSL, will retain their allocated IP addresses. For mobile terminals and for soft phones, the IP address may change, depending on the terminal type and depending on the degree of mobility. Examples are the following:

- Wireless LAN (WLAN) terminal; the IP address will change when a subscriber attaches to different WLAN access points at different locations.
- Cellular mobile phone; the IP address may change when a mobile data connection is released and re-established at another mobile access network.

Nomadic VoIP users may register to their VoIP network from different locations, in different access networks – for example, a VoIP subscriber from Sweden having the capability to, and being authorized to, register from a wired or wireless access point in France.

6.4 LOCATING THE REGISTRAR

Before VoIP subscribers can initiate a registration, they need to locate the registrar that they will be using, i.e. the registrar in which their binding will be deposited and that will handle the VoIP sessions initiated toward the subscriber. A VoIP subscriber may be registered to the VoIP network through more than one terminal. Multiple terminals are registered at the same registrar. Registration through multiple terminals will be dealt with in a later section.

The registrar, being a functional entity in the VoIP operator's network infrastructure, is hosted on one or more hosts in the network. The VoIP subscriber can forward a *registration*

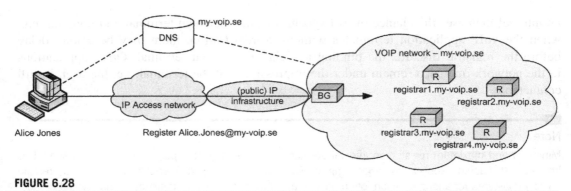

FIGURE 6.28

Sending a register message to a VoIP operator's network. BG, border gateway; R, registrar.

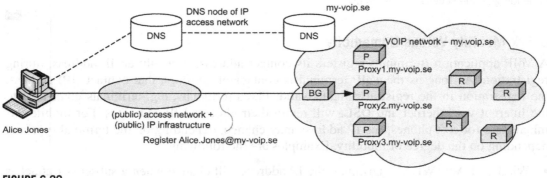

FIGURE 6.29

Sending a register message to a proxy in a VoIP operator's network. BG, border gateway; P, proxy (for control plane); R, registrar.

request message to this registrar by directing the request to the operator's *realm*, a domain name assigned to the VoIP operator, e.g. my-voip.se. The realm does not identify a designated registrar (see Figure 6.28).

When the VoIP subscriber directs the register request message to the VoIP operator's realm, the DNS service is used to route the message to that network, my-voip.se, as depicted by a dashed line in the figure. DNS constitutes, by design and by implementation, a global service. The VoIP user has an *entry* into this global DNS service, for obtaining information related to specific domains, this entry being the address to the local DNS node, provided by the access network provider. The required information (IP address of proxy of my-voip.se) may not be available in the local DNS node of the access provider. An iterative DNS query may be applied to obtain the required information.

VoIP-related signaling exchanged between VoIP subscribers and the VoIP network traverses a signaling proxy, the *user-to-network proxy*. Hence, the register request message is not directly sent to one of the registrars; it is sent to a user-to-network proxy first. This is depicted in Figure 6.29.

The blocks denoted "P" in Figure 6.29 represent a user-to-network proxy. As depicted in the figure, there may be several such proxies deployed in the VoIP network. User-to-network proxies may be associated with the access network or may be independent thereof. The VoIP terminal needs to determine which user-to-network proxy to use, prior to being able to carry out the registration. Various possibilities exist here:

- **Terminal configuration.** The VoIP terminal configures a proxy address. The proxy address may be configured in the form of an IP address (+port number) or in the form of a domain name associated with the proxy. In the latter case, DNS service is needed to obtain (i) the host name for the user proxy and (ii) the IP address for the proxy host. The method of using a configured domain name, as opposed to using a configured IP address, has the advantage that DNS-based load sharing over multiple proxies can be applied. In addition, DNS may provide information relating to a primary proxy node as well as information relating to a secondary proxy node, facilitating fallback when the primary proxy is determined to be unreachable.
- **DHCP query.** DHCPv6 may be used to request proxy node address. A designated DHCP message is specified here. Since DHCP is associated with the access network, obtaining a proxy node address through DHCP implies that an administrative relationship exists between the access network provider and the VoIP operator; the access network provider needs to provide the proxy node addresses.
- **GPRS access.** When a VoIP subscriber gains IP connectivity through GPRS access, then it may request a VoIP proxy in the same procedure as establishing the mobile data connection. This method requires that an administrative relationship exists between the GPRS operator, providing the mobile access, and the VoIP operator.

Besides using a user-to-network proxy in the VoIP operator's domain, the VoIP signaling to and from the VoIP user may traverse a proxy that is local to the host on which the VoIP terminal resides. This local proxy would be configured in the VoIP terminal, or might result from local (in terminal host) DNS configuration, and has no further relation with the VoIP network.

The DNS depicted in Figures 6.28 and 6.29 facilitates the forwarding, on a core network level, of the register request message up to a proxy of the VoIP operator's user-to-network proxies. A further step is needed to get the user-to-network proxy to forward the register request message to a registrar where the subscriber may be registered. Here the user-to-network proxy uses the services from an *interrogating proxy* (not to be confused with an *inbound proxy*). This is depicted in Figure 6.30.

Typically, the user-to-network proxy node forwards the register request message to *one of n* interrogating proxies, whereby the selection of interrogating proxy may be done by the user-to-network proxy node itself or may be based on load sharing from DNS (DNS internal to the VoIP network). Whereas a user-to-network proxy will, after the registration process, contain subscriber data, the interrogating proxy never contains subscriber data. Hence, any

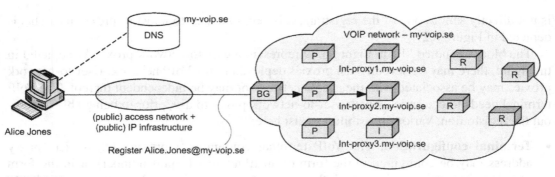

FIGURE 6.30

Sending a register request message through an interrogating proxy. BG, border gateway; I, interrogating proxy; P, proxy (for control plane); R, registrar.

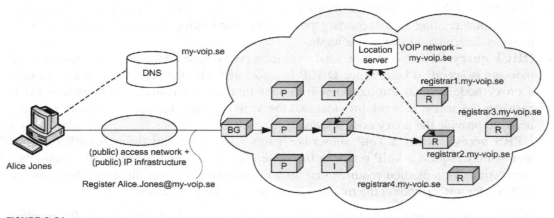

FIGURE 6.31

The role of a Location server in a VoIP network. BG, border gateway; I, interrogating proxy; P, proxy (for control plane); R, registrar.

of the available and operational interrogating proxies may, in principle, be used for the forwarding of the register request to a registrar.

As the role of interrogating proxy consists of selecting the registrar that will handle the registration, the following distinction needs to be made:

- The subscriber who is initiating the registration is not yet registered through any terminal and does not (temporarily) have a user profile available at a registrar. In this case, the interrogating proxy selects a registrar, based on internal configuration tables, and forwards the register request message to the selected registrar.
- The subscriber who is initiating the registration currently, already has a registrar assigned to him or her. In this case, the interrogating proxy forwards the register request to the registrar currently assigned to the subscriber.

The first situation – no registrar currently assigned to the subscriber – applies when the subscriber is not registered through any terminal at the time of registration. The latter situation, a registrar currently assigned to the subscriber, applies for example when the subscriber is already registered as a VoIP subscriber through at least one terminal. By design, a VoIP subscriber cannot be assigned to two or more registrars at the same time, which is why the registration request is forwarded to the currently assigned registrar.

The interrogating proxy does not hold subscriber data, yet it has to determine whether the subscriber is currently assigned to a registrar. For this purpose, a Location server is used. This is depicted in Figure 6.31.

The Location server[5] plays a pivotal role in the VoIP network. VoIP subscribers need to be provisioned (*administered*) in the Location server before they can register as VoIP users. The Location server maintains the following subscriber data (not exhaustive):

- **Static subscriber data:** service subscriber data, public user identity, authentication credentials.
- **Dynamic subscriber data:** address (e.g. domain name) of registrar currently assigned to this subscriber.

The aforementioned query from the interrogating proxy is used to obtain the address from the Location server, typically in the form of a domain name, of the registrar currently assigned to the subscriber. This query is depicted in Figure 6.31 as a dashed line (since it's a single database query, not message forwarding). The registrar that handles the registration request from the subscriber updates the Location server, informing the Location server that the subscriber is currently registered in that registrar. The next time that a registration occurs for this subscriber, the Location server can direct that registration request to the same registrar.

On receipt of the register request, the registrar will complete the registration by storing the *binding information*, consisting of public user identity, e.g. sip:alice.jones@my-company.se, and contact address, e.g. an IPv4 address.

6.5 REGISTRATION RELATIONSHIPS

The registration of a subscriber as a user in the VoIP network may be described as the establishing of relationships between entities in the network.

6.5.1 Subscriber Administered in VoIP Network, but Currently not Registered

The subscriber is provisioned in the Location server. The subscriber terminal is configured with information needed to initiate registration, such as public user identity and VoIP network domain (*realm*).

---

[5] A Location server may be used in other contexts as well, e.g. maintaining the network location (global cell identifier) or geographic location of a group of subscribers.

Before registration can be initiated, the subscriber terminal needs to obtain IP connectivity, including but not limited to, IP address.

A *provisioning relationship* exists between subscriber terminal and Location server.

6.5.2 Subscriber Administered in VoIP Network and Currently Registered

A registrar is allocated to the VoIP subscriber; the Location server has the address of the registrar for this subscriber. Registrar and user-to-network proxy contain subscriber data, resulting from the registration. The terminal has obtained the address of the user-to-network proxy as well as the address of the registrar in which it is registered. This results in the following additional relationships:

- *User-to-network relationship* between subscriber terminal and user-to-network proxy. All VoIP signaling to/from the subscriber terminal traverses this user-to-network proxy.
- *User–registrar relationship* between subscriber terminal and the registrar. VoIP sessions initiated by the subscriber are steered through this registrar; VoIP sessions destined for the subscriber are also steered through the registrar, for the purpose of forwarding the session establishment request to the subscriber's contact address.
- *Proxy–registrar relationship* between the user-to-network proxy and the registrar. VoIP signaling between subscriber terminal and registrar traverses the proxy.
- Relationship *between Location server and registrar*. The Location server stores the address of the registrar assigned to this subscriber, so VoIP sessions destined for this subscriber, as well as registration requests from the subscriber, can be forwarded to this registrar.

It is of utmost importance that these relationships are maintained. In the case of temporary outage, e.g. restart, of user-to-network proxy or registrar, for example, dynamic subscriber data would be lost, resulting in the loss of the above-mentioned *relationships*. The registration relation may be repaired only at the next registration. Networks will therefore take care that in the case of planned or unplanned node outage, subscriber data are moved to another node, thereby ensuring uninterrupted service availability.

6.6 NETWORK DOMAINS

A VoIP network has a particular domain associated with it, also referred to as a *realm*, e.g. my-voip.se. This domain is used for services and hosts associated with this VoIP network. The individual subscribers of this network, however, would not need to have my-voip.se as their domain name. Subscribers may have enterprise-specific or personal domain names for VoIP services. Examples, including the user part, include:

- alice.jones@my-company.se
- john.smith@another-company.org
- wendy@jackson.se

These subscribers may be served by, for example, my-voip.se. In all of these examples, the domain part of the user name would need to be provided in DNS with an *inbound SIP server* host name and address. When a VoIP session is established toward such a subscriber, the domain from where the VoIP session is established needs to be able to determine the VoIP network that is serving that subscriber. A call (VoIP session) to alice.jones@my-company .se would result in the domain from where the VoIP session is established receiving a host name and IP address of an inbound SIP server from my-voip.se. These enterprise-specific or personal domain names for VoIP services are hence only for subscriber identification, but do not represent a network domain.

The use of enterprise-specific or personal domain names for VoIP services poses the following interesting challenges:

1. When a VoIP session is established toward, e.g., sip:alice.jones@my-company.se, a DNS query is applied for the domain part of the public user identity, i.e. my-company.se, in order to obtain the host name and IP address of the inbound SIP server of the VoIP network serving users of the domain my-company.se. Since the DNS query is on the domain part only, and not on user@domain, it would be required that all VoIP users of my-company.se are subscribers to the same VoIP network. This might not be the case, for example due to company mergers or for strategic reasons. This situation would be even more pronounced for non-country-specific domains, such as another-company.org. Users of that domain may reside in different countries and are (inevitably) served by different VoIP operators.

2. The domain used for VoIP services may be the same as the domain used for other Internet services, such as HTTP or email. A DNS query for an HTTP host for a particular domain would be handled by the same DNS database as a DNS query for VoIP services for that same domain. This is due to the fact that the iterative process of directing a DNS query to the DNS node responsible for resource records (RRs) for that domain is based on the domain only. DNS information (RRs) related to VoIP services (specifically NAPTR RR) must therefore be administered in the same DNS node as DNS information relating to HTTP services (A record).

For this reason, the domain that enterprises may have allocated to them for a VoIP service is differentiated from other domains associated with that enterprise. For example:

- HTTP and email services: my-company.se
- VoIP services: my-company.voip.se

The domain my-company.voip.se is a *subdomain* of voip.se. Voip.se would be a domain owned by a VoIP operator in Sweden, my-voip.se. That operator assigns subdomains as <company>.voip.se. DNS queries related to any <company>.voip.se would be handled by a DNS node in the network my-voip.se, i.e. by that VoIP operator's own network.

Introduction to Session Initiation Protocol

7.1 INTRODUCTION

This chapter introduces the Session Initiation Protocol (SIP). SIP is the *control-plane* protocol used in the IMS network for registration, session establishment, message routing, terminal capability exchange, presence services, etc. Although SIP is not the only protocol used in IMS, it does constitute one of the cornerstones of IMS.

A thorough understanding of the capabilities of SIP is crucial for understanding IMS and how it may be applied for service development. In this chapter we will focus on the underlying principles of SIP, the concepts, and how it may be applied. We will regularly refer to IETF documents or other documentation for further, detailed information of SIP.

7.2 THE SIP STANDARD

The SIP model, developed by the Internet Engineering Task Force (IETF), was first described in IETF RFC 2543 in 1999, which was later revised as IETF RFC 3261 in 2002. Both IETF standards describe SIP "version 2.0". Standards preceding SIP, for establishing communication sessions over IP, include ITU-T H.323, ITU-T H.245, and ITU-T H.225. Many operational IP-based communication networks have been built around these standards. Gradually, SIP is taking over as the preferred communication standard for multimedia over IP. Reasons for this include the fact that SIP can be used for a wide variety of (multi)media communication, and that SIP is by design independent of access network technology, and also the flexibility and extensibility of SIP.

There are many enhancements defined for SIP; each enhancement is defined in a separate IETF RFC. SIP has a compatibility mechanism, allowing for new capability to be introduced without impacting operational systems that do not support the enhancement.

SIP is inspired by the Hyper Text Transfer Protocol (HTTP). Throughout the explanation of SIP, similarities with HTTP will be noticed, especially in the *transaction model* that is fundamental to SIP.

7.3 SIP SESSION VERSUS MEDIA SESSION

One of the main tasks of SIP is to facilitate the establishment of a *media session* between two end-points. A media session typically consists of real-time person-to-person communication,

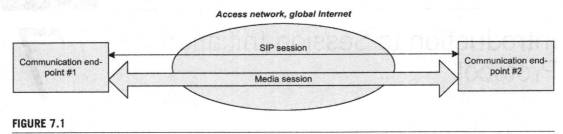

FIGURE 7.1

SIP session and media session.

such as voice call or video call. The respective end-points may be arbitrary users on the Internet. When establishing the media session between the two end-points, there will in fact be a SIP *session* and a *media session* (see Figure 7.1).

The two end-points establish a SIP session by exchanging *SIP messages*. One entity may initiate a SIP session by sending a session establishment request message to the other entity; the other entity may accept or reject the SIP session establishment request and signal the acceptance or rejection to the initiator of the SIP session. The SIP session is used to exchange information about the *media session* that needs to be established. The SIP session and the media session are related to one another. Establishment, modification, and termination of the media session (the user plane) are signaled between the two end-points, via the SIP session (the control plane).

A communication end-point that uses SIP for communication session establishment is known as a SIP user agent (SIP UA; denoted as "UA" for brevity). A SIP session between two communication end-points will typically not run directly point-to-point. There will be *SIP proxies* residing in the signaling path. These SIP proxies play a role in establishing the SIP session, e.g. routing a SIP session establishment request to the destination end-point, based on a public identifier of that communication end-point. For example, when a communication end-point wants to establish a SIP session with john.smith@my-company.se, then various SIP proxies aid in routing the SIP session establishment request message to the VoIP network serving this person (John Smith@...) and to the registrar holding that person's contact binding. The concept of a registrar will be explained in a later section. The media session, on the other hand, will typically traverse fewer nodes. This is accomplished by the end-points exchanging their contact addresses for media transfer, allowing the respective peers to send media directly to that contact address. This aspect of SIP is commonly depicted through the *SIP trapezoid* (see Figure 7.2).

Figure 7.2 shows that during SIP session establishment between UA1 and UA2, the SIP session is carried through a number of SIP proxies, labeled P1–P4. The *path* for the media session is defined during the SIP session establishment, but media transfer has not started yet (the concept of early media transfer will be dealt with in a later section). When the SIP session is established, it may, for the remainder of the life of this SIP session, be carried through a smaller number of SIP proxies. Media transfer between the two UAs may now

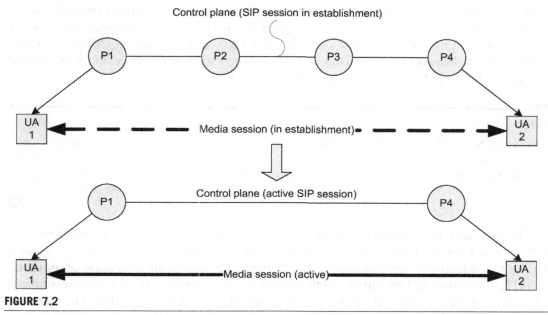

FIGURE 7.2

SIP trapezoid.

start. The relation between the SIP session and the media session remains for the entire communication session. The media session remains "under control" of the SIP session. A change in the definition of the media session, as well as the termination of the media session, is signaled between the UAs through the SIP session.

It is useful to give a brief description about the difference between UA (user agent) and UE (user equipment). The UA is an entity that has the capability to use SIP for communication. A UA may be found in a terminal (e.g. phone), but also in network nodes that are involved in SIP session establishment. A UE represents an end-user terminal. A UE contains a UA, but also contains other functional components that are needed in a phone, such as user-plane handling and user interaction. In this chapter, we deal with SIP signaling, so we refer to UA.

7.4 SIP TRANSACTION MODEL

SIP is a *transaction-based* protocol. The establishment of a SIP session between two UAs, e.g. two SIP phones, is done through the execution of one or more *transactions*. A transaction constitutes a request from one UA, the originator of the request, to another UA, the receiver of the request, to execute a particular task and to report the result of the execution of that task back to the originator of the request. There is an exception to that rule. The ACK transaction does not have a result associated with it. This will be further explained in a later section.

SIP capability, so far as the use of transactions is concerned, is described through a set of *methods*. A method constitutes a basic operation that may be applied between two UAs. A transaction constitutes the *execution* of a particular method. For example, INVITE is a particular method defined by SIP. When one UA wants to establish a SIP session with another UA, it initiates an INVITE transaction with that other UA.

Note

The transaction model may be compared with Remote Operations (ROS; ITU-T X.880), as applied by protocols like CAP (CAMEL Application Part; 3GPP TS 29.078) and MAP (Mobile Application Part; 3GPP TS 29.002). For ROS protocols, the basic operation capability is defined through a set of *operations*. One entity may apply a remote procedure call by sending an operation request to that remote entity.

The execution of a transaction between two UAs may result in the establishment of a SIP session between these UAs. This depends on the actual transaction that is applied and on the outcome of the transaction. For example, the successful execution of an initial INVITE transaction will lead to the establishment of a SIP session, whereas the execution of a REGISTER transaction will not lead to the establishment of a SIP session. The adjective "initial" for the INVITE transaction indicates that this particular INVITE transaction is used outside the context of an existing SIP session. A *re-INVITE*, on the other hand, refers to an INVITE transaction used within an existing SIP session.

The execution of a SIP transaction is done through a *client–server transaction state model*. A user agent client (UAC) initiates a SIP transaction; the SIP transaction is received and executed by a user agent server (UAS). UAC and UAS are *roles* that a UA can take, when initiating or receiving a transaction request (see Figure 7.3).

A SIP transaction consists of a request, zero or more provisional responses, and a final response. The execution of the INVITE transaction may also include the sending of an ACK request. This will be the case when the final response to the INVITE request is an *unsuccessful* final response. The ACK message in that case forms part of the INVITE transaction. When the INVITE transaction leads to a *successful* final response, it will also be followed by an ACK request. That ACK request constitutes a separate transaction, without final response. Each of these components of a transaction is carried in a SIP message. The role of each category of SIP message is as follows:

- **Request message:** Indicates the transaction that is requested. The request message contains various information elements (a collection of *SIP headers* and optionally a *message body*). These information elements are needed by the receiver of the request message to be able to execute the requested task. Some of these information elements are needed by the network to route the message to the intended receiver.
- **Provisional response:** Provides the initiator of the transaction with information about the progress of the execution of the requested task, whilst the execution of the task is

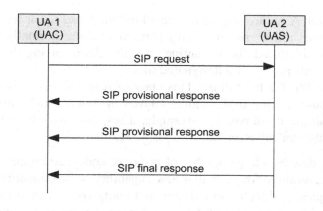

FIGURE 7.3

SIP transaction signaling.

not yet finished. A provisional response may, in turn, also contain various information elements.

- **Final response:** Indicates the outcome of the execution of the requested task (successful, unsuccessful). Again, various information elements may be carried in the final response.

The responses that a UAS may send to the UAC, in response to a *request message*, are divided into response classes, each class having its own meaning. The responses are defined in terms of a *response code*. The response codes that are defined for SIP are a subset of the response codes that are defined for HTTP/1.1, suitably extended for SIP. The following response classes are defined for SIP. Not all response code values in the given ranges are defined.

- **1xx-class (100–199):** Provisional responses. The UAS informs the UAC about the progress of the execution of the transaction request. The provisional response may contain additional information required by the UAC for the continuation of the transaction.
- **2xx-class (200–299):** Successful final response. The transaction is successfully executed.

All other response code values represent unsuccessful execution:

- **3xx-class (300–399):** Redirection. A response of this class informs the UAC that the destination, as indicated in the invite request message, should be contacted at an alternative address or that an alternative service should be requested from the destination subscriber.
- **4xx-class (400–499):** Request failure. The UA that received the request and that had taken the role of UAS for this request has indicated that it cannot process the request. The action to be taken by the UAC depends on the specific failure that is reported – for example, re-attempt the request at a later time.
- **5xx-class (500–599):** Server failure. A response within this class indicates that a system failure has occurred, e.g. in a proxy server. This system failure may, for example, be an

overload condition. A failure response of this kind may also result from an IP infrastructure failure, such as router overload. This particular failure normally has the effect that a particular proxy is placed "in quarantine", i.e. the UA receiving this response will not send a request to this proxy for a designated time.

- **6xx-class (600–699):** Global failure. This class of response codes is specific for SIP. It provides a response relating to the addressed user in general, including the situation that the user has multiple terminals. For example, a response may be given that the called user is not available on any terminal.

IETF RFC 3261 defines a large number of response codes; additional IETF RFCs define new response codes, required for particular new capability or functionality.

While a SIP request is directed to a designated entity (person or system), an intermediate proxy may provide an unsuccessful final response, i.e. a proxy between the initiating UA (acting as UAC) and the addressed entity. For example, an intermediate proxy may be in overload condition, preventing it from processing (proxying) the request. In that case, the intermediate proxy takes the role of UAS and sends a final response, e.g. 503 Service unavailable. A proxy receiving a 503 Service unavailable final response maps this to 500 Server Internal Error before forwarding the final response upstream. The reason is that the proxy receiving the 503 Service unavailable final response applies quarantining for the host relating to the 503 Service unavailable.

In the case of the INVITE transaction, some of the information elements contained in the request message and (provisional) response message are for the purpose of maintaining the SIP session after the INVITE transaction has completed successfully. Hence, they are not needed for the execution of the actual transactions, but will be used for subsequent transaction within the SIP session that was established. Figure 7.4 shows two examples of an INVITE transaction.

The left side of Figure 7.4 shows successful session establishment. The response codes, such as 180 and 200, as well as the associated labels "Ringing" and "OK", are described in a subsequent section. When UA1 has received the successful final response (200 OK), the SIP session between UA1 and UA2 is established. UA1 then initiates a new transaction, the ACK (*acknowledgement*) transaction, to UA2, to confirm to UA2 that the final response is received. In the case of unsuccessful call establishment (right side of Figure 7.4), when UA1 receives the unsuccessful final response (486 Busy Here), there is no SIP session established. UA1 then initiates an ACK message to UA2 to confirm to UA2 that the final response is received.

A common mistake is to associate UAC and UAS with a node or entity, such as a SIP phone. For example, a SIP phone that establishes a call (SIP session) is often considered to be a UAC, and by the same token a SIP phone that receives a call would be considered to be a UAS. However, UAC and UAS represent *roles* that a UA may take, depending on whether that UA is the initiator of a transaction or the receiver of a transaction request. This is depicted in Figure 7.5.

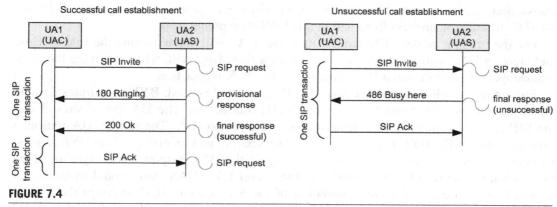

FIGURE 7.4

SIP INVITE transaction.

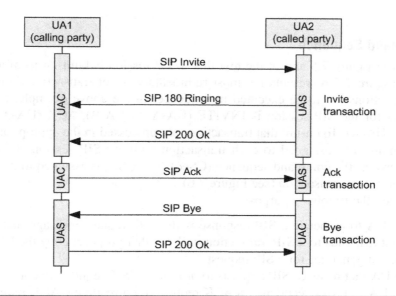

FIGURE 7.5

UAC versus UAS.

When UA1 establishes a SIP session with UA2, it creates a UAC state model for the INVITE transaction, i.e. it assumes the role of UAC. It is from within this UAC state model instance that the INVITE request message is sent. Since the INVITE request message is for the purpose of initiating a new SIP session, this INVITE request is referred to as *initial INVITE*.

UA2, meanwhile, is prepared to receive an initial INVITE request message. When UA2 receives the INVITE request from UA1, it creates a UAS state model for the INVITE transaction, i.e. it assumes the role of UAS. Provisional response and final response for the INVITE transaction are sent from within this UAS state model instance.

For the sending of the ACK request message, UA1 will again assume the role of UAC and create a UAC state model for the ACK transaction. UA2 will, when receiving the ACK request message, again start a UAS state model for ACK transaction.

For the termination of a SIP session, the BYE method is used. BYE constitutes a transaction in a manner that compares with INVITE transaction. The UA that wishes to end the SIP session initiates a BYE transaction to the remote UA. The remote UA returns a final response to the BYE request message. The sending and receiving of the BYE request message, as well as the sending and receiving of the BYE response message, are done from within a BYE UAC state model instance and BYE UAS state model instance. As depicted in Figure 7.5, for the termination of the SIP session, UA2 assumes the role of UAC and UA1 assumes the role of UAS. This shows that UAs can have alternating UAC and UAS roles.

7.4.1 Command Sequence

We have seen in Figure 7.5 and in the text that a SIP session is built up from a number of transactions. Figure 7.5 represents the most rudimentary set of transactions forming a SIP session. Transactions are to be executed in sequence. In the above example, the sequence of transactions for the SIP session is INVITE (UA-A → UA-B), ACK (UA-A → UA-B), BYE (UA-B→UA-A). To ensure that transactions are processed in the appropriate sequence, a sequential number is assigned to each transaction within a SIP session. This sequential number is known as the command sequence (CSeq). The CSeq is assigned to the transaction by the initiator of that transaction (see Figure 7.6).

The CSeq has the following purpose:

1. It allows a UA to associate a SIP response with a SIP request message that it had sent. The UA that responds to a SIP transaction will, for that purpose, copy the CSeq value in any response it generates for a SIP request.
2. It allows a UA that receives SIP requests to process the SIP requests in correct order.
3. It allows a UA or proxy receiving an ACK request to correlate the ACK request with the INVITE transaction.
4. It allows a UA or proxy receiving a CANCEL request to correlate the CANCEL request with the INVITE transaction.

Later in this chapter, we will describe the branch-ID. The branch-ID identifies a transaction branch between two SIP entities. The concept of *branch* is, likewise, described in a later section. The CSeq also identifies a transaction. A difference between CSeq and branch-ID is that the CSeq is used end to end between two SIP entities, i.e. the CSeq is carried over intermediate proxies. In addition, the CSeq may be used for *transaction correlation*, as described above.

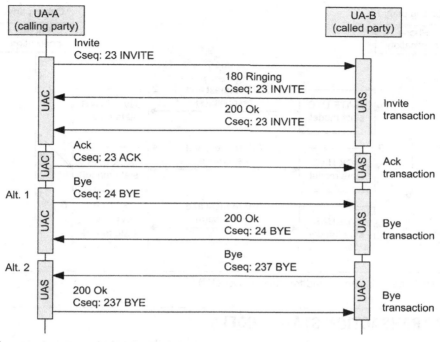

FIGURE 7.6

Use of Command sequence (CSeq) in a SIP session.

The CSeq is included in a SIP message in the form of a *header*. For example:

```
INVITE sip:john.smith@company-x.se SIP/2.0
CSeq: 23 INVITE
...
SIP/2.0 200 OK
CSeq: 23 INVITE
```

The CSeq header carries both a numerical value and the name of the method. For a SIP request message, the explicit mention of the method name in the CSeq header may be considered superfluous, as the method is also reflected in the request line. For a SIP response message, on the other hand, the status line (first line of the message) does not contain the method name. So, for a response message, the explicit mention of the method name in the CSeq header helps the reader to understand the message. So, once again, SIP is very readable!

The ACK transaction has special handling, as far as the CSeq header is concerned. An ACK transaction is closely related to an INVITE transaction; it acknowledges the final response of the INVITE transaction. This close relation is reflected in the fact that the ACK transaction contains the same CSeq value as the INVITE transaction it belongs to.

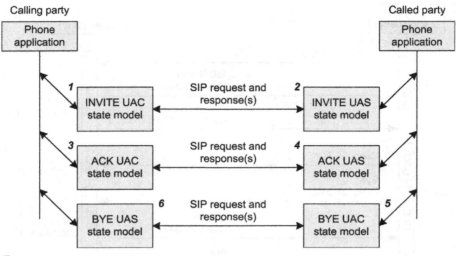

FIGURE 7.7

Transaction state model instances and transaction user (TU).

7.5 SIP TRANSACTION STATE MODELS

The transaction instance constitutes a process that may be created by a UA. In a way, the UA uses the *services* from a transaction instance. The UA creates a request message, starts a corresponding transaction client state model instance, and passes the request message to that client state model instance, with the request to further handle that transaction toward the remote end. Similar interaction between UA and transaction instance occurs at the receiving side. This is depicted in Figure 7.7. The phone application, being the user of the services from a transaction model instance, is referred to as the transaction user (TU).

At numeral **1**, the phone application of the calling party has prepared an INVITE request message and creates an INVITE UAC state model instance. It then passes the INVITE request message to that state model instance, for transfer of that message to the intended receiver. At numeral **2**, an INVITE UAS state model instance is created within the phone application. The received INVITE request message is passed on to the phone application. Since the INVITE request is an *initial INVITE*, the forwarding of the INVITE request to the phone application results in a new session handling process being initiated. (The session may be a phone call, but may also be a *chat session*.) A relation now exists between the phone application of the calling party and the INVITE UAC transaction model, as well as a relation between the phone application of the called party and the INVITE UAS transaction model. The phone application of the called party can now provide provisional response(s) and final response to that INVITE UAS transaction model instance, which will transfer the response(s), as they are provided, toward the INVITE UAC transaction model, which will offer the response to the phone application through the aforementioned relationship.

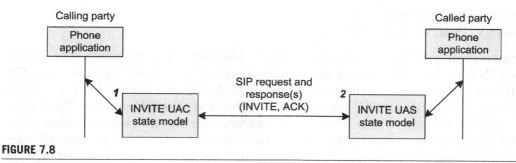

FIGURE 7.8

State model usage for unsuccessful call establishment.

At number **3**, the phone application of the calling party creates an ACK UAC state model instance and passes the ACK request message to it. At **4**, an ACK UAS state model instance is created, which passes the received ACK on to the phone application. The ACK request message relates to a session handling process that was started with a previous INVITE transaction. The ACK request message will therefore be forwarded to that existing call handling process.

At **5**, the phone application of the called party creates a BYE UAC state model instance and passes the BYE request message on to it. At **6**, a BYE UAS state model instance is created in the phone application of the calling party. That state model instance passes the BYE request on to the application. Similar to **4**, the BYE request is related to the ongoing call handling process, so the BYE request is passed on to that process. That call handling process can now return a final response on the BYE request; that final response will be handled by the BYE UAS transaction instance and the BYE UAC transaction instance.

UAC and UAS transaction model instances can be "destroyed" as soon as the transaction handled by that UAC instance and UAS instance is complete. It will be described in a later section that, due to potential retransmissions, transaction model instances may have to be kept alive for a specified duration after the last message is passed through that state model instance.

Figure 7.8 depicts the transaction usage for an unsuccessful call setup. When the INVITE transaction is not successful, e.g. a 486 Busy here is returned by the UA of the called party, the ACK request message is sent from the same UAC state model instance of the calling party's UA from which the INVITE request was sent. Likewise, the ACK request message is received in the same UAS state model instance of the called party's UA in which the INVITE request was received. So, this ACK request message does not lead to a new UAS being created. The ACK request message contains an identifier associating it with the existing UAS state model instance.

There are various significant reasons for this differentiated handling of the ACK transaction, as will be described below.

SIP makes a distinction between the following four transaction state models:

1. INVITE client transaction
2. INVITE server transaction

3. Non-INVITE client transaction
4. Non-INVITE server transaction.

The reason for this distinction between INVITE transaction client and server state models and non-INVITE transaction client and server state models is mainly the following. The INVITE transaction, if successful, leads to the establishment of a SIP session between the respective UAs (a SUBSCRIBE transaction may also lead to a SIP session). The further communication between these two UAs, e.g. to acknowledge the successful execution of the INVITE transaction or to request the termination of the SIP session, is done through independent transactions. These independent transactions are associated with the SIP session that is established between the two UAs.

When the INVITE transaction is not successful, there is no SIP session established between the respective UAs. In fact, an intermediate SIP proxy may generate the final response, e.g. 500 Server Internal Error. Therefore, in the unsuccessful INVITE transaction case, the ACK message that is used to acknowledge the reception of the final, non-2xx response is sent as part of the transaction. The ACK message traverses the same chain of SIP proxies that the INVITE request had traversed. The respective SIP proxies, as well as the UA acting as UAS for this request, destroy the transaction process as soon as the ACK is received.

The reason for the distinction between client transaction state model and server transaction state model is the following. A UAC may retransmit a request message when it has not received a (provisional) response for the request. The retransmission may be repeated according to a defined pattern. For a UAS, retransmission may be applied for final response.

The above-described differentiated usage of the ACK message also reflects on the UAC and UAS state model. An INVITE client transaction model has the ability to *send* an ACK message, whereas the INVITE server transaction model has the ability to *receive* an ACK message.

Figure 7.9 shows a graphical representation of an INVITE transaction between two UAs. The timers that are mentioned in the two state models in Figure 7.9 are used to govern the transaction signaling. They are used, for example, to trigger retransmission. Timers are described in IETF RFC 3261.

It was explained that the successful execution of a SIP transaction might result in the establishment of a SIP session. This applies specifically for the INVITE transaction. An INVITE transaction that is used to establish a SIP session is therefore referred to as an *initial INVITE transaction*. Within the SIP session, there will be additional transactions, notably the ACK transaction and the BYE transaction. These transactions do not lead to the establishment of a SIP session. Rather, these transactions are used *within* a SIP session. These transactions are referred to as *subsequent transactions* or *in-session transactions*. The INVITE transaction may also be used to *update* designated parameters of an existing SIP session, in which case it is referred to as *subsequent INVITE* or *re-INVITE*.

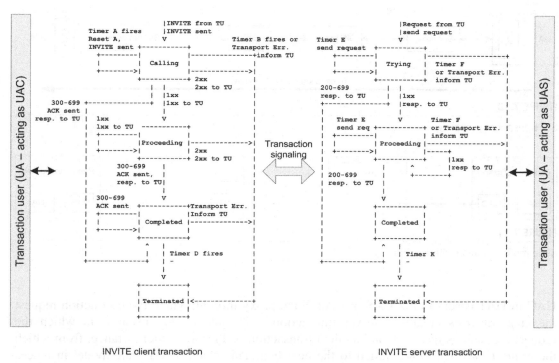

FIGURE 7.9

INVITE transaction signaling.
Source: IETF RFC 3261, figure 5 and figure 7 – adapted.

7.6 **PROXY ROLES**

Before we look into end-to-end SIP signaling between two user agents, we need to build an understanding of the various forms of SIP proxy that may be present in a SIP signaling chain. Two user agents involved in a SIP transaction form a SIP signaling chain, with zero or more SIP proxies in the chain. Depending on the type of transaction (session establishing or non-session establishing) and depending on the outcome of the execution of the transaction (successful, unsuccessful), the SIP chain may persist after the transaction, for the remainder of the SIP session.

Figure 7.10 depicts a SIP chain that may be applicable for a particular transaction. The transaction request, provisional response(s), and final response traverse a designated set of SIP proxies. The exact set of SIP proxies that is traversed by the transaction signaling is partly determined by the UAC at the time of transaction initiation and partly by the respective proxies, as the request message is being forwarded to the intended receiver. The building up of the SIP signaling chain starts with the creation of a UAC state model in UA1 and ends with the creation of a UAS state model in UA2. The request message that is sent from the

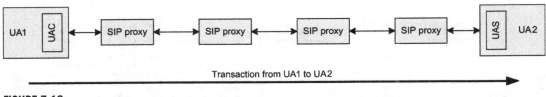

FIGURE 7.10

SIP transaction signaling.

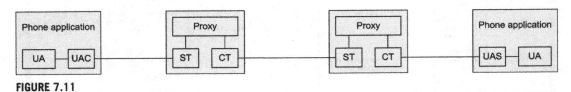

FIGURE 7.11

State model creation within SIP proxy.

UAC in UA1 is sent to the UAS in UA2. Each proxy through which the transaction request message traverses creates a server transaction (ST) state model instance, in which the request message is received, and a client transaction (CT) state model instance, from which the request message is forwarded to the next hop. This chain of CT state model instances and ST state model instances is depicted in Figure 7.11.

The following proxy roles are defined:

- Stateless proxy
- Stateful proxy
- Back-to-back user agent – this is not a proxy in the strict sense of the word, but may provide comparable functionality.

7.6.1 Stateless Proxy

A stateless proxy receives a SIP request message and forwards the request message to the next hop. The stateless proxy does not instantiate ST and CT state models, hence its statelessness. The SIP proxy handles each SIP request and SIP response independently. A stateless SIP proxy may be used for SIP communication between a SIP application on enterprise premises and a destination outside the enterprise network. The stateless proxy may reside at the edge of the enterprise network.

7.6.2 Stateful Proxy

The "statefulness" of a stateful proxy pertains to the transaction handling. A stateful proxy keeps the state of a transaction of which the messages traverse this proxy. It can hence aid in message retransmission and transaction cancellation. A stateful proxy creates an ST state model when receiving a SIP request. The ST offers the request message to the TU, in this case the SIP proxy application. The SIP proxy processes the request in accordance with its role. The SIP proxy may, for example, be an interrogating CSCF (I-CSCF) as defined in the IMS network (see Chapter 8). The processing of the request may result in modification of the request message, such as the addition of one or more Route headers, as will be elaborated upon in a later section. For the forwarding of the request message to the next hop, the SIP proxy application creates a CT state model and instructs the CT state model to dispatch the request message. For the remainder of the SIP transaction, there will be SIP transaction signaling between the sender of the request (or the previous hop) and the ST in the SIP proxy, and SIP transaction signaling between the CT in the SIP proxy and the destination of the request (or the next hop).

The distinction between record-routing stateful proxy and non-record-routing stateful proxy is explained as follows. An INVITE request message is sent from UA1, through SIP proxy A and SIP proxy B, to UA2. On the successful execution of the INVITE transaction, there will be a SIP session established between UA1 and UA2. Within this SIP session, subsequent transactions will take place between UA1 and UA2. These subsequent transactions may traverse one or more SIP proxies in the same manner as the initial transaction may have traversed one or more SIP proxies. Subsequent transactions that traverse a SIP proxy will, again, lead to creation of an ST state model and a CT state model. Whether subsequent transactions traverse a particular proxy depends on whether that proxy applies *record routing*.

A record-routing stateful proxy applies adaptation to the initial SIP request message indicating that it wants to be in the SIP transaction chain for all subsequent SIP transactions. The respective UAs between which the SIP session is established will then take appropriate action to send subsequent SIP transactions through this proxy. A non-record-routing SIP proxy does not make this adaptation to the initial SIP request message. As a result, the respective UAs between which the SIP session is established will not send the subsequent SIP transactions through this proxy.

A result of record routing by a SIP proxy is that the SIP session relationship is established *through* that proxy. This is depicted in Figure 7.12. In the example in Figure 7.12, SIP proxy A has applied record routing during the initial transaction. Hence, proxy A forms part of the SIP session relation between the phone application in SIP phone 1 and the phone application in SIP phone 2. This further implies that proxy A has a process active for handling subsequent SIP transactions within this SIP session. Subsequent SIP transactions within this SIP session traverse proxy A, but not proxy B. When the SIP session between the two phone applications is terminated, the transaction that is used to terminate the SIP session, namely the BYE transaction, also traverses proxy A. Proxy A will then terminate its proxy process.

Even though proxy A forms part of the SIP session between the two phone applications, there is still a *single* SIP session, traversing a SIP proxy.

Record routing relates to the routing of SIP transactions *within* a SIP session. Record-routing information is *exchanged* in an initial INVITE transaction and is then *applied* to the subsequent transactions in the SIP session.

Figure 7.12 does not reflect the user plane. The user plane may run directly between the SIP end-points or may be anchored in a media proxy, under control of proxy A.

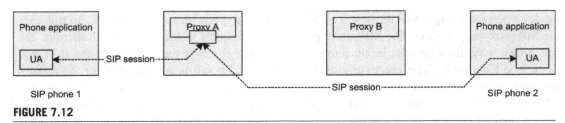

FIGURE 7.12

SIP session relationship for record-routing proxy.

7.6.3 Back-to-Back User Agent

A SIP node may act as a back-to-back user agent (B2BUA). A B2BUA is strictly not a proxy. A B2BUA is built up from two SIP signaling end-points. In some aspects, a B2BUA may offer the same capability as a proxy. A B2BUA is depicted in Figure 7.13. It should be emphasized that a transacting traversing a B2BUA consists of two transactions: one terminating at the B2BUA and another initiated by the B2BUA.

A B2BUA will, when it receives an initial SIP request message, specifically an initial INVITE request, act as UAS to the initiator of the INVITE request. The B2BUA will also act as UAC to the intended receiver of the INVITE request. The INVITE transaction that traverses the B2BUA (as explained above: there are *two* transactions) results in the establishment of a SIP session toward the initiator of the SIP INVITE transaction, as well as the establishment of a SIP session toward the intended receiver of the SIP INVITE transaction. These SIP sessions are maintained independent of one another. The media session (user plane), however, still runs end-to-end between the calling and called parties. That is, the B2BUA does not have to anchor the media plane in a media proxy, but rather the B2BUA acts as *end-point* for both sessions. When a SIP transaction is handled between a calling party (initiator of the SIP session) and called party (intended receiver of the SIP session), and the SIP session traverses a B2BUA, then the UAS and the UAC that are created by the B2BUA for this session do not have to run synchronously. For example, a provisional response received from the destination party does not have to be forwarded as a provisional response to the initiator of the SIP session.

A B2BUA has more capabilities than a record-routing proxy. A B2BUA may, for example, terminate a SIP session; this may be required when the B2BUA forms part of an online charging application and it has been determined that the available credit for the calling subscriber is depleted. Another use-case for B2BUA is a SIP application that wants to interrupt a speech session to deliver an announcement (connect the respective parties to an announcement device).

A disadvantage of B2BUA compared to proxy behavior is the additional processing cost. A B2BUA has to maintain two SIP sessions: one session to the initiator of the SIP session (the entity that generated the initial INVITE transaction) and another session to the receiver of the SIP session. In addition, it has to apply mapping of SIP request and response messages received within one SIP session to corresponding SIP request and response messages in the other SIP session. So, the implementation and testing of a B2BUA-based service is generally more expensive.

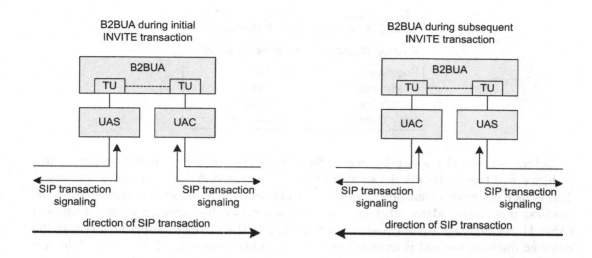

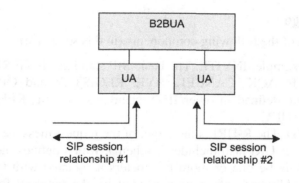

FIGURE 7.13

Back-to-back user agent cases.

7.7 SIP SESSION ESTABLISHMENT

In this section we will look into the establishment of a SIP session between two user agents. In this example, the SIP session will be established through a number of SIP proxies. Specifically, the following aspects will be considered:

- Routing of initial SIP request message from UAC to UAS
- Routing of SIP response message(s) from UAS to UAC
- Building a SIP routing path for subsequent SIP transactions
- Routing of subsequent SIP request messages between respective UAs.

Table 7.1 Structure of a SIP Message	
Request Message	**Response Message**
Request line	Status line
Header lines	Header lines
<empty line>	<empty line>
Body (optional)	Body (optional)

Before we consider a step-by-step method to build up a SIP request message routing path, we will briefly discuss the structure of a SIP message. A SIP message is text based. This is nice for human readability, but not very efficient where it comes to transmission and machine processing. Many other protocols use a notation like abstract syntax notation 1 (ASN.1) in combination with Basic Encoding Rules (BER), which results in very efficient message transmission and is more effective for machine processing. This is one of the reasons that many initiatives have been undertaken to apply SIP compression, in various forms.

A SIP message has the structure shown in Table 7.1.

7.7.1 Request Message

The request line consists of the following components (in this sequence):

- **Method name.** For example, INVITE. The basic SIP standard (IETF RFC 3261) defines six methods: INVITE, ACK, CANCEL, BYE, REGISTER and OPTIONS. Various additional methods are defined in other IETF standards, such as REFER, MESSAGE, SUBSCRIBE and NOTIFY.
- **Request URI (R-URI).** The R-URI is the target of the request message. It may be a URI or an IP address. The R-URI also includes a schema, which defines the type of address that follows. There may be one or more parameters associated with the R-URI. These parameters provide additional information needed by the network for processing the request message.
- **SIP protocol version.** The current version of SIP is 2.0. This is denoted by adding SIP/2.0 to the request line – for example, INVITE sip:john.smith@company-x.se SIP/2.0.

Header lines are lines of information that provide additional information to the request message. Header lines may be used for purposes such as message routing, transaction identification, branch identification, SIP path building, dialog identification, contact address exchange, and subscriber identification. We will discuss all relevant headers as we go along. The concept of dialogs is explained in a later section.

Certain SIP headers may contain a URI. Examples of such headers are Route, Record-Route, and Contact. There may be one or more parameters associated with a URI in a header, as is the case for the R-URI. There may also be one or more parameters associated with the header line as a whole. To identify parameters as *URI parameters* or *header parameters*, angle brackets (<>) are used in a header line. Examples are as follows:

```
Contact: <sip:john.smith@company-x.se>;expires=3600
Contact: <sip:john.smith@company-x.se;transport=udp>
```

The header parameter expires=3600 indicates that the contact address has a validity of 3600 s. The parameter transport=udp is directly associated with the URI; it indicates that requests destined for john.smith@company-x.se should be sent over UDP (for the SIP branch arriving at the host of company-x.se).

The body (which is optional) within a request message contains the *payload* that the SIP message transfers. Generally, the body of a SIP message is application information, exchanged between end-points; it is not used for message routing, transaction handling, etc. One prominent example of a message body is the Session Description Protocol (SDP). SIP messages may be used to exchange information about the media session to be established. This information is exchanged through the SDP, in the message body.

7.7.2 Response Message

A SIP response message has a structure that is similar to the request message. Instead of a request line, it contains a status line. The status line provides the (provisional) outcome of the execution of the request. The status line contains the following components (in this sequence):

- **SIP protocol version.** This will be set to SIP/2.0.
- **Status code.** This is a numerical representation of the (provisional) outcome of the execution of the request. It is fit for machine processing.
- **Reason phrase.** This is a text string accompanying the status code. It expresses the status code in a readable format – for example, SIP/2.0 183 Session progress.

The function of the header lines in the response message is identical to the function of the header lines in the request message. However, there will typically be different header lines present in a response message than in the request message the response message relates to. Some headers found in a response message are copied from the request message. This may particularly be the case with certain *system headers*, such as CSeq, Call-ID, and Via. System headers are the headers that are used for SIP message routing and transaction handling.

The body within a response message has the same function as the body within a request message. The actual presence of a body in a response message and its use depends on the application.

7.7.3 Initial Request Message Routing

The routing of the initial request message is explained with reference to Figure 7.14. The sequence shows an example of message flow. Not all the Route headers shown have to be

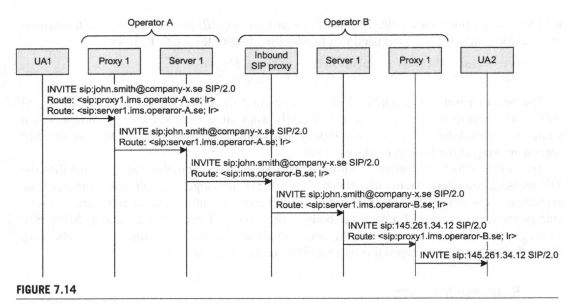

FIGURE 7.14

Initial request message routing.

always present. For example, the INVITE request message between server 1 from operator A and inbound SIP proxy from operator B may be sent without a Route header.

A SIP request message contains two items of information that are used for routing:

- Request URI (R-URI). The first line of a SIP request message is referred to as the *request line*. The request line contains an R-URI; the R-URI identifies the destination of the SIP request.
- The SIP request message may contain one or more *Route headers*. A Route header is a header line containing information about the required routing of the request message. This information is contained in the form of a URI, identifying a required next hop (a *route*). The URI in the Route header normally contains a host name or a domain name. There may also be a user part included in the URI in the Route header.

The routing of a SIP request message by an entity (user agent or proxy) is based on the following principles:

- If the request message contains one or more Route headers, then use the topmost Route header to determine the next hop (address + transport) for the message.
- If the request message contains no Route headers, then use the R-URI to forward the message.

Through the use of Route headers, the routing of a request message is done without affecting the *destination* of the message, provided that loose routing is applied. In the example in Figure 7.14, the destination of the message is sip:john.smith@company-x. se. Meanwhile, the message will be sent through various SIP proxies. The set of proxies through which a request message shall traverse may be (partly) known by the sender of the message, i.e. by the UA acting as UAC, or may be determined as the request message is being routed to the destination. In Figure 7.14, the UA sending the INVITE request had previously learned two proxy addresses, to be used for sending SIP requests:

```
Route: <sip:proxy1.ims.operator-A.se; lr>
Route: <sip:server1.ims.operaror-A.se; lr>
```

The first route address, *sip:proxy1.ims.operator-A.se; lr*, constitutes the user-to-network proxy through which this subscriber had registered his VoIP terminal in the network (i.e. place *contact binding* in a registrar in the operator's VoIP network). This user-to-network proxy address may be configured in the VoIP terminal. The second route address, *sip:server1.ims.operator-A.se;lr*, constitutes the server (registrar) in which this subscriber is registered. The subscriber had received that route address from the registrar, during registration.

A URI may contain one or more parameters. One example is the 'lr' URI parameter in the Route header. "lr" stands for *loose routing*. Loose routing implies the above-described method whereby the route to be followed by a request message is strictly defined through a set of Route headers ("Route set") without affecting the R-URI.

Note

Loose routing replaces the strict routing method. Strict routing is defined in the predecessor of IETF RFC 3261, IETF RFC 2543.

Route headers may be added and removed from a SIP request message as the message is being routed to its destination. Adding a Route header is done by "pushing" the Route header on top of the existing Route set, or by adding the Route header if none is present at that moment. Removing a Route header is done by "popping" the Route header from the top of the Route set.

When a SIP proxy receives a request message it will check the Route header present in the message. If the proxy determines from the Route header that this message is destined for this proxy, then the proxy removes the Route header and processes the message. If a SIP proxy determines from the Route header that this message is not destined for this proxy, then it will forward the message based on the above-described rule (loose routing).

It should be borne in mind that the URI contained in a topmost Route header does not have to be equal to the SIP proxy's own host name, in order for that proxy to feel *it is being addressed*. For example:

```
INVITE sip:john.smith@company-x.se SIP/2.0
Route: <sip:inbound-sip-proxy.ims.operator-B.se;lr>
```

Let this INVITE be generated by a SIP application, for the establishment of a SIP session, toward sip:john.smith@company-x.se, whereby the INVITE is sent into the IMS network ims.operator-B.se. The entity (SIP proxy) that is routing this message derives an address of the next hop to send this message to. Incoming SIP traffic for a particular operator may be dynamically distributed, by means of DNS-based load sharing, over this multitude of inbound SIP proxies. The next hop for this INVITE message may therefore be one of a multitude of inbound SIP proxies. Each inbound proxy may have its own host name, but each of these inbound proxies will be addressed by sip:inbound-sip-proxy.ims. operator-B.se.

The processing of a SIP request by a SIP proxy may include the addition of one or more Route headers. One example of adding a single Route header is the case where a SIP server needs to forward a SIP request to an inbound SIP server of the domain of the destination subscriber. An example of adding two Route headers is the forwarding of a SIP request message from a SIP registrar to a SIP application server.

For the example in Figure 7.14, the end-to-end routing of the INVITE request message is described in Table 7.2.

It follows from Figure 7.14 and the description in Table 7.2 that the hop between the user-to-network proxy (proxy 1) in the destination subscriber's network represents the only SIP routing case whereby routing takes place on R-URI. One of the implications of this is the fact that the destination subscriber's contact address could be a *local IP address*. None of the SIP proxies and servers, except for the destination subscriber's own proxy, needs to be able to route on that IP address.

The above-described routing of a SIP request from one subscriber (calling party) to another subscriber (called party) is a representative example of communication between two subscribers. However, many deviations are possible, such as (not exhaustive):

- The request message is routed through a stateless proxy (in addition to being routed through the proxies and servers described above).
- One or more intermediate SIP nodes take the role of B2BUA; in this case, the SIP session from calling party to called party is divided into multiple SIP sessions, each spanning a section of the SIP chain, between calling party and called party.
- The request message traverses an additional SIP proxy at the border between two interconnected VoIP networks. This routing through a border SIP proxy may involve the use of internal versus external DNS.
- The request message is routed through one or more SIP application servers; this routing would be done by the service node/registrar serving the calling or called subscriber.
- The registrar of the destination subscriber or a SIP application server of the destination subscriber forwards the request message to two or more terminals associated with the destination subscriber (parallel/sequential alerting).

Table 7.2 Example of SIP Request Message Routing

Sending Entity	Action	Request Message
UA1	Set R-URI to destination subscriber. Set Route set.	INVITE sip:john.smith@company-x.se SIP/2.0 Route: <sip:proxy1.ims.operator-A.se; lr> Route: <sip:server1.ims.operaror-A.se; lr>
Proxy 1 (operator A)	Remove topmost Route header.	INVITE sip:john.smith@company-x.se SIP/2.0 Route: <sip:server1.ims.operaror-A.se; lr>
Server 1 (operator A)	Remove topmost Route header. Add Route header for target VoIP domain.	INVITE sip:john.smith@company-x.se SIP/2.0 Route: <sip:ims.operaror-B.se; lr>
Inbound SIP proxy (operator B)	Remove topmost Route header. Add Route header for server (registrar) of destination subscriber.	INVITE sip:john.smith@company-x.se SIP/2.0 Route: <sip:server1.ims.operaror-B.se; lr>
Server 1 (operator B)	Remove topmost Route header. Replace R-URI by stored contact address (contact binding). Add Route header associated with contact binding (*path*).	INVITE sip:145.261.34.12 SIP/2.0 Route: <sip:proxy1.ims.operaror-B.se; lr>
Proxy 1 (operator B)	Remove topmost Route header. Route based on R-URI.	INVITE sip:145.261.34.12 SIP/2.0

IETF RFC 3263 describes the methods for a SIP proxy to determine the address, transport protocol, and port number of a subsequent SIP server. One important tool in this process is DNS. SIP domains and SIP hosts will be configured in DNS with appropriate resource records.

We have seen in this section that a request message may span a number of branches between UAC and UAS. Obviously, we want to prevent a request message from continuing to be passed on *indefinitely* or the request message entering a *loop*. For that reason, the number of branches a request message may follow is restricted; for this purpose, a designated SIP header is used, Max-Forwards. When a request message is created by a UA, thus acting as UAC, the Max-Forwards header is included in the SIP message, with starting value 70. Every proxy that forwards the request message decrements the Max-Forwards value by 1. When it reaches 0, a proxy will no longer forward the message but will, instead, return unsuccessful final response message *483 Too Many Hops*.

Note

The Max-Forwards header in SIP may be compared with the Hop counter in ISUP, which has similar functionality.

For *in-session* SIP transactions, the Max-Forwards header could be loaded with a lower value than 70. The reason for this is that, for in-session SIP transactions, the request message may traverse fewer proxies, when one or more of the proxies do not record-route. However, for consistency, the Max-Forwards header is always started with the value 70.

7.7.4 Response Message Routing

Now that the request message has been routed from UAC (calling party) to UAS (called party), we will describe how that UAS may return the response message(s) to the UAC. The sending of the request message from UAC to UAS has resulted in a *transaction trail* being established (see Figure 7.15).

The ST and CT instances shown in parentheses are *intermediate* ST and CT instances. The transaction trail between UA1 and UA2 runs via a set of *transaction branches*. Each CT instance, from UA1 onward, sends or forwards the request message to the next hop. That next hop may be a SIP proxy or SIP server or may be the final destination. Each SIP proxy or SIP server determines the next hop. As depicted in Figure 7.15, there are four *branches* for this transaction. Each branch is associated with a specific CT instance (in the sending node) and a specific ST instance (in the receiving node). SIP requests and responses carry a *branch identifier*. The branch identifier is generated by the node in which the CT resides and will be unique within that node. The branch identifier may be constructed, with an algorithm, from various system headers in the request message, such as From tag, To tag, and Call-ID. Call-ID, From tag, and To tag will be explained in the section on dialogs.

Each entity that generates or forwards a request message relating to a new transaction creates a new branch and adds the branch identifier to the request message. In this manner, a SIP transaction trail is built up as the request message travels toward its destination. The trail is hence a concatenation of branch identifiers. Considering the example in Figure 7.14, the INVITE request message from UA1 to its adjacent proxy has the following structure:

```
INVITE sip:john.smith@company-x.se SIP/2.0
Route: <sip: proxy1.ims.operator-A.se; lr>
Via: SIP/2.0/UDP 164.45.63.125:44234;branch=z9hG4bK9876uytr6543refd
```

Whereas the INVITE request message from proxy 1 of UA2 to UA2 may have the following structure:

```
INVITE sip:145.261.34.12 SIP/2.0
Via: SIP/2.0/UDP 164.48.60.241:55034;branch=z9hG4bK-d87543
```

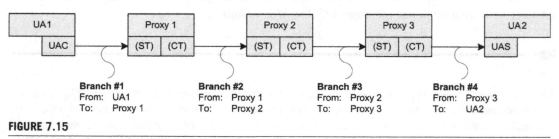

FIGURE 7.15

Signaling trail from UAC to UAS.

```
Via: SIP/2.0/UDP 164.48.60.242:55323;branch=z9hG4bK-12qwas
Via: SIP/2.0/UDP 164.48.60.243:56344;branch=z9hG4bK09iujhbv87ytgfvc
Via: SIP/2.0/UDP 164.45.63.123:34555;branch=z9hG4bK23wesdxc45rtfgvb
Via: SIP/2.0/UDP 164.45.63.124:12456;branch=z9hG4bK45rtfgvb67ghbn
Via: SIP/2.0/UDP 164.45.63.125:44234;branch=z9hG4bK9876uytr6543refd
```

A Via header in the request message identifies a transaction branch. The *set of Via headers* in the request message represents the transaction trail for this transaction. Each entity that sends or forwards a request message (relating to a new transaction) adds a Via header to the request message, containing (as a minimum) the following information:

- **Transport indicator.** This information element indicates the transport protocol used for this branch. This allows the receiver of the request message to use the same transport protocol for any response on this request message – for example, *SIP/2.0/UDP*, to indicate that UDP is used for this branch.

  If the transport indicator indicates TCP, then it is not explicitly indicated *which* TCP connection is used. There may be multiple TCP connections open between communicating entities (hosts).

- **Address of sending entity.** This information is needed by the receiver of the request message (i.e. the next hop) to send response messages to the entity (host) that initiated this branch, i.e. the host where the UAC instance for this branch was created. The address may be an IP address or a domain name. However, using a domain name will require additional DNS lookup (if not cached). A port number may be included in the address. Without a port number, the default port for SIP is used. For UDP, TCP, and SCTP, this is port 5060. It is also possible to receive the port number in the SRV record from DNS, if a domain name is used for the Via header.

  The address in the Via header may also contain one or more parameters, such as *maddr*, *ttl*, and *sent-by*. Refer to IETF RFC 3261 for more information on that.

- **Branch value.** The branch identifier is a unique character string within the SIP application on the host (IP address + port number) that created the transaction branch – for example, z9hG4bK-d87543-626b297a31680e08-1--d87543-.

Note

The branch value always starts with the characters "z9hG4bK". This is for historical reasons.

One guiding principle in SIP is that information that is required for SIP messages is, as much as possible, maintained in the end-points of the SIP session. This principle is clearly visible in this building up of the set of Via headers (*transaction trail*) in the SIP request message. The set of Via headers grows as the request message traverses through successive hops.

A SIP response message follows the same signaling path as the request message to which this response message relates. The set of Via headers is used here. When a UA receives a request message, hence taking the role of UAS, it stores the received set of Via headers. This set of Via headers is stored in the server transaction state model instance; it is needed only for the processing of this particular transaction. The UA acting as UAS for this transaction will send zero or more provisional responses and will send a final response. For the sending of a response, the UAS uses the set of Via headers associated with this transaction. In the example in Figure 7.14, the following Via header set would be used in the response message:

```
Via: SIP/2.0/UDP 164.48.60.241:55034;branch=z9hG4bK-d87543
Via: SIP/2.0/UDP 164.48.60.242:55323;branch=z9hG4bK-12qwas
Via: SIP/2.0/UDP 164.48.60.243:56344;branch=z9hG4bK09iujhbv87ytgfvc
Via: SIP/2.0/UDP 164.45.63.123:34555;branch=z9hG4bK23wesdxc45rtfgvb
Via: SIP/2.0/UDP 164.45.63.124:12456;branch=z9hG4bK45rtfgvb67ghbn
Via: SIP/2.0/UDP 164.45.63.125:44234;branch=z9hG4bK9876uytr6543refd
```

Response message routing is hence done *top to bottom* through the Via header set. The UAS and intermediate proxy route the response message to the transport address (including IP address, transport protocol, and port number) indicated in the topmost Via header. Every proxy that receives a response message passes it to the transaction process indicated by the branch. The proxy handling the response message removes the topmost Via header. If there are one or more Via headers left, the message must apparently be forwarded further to the UAC. In that case, the proxy uses the transport address of the topmost Via header. Otherwise, if there are no more Via headers left, the response message has apparently arrived at the UAC that initiated the transaction, in which case no further routing of the response message, to UAC, is needed.

In the above-described manner, response messages traverse the same path as the initial request message that the response messages belong to. Figure 7.16 shows a graphical representation of this request and response routing.

The message part in Figure 7.16 marked in bold indicates the information in the message that is used for routing that message. For a request message, routing is based on Route header and R-URI; for a response message, routing is based on Via header. As is also clear from Figure 7.16, the routing of the INVITE request message requires successive "next hop decisions"; various entities have to take action to determine the next hop for the request message. This is shown by the iterative addition and removal of Route headers. For the return path, followed by the response messages, no such successive next hop decision has to be made. The entire return path is built up as the request message travels from UAC to UAS.

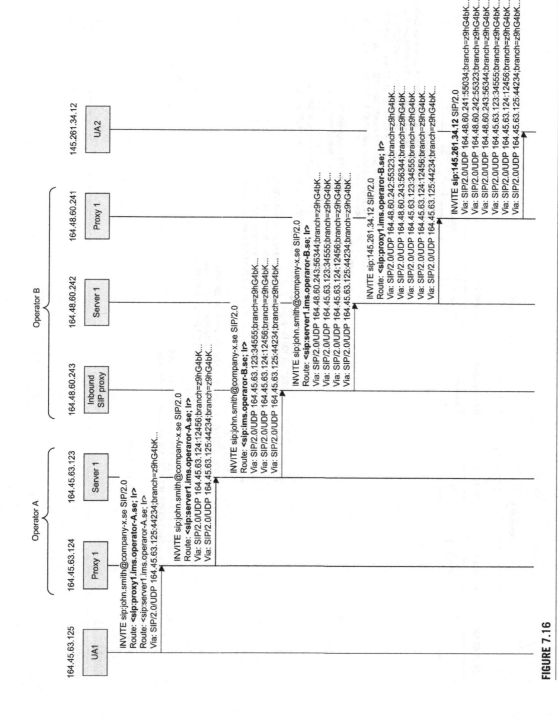

FIGURE 7.16

Example SIP request and response routing (parts 1 and 2).

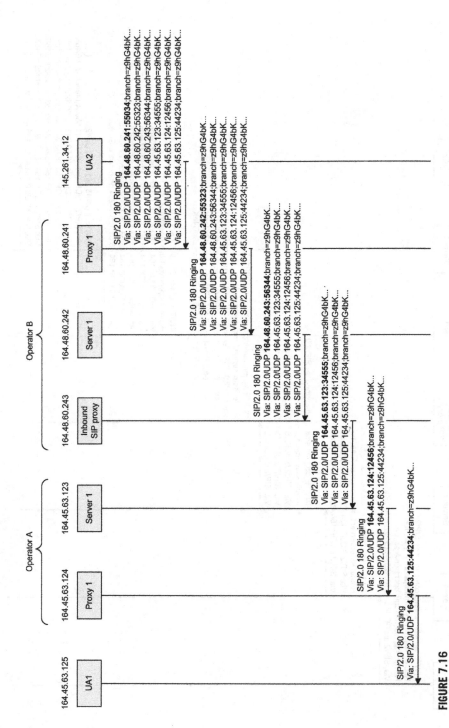

FIGURE 7.16

(Continued).

The entire Via header set is stored by the UAS and is used for every response message to be returned for this transaction, including response retransmission. There are cases where a branch identifier is also reused in the direction from UAC to UAS. These include:

- When UAC (or proxy) sends an ACK request in response to a non-2xx final response, this ACK request forms part of the INVITE transaction that led to this non-2xx final response. The ACK shall hence be sent *over the branch* between this UAC and the next hop. So, the Via header in the ACK request will contain the same branch value as the corresponding INVITE request.
- When a UAC (or proxy) retransmits a request, due to expiry of a UAC state model-related timer, the retransmitted request will be directed to the UAS instance that *was supposed* to have been created as a result of the initial request message. The proxy that receives a retransmitted request message can therefore easily determine from the branch identifier that the request message relates to an existing UAS instance and that the request message represents a retransmission.
- When a UAC cancels an INVITE transaction, it uses the same branch ID for the CANCEL transaction as was used for the INVITE transaction that it wants to cancel.

We will see in the following sections that, *practically*, SIP transactions will span no more than three or four branches, and not six branches, as in the example in Figure 7.16.

As discussed earlier, the INVITE transaction has an ACK request message associated with it. When the UAC has received the final response, an ACK message is sent from UAC to UAS. We will first explain how a SIP routing path is established for subsequent SIP requests within a SIP session. This SIP routing path will then be used for the sending of the ACK message in the case of a successful final response.

7.7.5 Building a SIP Routing Path for Subsequent SIP Requests

When a SIP transaction occurs between two user agents (UA), one acting as UAC and the other one acting as UAS, this may lead to the establishment of a SIP session. INVITE is one method that may result in a SIP session between the respective UAs. More precisely, the SIP session is established only when the INVITE request leads to an affirmative final answer, i.e. a final answer in the 2xx-class, such as 200 OK. When a SIP session is established between two UAs, *subsequent* transactions may be applied between these UAs, also referred to as *in-session* transactions (as opposed to *initial transaction*). For the initial transaction, the routing of the request message is done iteratively; successive proxies take specific action to determine the next hop for the request. For subsequent transactions, no such iterative process is needed. The path that will be followed by an in-session request message is known at the very moment that one UA initiates an in-session transaction. There are two reasons for this:

1. Remember the description of the routing of the request message of an initial transaction; proxies through which the request message traverses take action to determine the next hop for the request message. As part of this process, each proxy builds up a *transaction* *trail*, representing the path followed by the request message. This transaction trail, formed by the set of Via headers, is used for the routing of response message(s). In addition, proxies will build up a *subsequent transaction path*. This subsequent transaction path is formed by a set of headers, called *Record-Route* headers, representing the set of proxies through which a subsequent transaction will be routed.

2. The subsequent transaction path does not have to be identical to the initial transaction path. The subsequent transaction path does not have to traverse the same set of proxies as the initial transaction path; some proxies that formed part of the initial transaction path may opt to not form part of the subsequent transaction path. When a proxy wants to be able to apply charging for a media session, it has to remain in the SIP session.

When an INVITE transaction is completed successfully, a SIP session is established between the two UAs between which the INVITE transaction was executed. This SIP session, a *relationship* if you will, may span one or more proxies that the SIP INVITE request and response messages traversed. Such a proxy may want subsequent transactions within this SIP session to also traverse this proxy. Here, a proxy uses the record-route mechanism. Let's have a brief look how this *record routing* is applied (see Figure 7.17).

The INVITE request and response messages in Figure 7.17 show only the request line and the Record-Route headers. The Route headers and Via headers, used for transaction message routing, are shown in Figure 7.16. When the INVITE request message traverses a proxy, the proxy may add a *Record-Route* header to the INVITE request message, including its own address. In this manner, a set of Record-Route headers is constructed, as the INVITE request message travels toward the destination UA, acting as UAS for the INVITE transaction. When a proxy does not wish to form part of the signaling path for subsequent transactions in this SIP session, it does not add a Record-Route header to the INVITE request message. The UAS copies the entire Record-Route set in the response message. The Record-Route set is then conveyed to the UAC. The UAC and UAS now both have a Record-Route set that may be used for subsequent transaction routing.

The Record-Route set may be included, from UAS to UAC, in a *provisional* response or in the *final* response. If a *reliable* provisional response is applied, the Record-Route set will be included in the provisional response, as will be explained further in a later section. The subsequent transaction path is defined during the initial INVITE transaction and does not change for this SIP dialog. It is not possible for a proxy to *remove* itself from a SIP signaling path once the SIP session is established. Neither is it possible for a proxy to be added later on. For this reason, the building of the Record-Route set and sharing this between the two UAs needs to be done only in the initial INVITE transaction.

The address of the proxy in the Record-Route header typically has the form of a fully qualified domain name (FQDN). We also see in the example in Figure 7.17 that the Record-Route headers have the "lr" parameter included. This indicates that the proxy supports loose routing, i.e. loose routing is applied for subsequent transaction signaling through this proxy.

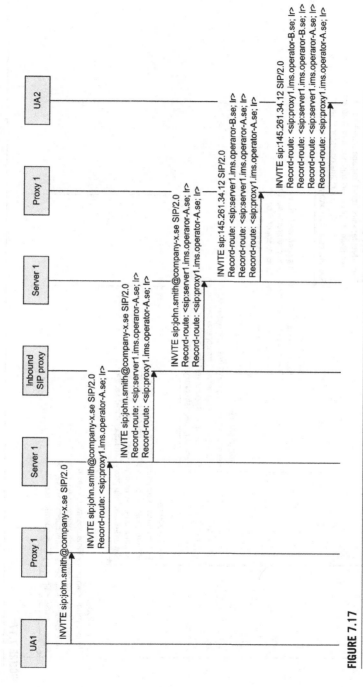

FIGURE 7.17

Record routing (parts 1 and 2).

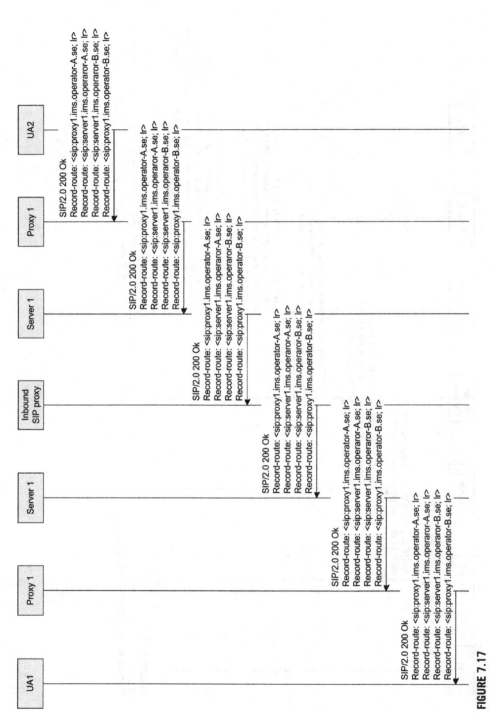

FIGURE 7.17

(Continued).

The FQDN that is used in the Record-Route header will be a domain name associated with the *specific* host through which the SIP session runs.

We also notice that the *order* of the Record-Route headers in the response message is reversed, compared to the order of the Record-Route headers in the request message. This is because request messages relating to a subsequent request travel in the reverse direction, depending on which UA initiates the subsequent transaction. So, the UAC needs to receive (and store) the Record-Route set in reverse order, compared to how the set was built up in the INVITE message.

A further noteworthy aspect of the Record-Route header is the fact that a proxy may alter the Record-route header, before forwarding it to the UA (see Figure 7.18).

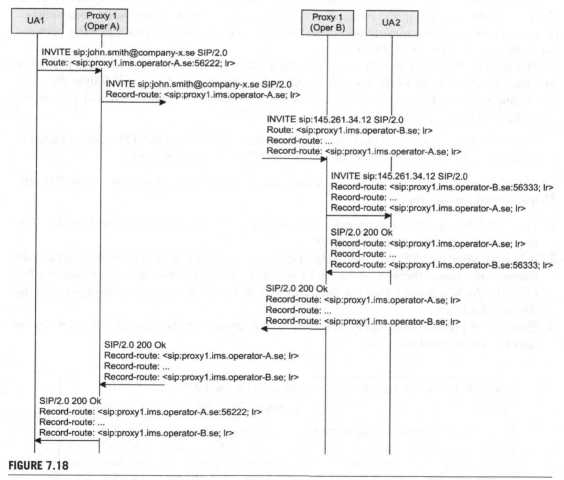

FIGURE 7.18

Record-route manipulation by proxy.

The signaling sequence of Figure 7.18 shows only the INVITE transaction signaling through the proxy, not through other entities in the network. The rationale of the method depicted in Figure 7.18 is that a proxy node may want to be addressed in different ways, depending on the direction from which it is addressed. The addressing from user to proxy may run over a secure connection that uses a designated port number, assigned specifically for that secure connection, e.g. port 56222. When the proxy forwards the INVITE message into the core network, it uses a standard port number for the Record-Route header, e.g. port 5060; however, port 5060 may be omitted, since it is the default port for SIP over UDP, TCP, or SCTP. On the called party's side, the proxy uses the designated port number assigned for the secure connection with the called party to set the Record-Route header, e.g. 56333. In the reverse direction, the respective proxies apply reverse mapping of the Record-Route headers.

The resulting stack of Record-Route headers for UA-A and UA-B will therefore be as shown in Table 7.3. As in the example in Figure 7.18, the node in the VoIP core network resides in a trusted environment and doesn't need a secure connection for signaling between each other, so standard port 5060 is used.

The Record-Route set that the two peers in a SIP session exchange relates to the entire SIP session, after the INVITE transaction. All subsequent transactions will follow the same SIP signaling path. Hence, the Record-Route set needs to be constructed and exchanged only once. To be more specific:

- Record-Route set is constructed during the sending of the initial INVITE request message
- Record-Route set is included in the initial INVITE transaction response.

The SIP session responder may typically include the Record-Route set in the 200 OK. There are two reasons for this:

1. The 200 OK is a *reliable response*. The reliability of 200 OK is constituted by the ACK request message following the 200 OK.
2. It is only after the 200 OK final response that a requirement arises for subsequent SIP transactions, hence there is no reason to return the Record-Route set earlier than the 200 OK. The ACK request following the 200 OK will be the first request message using the Record-Route set.
3. However, if preconditions apply or early media is used, then the Record-Route set will be sent in a provisional response.

Table 7.3 Stack of Record-Route Headers for UA-A and UA-B

UA-A	UA-B
<sip:proxy1.ims.operator-A.se:56222; lr>	<sip:proxy1.ims.operator-B.se:56333; lr>
<sip:server1.ims.operator-A.se; lr>	<sip:server1.ims.operator-B.se; lr>
<sip:server1.ims.operator-B.se; lr>	<sip:server1.ims.operator-A.se; lr>
<sip:proxy1.ims.operator-B.se; lr>	<sip:proxy1.ims.operator-A.se; lr>

If the final response to the initial INVITE request message is a response in the 3xx to 6xx-class, then there is strictly no need to return a Record-Route set. This is because the ACK message following the unsuccessful final response does not use the in-session signaling path as defined by the Record-Route set. Instead, the ACK message in this case follows the same path as the initial INVITE request. However, practical implementation of a SIP client may always return a Record-Route set, even for an unsuccessful final response. The UA that had initiated the SIP session establishment will ignore the Record-Route set.

7.7.6 Exchanging Contact Addresses for Subsequent SIP Requests

In previous sections, we have seen:

- How a SIP signaling path is defined for an initial SIP transaction, i.e. a SIP transaction that is used for establishing a SIP session.
- How a SIP signaling path is defined that will be used for *subsequent* SIP transactions, i.e. SIP transactions that are used within an established SIP session.

We will now describe how the signaling for such subsequent SIP transaction is applied.

For the initial SIP transaction, the destination entity, e.g. the called party for a call, is identified with a public user identity in the form of a SIP URI, e.g. sip:john.smith@company-x.se. The routing of the initial request message is based on the successive addition and removal of Route headers; various proxies and servers in the SIP signaling path take designated action to determine the next hop for the request message. Meanwhile, the R-URI, identifying the destination of the request message, remains sip:john.smith@company-x.se. At some point in the signaling, the R-URI is replaced by the called party's contact address, as registered in that called party's registrar. An optional additional Route header may be added for routing the request message through a proxy associated with the called party.

When the two parties involved in the SIP session apply a *subsequent* transaction, how will they address the remote party? For example, Wendy Jones (sip:wendy.jones@company-y.se) has established a call with John Smith (sip:john.smith@company-x.se); then how does Wendy address John for a subsequent transaction, and vice versa. Using the peer's public user identity would not be appropriate. One reason, to start off with, is the fact that a subscriber's public user identity identifies the person, not a specific device. A call to John Smith might have been answered on John's mobile phone, so a subsequent transaction for that SIP session must be directed to John's *mobile phone* and not to John *as a person*.

To facilitate transparent and convenient addressing, the parties involved in the establishment of a SIP session exchange their *contact addresses*. The contact address constitutes the address that may be used for contacting the peer within a SIP session. The exchange of contact addresses is shown below:

```
/*Initial INVITE request, sent from UA-A, Wendy Jones */
INVITE sip:john.smith@company-x.se SIP/2.0
Route: <sip:proxy1.ims.operator-A.se:56222; lr>
Contact: <sip:wendy.jones@164.45.63.125:55123>
```

```
/*200 OK response on the initial INVITE request, sent from UA-B, John Smith */
SIP/2.0 200 OK
Contact: <sip:john.smith@145.261.34.12:55456>
```

The user agent that initiates the SIP session, by sending the initial INVITE transaction, includes a Contact header in the initial INVITE request message. The remote party of this SIP session, i.e. the user agent that responds to the SIP session establishment request, includes a Contact header in a response message, e.g. in the 200 OK. As shown above, subsequent SIP transactions to Wendy Jones for this SIP session will be directed to sip:wendy.jones@164.45.63.125:55123. Subsequent SIP transactions toward John Smith for this SIP session will be directed to sip:john.smith@145.261.34.12:55456. The contact address that is exchanged typically has the form of an IP address. It may contain a user part; this may be used when there are multiple SIP applications (user agents) installed on one host, or when multiple accounts are registered via one user agent.

The contact addresses that the two peers in a SIP session exchange relate to the SIP dialog. If that SIP dialog becomes a confirmed dialog (when the SIP session becomes active), then that contact address applies to the entire SIP session after the INVITE transaction. Hence, the contact address needs to be provided only once. To be more specific:

- The contact address of the SIP session initiator needs to be present in the initial INVITE transaction request.
- The contact address of the SIP session responder needs to be present in a provisional response (except 100 Trying) or in the final response (2xx-class of final response).

It *is* possible to modify the *remote target* during a SIP session, which is referred to as target refresh. This effectively changes the contact address that will be used for contacting that remote party from that moment onward.

The SIP session responder may typically include the contact address in the 200 OK. Refer to the description of the Record-Route header for rules with respect to the exchange of contact addresses for different cases.

The contact address that a user agent provides to a peer, during SIP session establishment, may differ from the contact address that the UA had previously registered at the registrar (see Figure 7.19). In the example in Figure 7.19, the UA has registered contact address 145.261.34.12:5060. Hence, new SIP sessions will arrive at that IP address and port number. However, the UA wishes to receive subsequent SIP requests on a different port, e.g. port 55333. Subsequent SIP addressing, within a SIP session, does not take cognizance of the registered contact address, which is why they may differ.

The signaling between the UA and its user-to-network proxy may run via a secure connection. In that case, it will be common that the same port number, associated with the secure connection, is used for addressing the UA for *initial SIP requests* and addressing the UA for *subsequent SIP requests*. Hence, the UA will in that case report a contact address in the initial INVITE response message that is the same as the contact address that that UA had registered with the registrar.

7.7.7 Subsequent Request Message Routing

Now that we have seen how the two user agents exchange contact addresses (to address each other for subsequent transactions) and exchange Record-Route sets (indicating what path to follow for these subsequent transactions), we will take a closer look at a typical subsequent transaction.

Figure 7.20 shows a limited SIP signaling sequence for an initial INVITE transaction, showing only the request line, status line, Route headers, and Record-Route headers, as

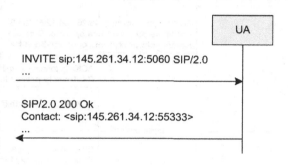

FIGURE 7.19

Initial contact address versus *in-session* contact address.

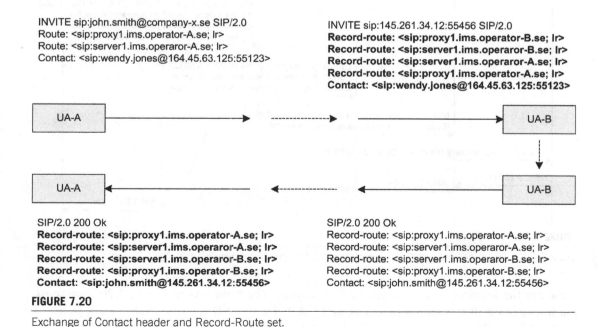

INVITE sip:john.smith@company-x.se SIP/2.0
Route: <sip:proxy1.ims.operator-A.se; lr>
Route: <sip:server1.ims.operaror-A.se; lr>
Contact: <sip:wendy.jones@164.45.63.125:55123>

INVITE sip:145.261.34.12:55456 SIP/2.0
Record-route: <sip:proxy1.ims.operator-B.se; lr>
Record-route: <sip:server1.ims.operaror-B.se; lr>
Record-route: <sip:server1.ims.operaror-A.se; lr>
Record-route: <sip:proxy1.ims.operator-A.se; lr>
Contact: <sip:wendy.jones@164.45.63.125:55123>

SIP/2.0 200 Ok
Record-route: <sip:proxy1.ims.operator-A.se; lr>
Record-route: <sip:server1.ims.operaror-A.se; lr>
Record-route: <sip:server1.ims.operaror-B.se; lr>
Record-route: <sip:proxy1.ims.operator-B.se; lr>
Contact: <sip:john.smith@145.261.34.12:55456>

SIP/2.0 200 Ok
Record-route: <sip:proxy1.ims.operator-A.se; lr>
Record-route: <sip:server1.ims.operaror-A.se; lr>
Record-route: <sip:server1.ims.operaror-B.se; lr>
Record-route: <sip:proxy1.ims.operator-B.se; lr>
Contact: <sip:john.smith@145.261.34.12:55456>

FIGURE 7.20

Exchange of Contact header and Record-Route set.

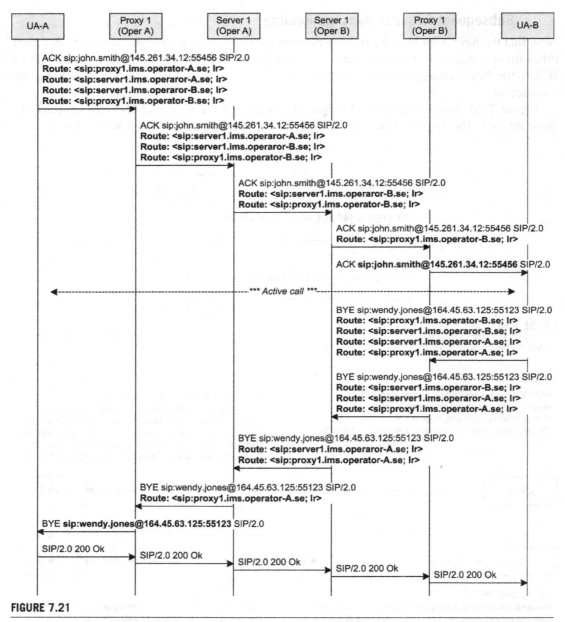

FIGURE 7.21

Using Record-Route header and Contact header for subsequent transaction.

these are the headers that are used for SIP routing. The bold text in Figure 7.20 represents the information that is stored in UA-A and UA-B for this SIP session. Subsequent transactions will use that information, as depicted in Figure 7.21. In Figure 7.21 UA-A uses the stored Record-Route set and contact address to set the Route headers and R-URI respectively, in the ACK request to UA-B. UA-B uses the stored Record-Route set and contact address to set Route headers and R-URI in the BYE request to UA-A.

Subsequent transactions are, in many aspects, regular transactions and follow generic transaction handling rules, such as:

- Routing the request message is based on Route header and R-URI.
- A transaction trail is built up, consisting of a concatenation of branches, reflected in a set of Via headers.
- A transaction (except ACK) has one final response and zero or more provisional responses.
- Response message(s) follow(s) the same path as the request message.
- Reliable provisional response may be applied.
- Retransmission of the request message (if needed) and of the final response message (if needed) may be applied.

7.8 SIP TRANSPORT CONSIDERATIONS

Now that we have seen how a SIP session can be established (with INVITE transaction) and terminated (with BYE transaction), let's have a closer look at the transport-related matters for SIP signaling.

As described in an earlier section, SIP, being an OSI application protocol, runs on top of a transport layer, such as UDP, TCP, or SCTP. For every branch that is established by a SIP UA, i.e. for every transaction that is initiated and for which a request message is sent from the UAC state model instance, the transport mode has to be determined. The transport method used for sending a request message may be determined from the Route header or from the R-URI. Consider, for example, the following INVITE request message:

```
INVITE sip:john.smith@company-x.se SIP/2.0
Route: <sip:proxy1.ims.operator-A.se; lr>
Route: <sip:server1.ims.operaror-A.se; lr>
```

The routing of this request message will be based on the topmost Route header. This Route header has the form of a domain name (proxy1.ims.operator-A.se). Hence, the sender of this request message has to apply DNS resolution in order to determine transport protocol, IP address, and port number. The following DNS services (resource records) are required here:

- **NAPTR record query.** This is used to obtain an overview of supported services for this domain (proxy1.ims.operator-A.se) – for example, SIP over UDP (_sip._tcp.proxy1.ims.operator-A.se) or SIP over TCP (_sip._tcp.proxy1.ims.operator-A.se).
- **SRV record query.** This is used to obtain host name (or host names) of the SIP server(s) associated with the selected SIP service.

- **A record query or AAAA record query.** This is used to obtain IP address (or IP addresses) of the host (or IP address(es) of one of the multiple hosts). An A record query is used to obtain IPv4 address; an AAAA record query is used to obtain an IPv6 address.

When the domain (proxy1.ims.operator-A.se) supports multiple SIP services, e.g. SIP over UDP and SIP over TCP, the sending entity will select one of the services, based on the priority indicated in the NAPTR record. It is, however, also possible that a transport indicator is already associated with the domain in the Route header. In this case, preference will be given to that transport protocol. IETF RFC 3263 provides guidelines on the selection of transport protocol.

The port number to use for the routing may be received as part of the SRV record. If the port number to use is not provided as part of SRV record and is not explicitly included in the Route header, then the sender should use the default port number for SIP.

The second "hop" for the above example INVITE message in this section is directed to server1.ims.operaror-A.se. Here, the proxy node, which received the INVITE message from the UA, applies similar DNS queries as the UA had applied, to determine SIP service, transport protocol, IP address, and port number. Of course, it may have this information in cache.

When the INVITE request message is finally routed to the destination terminal, based on the registered (in registrar) contact address, then typically an IP address is already available, since the registered contact address is typically an IP address. By default, the request message is sent over UDP to the SIP terminal, unless explicitly indicated by the terminal during registration (as a URI transport parameter added to the contact address) that TCP (or another transport protocol) should be used.

Response messages follow the same path, in reverse order, as the request message. In addition, response messages shall use the same transport protocol as used for the respective "hops" (branches). Here, the transport protocol used for the request message is recorded in the *transaction trail*, i.e. in the set of Via headers. For example:

```
Via: SIP/2.0/UDP 164.48.60.241:55034;branch=z9hG4bK-d87543
Via: SIP/2.0/UDP 164.48.60.242:55323;branch=z9hG4bK-12qwas
Via: SIP/2.0/UDP 164.48.60.243:56344;branch=z9hG4bK09iujhbv87ytgfvc
Via: SIP/2.0/UDP 164.45.63.123:34555;branch=z9hG4bK23wesdxc45rtfgvb
Via: SIP/2.0/UDP 164.45.63.124:12456;branch=z9hG4bK45rtfgvb67ghbn
Via: SIP/2.0/UDP 164.45.63.125:44234;branch=z9hG4bK9876uytr6543refd
```

For each hop, the transport protocol is explicitly indicated. When the response message is sent, the UAS as well as each intermediate proxy knows which transport protocol to use.

As follows from the above, response message routing is typically easier than request message routing, since the Via headers contain, for each hop, the required IP service (e.g. SIP over UDP), the IP address, and the port number. Hence, no DNS query would be needed for response message routing. If the Via header contains a domain name, then DNS query will be needed.

DNS resource records may be cached by the DNS client. Addressing information in a VoIP network, such as host names and IP addresses, is reasonably static. The time-to-live (TTL) value of these records may therefore be set to a long value, thereby reducing the number of DNS queries required.

7.8.1 Internal DNS Versus External DNS

When an entity, e.g. a SIP terminal or a SIP proxy, has to route a SIP request message, the request message may contain a multitude of addressing information. For example:

```
ACK sip:john.smith@145.261.34.12:55456 SIP/2.0
Route: <sip:proxy1.ims.operator-A.se:56222; lr>
Route: <sip:server1.ims.operaror-A.se:5060; lr>
Route: <sip:server1.ims.operaror-B.se:5060; lr>
Route: <sip:proxy1.ims.operator-B.se:5060; lr>
```

This SIP request message is sent by a SIP terminal. The SIP terminal has to send the request message to proxy1.ims.operator-A.se. This proxy may have to be accessible by the SIP terminal via public IP infrastructure. Hence, the DNS resource records associated with proxy1. ims.operator-A.se should be available from public DNS, *external DNS*. The other proxies in this example (server1.ims.operaror-A.se, server1.ims.operaror-B.se and proxy1.ims.operator-B.se) are accessed from within the VoIP network. Availability of DNS resource records relating to these proxies and servers may therefore be restricted to entities within the VoIP network, i.e. through *internal DNS*.

Note

A user-to-network proxy resides on the VoIP network boundary. This proxy is accessed by a SIP terminal as well as by other proxies in the VoIP network. When a SIP terminal accesses the user-to-network proxy, it is accessed over public IP infrastructure. When other proxies in the VoIP network access the user-to-network proxy, it is accessed from a trusted environment.

For access over public IP infrastructure, required DNS resource records have to be available through external DNS. For access from a trusted environment, DNS resource records may be restricted to internal DNS.

7.8.2 Reliability of SIP Requests and SIP Responses

SIP runs over IP infrastructure, which is inherently "unreliable". IP packets may get lost in the transmission from sending host to destination host. Such transmission failure is predominantly caused by congestion in the network, specifically overflow at input queues in IP routers. Yet we need a mechanism to guarantee the SIP signaling. This "guarantee" relates to the transaction signaling, including:

- A guarantee that a request message has arrived at the destination UAS
- A guarantee that the (final) response from the UAS has arrived at the UAC.

One method for achieving this guarantee is the use of TCP as transport protocol for SIP. TCP entails the establishment of a *data connection* between two hosts, commonly referred to as the "TCP socket" (the "socket" is actually the entrance, at the respective hosts, into said data connection). One of the tasks carried out by TCP is retransmission of IP packets in cases where it is detected by the sender that an IP packet was lost. In this way, TCP provides the required reliability.

TCP is also needed when a SIP message exceeds the maximum transmission unit (MTU) size applicable for the connection between two hosts. For Ethernet, the MTU is 1500 bytes. If the MTU is known to the sending entity, TCP will be used when the message is larger than MTU – 200. This buffer of 200 bytes allows for a response message that is larger than the request message.

SIP also has its own reliability mechanism, which may be used in conjunction with, and independently of, TCP or UDP. We will describe the following reliability mechanisms:

- Reliability of request message
- Reliability of successful final response
- Reliability of unsuccessful final response
- Reliability of provisional response.

7.8.2.1 Reliability of Request Messages

Remember that a request message is transferred from UAC to UAS in a hop-by-hop fashion. The request message is sent from the host of the UAC to the next host, and then to another host until it arrives at the host of the UAS (see Figure 7.22). P1–P3 represent intermediate SIP proxies in the signaling path for this transaction.

All of the "hops" depicted in Figure 7.22 run over IP infrastructure so the respective INVITE request message may be lost, resulting in the INVITE message not arriving at the intended destination. To overcome this dilemma, a hop-by-hop reliability mechanism is implemented. This reliability mechanism ensures that a request message arrives at the next hop, and at the hop after that, etc., until it has arrived at the UAS (see Figure 7.23).

The square blocks in Figure 7.23 represent IP routers. The INVITE message may get "stuck" (and eventually dropped) in a queue in an IP router. When the UA acting as UAC sends the request message, it starts a timer that governs the transmission of this request message to the next entity, in this example proxy P1. This timer, referred to as "timer A", is loaded with a default value of 500 ms.[1] If no response is received within 500 ms, UAC determines that the request message did not arrive at the next entity, so the UAC will retransmit the request message. The arrival of the retransmitted request message is again verified through timer A, set to twice the default value, i.e. 2 × 500 ms. Once a provisional (or final) response is received, the UAC stops timer A and will not further retransmit.

In order to get a *fast response* on the request message, a special provisional response is used. The request message is *destined* for a particular remote party, with a number of proxies on the signaling path. In order to have request message reliability "per hop", a UAS state model instance generates a provisional response. The state model generates this provisional response, not the SIP application. This provisional response is the 100 Trying. Unlike the

---

[1] 500 ms is considered to be the typical round-trip time for SIP request and response between two SIP entities.

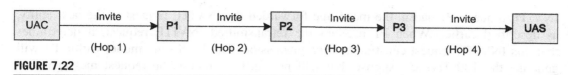

FIGURE 7.22

Hop-by-hop transmission of request message.

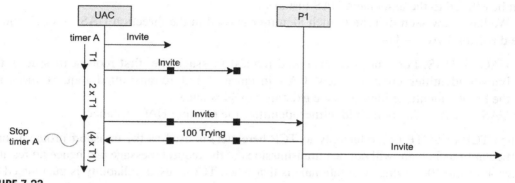

FIGURE 7.23

Retransmission of request message due to request message being lost.

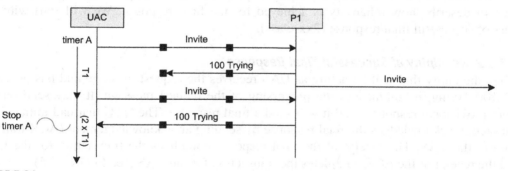

FIGURE 7.24

Reliability of 100 Trying.

regular provisional responses, the 100 Trying is not propagated further to the initiator of the INVITE transaction. When the UAC receives the 100 Trying, it knows that the next entity has received the INVITE. The sending of the 100 Trying is based on the topmost Via header (to determine IP address, transport protocol, and port number).

When the UAC does not receive a response within $64 \times T1$ (=32 s), the transaction is considered to have failed and will be aborted.

Figure 7.24 shows the reliability of the 100 Trying response. The UAC did not receive a provisional response (in the form of 100 Trying) within T1 (500 ms), so it retransmits the

INVITE request. P1 had in the meantime forwarded the INVITE request to the next entity, as described earlier. When P1 receives the retransmitted INVITE request, it determines that this INVITE request constitutes a retransmission. The UAS state model within P1 will generate the 100 Trying response, but will not further process the request message. The retransmitted request message is sent over the same *branch* as the initial request message, so carries the same branch identifier in the Via header. That is how the retransmitted request can be offered to the associated UAS instance.

We have now seen that the branch identifier is used in the direction UAS → UAC and in the direction UAC → UAS:

- UAC → UAS. For a non-retransmitted request message (i.e. first request message), the branch identifier creates a new UAS instance; for a retransmitted request message, the branch identifier identifies the existing UAS instance.
- UAS → UAC. The branch identifier identifies the existing UAC instance.

When TCP or SCTP (we refer only to TCP hereafter) is used as the transport protocol for a particular branch, there will not be retransmissions of the request message and hence no retransmission of the 100 Trying. The rationale is that when TCP is used, reliability is guaranteed by the transport protocol. The 100 Trying in response to the first (non-retransmitted) request message will still be sent, but has no effect on the UAC. Timer A will not be started in the UAC.

Now that we have seen how reliability is provided for the request message per branch, we will describe how reliability is achieved for the final response. We will start with the class of successful final response (2xx-class).

7.8.2.2 Reliability of Successful Final Response

When the entity that will be acting as UAS receives the request message and has provided the 100 Trying, it commences the processing of the request message. It may send one or more provisional responses and it will send a final response. The UAC that had initiated the transaction acknowledges the final response by sending an acknowledgement (ACK transaction) to the UAS. The receipt of the final response completes the transaction for the UAC and the receipt of the ACK completes the transaction for the UAS (see Figure 7.25).

The provisional response, in this example 180 Ringing, is not sent reliably. Hence, the UAS takes care not to include information in the provisional response that is necessary for the further handling of the transaction or for the further handling of the SIP session.

The UAS applies a similar retransmission method for the final response as the UAC applies for the request message. When the UAS has sent the final response, it starts a timer that monitors the receipt of the ACK message for this transaction (see Figure 7.26).

The timer that oversees the successful transmission of the successful final response, timer G in the UAS, is started at T1 (500 ms) and doubles at every retransmission with a maximum of 4 s.

Figure 7.27 shows the case where the ACK request message from UAC was lost. The UAS did not know that the non-arrival of ACK, resulting in expiring of timer G, was the result of failure in transmission of the successful final response or failure in transmission of

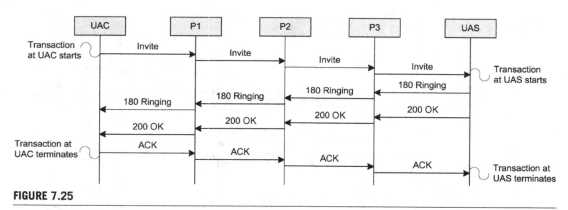

FIGURE 7.25

End-to-end confirmation of successful transmission of successful final response.

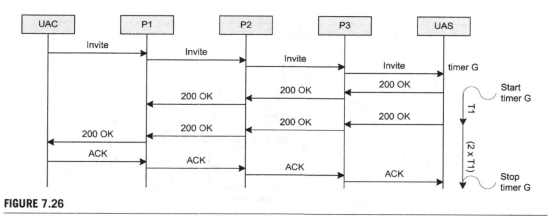

FIGURE 7.26

Retransmission of successful final response.

the ACK message. In this example, where the ACK message transmission failed, the UAC will receive a retransmission of the final response. The UAC will then retransmit the ACK message. This also shows that the UAC should remain active for a finite duration, in order to be able to receive a possible retransmission of the final response.

7.8.2.3 Reliability of Unsuccessful Final Response

The reliability of the unsuccessful final response is provided in a hop-by-hop manner between two SIP entities. The rationale of this approach is that when the INVITE transaction is "unsuccessful", it does not lead to the establishment of a SIP session. The INVITE transaction had resulted in the establishment of a transaction trail. The unsuccessful result of the INVITE transaction leads to "cleaning up the trail". The unsuccessful final response is transmitted to the UAC and traverses zero or more SIP proxies, i.e. is transmitted over one or more branches. Each time the response traverses a branch, the branch can be released (see Figure 7.28).

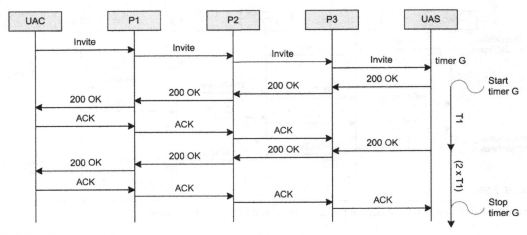

FIGURE 7.27

Retransmission of successful final response and ACK.

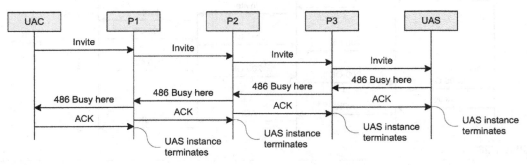

FIGURE 7.28

Hop-by-hop termination of UAS and proxy.

For an unsuccessful final response, transmission and retransmission is hop-by-hop. For example, whilst P3 has sent ACK to UAS and has forwarded the 486 Busy to the UAC, it may receive a retransmission of the 486 Busy from the UAS, due to the ACK message to the UAS being lost. P3 in this case will not retransmit the 486 Busy to the UAC; it has already done so. Instead, it will retransmit the ACK to the UAS.

7.8.2.4 Reliability of Provisional Response

We have seen in the above description that the *final* response is sent reliably, by virtue of the ACK message. If no ACK message arrives at the UAS within a designated time, then the UAS will re-send the final response. We have also seen that *provisional responses* are sent unreliably. If a provisional response does not arrive at the UAC, then the UAS will be unaware and the provisional response will not be re-sent.

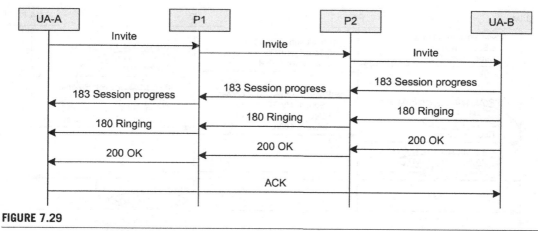

FIGURE 7.29

INVITE transaction with 183 Session Progress.

There may, however, be situations where the reception of a provisional response by the UAC is essential for the completion of the INVITE transaction. One example is a two-step offer-answer message sequence for codec negotiation. In these situations, a mechanism may be applied for reliable transmission of a provisional response. Let's have a look again at signaling (example) associated with an INVITE transaction (see Figure 7.29).

In the example in Figure 7.29, UA-B sends provisional response 183 Session progress, prior to 180 Ringing. The vertical spacing between the successive messages is not interpreted as linear in time. As can also be seen from the figure, SIP proxies P1 and P2 do not record route; the ACK message does not traverse these proxies.

183 Session progress is a generic provisional response, indicating that the handling of the INVITE transaction is in progress. It may, for example, be used to convey information to the calling party, indicating to the calling party that *early media* will be played (to the calling party). Early media, for example, may be a call progress announcement. Early media is described in detail in a later section. If UA-B intends to send early media to the calling party, it may want to defer the sending of this early media to the calling party until it has received confirmation that the 183 Session Progress has arrived at UA-A.

For this purpose, a *provisional acknowledgement* may be used. The provisional acknowledgement is defined in IETF RFC 3262. The provisional acknowledgement is done through a designated SIP method, the PRACK. When UA-B wants to receive confirmation of the successful arrival of the provisional response at UA-A, then UA-B may request an acknowledgement from UA-A. This acknowledgement is sent from UA-A to UA-B in the form of a PRACK transaction. The PRACK transaction may be compared with the ACK transaction. Some essential differences are:

- The PRACK transaction consists of a request message and final response message (for an ACK transaction, there is no response).

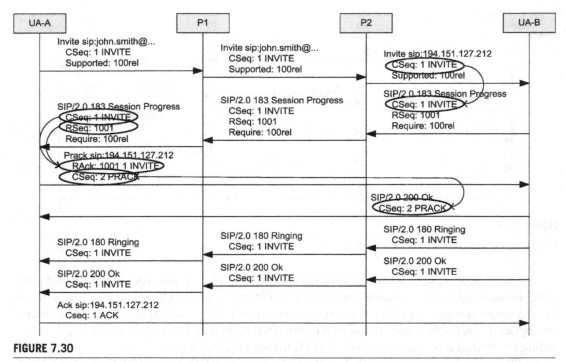

FIGURE 7.30

Message sequence for reliable provisional response.

- The support of the PRACK method is optional for a UA, whereas the support of the ACK method is mandatory (it forms part of standard SIP functionality).
- PRACK uses its own branch-ID and CSeq value.

The use of PRACK is explained with reference to Figure 7.30.

There may be multiple provisional responses for a transaction; each provisional response may have a provisional acknowledgement associated with it. For this reason, a provisional acknowledgement needs double identification: (i) the transaction (within the SIP session) it belongs to; and (ii) the provisional response within that transaction it belongs to. The following SIP headers are used for the provisional acknowledgement mechanism:

- **RSeq.** Response sequence, used in a provisional response when that provisional response is sent reliably. The RSeq header is a numerical value identifying the response within a transaction. It may start at an arbitrary value and has meaning within a transaction only.
- **RAck.** Response acknowledgement, used in a PRACK request message. The RAck header is a combination of two numerical values: (i) the Response sequence (RSeq) value for which the PRACK is requested; and (ii) the Command sequence (CSeq) value of the transaction of this provisional response. The combination of CSeq and RSeq uniquely identifies a particular provisional response of a particular transaction.

When UA-B receives a PRACK request, it can associate the PRACK request with the corresponding provisional response. UA-B now knows that the 183 Session Progress has arrived, so can take further action like the playing of an announcement.

A PRACK transaction is a transaction in its own right. This implies that the request message routing, response message routing, CSeq stepping, etc. follow the rules as described earlier. Also, retransmission of the PRACK request message and the PRACK response message may be applied.

You will have noticed in Figure 7.30 that the PRACK transaction does not traverse SIP proxies P1 and P2. In addition, the PRACK request message is directed to UA-B's contact address (sip:194.151.127.212) as opposed to UA-B's public identity (sip:john.smith@...). The provisional response for which the PRACK is requested (183 Session progress) contains Record-Route headers and the contact address of UA-B. UA-A uses these Record-Route headers and contact address for the routing and addressing respectively of the PRACK request to UA-B. The intermediate proxies, so far as they do not record route, need not be aware of the PRACK transaction. The non-record routing proxies will handle just the INVITE transaction.

As stated earlier, reliable provisional response is an optional feature. A UA, acting as a UAS, may apply reliable provisional response only when both the INVITE initiator and the INVITE responder support this feature. This support is indicated through the *Supported* header. For example:

```
INVITE sip:john.smith@company-x.se SIP/2.0
Supported: 100rel
```

The INVITE request message may include the Supported header with designated *option tags*. Support for optional SIP capability is indicated in one or more tags in the *Supported* header. The option tag "100rel" is derived from "100 response class" (i.e. the class of provisional responses, excluding 100 Trying) and "rel" (reliable transmission). Each newly standardized optional SIP functionality has an option tag defined for it (in the IETF RFC that defines this optional SIP functionality). The INVITE responder may indicate in the provisional response that a reliable provisional response is required. This is indicated through the *Require* header. For example:

```
SIP/2.0 183 Session progress
Require: 100rel
```

Both the INVITE initiator or INVITE responder may request a reliable provisional response. When the INVITE request contains Require: 100rel, the request for a reliable provisional response applies to any of the provisional responses (except 100 Trying) for this transaction. If, however, the INVITE request does not contain Require: 100rel, but contains Supported: 100rel, then the INVITE responder may determine per provisional response whether reliable transmission applies.

When a request message or response message indicates Require: 100rel, it implicitly indicates also that the sender of this message *supports* this functionality. Hence, it would not be required to also explicitly indicate Supported: 100rel in this case.

The use of reliable transmission of a provisional response applies both to UDP and TCP. It offers end-to-end reliable transmission of a provisional response; the provisional response may traverse a mix of reliable transport (TCP) and unreliable transport (UDP).

7.9 CANCELING A SIP TRANSACTION REQUEST

A SIP transaction constitutes a request from one SIP entity to another SIP entity to execute a particular task. An INVITE transaction is a prominent example: one SIP entity confers a request to another SIP entity to establish a SIP session. The establishment of the SIP session implies engagement in media transfer, in accordance with a set of codecs. In other words, the INVITE transaction constitutes a call setup process.

Call setup includes, by (human) nature, an alerting phase; the called party is alerted and may answer or decline the incoming call. Whereas the called party may *decline* the call, the calling party may *abandon* the call attempt. Abandoning the call attempt means that the processing of the SIP INVITE transaction shall be canceled. A special SIP method is used here, the CANCEL method. A CANCEL transaction is sent to the SIP responder, requesting the SIP responder to cancel the execution of the INVITE transaction. One issue here is: How can we get this CANCEL transaction sent to the exact same entity that has received the INVITE request? We have seen in earlier sections that when the SIP transaction is initiated, a SIP trail is established, the path along which the request message travels toward its destination. To cancel the INVITE transaction, it is necessary to give the INVITE responder an indication that this INVITE transaction should be canceled. However, the INVITE initiator does not have knowledge about this SIP trail along which the INVITE request message has traveled. The exact SIP trail is determined on a hop-by-hop basis. A proxy receives information about the transaction trail built up so far, but does not receive information about the transaction trail that will be built ahead of it. In order to send a CANCEL instruction to the INVITE responder, the CANCEL request is sent by the INVITE initiator and by each intermediate proxy, to the same transport address as the corresponding INVITE was sent. This same transport address includes IP address, transport protocol, and port number (see Figure 7.31).

A CANCEL transaction is a regular transaction in the sense that it consists of a request message and a final response message. Unlike transactions of other SIP methods, each intermediate proxy sends the final response on a CANCEL transaction. The rationale is that the CANCEL request is not meant to request a particular functionality from a remote end; instead, the CANCEL request is used only to cancel the processing of a particular INVITE transaction. Hence, there is no need to provide any response to the CANCEL request, other than a 200 OK, indicating that the request is received and will be forwarded to the next hop.

Figure 7.32 shows a signaling sequence for an INVITE transaction that is canceled. When a CANCEL request arrives at an intermediate proxy, the proxy needs to match this CANCEL request with the associated INVITE transaction. This is done by using the same branch identifier for the CANCEL request as for the INVITE request (see Figure 7.33).

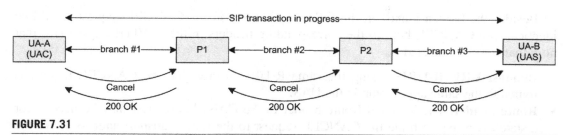

FIGURE 7.31

Hop-by-hop sending of and responding to CANCEL request.

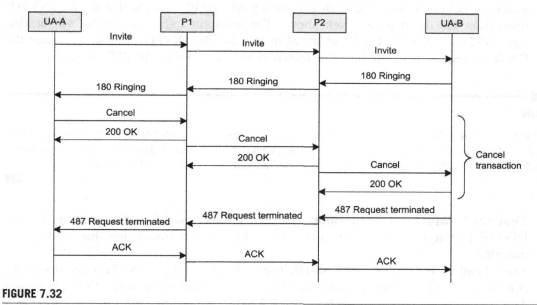

FIGURE 7.32

Signaling sequence for CANCEL transaction.

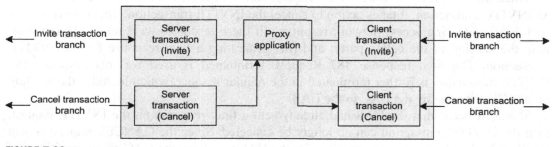

FIGURE 7.33

Matching CANCEL request with INVITE request.

Besides the branch identifier, the following headers in the CANCEL request should be identical, per CANCEL hop, to the corresponding headers in the INVITE request (for that hop) that is to be canceled:

- **Request URI (R-URI).** Using the same R-URI ensures that the CANCEL request is routed to the same application in the UAS.
- **Route header.** The identical Route header in the CANCEL request may be needed for a stateless proxy to route the CANCEL request to the same transport address as for the INVITE request.
- **Via header.** Since CANCEL is a hop-by-hop transaction, the CANCEL request contains a single Via header only. The branch identifier will be identical to that for the INVITE transaction for the reasons given above. The transport indication in the Via header, e.g. SIP/2.0/UDP, will also be identical to that for the INVITE transaction, since the CANCEL request is sent to the same transport address as for the INVITE request.

Note

An entity sending a CANCEL request shall not apply a DNS query or use cached DNS resource records for routing of the CANCEL request. Such behavior may result in the CANCEL request being sent to a different host. Instead, the entity should use the same transport address as used for the INVITE transaction.

- **Command sequence (Cseq) value.** The CSeq value should be the same as for the INVITE transaction. This allows the UAS to identify the transaction that needs to be canceled.
- **From header, To header, and Call-ID.** The To header in the CANCEL request will not contain a tag. This follows from the fact that the To header in the initial INVITE request does not contain a tag either. If, however, CANCEL is used to cancel the execution of a re-INVITE, then a tag will be present in the CANCEL request.

When the UAS receives a CANCEL request and has matched the CANCEL request with an INVITE transaction, it takes action to cancel the INVITE transaction. This means actions such as: (i) releasing access network resources (if these resources were reserved); (ii) stopping the alerting of the called party; and (iii) generating a final response for the INVITE transaction. The final response 487 Request Terminated is used for this purpose. The INVITE transaction is further terminated in the regular manner, which includes the sending of an ACK message by the UAC to the UAS.

Should it occur that the UAS had already sent a final response on the INVITE request, then the INVITE transaction can no longer be canceled. Since the CANCEL request is sent on a hop-by-hop basis, it is not possible for the UAS to inform the UAC about the failure of canceling the INVITE transaction. The UAC will in this case receive the final response on

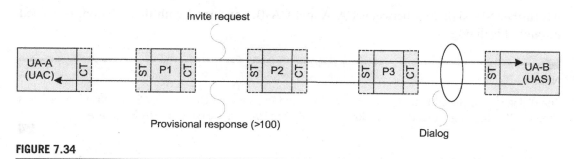

FIGURE 7.34

Dialog establishment between UAC and UAS. CT, client transaction; P, proxy; ST, server transaction.

the INVITE request. If the final response is a 200 OK, the UAC will have to acknowledge the 200 OK in the regular manner, and then terminate the SIP session in the usual fashion by initiating a BYE transaction.

The cancellation of a transaction would not be possible prior to receiving the first provisional response for that transaction. The first provisional response would typically be 100 Trying. This is because in this case the CANCEL request might arrive at the next hop before the INVITE request has arrived. The next hop would in that case not be able to process this CANCEL request. The UAC that intends to cancel an INVITE transaction, e.g. because the calling party has abandoned the call attempt, must wait until the first provisional response is received and may then send the CANCEL request. This wait for the first provisional response before sending the CANCEL request has to be done by each intermediate proxy in the SIP chain up to the UAS.

A CANCEL request may be issued before the dialog was established, i.e. before a To tag was received. The routing of the CANCEL message and the handling of the CANCEL message at the next hop are as described above.

7.10 SIP DIALOGS

This section introduces the concept of SIP dialog. SIP dialog has a specific relation to SIP session and SIP transaction. A SIP session is established by means of an INVITE transaction. The establishment of the SIP session is initiated by a particular UA, acting as UAC for the INVITE transaction. The INVITE request is propagated to the intended destination, acting as UAS for the INVITE transaction. The path toward the destination, as well as the actual destination of the INVITE request, is determined through an iterative process. Successive SIP proxies along the path from UAC to UAS determine the next receiver of the INVITE request (see Figure 7.34).

Figure 7.34 depicts SIP transaction initiation between UAC and UAS. As soon as UA-A has received a provisional response (except 100 Trying) from UA-B, a *dialog* is established between UA-A and UA-B. The dialog constitutes a relation between the respective UAs.

All further SIP signaling between UA-A and UA-B, concerned with this relation, is carried through this dialog.

Note

The dialog is established between UA-A and UA-B and not between UAC and UAS, because this dialog is used for SIP signaling, i.e. SIP transactions, from UA-A to UA-B as well as from UA-B to UA-A.

When the INVITE transaction between UA-A and UA-B is successfully completed (2xx-class final response), the dialog between UA-A and UA-B transits to the *confirmed* state, i.e. becomes a *confirmed dialog*. Prior to the final response, i.e. as long as the execution of the INVITE transaction is in progress, the dialog between UA-A and UA-B is referred to as *early dialog*.

A SIP message carries a dialog identifier; this enables a UA to associate an incoming message with a particular SIP dialog. The dialog identifier is composed of three elements:

- **From tag.** An INVITE request message contains a From header. This header identifies the party who initiated the INVITE transaction. The From header also includes a *tag*. This tag is used for dialog identification. The From tag is generated by the initiator of the INVITE transaction; it is unique within the UA that generates this tag. All SIP messages within this dialog contain a From header with this tag value.
- **To tag.** An INVITE request message also contains a To header. This header identifies the party or entity for whom the INVITE transaction is destined. When the UAS receives the INVITE request, it allocates a tag, to be included in the To header. The To tag is a random value generated by the receiver of the INVITE transaction; it is unique within that UA. All SIP messages within this dialog contain a To header. The first $>$ 100 provisional response messages ($>$100 meaning any response in the range 101–199), as well as all subsequent transaction messages within this dialog, will have the To tag in the To header.
- **Call-ID.** The INVITE request message further contains a Call Identifier, contained in the Call-ID header. The Call-ID is generated by the initiator of the INVITE transaction and is used in all SIP messages within this SIP session.

Entities allocating From tag, To tag, or Call-ID may use the tag or Call-ID to embed process-related information or subscriber-related information. Such information has no meaning for other entities in the system. When the entity that allocated, for example, a To tag receives a request or response message, it may use the To tag to facilitate retrieving subscriber data or to locate the software process within the node.

An example INVITE transaction message sequence, including dialog identification, is the following:

```
INVITE sip:john.smith@company-x.se SIP/2.0
From: "Wendy Jones"<sip:wendy.jones@company-y.se>;tag=64232014
```

```
To: "John Smith"<sip:john.smith@company-x.se>
Call-ID: 12rjhafvsnadf8wyt5njsdffgmsdfngowe5iu

…

SIP/2.0 183 Session progress
From: "Wendy Jones"<sip:wendy.jones@company-y.se>;tag=64232014
To: "John Smith"<sip:john.smith@company-x.se>;tag=15623985
Call-ID: 12rjhafvsnadf8wyt5njsdffgmsdfngowe5iu
```

The From header and To header are used for dialog identification, not for routing. This is also evident from the fact that the response message, the 183 Session Progress, is sent from a UAS of sip:john.smith@… to the UAC of sip:wendy.jones@…, but still the message includes Wendy Jones in the From header and John Smith in the To header. The reason for this is that the From header and To header are set per transaction. The INVITE transaction in this example is sent from Wendy Jones to John Smith. The response message routing is based on the Via header.

The URI contained in the From header will typically be public identification of the initiator of the INVITE transaction. Its purpose is (i) identifying the calling party (calling line presentation) and (ii) allowing the destination party to store the identity and, later, establish a communication session towards that calling party, using the stored public identity. The From header may contain a *Display name*. It may be a freely chosen text string to identify the calling party. SIP phones typically allow the user to configure a user name that is used in the SIP signaling.

The To header contains an identification of the destination party. It does not have to be equal to the R-URI in the INVITE request message. SIP phones may, for example, set the URI in the To header equal to the R-URI or equal to the user part of the R-URI. Whereas the R-URI in the INVITE transaction will at some point be set to the contact address of the destination party (e.g. INVITE sip:john.smith@company-x.se SIP/2.0 becomes INVITE sip:145.261.34.12 SIP/2.0), the URI in the To header may remain unmodified up to the UAS. It indicates to the UAS *how* that UAS was identified in the establishment of the call. This may be useful when the destination party has multiple public user identities, e.g. sip:john.smith@company-x.se and sip:+46104812211@company-x.se. The destination subscriber, or an application server acting on behalf of the destination subscriber, may apply different call acceptance policies, based on how the person is called. The To header may further contain a display name. This display name is analogous to the display name in the From header, using free text to identify the destination party.

When the UA-B in this example intends to terminate the SIP session, it initiates a BYE transaction:

```
BYE sip:wendy.jones@company-y.se SIP/2.0
From: "John Smith"<sip:john.smith@company-x.se>;tag=15623985
To: "Wendy Jones"<sip: wendy.jones@company-y.se>;tag=64232014
Call-ID: 12rjhafvsnadf8wyt5njsdffgmsdfngowe5iu

…

SIP/2.0 200 OK
```

```
From: "John Smith"<sip:john.smith@company-x.se>;tag=15623985
To: "Wendy Jones"<sip: wendy.jones@company-y.se>;tag=64232014
Call-ID: 12rjhafvsnadf8wyt5njsdffgmsdfngowe5iu
```

For the transaction from John Smith to Wendy Jones, the respective tags are swapped, compared to the tags in the initial INVITE transaction. The From header in the BYE request message contains the From header that was allocated by the UA from John Smith. The display name in the To and From headers would typically not be used other than in the initial INVITE transaction. However, the display names are normally retained in all SIP messages.

The URIs in the To header and in the From header require special attention for application developers. The URI in the From header may be used for display purposes. A SIP application may want to modify the URI in the From header, for example by an enterprise-specific number:

```
From: "Wendy Jones"<sip:146104812211@company-y.se>;tag=64232014
```

into

```
From: "Wendy Jones"<sip:2211@company-y.se>;tag=64232014
```

2211 would, in this example, be the corporate number.

According to SIP as defined in IETF RFC 3261, a SIP dialog is strictly identified by the From tag, the To tag, and the Call-ID. From that point of view, it would be permissible for a SIP application server to modify the URI in the From header. This is depicted in Figure 7.35.

The receiver of the INVITE request message, UA-B, copies the From header into the provisional response message. UA-A will now receive a provisional response message containing a From header with a different URI than it had sent out. Since, according to IETF RFC 3261, SIP dialog is identified by the two tags and the Call-ID, UA-A should be able to associate this provisional response message with the appropriate SIP dialog. However, the predecessor of IETF RFC 3261, IETF RFC 2543, defined that the URI in the From header and To header are used for dialog identification. Some SIP entities in operation still require that the URI in the From header and To header is not modified for a dialog. In order to guarantee correct interworking with these SIP entities, a SIP application server that wants to modify the URI in the From header or in the To header will have to do so in all SIP messages for this SIP dialog. This implies, for example, that such a SIP application has to apply record routing.

When an application server modifies the From header, it takes care that this modification does not affect subsequent application server(s) that use the From header within their service logic processing. This dilemma shows that when multiple application servers act on one SIP session, the operator needs to ensure the correct behavior of these application servers, in order to prevent unwanted service interference.

IETF RFC 4916 defines a mechanism through which a user agent (UA) can modify its identity once a SIP dialog is established. More specifically, that UA signals a modified From URI to the peer. This modified From URI will be the URI identifying this UA from that moment onward.

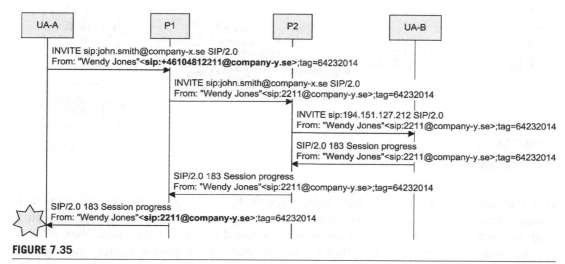

FIGURE 7.35

Modifying the From header URI.

7.10.1 Multiple Early Dialogs

We now encounter a very powerful aspect of SIP, namely the existence of *multiple* early dialogs for an INVITE transaction. As explained earlier, the execution of an INVITE transaction results in the establishment of a SIP dialog between UAC and UAS. Figure 7.36 depicts an INVITE transaction between two UAs, with no proxy in between.

A SIP proxy or SIP application server on the SIP path between UAC and UAS may forward the INVITE request message to two or more destinations. This is depicted in Figure 7.37. The proxy in Figure 7.37 may be a *forwarding proxy* or a *forking proxy*. We will describe the establishment of multiple early dialogs for the INVITE transaction based on a forking proxy.

Consider the routing of the INVITE request message destined for a particular destination, e.g. sip:john.smith@company-x.se. The INVITE request message would be forwarded to the registrar where John Smith is currently registered. That registrar holds the binding of John Smith's SIP terminal, i.e. it knows the contact address to forward the INVITE request message to. John Smith may, however, be registered at *two* SIP terminals

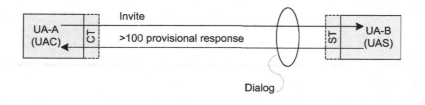

FIGURE 7.36

Single dialog for INVITE transaction.

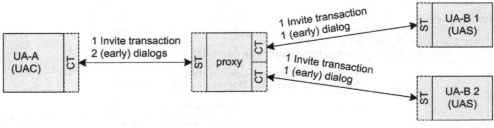

FIGURE 7.37

Multiple dialogs for INVITE transaction.

at that moment. Both terminals have gone through the registration process, whereby each terminal has deposited its contact address in the registrar. The registration process in a VoIP network ensures that, once a registrar is assigned to a particular subscriber, all registration requests from that subscriber are sent to that same registrar. Hence, the registrar may forward the INVITE to two, or more, contact addresses. This action is known as *forking*. When forking occurs, the INVITE request is received on two (or more) terminals. This forwarding of the INVITE request to the two or more terminals may be done in parallel or sequentially. The effect is that the call is offered to two or more phones: parallel alerting or sequential alerting. Whether parallel alerting or sequential alerting is applied depends on the *priority* value assigned to the respective contact addresses that are registered in the registrar. The priority of a contact address is denoted by a parameter q; the value of q indicates the relative priority.

Example 7.1

The following contact addresses are registered for John Smith:

  sip:145.261.36.14; q = 1.0
  sip:145.261.34.12; q = 0.5

 Contact address sip:145.261.36.14 has a higher priority than contact address sip:145.261.34.12. The INVITE will first be offered to the first contact address for a configurable (in the registrar) period, e.g. 15 s. If there is no final response within this period, or a non-2xx final response is received, the INVITE is offered to the second contact address.

Example 7.2

The following contact addresses are registered for John Smith:

  sip:145.261.36.14; q = 1.0
  sip:145.261.34.12; q = 1.0

The contact addresses have equal relative priority. The INVITE is offered to both contact addresses at the same time. If a contact address does not have an explicit relative priority specified, q = 0 is assumed.

The forking proxy creates a client transaction for each INVITE request message to a contact address. There may be one (typical) or more proxies in the SIP path from registrar to the respective UA-B terminals.

The two INVITE request messages sent by the registrar to UA-B(1) and UA-B(2) are identical, except for:

- **R-URI.** The R-URI contains the contact address of the terminal.
- **Route header.** The Route header identifies the path that shall be followed by the INVITE request message, from registrar to UA-B terminal. Different terminals of the same subscriber may be registered via different proxies. Hence, the Route headers in the INVITE messages may indicate different proxies.
- **Via header (branch).** For each INVITE request message sent from the registrar, a client transaction is initiated; for each client transaction, a designated branch identifier is used. The IP addresses in the Via header should be identical, since the INVITE messages are sent from the same host and application.

Each UA-B may generate provisional responses (the respective UA-B instances are not aware that the INVITE they are receiving is also being received by another entity). Each UA-B generates a To tag, to be used in the further signaling between this UA-B and UA-A. So, there will be a dialog established by each UA-B that receives the INVITE request. Provisional response messages that are returned by the respective UA-Bs, in this case UA-B(1) and UA-B(2), are forwarded transparently to UA-A. This has the effect that from the forking proxy up to UA-A, there are *two* dialogs established. Each response message received by UA-A relates to a specific dialog. For each dialog, different provisional responses may be received; the forking proxy does not coordinate this. The provisional responses received by UA-A, for the different dialogs, will differ in at least the following SIP headers:

- **To tag.** The To tag is generated by the respective UA-B. UA-A will determine from the combination of From tag, To tag, and Call-ID that a response message relates to another dialog.
- **Contact address.** The respective UA-Bs will provide their own contact address in the provisional response. Alternatively, a UA-B provides its contact address only in the 200 OK final response.
- **Record-Route.** Each UA-B returns the Record-Route set, if applicable. Since the UA-B terminals may be registered via different proxies, the Record-Route set may be different. Alternatively, a UA-B returns the Record-Route set only in the 200 OK final response.
- **Content-Type and Content-Length.** The different UA-B terminals may provide different SIP bodies in the response message, such as different SDP answers. Possibly, a UA-B may return an SDP response only in the 200 OK final response.

There may be additional SIP headers that differ for the respective dialogs received by UA-A, such as Supported, Required, and User Agent. Each of these headers shall be processed by the UAC in accordance with the rules applicable for that header. For example, a reliable provisional response may be requested for a particular dialog. UA-A will in that case apply the PRACK transaction for that dialog, using the contact address and Record-Route set received for that dialog.

The general behavior of the forking proxy is as follows:

- Provisional responses from the respective user agent servers (UASs) will be forwarded transparently to the user agent client (UAC).
- When the UAC has received a 180 Ringing response and then receives a second 180 Ringing response, relating to a second dialog, this has no effect on the alerting state of the terminal. That is, the terminal of the calling party remains in an alerting state.
- The receipt of multiple early dialogs by the UAC (calling party) has the side-effect that the UAC is aware that multiple terminals of the called party are being alerted. The arrival of a second 180 Ringing (relating to another dialog) is normally not shown to the calling party.
- When a non-2xx-class final response is received from a particular UAS, the INVITE transaction with that UAS terminates (and the forking proxy will send ACK to that UAS). The forking proxy will store the final response received from that UAS and will continue to send INVITE to other available contact addresses (this may have been done in parallel), until a 2xx-class final response is received from a particular UAS. That final response is forwarded to the UAC. The forking proxy cancels any INVITE transaction still ongoing to other contact address(es), in a manner that is similar to the canceling of an INVITE transaction by a UAC.
- When none of the contact addresses returns a successful final response, the forking proxy determines "the best final response" from the set of received final responses and returns that best final response to the UAC. IETF RFC 3261 describes the rule for determining the best final response.

Note

When one of the terminals provides additional information in the unsuccessful final response, in the form of a SIP header or SIP body, but another final response is determined to be the best final response, this additional information will not be conveyed to the UAC.

When a UA initiates an INVITE transaction, it will not know whether this INVITE may be subject to forking. This implies that a UA should be prepared to receive multiple early dialogs in response to initiating an INVITE request. There is no defined possibility for a UA to indicate in the INVITE request whether or not it supports multiple early dialogs (other than using the *no-fork* option; see description of Request-Disposition header). If a UA does not support

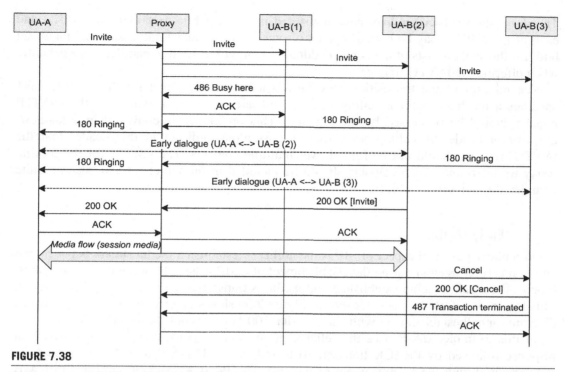

FIGURE 7.38

Forking scenario for three destination terminals.

multiple early dialogs, then the VoIP network may overcome this dilemma by converting the multiple early dialogs from the two or more UASs into a single dialog to the UAC.

Figure 7.38 shows an example of a typical parallel alerting service, based on the forking proxy principle. In the example in Figure 7.38, the proxy is acting on behalf of the destination subscriber, who is registered at three SIP devices: UA-B(1), UA-B(2), and UA-B(3). UA-B(1) is determined to be busy (response: 486 Busy Here). UA-B(2) and UA-B(3) are available for communication, so respond with 180 Ringing. UA-B(2) answers the call; the early dialog between UA-A and UA-B(2) transits to the confirmed state. The proxy cancels the INVITE transaction with UA-B(3). UA-A will, for a certain duration (64 × 500 ms), keep information related to the early dialogs with UA-B(1) and UA-B(3). This is because, due to race condition, a 200 OK may arrive from UA-B (3). In such case, UA-A acknowledges that 200 OK with an ACK message and after that issues a BYE request to UA-B(3), in order to release the SIP session from UA-B(3).

7.10.2 Target Set

The process in the registrar of matching the R-URI of the incoming INVITE request with one or more registered contact addresses, for that R-URI, is known as building the *target set*.

The destination subscriber may have registered multiple public user identities (PUIs) and each of these PUIs may be bound to a set of one or more contact addresses. The target set built for this call consists of the contact addresses that will be tried in parallel or sequentially when offering the INVITE request.

The initiator of the transaction may further influence the target set. IETF RFC 3841 specifies a mechanism for a calling party to indicate *preferences* as to how the INVITE request should be processed by the registrar. This preference is conveyed in a *Request-Disposition* header. A calling party may, for example, indicate in this header that the INVITE request should not be forked over multiple contacts. The process for the forking proxy to determine *which* contacts should be included in the target set can become quite complex through this flexibility.

7.10.3 Early Media

A particularly powerful aspect of SIP is the ability to establish a media stream between two end-points (user agents) during the establishment of a SIP session between these two agents. Practically, a call is being established and media is transferred before the call is answered. This feature is referred to as *early media*. Figure 7.39 shows an example message sequence diagram for call establishment with early media. 100 Trying is not shown in the figure.

In this example, UA-B uses the reliable provisional response mechanism (provisional response followed by PRACK transaction) to indicate to UA-A that it (UA-B) intends to send media during the establishment of the call. The PRACK message indicates to UA-B that UA-A has received the 183 Session Progress and that UA-A is ready to receive (early) media. This early media may, for example, be a personalized ringback tone generated by the terminal. The 180 Ringing provisional response gets the SIP phone to transit to the alerting state; the subscriber may then see an indication on the display that the call has reached the destination subscriber and that alerting has started. There is already early media flowing to the calling party, so the terminal should not generate local ringtone.

In the example of Figure 7.39, the early media is transferred as a media stream associated with the (early) dialog between UA-A and UA-B. The early media may, in this case, be sent using the media transmission parameters that will apply for the established call. UA-A and UA-B may have one negotiation iteration for the media transmission parameters, such as codec IP address and port number. When the call is answered, UA-B may continue to send media using the agreed media transmission parameters, with the difference being that the media sent from UA-B will be speech from the called party, instead of personalized ringback tone.

Early media may take one of the following directions: (1) media from UA-A to UA-B (*forward media*); (2) media from UA-B to UA-A (*backward media*); and (3) media between UA-A and UA-B in both directions (*bothway media*). If the call crosses the CS–IMS domain boundary, then controlling early media as described here requires support of early media

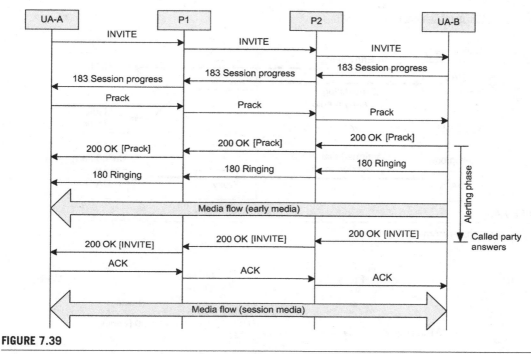

FIGURE 7.39

SIP session establishment with early media.

in the media gateway and media gateway controller at the boundary between CS and IMS networks.

A more common case of early media than that depicted in Figure 7.39 is the situation where the early media transfer is associated with a designated dialog with the initiating party. This is illustrated in Figure 7.40. The proxy shown in Figure 7.40 acts as forking proxy. It forks the SIP INVITE request to the intended receiver (UA-B) and to a media server. Both UA-B and the media server establish a SIP dialog with UA-A. At this point, both dialogs are early dialog. The media server uses the early dialog with UA-A to transfer media to that UA-A. When UA-B answers the call, the early dialog between UA-B and UA-A transits to the confirmed state. UA-A will discard the early dialog with the media server and will, from that point onward, accept only media from UA-B. Figure 7.41 shows an example signal sequence for this use-case of early media.

When UA-B answers the call, the proxy will cancel the SIP transaction that was initiated toward the media server. As is reflected in Figure 7.41, the early media transfer between media server and UA-A is in the backward direction only. The reliable provisional response between media server and UA-A may include information indicating that media transfer will be in the backward direction only.

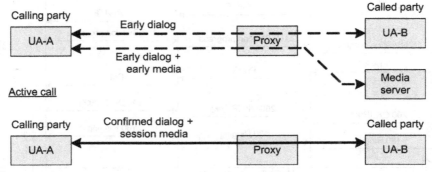

FIGURE 7.40

Early media transfer from media server.

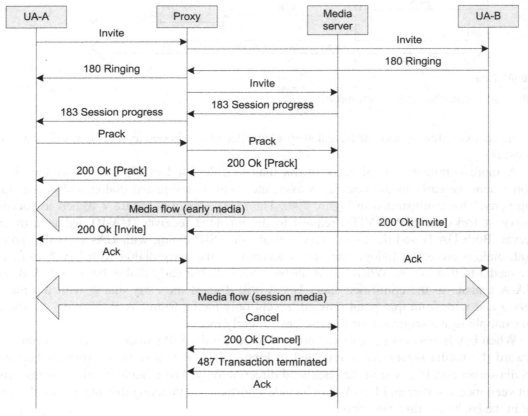

FIGURE 7.41

Early media from media server.

Note

Early media transfer from the media server to UA-A implies that RTP messages are sent from the media server to UA-A. The RTCP message transfer, however, may be bidirectional. The media server has provided UA-A with relevant information for RTCP transfer to the media server, such as IP address and port number.

In the above example, the media server sends the media through early SIP dialog. If the media server, instead, responds with 200 OK, then we need the aid of an application server to map this 200 OK to 183 Session Progress. Otherwise, the call from the calling party would be answered and charging for the calling party would start.

In the case where two early dialogs are established to a calling party's terminal and early media is provided for *both* dialogs, the calling party's terminal has to decide which media it will render. In the IMS network, this decision on which early media to provide to the calling party is taken by the network instead of by the calling party's terminal.

When early media transfer ceases (e.g. the playing of an alerting announcement is finished), the early dialog used for the transfer of that early media is no longer needed. An application server acting on behalf of the calling or called party may send an explicit indication to the calling party that this dialog is terminated. This indication is done through the 199 Early Dialog Terminated provisional response. The calling party's terminal may then discard any resource related to that early dialog. The 199 Early Dialog Terminated response is described in IETF RFC 6228.

7.11 MEDIA TRANSMISSION: OFFER–ANSWER MODEL

In the descriptions provided so far, we have seen how one SIP user agent (UA) establishes a SIP session with another SIP UA and that media may be exchanged between the respective SIP UAs once the SIP session is established. The flow of media requires the exchange of user-plane details. Put differently, the respective UAs have to exchange a set of information that is needed for the sending and receiving of the media. In this section, we will therefore look in detail into the media transmission aspects of SIP. In particular, we will describe the mechanism for SIP user agents to negotiate the characteristics of media transfer. This mechanism is known as the *offer–answer model*. The offer–answer model entails that, during the establishment of a SIP session or during an active SIP session, one user agent provides an *SDP offer* and the other user agent responds with an *SDP answer*. This process may be repeated one or multiple times (see Figure 7.42).

In the examples in Figure 7.42, the SDP offer is always generated by UA-A. There may also be a mix of SDP offer–answer direction. For example, UA-A provides an SDP offer, followed by UA-B providing an SDP answer. Subsequently in the (early) dialog, UA-B provides an SDP offer and UA-A provides an SDP answer.

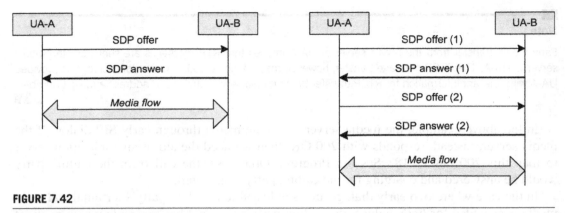

FIGURE 7.42

Basic offer–answer model signaling.

The offer–answer model used by SIP is described in IETF RFC 3264. The offer–answer model uses the Session Description Protocol (SDP) as defined in IETF RFC 4566. It is important to apply a strict distinction between these two:

- SDP (IETF RFC 4566). SDP is a notation (protocol) for describing a media session.
- Offer–answer model (IETF RFC 3264). The offer–answer model describes how two user agents agree on a media session. The SDP forms an integral part of the offer–answer model; the SDP is used to describe a media session offer or a media session answer. Media session offer is commonly referred to as SDP offer; media session answer is commonly referred to as SDP answer.

The media session offer and media session answer are exchanged between respective SIP user agents via SIP messages. There is no one-to-one relationship between SDP offer and SDP answer and designated SIP messages. An SDP offer and answer sequence may occur in various stages of SIP session establishment or during an established SIP session. A few basic rules for offer–answer sequence are:

- The SDP is defined in terms of a number of media streams. The SDP offerer proposes a set of media streams. The SDP answerer may accept one or more of the proposed media streams.
- Each media stream is described in terms of a set of parameters, such as codec(s), (optional) IP address, and port number.
- SDP offer and SDP answer will be sent reliably. It is not prohibited that an SDP answer is provided in an unreliable provisional response. But then the same SDP answer will also be provided in a reliable response (which may be a provisional or a final response).
- An SDP offer–answer process may consist of an iterative sequence of SDP offer and SDP answer exchange.

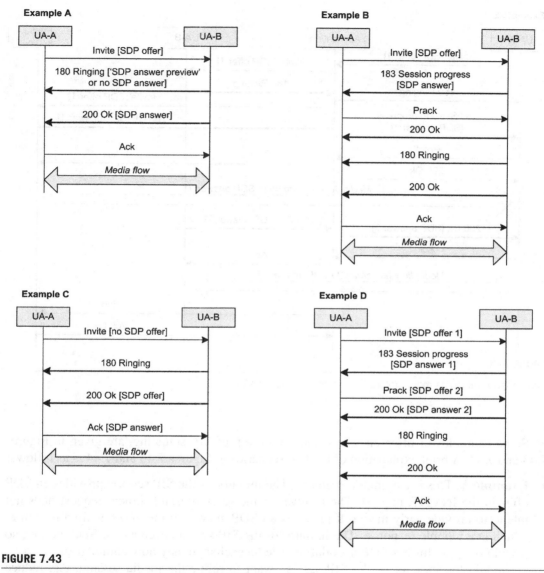

FIGURE 7.43

Offer–answer model as meta protocol on top of SIP (examples A–D).

- When a media session is established, either one of the involved user agents may start a media update procedure, for example to add or remove a media stream. The procedure for updating a media stream follows the procedure for initial offer–answer.
- There will be at most one unanswered SDP offer outstanding per SIP client.

Example E

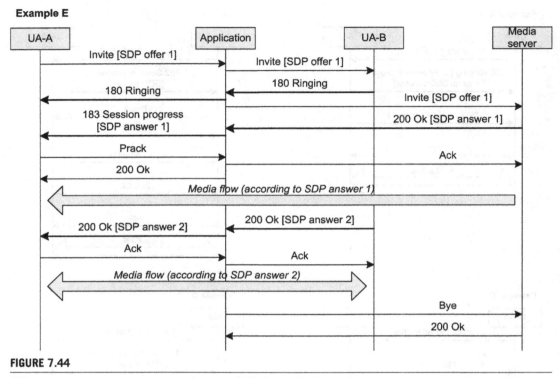

FIGURE 7.44

Offer–answer model as meta protocol on top of SIP (example E).

Some examples of offer–answer sequence, on top of SIP signaling, are given in Figures 7.43 and 7.44. A brief explanation of the five scenarios in Figures 7.43 and 7.44 is as follows:

- **Example A.** This is a common scenario. The initiator of the SIP session provides an SDP offer in the INVITE request. The receiver of the session establishment request does not intend to provide early media. It provides an SDP answer in the 200 OK final response, which is a reliable response. The initiator of the SIP session accepts the SDP answer, so it will not apply further SDP negotiation. Media exchange may now commence.

  Should the initiator of the SIP session want to refine the media session, e.g. reduce the set of accepted codecs for a particular accepted media stream, then the initiator may apply in-session codec negotiation, straight after the ACK message.

- **Example B.** In this example, the SDP negotiation is initiated and completed prior to session answer. This model may be useful when, for example, the called party intends to provide early media, in which case the SDP answer has to be available during the session establishment process. Since SDP has to be conveyed reliably, 183 Session Progress with reliable provisional response is used. The 200 OK final response in this case does not need to contain an SDP answer; it was already provided reliably.

- **Example C.** The initiator of the SIP session provokes the responder to propose a media description. It may do so by sending an INVITE request without SDP offer. The responder provides the SDP offer in the 200 OK final response. The SDP answer must now be included in the ACK message.

  This sequence may be applied in service-initiated call establishment – for example, a click-to-call service, where a network service establishes a call between two remote parties. The service provokes one remote party to initiate offer–answer signaling and relays the offer–answer signaling to another remote party.

- **Example D.** In this example, a first SDP offer–answer sequence occurs in the INVITE message and provisional response message. The initiator of the SIP session wants to narrow down the list of accepted codecs, so provides a second SDP offer; this second SDP offer is included in the PRACK request message. Every SDP offer results in an SDP answer, so the 200 OK final response on the PRACK request includes an answer to this second SDP offer. The second SDP answer is acceptable for the initiator of the SIP session, so no further offer–answer signaling takes place.

- **Example E.** In this example, the INVITE from UA-A is forked by an application to UA-B and to a media server. The media server provides an SDP answer (SDP answer 1). That SDP answer is conveyed reliably to UA-A, through a reliable provisional response method. It is used for early media transfer, e.g. a personalized greeting. When UA-B answers the call, it provides an SDP answer as well, SDP answer 2. That SDP answer is also conveyed reliably to UA-A, in a 200 OK. UA-A will, from now onward, apply media transfer in accordance with SDP answer 2.

Offer–answer signaling may also be initiated when the SIP session is established. Two typical examples will be described.

Example 7.3: New Offer by Calling Party in Response to SIP Session Establishment With SDP answer

In the example of Figure 7.45, UA-B has provided an SDP answer in the 200 OK final response. The offer–answer sequence is now complete and the SIP session is established. UA-A may want a refinement of the set of negotiated codecs. The SIP session is already established, since the SDP answer came in the final response. Therefore, UA-A will provide a new SDP offer immediately following the ACK message.

This situation may occur when the SDP offer includes multiple codecs for a particular media stream, e.g. G.711 and G.722, and the SDP answer indicates that both codecs are accepted for this media stream. Such an SDP answer mandates UA-A (and UA-B) to accept both G.711-based media packets and G.722-based media packets for this media stream. UA-A may therefore send a new SDP offer, this case containing a single codec only.

UA-B may, briefly, have started sending media in accordance with SDP answer 1. UA-A may, until the new offer–answer sequence is complete, drop these media packets or render them.

The sending of an UPDATE request following the ACK request *may* lead to race conditions. Specifically, the UPDATE request may arrive earlier than the ACK request. IETF RFC 5407 contains examples of how a SIP UA may deal with such race conditions.

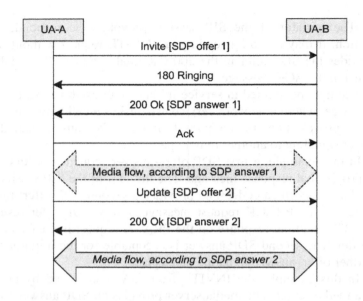

FIGURE 7.45

Re-initiation of offer–answer after SIP session establishment.

The use of UPDATE versus re-INVITE for changing the SDP needs some explanation. When an SDP update is needed before a SIP session is established, the UPDATE method is used. For changing the SDP during an established SIP session, the UPDATE method or the INVITE method (re-INVITE) may be used. When re-INVITE is used during an established SIP session, the remote party could be "prompted" to accept the modified SDP. When the remote party accepts the modified SDP it sends a 200 OK (followed by an ACK request from the initiator of the re-INVITE). Using the UPDATE method requires that the remote party has indicated (in the Allow header) that it supports the UPDATE method.

Example 7.4: Adding or removing codecs during a call

The example in Figure 7.46 shows how one of the peers involved in a media session may initiate a new SDP offer–answer sequence during a call. One of the peers may, for example, activate the camera function in the SIP phone. That SIP phone will offer video capability to the peer. If the peer phone accepts the video capability, the two phones can start exchanging video media streams, in addition to the voice media stream. A similar SIP message sequence applies when removing codecs from the call, e.g. deactivating the video capability of the phone.

Another common use of re-INVITE or UPDATE in an active SIP session is placing a call on hold and, later on, retrieving that call.

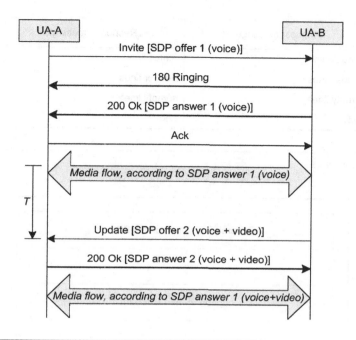

FIGURE 7.46

Adding video codec during a call.

7.11.1 A Closer Look at the SDP Structure

This section provides a closer look at the signaling aspects of transporting the SDP through SIP. Two aspects will be looked into:

1. Position of an SDP in a SIP message.
2. Structure of the SDP.

7.11.1.1 Position of an SDP in a SIP Message

A SIP request message and a SIP response message may contain a *body*. The body within a SIP message constitutes the *payload* that the SIP message transfers. The position of the body within a SIP message is shown in Figure 7.47.

Header lines and body are both optional components in a SIP message. In a basic voice or video call, there will be only a few SIP messages carrying a body (SDP). A SIP message without header(s) is, on the other hand, very unlikely for any SIP session and therefore more theoretical.

The presence of a SIP body in a SIP message is indicated through designated SIP headers, Content-Type and Content-Length:

- **Content-Type.** The Content-Type includes a *media type* and a *media subtype*, separated by a forward slash. The media type indicates what kind of media is included in the body part, such as text, image, audio, video, or application. Media type "application"

Request message	Response message
Request line	Status line
Header lines	Header lines
<empty line>	<empty line>
Body	Body

FIGURE 7.47

SIP body within SIP message.

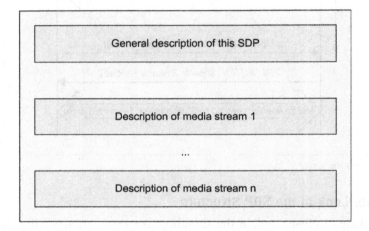

FIGURE 7.48

Overall structure of SDP.

indicates that the body contains a designated application protocol. Media subtype "SDP" indicates that the designated application protocol contained in this body is SDP. Media type "application" and media subtype "SDP" are defined in IETF RFC 4566, which is the RFC describing the SDP. For example:

```
Content-Type: application/sdp
```

- **Content-Length.** The Content-Length indicates the length of the SDP in the body, expressed in octets. A SIP message always contains the Content-Length header. If the SIP message does not contain a body, then the length is set to 0. For example:

```
Content-Length: 142
```

7.11.1.2 Structure of the SDP

The overall structure of an SDP offer or SDP answer is illustrated in Figure 7.48. The SDP is built up from a number of text lines. Each text line contains either general information about the SDP or information about a specific media stream. We will look at the various

Table 7.4 General Description of the SDP

v	Protocol version	This line indicates the version of the SDP. Version 0 is the only SDP in use at the moment.
o	Originator and session identifier	This line consists of a number of information fields, including: <ul><li>Name of the session creator (representing the entity that has created this SDP).</li><li>Session identifier. The inclusion of a session identifier ensures that each SDP generated by this entity is unique.</li><li>Session description version. Whenever an entity provides a new session description for a media session, e.g. to add or remove a codec, the version of the session will be incremented.</li><li>Network type identifier, indicating the type of network that is used for media transportation, such as "IN" indicating Internet.</li><li>IP version, such as "IP4" or "IP6".</li><li>Address of the host from where the session description is created. This may be a fully qualified domain name or an IP address (IPv4 or IPv6).</li></ul>
s	Session name	This text line allows the generator of the media session to give a descriptive name to the media session, such as "private conversation".
c	Connection information	(optional) If this field is present, then it applies to all media descriptions in the SDP, except for those media descriptions that have designated connection information associated with it. It consists of the following information fields: <ul><li>Network type identifier, indicating the type of network that is used for media transportation, such as "IN" indicating Internet.</li><li>IP version, such as "IP4" or "IP6".</li><li>Address of the host from where the session description is created. This may be a fully qualified domain name or an IP address (IPv4 or IPv6).</li></ul>
t	Time	The time line may be used to provide a start time and a stop time (in absolute time stamp), specifying the validity of the media session. Alternatively, an unbound session may be provided (both start time and stop time set to 0). When SDP is used with SIP, the media session described by the SDP will be bounded by the SIP session or dialog.

text lines. Some SDP lines may be used in both the general description part and the media description part of SDP.

General Description of SDP

The general description of the SDP consists of the descriptor lines given in Table 7.4. Only the most commonly used descriptors are listed (refer to IETF RFC 4566 for a complete set).

An example of the general description of SDP is as follows:

```
v=0
o=- 6 2 IN IP4 164.48.60.241
s=CounterPath X-Lite 3.0
c=IN IP4 164.48.60.241
t=0 0
```

Table 7.5 Media Stream Description

m	Media	The media line is built up from a number of information fields.

- Media type. This field may indicate, among others, audio, video, and text.
- Port. This field indicates to which port the media of this media stream shall be sent. If there is no designated Connection information defined for this media stream, the media shall be sent to the IP address as defined by the Connection information in the SDP general description part.
- Protocol. This field identifies the transport protocol that is used for transporting the media stream. Two prominent examples are:
  1. RTP/AVP, which implies that RTP/UDP is used with codec definition from IETF RFC 3551.
  2. RTP/SAVP, which implies the use of SRTP/UDP in accordance with IETF RFC 3711 (Secure RTP).
- Media format. There may be one or more media format fields present in the media line. For protocol RTP/AVP and RTP/SAVP, each media format denotes a specific codec (payload). A media format may refer to static payload type or dynamic payload type. Static payload types are defined in the corresponding application (AVP or SAVP). Dynamic payload types are defined by means of attribute description lines following the media description line.

b	Bandwidth	The bandwidth line defines the proposed bandwidth, in kb/s, for the media stream.
a	Attribute	(optional) Various media attribute descriptions are defined. We will mention the most important and commonly used ones.

- ptime. This attribute defines the packetization time to be applied by the codec (milliseconds represented by the media in a packet). If no ptime is provided for a codec, then the packetization shall follow the rules of the application (e.g. IETF RFC 3551 for RTP/AVP).
- rtpmap. A dynamic payload type definition in the m-line is defined through RTP mapping. The RTP mapping defines the payload type in terms of codec type and clock rate (sampling rate).
- fmtp. A media type may require additional format information. An example is the following:

  a = rtpmap:101 telephone-event/8000
  a = fmtp:101 0–15

  Payload type 101 is dynamically defined as telephone-event (including DTMF tones) at 8000 Hz sampling rate. The fmtp attribute defines that DTMF tones in the range 0–15 may be conveyed (0–15 corresponding to the keypad position on a phone).

The following four attributes define the *stream direction* of a media stream:

- sendrecv. The media stream may be used in both directions.
- recvonly. The media stream may be used only in the receiving direction, relative to the entity that has generated this SDP.
- sendonly. The media stream may be used only in the sending direction, relative to the entity that has generated this SDP. This may be used for a unidirectional media stream such as announcement device.
- inactive. No media shall flow for this stream.

These attributes may be used, among others, to put a media session on hold and, at a later time, to resume the media session.

(Continued)

Table 7.5 (Continued)		
		Whereas these attributes affect an RTP session, the RTCP session associated with the RTP session is not affected. Hence, RTCP messages may still be sent when an RTP session is suspended.
c	Connection information	A media description may have connection information associated with it. This allows for different media transport addresses for the different media components of a multimedia call. For example, different IP addresses may be used for the audio stream and the video stream in a call.

Media Stream Description

A media stream is described essentially through the following descriptor lines (Table 7.5):

- a media descriptor line;
- (optionally) a bandwidth line; and
- a set of attribute lines.

The attributes applicable to a media stream may be defined specifically for that media stream (media-level attribute) or may be defined for the SDP in general (session-level attribute). In the latter case, the attribute applies to all media descriptions in this SDP. For a complete list of media stream descriptor lines, refer to IETF RFC 4566.

The media streams that are defined with the various media lines relate to RTP sessions, carrying media in accordance with the media description. Each RTP session has an RTCP session associated with it. For the RTCP session, there is no special definition included in the SDP. A general rule is that the RTCP session should use the same network connection (IP address) and transport protocol as the corresponding RTP session. A further general rule is that an RTP session should use an even port number and the corresponding RTCP session the next higher (odd) port number. Deviations from this rule are allowed, by explicitly defining the port to be used for the RTCP session. That method is defined in IETF RFC 3605.

7.11.2 Some SDP Examples

We will show a few examples of typical SDP offer–answer message sequences. Only those SIP headers that are relevant for the SDP offer–answer are shown. The calls are established between X-Lite PC-based soft SIP phone and Grandstream GXV3000 desktop-based video phone. Protocol capture is done with Wireshark.

Example 7.5: Basic Voice Call Establishment

```
INVITE sip:7921@test-ims.network.nl;transport=udp SIP/2.0
Content-Type: application/sdp
Content-Length: 464
```

```
Session Description Protocol Version (v): 0
Owner/Creator, Session ID (o): 12937078313661371 1 IN IP4 164.48.50.132
```

Session Name (s): CounterPath X-Lite 4.0
Connection Information (c): IN IP4 164.48.50.132
Time Description, active time (t): 0 0
Session Attribute (a): ice-ufrag:212bd4
Session Attribute (a): ice-pwd:98a0c0f8e76b172993b3f8023166047f
Media Description, name and address (m): audio 60080 RTP/AVP 107 0 98 8 102 101
Media Attribute (a): rtpmap:107 BV32/16000
Media Attribute (a): rtpmap:98 iLBC/8000
Media Attribute (a): rtpmap:102 L16/16000
Media Attribute (a): rtpmap:101 telephone-event/8000
Media Attribute (a): fmtp:101 0-15
Media Attribute (a): sendrecv
Media Attribute (a): candidate:1 1 UDP 659136 164.48.50.132 60080 typ host
Media Attribute (a): candidate:1 2 UDP 659134 164.48.50.132 60081 typ host

...
SIP/2.0 200 OK
Content-Type: application/sdp
Content-Length: 157

Session Description Protocol Version (v): 0
Owner/Creator, Session ID (o): 7921 8000 8000 IN IP4 164.48.50.245
Session Name (s): SIP Call
Connection Information (c): IN IP4 164.48.50.245
Time Description, active time (t): 0 0
Media Description, name and address (m): audio 5004 RTP/AVP 0
Media Attribute (a): sendrecv
Media Attribute (a): rtpmap:0 PCMU/8000
Media Attribute (a): ptime:20

In this example, the calling party offers in the INVITE request six media formats from the RTP/AVP application:

Payload type 107   BV32 (dynamic payload type) at 16 kHz sampling
Payload type 0   PCM μ-law (static payload type)
Payload type 98   iLBC (dynamic payload type) at 8 kHz sampling
Payload type 8   PCM A-law (static payload type)
Payload type 102   L16 (dynamic payload type) at 16 kHz sampling
Payload type 101   Telephone event at 8 kHz sampling; phone keys 0–15

The SDP answer in the 200 OK indicates that the following codecs are accepted:
Payload type 0 PCM μ-law (static payload type).

Example 7.6: Basic Voice and Video Call Establishment

In this example, we show only the SDP offer and answer.

SDP offer
>   Session Description Protocol Version (v): 0
>   Owner/Creator, Session ID (o): 12937080002117200 1 IN IP4 164.48.50.132
>   Session Name (s): CounterPath X-Lite 4.0
>   Connection Information (c): IN IP4 164.48.50.132
>   Time Description, active time (t): 0 0
>   Session Attribute (a): ice-ufrag:d61a3a
>   Session Attribute (a): ice-pwd:7b9090627e1a0d3c065c5d7c634bd306
>   Media Description, name and address (m): audio 60056 RTP/AVP 107 0 98 8 102 101
>   Media Attribute (a): rtpmap:107 BV32/16000
>   Media Attribute (a): rtpmap:98 iLBC/8000
>   Media Attribute (a): rtpmap:102 L16/16000
>   Media Attribute (a): rtpmap:101 telephone-event/8000
>   Media Attribute (a): fmtp:101 0-15
>   Media Attribute (a): sendrecv
>   Media Attribute (a): candidate:1 1 UDP 659136 164.48.50.132 60056 typ host
>   Media Attribute (a): candidate:1 2 UDP 659134 164.48.50.132 60057 typ host
>   Media Description, name and address (m): video 56076 RTP/AVP 34 115
>   Media Attribute (a): rtpmap:34 H263/90000
>   Media Attribute (a): fmtp:34 QCIF=1;CIF=1;VGA=2
>   Media Attribute (a): rtpmap:115 H263-1998/90000
>   Media Attribute (a): fmtp:115 QCIF=1;CIF=1;VGA=2;I=1;J=1;T=1
>   Media Attribute (a): sendrecv
>   Media Attribute (a): candidate:1 1 UDP 659136 164.48.50.132 56076 typ host
>   Media Attribute (a): candidate:1 2 UDP 659134 164.48.50.132 56077 typ host

SDP answer
>   Session Description Protocol Version (v): 0
>   Owner/Creator, Session ID (o): 7921 8000 8000 IN IP4 164.48.50.245
>   Session Name (s): SIP Call
>   Connection Information (c): IN IP4 164.48.50.245
>   Time Description, active time (t): 0 0
>   Media Description, name and address (m): audio 5004 RTP/AVP 0
>   Media Attribute (a): sendrecv
>   Media Attribute (a): rtpmap:0 PCMU/8000
>   Media Attribute (a): ptime:20
>   Media Description, name and address (m): video 5006 RTP/AVP 34
>   Media Attribute (a): sendrecv
>   Media Attribute (a): rtpmap:34 H263/90000
>   Media Attribute (a): fmtp:34 CIF=3 MaxBR=3840
>   Media Attribute (a): framerate:10

The offered audio media formats and the accepted media formats are identical to those for the voice call. For the video component of this call, the following codecs are offered:

Payload type 34	H.263 (static payload type)
Payload type 115	H.263-1998 (dynamic payload type) at 90 kHz clock rate

For both offered codecs, a set of additional qualifiers is provided, in the form of format attributes (a=fmtp:34 QCIF=1; …).

The accepted video codecs are:

Payload type 34	H.263 (static payload type).

Introduction to the IMS Network

8.1 INTRODUCTION

This chapter provides a methodological description of the IMS network architecture. IMS service development requires thorough understanding of the principles and protocols of multimedia over IP and of the architecture of the IMS network. Principles and protocols of multimedia over IP have been explained in previous chapters. This chapter builds on that; it focuses on the principles and protocols in applied architecture, and on the IMS network, and explains the reasoning behind the IMS architecture. This involves functional decomposition of the IMS network into the core network (control plane, user plane, Location register), media plane (media transmission), access network (mobile, wireless, wireline), and application layer.

After explaining the IMS network architecture and protocols, the chapter then explains IMS procedures, including registration, session establishment, subscriber mobility, messaging, service invocation, and session control. The focus is on how these procedures relate to application development.

8.2 OVERVIEW OF IMS STANDARDS AND RELEASES

Let's first take a look at the formal documents specifying the IMS network. IMS was introduced in 3GPP Release 5 (Rel-5) as the standard *multimedia communication system*. After its introduction in Rel-5, 3GPP has continually added new functionality to IMS in subsequent releases.

IMS is fully specified in 3GPP technical specifications (TSs). TSs may include reference to standard documents from other standardization organizations. Those other standard documents then become, in full or in part, an integral part of the IMS specification. One prominent example is the reference to IETF RFCs. Table 8.1 provides an overview of the main 3GPP TSs that specify IMS. Some of these TSs are IMS specific; others also specify functionality relating to other systems, such as the 3G mobile network.

In various sections in this book, we will refer to these or other 3GPP TSs, where applicable. A full listing of IMS-related 3GPP TSs can be obtained from www.3gpp.org.

Table 8.2 provides an overview of some of the IETF RFCs that are referred to in the 3GPP IMS specifications. A full list of IETF RFCs can be obtained from www.ietf.org.

Table 8.1 3GPP Technical Specifications for IMS (not exhaustive)

3GPP TS	Title and Description
22.228	Service requirements for IMS; Stage 1
	Describes the service requirements based on which the IMS network architecture and protocols are defined.
22.340	[Rel-6] IP Multimedia Subsystem (IMS) messaging; Stage 1
	Describes the service requirements for IMS-based messaging.
23.002	Network architecture
	Describes total network architecture for the "3GPP System", including IMS. Describes the interfaces (reference points) between the respective entities in the 3G network.
23.003	Numbering, addressing, and identification
	Describes the formats of the various identifiers used in IMS.
23.008	Organization of subscriber data
	Describes subscription data for, among others, IMS services.
23.167	[Rel-7] IP Multimedia Subsystem (IMS) emergency sessions
	Describes how emergency sessions may be established from SIP terminals.
23.218	IP multimedia (IM) session handling; IM call model
	Describes the session establishment within IMS, specifically the processes in S-CSCF.
23.228	IP Multimedia Subsystem; Stage 2
	Describes network architecture, subscription data, and session procedures.
24.229	IP multimedia call control protocol based on SIP and SDP; Stage 3
	Describes how IMS uses SIP for session establishment and other services.
29.228	IMS Cx and Dx interfaces; signaling flows and message contents
	Describes the usage of Diameter between CSCF and HSS, as well as other Diameter-based interfaces.
29.229	Cx and Dx interfaces based on the Diameter protocol; protocol details
	Describes the protocol details of the Diameter messages described in 3GPP TS 29.228.
32.260	IP Multimedia Subsystem (IMS) charging
	Describes the charging principles of IMS

The indication [Rel-x] in this table indicates that that TS was introduced in that 3GPP release.

Table 8.3 provides an overview of some of the main IMS-related functional enhancements introduced in the respective 3GPP releases.

Where applicable, we refer in the text to a specific 3GPP release for designated IMS functionality.

8.3 IMS NETWORK ARCHITECTURE – A GLOBAL VIEW

This section introduces the IMS network architecture. Figure 8.1 shows the architecture of the IMS network as defined in 3GPP technical specifications.

Rather than throwing you, the poor reader, in at the deep end or into the dark (bombarding you with boxes and reference points), we will provide a stepwise introduction to the

Table 8.2 IETF RFCs Referenced by IMS (not exhaustive)

IETF RFC	Title and Description
3261	Session Initiation Protocol
3262	Reliability of provisional responses in SIP
3263	Locating SIP servers
3264	Offer–answer model with the Session Description Protocol
3323	Privacy mechanism for Session Initiation Protocol
3325	Private extensions to Session Initiation Protocol for asserted identity within trusted networks
3326	Reason header field for the Session Initiation Protocol
3455	Private header (P-header) extensions to SIP for 3GPP
3588	Diameter Base Protocol
3608	SIP extension header field for Service-Route discovery during registration
3841	Caller preferences for the Session Initiation Protocol
4028	Session timers in the Session Initiation Protocol
4240	Basic network media services with SIP
4244	Extension to Session Initiation Protocol for request history information
4457	SIP P-user-database private header
5002	P-profile-key private header
5009	Private header (P-header) extension to SIP for authorization of early media
5502	P-served-user private header for the IMS network

Table 8.3 IMS Functionality for 3GPP Releases

Release	Year[a]	Functional Description
Rel-5	2001	Introduction of IMS
Rel-6	2003	IMS emergency services
		Combinational services
		Voice call continuity
Rel-7	2005	Single radio voice call continuity (SR-VCC)
		Multimedia Telephony
Rel-8	2007	IMS centralized services
		IMS service continuity
		Multimedia interworking between IMS and CS networks
		IMS Multimedia Telephony and supplementary services
Rel-9	2009	IMS emergency calls over GPRS and enhanced packet system (EPS)
		Enhancements of IMS customized alerting tone service
		IMS restoration procedures
Rel-10	2010	IMS service continuity – inter-device transfer enhancements

[a]The indicated year is an approximate date of release of the first baselined version of the IMS specifications of that 3GPP release.

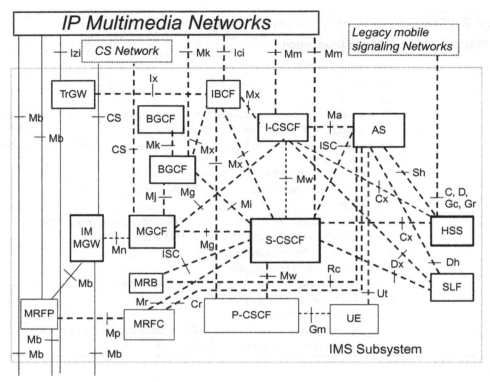

FIGURE 8.1

IMS network architecture...

Source: 3GPP TS 23.228 v8.11.0, figure 4.0. © 2010 3GPP™ TSs and TRs are the property of ARIB, ATIS, CCSA, ETSI, TTA and TTC who jointly own the copyright in them. They are subject to further modifications and are therefore provided to you "as is" for information purposes only. Further use is strictly prohibited

various components (functional entities) from which the IMS network is built up. We will apply the following approach:

- Logical separation of the IMS network into:
  - User equipment (UE)
  - Access network
  - Core network
  - Application layer.
- Strict separation between control plane and user plane.

This section restricts itself to giving an overview of the high-level architecture. Detailed descriptions of the various functional entities are given in subsequent sections.

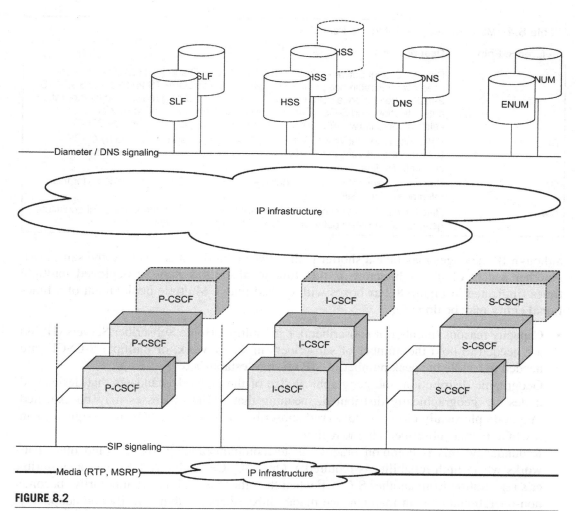

FIGURE 8.2

IMS core network main functional entities.

8.3.1 IMS Core Network

The IMS core network consists of a limited number of functional entities (see Figure 8.2). Figure 8.2 shows the most important functional entities for basic SIP session establishment. Other entities, fulfilling a specific role, are dealt with separately. Functional entities (proxy call state control function, P-CSCF, etc.) constitute applications, deployed on an IP host, with the IP hosts connected to the operator's IP infrastructure. One host may contain more than one functional entity, i.e. functional entities may be co-located. A typical example is that S-CSCF, P-CSCF, and I-CSCF are co-located. Co-location does not fundamentally alter the signaling in the network; the same procedures and signaling are applied,

Table 8.4 IMS Core Network Reference Points

Reference Point	Description
Cx	Reference point between CSCF (I-CSCF or S-CSCF) and the HSS. The protocol used over this reference point is Diameter. See also 3GPP TS 29.228 and 3GPP TS 29.229. The reference point is used, among others, during registration (authorization, authentication, and S-CSCF assignment) and during terminating session establishment (forwarding session establishment request to correct S-CSCF).
Dx	Reference point between CSCF (I-CSCF or S-CSCF) and SLF; used by CSCF to obtain the address of the HSS containing (or having access to) a particular subscriber's data.
Gm	Reference point between user equipment (UE) and P-CSCF. The protocol for this reference point is SIP.
Mw	Reference point between a CSCF and another CSCF. This reference point constitutes general SIP signaling between CSCFs in the IMS core network.

although IP messages may take a shorter path between entities that reside in the same host. Another aspect of the architecture is that functional entities may be deployed multiple times (indicated in Figure 8.2 as boxes with dashed lines). Multiple deployment of a functional entity may be done for:

- Capacity reasons (number of subscribers), e.g. multiple Home Subscriber Servers (HSSs) are needed to hold the number of subscribers in the network, or multiple S-CSCFs are needed to handle the peak number of SIP sessions established.
- Geographic distribution; the geographical size of the network mandates that designated nodes are geographically distributed, meaning that traffic processes may be executed in a node physically close to the served subscriber (e.g. subscriber being assigned to an S-CSCF in that subscriber's home region).
- Redundancy; this is common practice in telecommunication networks (and other networks where high reliability is required[1]). One S-CSCF is able to take over a subscriber's registration from another S-CSCF when the other S-CSCF has (temporarily) become non-operational or is in maintenance mode. Subscribers are dynamically and adaptively assigned to an available S-CSCF during registration.

The various entities shown in Figure 8.2 have interfaces with other specific entities, which are interfaces known as "reference points". Showing all reference points would clutter up the diagram (such as in Figure 8.1), so only the main ones have been illustrated. Table 8.4 lists some relevant reference points for the IMS core network. The table focuses on reference points on the control plane. For a complete overview of IMS reference points, refer to 3GPP TS 23.002 (network architecture) and 3GPP TS 23.228 (IP Multimedia Subsystem).

Various other reference points will be described in designated sections, such as reference points between the IMS core network and border nodes.

---

[1] Uptime of 99.999%, also referred to as 9^5 or "five nines".

It is not possible to reflect the "typical" message flow in an illustration of the IMS architecture, such as SIP signaling between CSCFs and Diameter signaling between CSCF and HSS. Different call cases have their own message sequence associated with them and will involve different functional entities. All entities are connected to the IP infrastructure (details not shown).

The CSCFs "talk SIP" to each other (Mw reference point), as well as to access gateways and network gateways. The I-CSCF and S-CSCF "talk Diameter" with SLF (Dx reference point) and HSS (Cx reference point). These reference points (related to SIP and Diameter) will be described in detail in subsequent sections.

Figure 8.2 shows DNS. All entities in the IMS network require services from DNS in order to communicate with other entities. "DNS" in the IP infrastructure generally represents *access to global DNS*. DNS clients access DNS through a local DNS server. That local DNS server has a functional connection with other DNS server(s). Obtaining resource records from DNS is an iterative process. A DNS server may forward the DNS query from a DNS client to another DNS server of another domain, etc.

A DNS server in an IMS network may not give unconditional access to global DNS for entities in the IMS network. DNS service for entities in an IMS network may be restricted to those resource records that are needed for communication within the IMS network. When communication is to be sent from an entity in the IMS network to an entity outside the IMS network, then this communication will typically be routed via a border gateway (e.g. an interconnect border control function, IBCF, for IMS interconnect). The border gateway will have access to resource records outside its own IMS network. This distinction may be made through *internal DNS* versus *external DNS*.

Figure 8.2 further shows the media plane through a single line, representing the transmission of RTP-based media and Message Session Relay Protocol (MSRP). RTP is used in this figure as representative of a common media stream for communication over IP. MSRP is used for session based message transfer. Within the IMS core network, media transfer between two IMS users runs directly end to end, without traversing intermediate media proxies. At the border of an IMS network, however, media is routed through a media proxy. In addition, media streams may be routed through special media handlers, for example when media transcoding is required.

8.3.2 IMS Access Network

IMS is, in principle, "access independent". This means that an IMS network can be accessed through IP carrier access network (IP-CAN) of different types. The IP-CAN provides the IP connectivity for the terminal, as well as mobility. The IMS terminal applies control-plane signaling and media transfer to the IMS core network through the IP-CAN. This is seen in Figure 8.1, which shows a single interface between user equipment (UE) and P-CSCF, the Gm reference point. The Gm reference point represents SIP signaling between UE and P-CSCF. For data transfer to and from UE (user plane), the Mb reference point is defined.

The Mb reference point is formally specified only for IPv6. However, IPv4 may be used as well. The path followed by the media to and from UE depends on the access network. General principles for media transfer are:

- The media path will be as short as possible.
- The media will, as far as possible, be transferred end to end, without being routed through media proxies.

When IMS was introduced in 3GPP Rel-5, it was specified for use with the 3G mobile network. The 3G access network has evolved toward HSPA and LTE and, hence, HSPA and LTE may be used for access to IMS. 3G access (UMTS) does not provide sufficient capability for mobile voice based on IMS. However, it does provide the capability for IMS services such as messaging, presence, and Combinational services, such as Rich Communication Suite (RCS).

Over time, the use of other access networks for IMS has been specified as well, such as Ethernet LAN (in combination with ADSL), wireless LAN (in combination with ADSL), fiber to the home (FTTH), and cable.

The 3G (also HSPA, LTE) access network is fully specified by 3GPP. Other access networks use technology specified by other organizations, such as Ethernet, which is standardized by IEEE (IEEE standard 802.3), and ADSL, which is standardized by ITU (ITU-T recommendation G.992.1).

We will look briefly at wireline access as one example (see Figure 8.3). In the example in Figure 8.3, SIP terminals are connected to a corporate local area network (LAN), possibly to a metropolitan area network (MAN) or wide area network (WAN). Access to the IMS network is done through (public) IP infrastructure. SIP signaling to and from the SIP terminal runs via the P-CSCF, i.e. the P-CSCF is the point of contact for the SIP terminal. When a

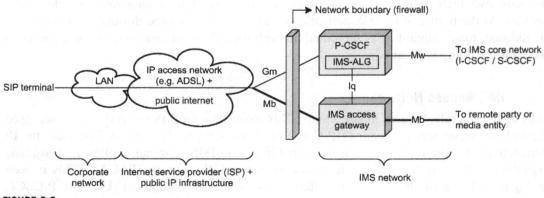

FIGURE 8.3

Wireline access to IMS.

SIP session is established, SIP signaling and media will follow different paths into the IMS network. The SIP signaling will be directed to P-CSCF and media will be directed to the IMS access gateway. The traversing of the media through the IMS access gateway is under control of the P-CSCF. The IMS application level gateway (IMS-ALG) may be linked in the control plane when the access network and core network use different IP versions or when the SIP terminal is using a local IP address.

Access to the IMS network for wireline access network is commonly done through an access session border gateway (A-SBG). This is reflected in Figure 8.4.

Session border gateway (SBG) is not formal IMS terminology, but is nevertheless commonly used. A-SBG comprises the following functional entities:

- **Session border controller (SBC).** The SBC is the control-plane entity of the A-SBG. It is the "SIP entry" for the end-user terminal towards the IMS network. It may perform various tasks on SIP signaling, e.g. validating and correcting SIP headers, modifying SIP headers to suit the IMS operator's requirements, adapting SIP headers to cater for known anomalies of certain SIP terminals. The SBC may also enforce frequent periodic registration by the SIP terminal, to keep pinholes open at the end-user's host. Additional tasks include topology hiding (hiding network addressing information between user and network), access security, such as transport layer security (TLS), and SIP firewall.
- **Session gateway (SG).** The SG is the user-plane entity of the A-SBG. The A-SBG may verify that for every media session traversing the SG, there is a corresponding SIP session. The media that is steered through the SG may be subject to bandwidth management. Also, RTP media streams may contain confidential information (carried in the RTP message header). The SG will, under control of the SBC, remove confidential information, subject to operator policy.

A-SBG may also contain P-CSCF. The IMS-ALG functionality that may be included in P-CSCF is in that case performed by the SBC. A-SBG is often very flexible and offers the operator a wide range of configuration options.

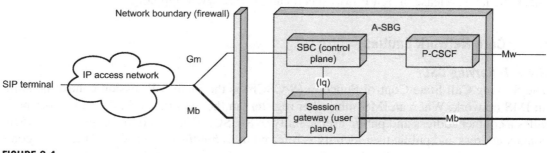

FIGURE 8.4

Access session border gateway (A-SBG).

As an implementation option, the user-plane component of the A-SBG may be separated from the control-plane component and placed on a different host. In that manner, operators may apply separate dimensioning (number of nodes and capacity per node) for the SBC and the SG, depending on the number of subscribers and the expected traffic. Whereas one SBC may, in this separated constellation, control multiple SGs, one SG is controlled by only one SBC. One reason for this is that an SBC needs to keep track of the available resources in the SG. An SG may be configured into multiple *logical SGs*. Each logical SG is controlled by just one SBC.

When A-SBG and P-CSCF are deployed as separate entities, this has impact on the ability to establish an IMS security association between UE and the IMS network. As the A-SBG normally forms the "entry point" into the IMS network, it makes sense to establish the IMS security association between UE and A-SBG, and not between UE and P-CSCF. However, if P-CSCF and A-SBG are separate nodes, then it is not possible to establish an IMS security association between UE and A-SBG. To have an IMS security association between UE and A-SBG, the IMS security association-related information contained in the 401 Unauthorized from S-CSCF would have to be forwarded from P-CSCF to A-SBG. P-CSCF functionality is then, if you will, split over P-CSCF and A-SBG. With separate A-SBG and P-CSCF, an operator may therefore apply TLS-based security (between UE and A-SBG) instead of establishing an IMS security association. TLS requires the use of TCP or SCTP as transport protocol.

8.4 IMS NETWORK ARCHITECTURE – A CLOSER LOOK

In this section, we take a closer look at the IMS network architecture and describe the individual functional entities. Here, we focus on entities that reside on OSI layer 5 and higher, i.e. applications on top of TCP/IP or UDP/IP. Hence, IP infrastructure is not shown. DNS is not specifically shown either, as DNS is required for all IP services and therefore forms part of the IP infrastructure.

As the IMS network standard is continuously evolving, it is recommended that the reader check the latest release of 3GPP IMS specifications (see www.3gpp.org).

8.4.1 Core Network Entities

8.4.1.1 Serving CSCF

The Serving Call State Control Function (S-CSCF) is the main SIP session control node in an IMS network. When an IMS subscriber registers in the IMS network, the subscriber provides a contact address and public user identity to S-CSCF. The functional coupling between contact address and public user identity is referred to as *binding*. The S-CSCF therefore constitutes the registrar (as defined in IETF RFC 3261) for the subscriber. There may be multiple S-CSCFs in the IMS network. When a subscriber undertakes initial registration in the

IMS network (i.e. at that moment there is no binding in an S-CSCF for that subscriber), then an S-CSCF needs to be selected for that subscriber. This selection is done by I-CSCF, as described later. Once an S-CSCF is assigned to a subscriber, the S-CSCF's address is stored in HSS. When an initial SIP request needs to be sent to that subscriber's S-CSCF, a query is sent to the HSS, to obtain the S-CSCF address.

Figure 8.5 shows the interfaces (reference points) between S-CSCF and other entities. Charging-related reference points are not shown in the figure.

When an IMS subscriber establishes a SIP session (originating SIP session), the signaling relating to that SIP session will traverse the S-CSCF assigned to that subscriber. (One exception is emergency calls; these may be handled by an emergency CSCF, E-CSCF, instead of by an S-CSCF). The S-CSCF acts as SIP proxy. The S-CSCF may apply various services to the SIP session establishment from this subscriber, including:

- Authentication (during registration)
- Invocation (also known as *triggering*) of an external service (residing in SIP-AS), through the ISC reference point
- Number normalization (when the SIP request is addressed to a phone number)
- Phone number to SIP URI mapping, with the aid of the E.164 number database (ENUM)
- Outbound routing, including finding the serving IMS network for the destination of the SIP request
- Generating charging record(s) for the SIP session.

The above-listed actions also apply when the IMS subscriber initiates a stand-alone SIP transaction, i.e. a SIP transaction that does not lead to the establishment of a SIP session – for example, MESSAGE or OPTIONS.

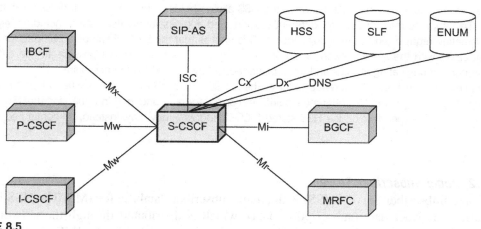

FIGURE 8.5

Reference points for S-CSCF.

When a SIP request message arrives for an IMS subscriber (terminating the SIP session), the signaling relating to that SIP session will, again, traverse the S-CSCF assigned to that subscriber. The S-CSCF acts as SIP proxy and applies the following services:

- Invocation of external service (residing in SIP-AS), through the ISC reference point
- Building a target set of contact addresses to which the SIP request will be sent
- Forwarding the SIP request to the contact address(es) present in the target set
- Generating charging record(s) for the SIP session.

As long as a subscriber remains registered with at least one contact address (binding), that subscriber remains assigned to the S-CSCF. When the last binding is removed from the S-CSCF for that subscriber, the subscriber is no longer assigned to any S-CSCF. A subsequent (temporary) registration by that subscriber may be directed to another S-CSCF.

The S-CSCF is always located in the Home IMS network. The "Home IMS network" is the IMS network where the subscriber has a subscription. The concept of *IMS roaming* has a different meaning than *GSM roaming*. When an IMS subscriber is establishing multimedia sessions from a remote location, the session signaling (SIP signaling) is always routed through the S-CSCF in that subscriber's Home IMS network.

The different protocols used to and from S-CSCF may use different port numbers, to differentiate between these protocols. The SIP signaling to and from S-CSCF, however, will typically use the same port number. Designated SIP header parameters are used to indicate to S-CSCF what kind of session case the signaling relates to, such as originating (SIP session originated by IMS subscriber) or terminating (SIP session established toward IMS subscriber).

Note

When an S-CSCF is assigned to a subscriber, the HSS always holds a single S-CSCF address for that subscriber. This address may be used for different session- and non-session-related cases, including registration, originating request, and terminating request. This implies that the S-CSCF will be addressed for these different cases with the same network and transport address. It is therefore not possible to use different port numbers for originating request and terminating request over the Mw reference point.

Signaling through the ISC reference point at the S-CSCF may, however, use a different port number than signaling through the Mw reference point. Transport address and port number for the ISC reference point are not dependent on the HSS stored S-CSCF address; they may, instead, be configured in the S-CSCF.

8.4.1.2 Home Subscriber Server

The Home Subscriber Server (HSS) is the main subscriber database for IMS. The HSS holds the static subscriber data (subscription data), which is distributed through the network, to designated functional entities (nodes). When a subscriber registers in the IMS network, subscription data are retrieved from HSS, by the S-CSCF that is assigned to that subscriber.

Some of the subscription data are also sent from S-CSCF to P-CSCF during the registration process (see Figure 8.6).

Strictly speaking, the HSS provides *access to subscriber data*. It does not necessarily have to *contain* the subscriber data. An HSS may act as *front-end* to a centralized subscriber database. A multitude of HSS front-ends may access that centralized subscriber database.

The arrow labeled (1) represents transfer of subscription data to S-CSCF, such as IMS public user identities (IMPU) and service trigger data (initial filter criteria, IFC). The IMPUs that are sent from the HSS to S-CSCF are public identifiers that may be used to identify the subscriber for originating and terminating multimedia sessions. Arrow (2) represents data that are sent from HSS to P-CSCF, via S-CSCF. These data comprise a set of IMPUs (which is also stored in S-CSCF) and charging subscription data. Arrow (3) represents the data that are forwarded to the UE. This comprises the set of IMPUs that is also kept in S-CSCF and P-CSCF. Arrow (4), finally, comprises data that are associated with the service subscription data (IFC) and that are included in the service invocation message to SIP-AS.

When an S-CSCF is assigned to a subscriber, the HSS stores the S-CSCF address (constituting dynamic subscriber data). SIP session establishment requests and stand-alone SIP transactions destined for a subscriber are routed via an I-CSCF. I-CSCF contacts the HSS, whereupon the HSS can inform I-CSCF about the address of the S-CSCF where the subscriber is registered.

The HSS may be used for *transparent* subscription data and *non-transparent* subscription data. Transparent subscription data are data that are not standardized; they are application specific. A SIP-AS may, for example, use the HSS for permanent storage of dynamic subscriber data. SIP-AS obtains the subscriber data from the HSS when the subscriber registers in the IMS network and in the SIP-AS. The reference point between SIP-AS and HSS is labeled Sh.

Non-transparent subscriber data are formed by standardized information elements, which may be transported over the Sh reference point by means of standardized parameters. The HSS has knowledge of the structure of the non-transparent subscriber data.

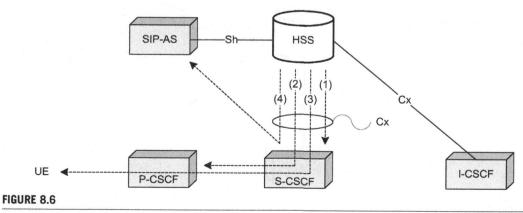

FIGURE 8.6

Reference points for HSS.

8.4.1.3 Subscriber Locator Function (SLF)

An IMS network may have multiple HSSs deployed, with each HSS serving a group of sub-scribers of that IMS network. Subscribers are distributed over these multiple HSSs. When an entity such as I-CSCF needs to contact the HSS, to obtain subscriber information, it would need to know which HSS to contact. To help entities to find the HSS for a particular sub-scriber, a subscriber locator function (SLF) may be used. SLF is a database containing a record of all subscribers in the operator's IMS network. Each subscriber record in SLF con-tains a reference (pointer) to the HSS for that subscriber. Alternatively, SLF uses a hash algo-rithm to determine the HSS for a subscriber. Connections to SLF are shown in Figure 8.7.

An IMS network may have multiple SLF nodes deployed, for redundancy and for load distribution. An entity using the services from SLF has the address of the SLF entity/entities (e.g. primary SLF and secondary SLF) configured. When multiple SLFs are deployed in an IMS network, then each SLF will have an entire copy of the set of subscriber records. The subscriber data may be located in an external database, as opposed to being contained in the SLF itself. The SLF then merely *provides access* to that subscriber data.

The protocol used over the Dx reference point (CSCF–SLF) and over the Dh reference point (SIP-AS–SLF) is Diameter. A designated Diameter message may be used by CSCF or SIP-AS to query SLF.

The subscriber database in SLF typically uses IMPU as search key: queries from CSCF or SIP-AS for a particular subscriber will always be based on (one of the) IMPU(s) from that subscriber.

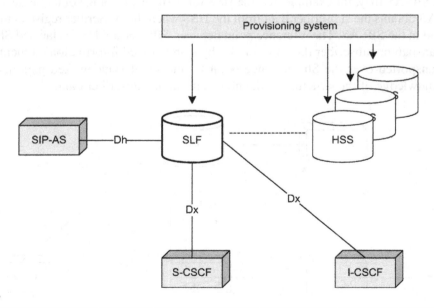

FIGURE 8.7

SLF reference points.

8.4.1.4 P-CSCF

The proxy call state control function (P-CSCF) is the user-to-network proxy. All SIP signaling to and from the end-user runs via a P-CSCF of the IMS network (Home IMS network or Visited IMS network; see below) – see Figure 8.8. Not all reference points of the P-CSCF are shown.

When a subscriber registers in the IMS network with a particular terminal, the registration signaling will traverse a P-CSCF. When the subscriber registers through more than one terminal, then these multiple terminals may register through different P-CSCFs (but the registrations will be done at the same S-CSCF for the subscriber). Some of the tasks of the P-CSCF are applying *assertion* of SIP headers received from a terminal and establishing a security association. When a subscriber registers, the subscriber provides, among others, identification of him or herself, e.g. sip:john.smith@abc-company.se. This identification is (one of) the IMPU(s) for that subscriber. In addition, the subscriber provides an IMPI, an identifier of the (type of) IMS terminal. This registration process will be subject to authentication by S-CSCF (for selected registration cases, authentication may be disabled). When the registration process is complete, the P-CSCF receives from the HSS, via S-CSCF, a set of IMPUs for this subscriber. When the subscriber subsequently establishes a SIP session, the P-CSCF uses the default IMPU, of the set of IMPUs, as an *asserted* identity of the subscriber, the *P-asserted-identity* (PAI). The PAI is included in the SIP request that is sent to the IMS network; the PAI then serves as network-asserted (and hence trustworthy) indication of the party sending the SIP request. The subscriber can select another one of the available IMPUs to be used as PAI. This may be done by including a P-preferred-identity header in the SIP INVITE request or other SIP request message. The use of the P-preferred-identity is explained in a later section.

The P-CSCF may establish a security association with the UE, using security association information received from the HSS, via S-CSCF. All SIP signaling will subsequently be transferred over this security association. As explained, instead of having a security association between UE and P-CSCF, TLS between UE and A-SBG may be used.

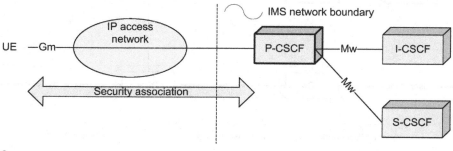

FIGURE 8.8

P-CSCF reference points.

The P-CSCF resides in the Home IMS network or in a Visited IMS network. The latter may be the case when registration is initiated from an IP access network in a foreign location/country. An application in the terminal may select a local P-CSCF for registration, based on the location (e.g. IP address) from where the registration is initiated (see Figure 8.9).

In the case of P-CSCF in a Visited IMS network, referred to as IMS roaming, the interconnect border control function (IBCF) and transition gateway (TrGW) will be used at the border of the respective IMS networks. IBCF and TrGW will be described in more detail in a subsequent section. When the P-CSCF in the Visited IMS network addresses I-CSCF or S-CSCF in the Home IMS network, the SIP signaling will be force routed via IBCF of the respective IMS network. This force routing may be based on, for example, internal vs. external DNS. The respective IBCFs may, in turn, force the media to be routed via TrGW. This has the effect that the media stream is routed via the Home IMS network as well.

8.4.1.5 I-CSCF

The Interrogating Call State Control Function (I-CSCF) is used for forwarding an initial SIP request to S-CSCF when the initiator of the request does not know which S-CSCF should receive the SIP request. Generally, I-CSCF contacts the HSS to obtain the address of the S-CSCF that will receive and process the SIP request (see Figure 8.10).

The protocol used over the Cx and Dx reference points is Diameter. An initial SIP request is routed via I-CSCF to S-CSCF in the following main use-cases:

1. During registration. P-CSCF forwards a registration request (SIP REGISTER message) to I-CSCF, whereupon I-CSCF contacts the HSS, to find out which S-CSCF should receive the SIP REGISTER message.
2. During SIP session establishment or stand-alone SIP transaction (other than REGISTER). The SIP request is sent to I-CSCF, whereupon I-CSCF contacts the HSS to find out which S-CSCF should receive the SIP request message.

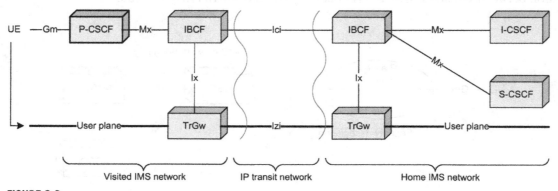

FIGURE 8.9

IMS roaming.

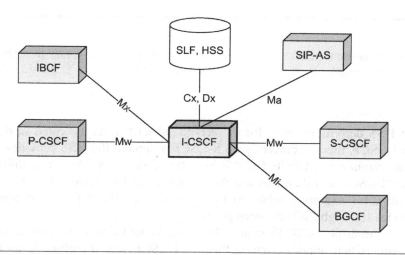

FIGURE 8.10

I-CSCF reference points.

The second of the above-listed main set of use-cases, SIP session establishment or stand-alone SIP transaction, may be further subdivided as follows:

a. Terminating request routing. A SIP request message is routed to I-CSCF of the serving IMS network. The "serving IMS network" is derived from the domain of the destination (Request URI, R-URI) of the SIP message. I-CSCF contacts the HSS based on the R-URI of the request message. The HSS provides I-CSCF with the address of the S-CSCF that is currently assigned to the indicated subscriber. For example:

```
INVITE sip:john.smith@abc-company.se SIP/2.0
```

I-CSCF requests the HSS to provide the S-CSCF address related to sip:john.smith@abc-company.se.

b. SIP-AS originated request routing. In this case, the request message relates to the originating SIP session establishment or stand-alone transaction. I-CSCF should in this case obtain from HSS the address of the S-CSCF serving the *originating* party, as opposed to the *destination* party. This requirement on I-CSCF is indicated in the Route header. For example:

```
INVITE sip:john.smith@abc-company.se SIP/2.0
Route: sip:icscf.ims-operator.se;lr;orig
P-asserted-id: sip:wendy.jones@my-company.se
```

The "orig" parameter in the Route header is, for example, added by a SIP-AS during establishment of the call. I-CSCF requests the HSS to provide the S-CSCF address relating to sip:wendy.jones@my-company.se. Originating session routing through I-CSCF may occur when the session is, for example, initiated by a SIP application server that does not have knowledge of the address of the S-CSCF of the calling party.

Note

For originating SIP session establishment or stand-alone transaction handling through I-CSCF, the P-served-user header might be used instead of P-asserted-identity. This is, however, not yet formally standardized.

When the HSS determines that, for the served subscriber, as indicated in the Diameter request from I-CSCF, there is currently no S-CSCF assigned, the HSS provides an *S-CSCF capabilities description* to I-CSCF. I-CSCF may, based on the S-CSCF capabilities description, select an S-CSCF and then forward the request to that selected S-CSCF. The S-CSCF capabilities may, for example, relate to the capability of the S-CSCF to invoke a SIP-AS-based service (through ISC reference point).

There may be multiple I-CSCFs in an IMS network. An I-CSCF does not hold subscriber data. A SIP request that requires routing through I-CSCF may therefore be routed through any of the available I-CSCFs in the network. Incoming SIP request messages from other IMS networks will arrive at the IBCF, from where they will be forwarded to an I-CSCF.

8.4.1.6 DNS

DNS in the IMS network has a special role. All entities in the IMS network that are addressable through SIP signaling (or other IP-based services) need to be defined in DNS, in so far as *domain names* are used for addressing these entities. For the various entities in the IMS network, the following resource records are defined in DNS:

- Naming Authority Pointer (NAPTR) record (IETF RFC 3403). NAPTR records are used to obtain an overview of supported IP services for a particular domain.
- Service (SRV) record (IETF RFC 2782). SRV records are used to obtain host name(s) and port number(s) for a specific service of a domain (a service received in response to the NAPTR query).
- A record/AAAA record (IETF RFC 3404). A/AAAA records are used to obtain IP address(es) for a particular host (a host received in response to the SRV query).

Most entities in the IMS network, such as S-CSCF, HSS, MRFC, MRFP, and I-CSCF, would not need to be accessible by entities *outside* the IMS network. These entities need to be addressable only by other entities *inside* the IMS network. When the DNS node in the IMS network is accessed with a request for information (NAPTR record, SRV record, or A/AAAA record), DNS will behave as *internal DNS* (iDNS) or *external DNS* (eDNS). Internal DNS is accessible only by entities within the IMS network; it is used for domain resolution for IP routing within the IMS network. External DNS is also accessible by entities outside the IMS network; it is used for domain resolution for IP routing that crosses the IMS network boundary. Internal DNS and external DNS may be deployed in physically different nodes.

In Figure 8.11 are listed a few entities that are provided in iDNS and eDNS; these lists are not complete.

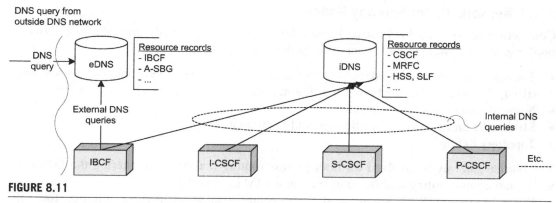

FIGURE 8.11

Internal DNS versus external DNS.

Besides resource records related to nodes in the IMS network, DNS also contains resource records relating to enterprise domains, such as abc-company.se, being the domain part of sip:john.smith@abc-company.se. A domain like abc-company.se does not relate to a host. The domain is "owned" by a company (organization, etc.) and is administered by an IMS operator. The IMS operator offers services (IP multimedia calls, etc.) to users of that domain (user@abc-company.se). When a call is established to sip:john.smith@abc-company.se, the S-CSCF handling this call would need to obtain a NAPTR record for abc-company.se, in order to find the supported SIP services of that domain. If the call to sip:john.smith@abc-company.se is established from outside the IMS network, then the NAPTR record for abc-company.se needs to be accessible via eDNS. The SRV record(s) associated with the supported SIP services of that domain, as well as the A records associated with the hosts supporting those SIP services (for external calls), also need to be accessible via eDNS. The SRV record(s) and A record(s) received through eDNS for this example would typically point to an IBCF. The rationale is that SIP signaling from outside the IMS network should enter the IMS network through IBCF. IBCF may, in turn, use iDNS to obtain resource records needed to forward the SIP request to I-CSCF.

When the call to sip:john.smith@abc-company.se is established from within the IMS network serving the domain abc-company.se, the NAPTR record query for that domain may be directed to iDNS or eDNS. For the former method (using iDNS for this query), the S-CSCF would need to be configured with the supported enterprise domains. In addition, both iDNS and eDNS would need to contain the required NAPTR record for this domain. For the latter method (using eDNS for this query), the NAPTR record for this domain is needed in eDNS only. To resolve this, there may be a functional connection between iDNS and eDNS. iDNS may forward a DNS query to eDNS.

8.4.2 **Network Border Gateway Nodes**

Connection between an IMS network and other networks is generally done through network border gateways. Main tasks for network border gateways comprise:

- Transport and signaling adaptation, including adapting between different IP versions (IPv4, IPv6) and adapting between different transport protocols (UDP, TCP, SCTP)
- Network protection (SIP firewall)
- SIP header and parameter verification and manipulation
- Topology hiding.

The border gateway is involved for both ingress routing (call/session entering the IMS network) and egress routing (call/session leaving the IMS network).

Refer to Figure 8.12 for network architecture related to ingress routing, for SIP interworking.

The interconnect border control function (IBCF) is used for interconnection between an IMS network and another IMS network or non-IMS IP network. The IBCF acts as SIP proxy or as B2BUA. It may remove or adapt certain SIP headers, based on inter-operator agreement. The IBCF may force the user plane of a SIP session to be routed through the IMS network. This is shown in the figure as the transition gateway (TrGW). IBCF has, in this example, adapted the end-point addresses (for RTP and RTCP termination) in the SDP offer and SDP answer, so as to route the media through the TrGW. The combination of IBCF and TrGW is commonly referred to as the network session border gateway (N-SBG).

Topology hiding would normally be applied by IBCF. By applying topology hiding, the IBCF shields any network-internal-topology-related information between the two networks. Various SIP headers may contain topology-related information, such as:

- Via header
  - IP address and port number of SIP proxies
  - Branch-ID (as proprietary implementation option, implementation-specific information, such as process identifier or blade identifier, may be embedded in branch-ID)
- Record-Route header (contains domain name of SIP proxies in the network)
- Call-ID, From tag, To tag (host IP address/domain name is often embedded in these identifiers; in addition, they may contain process identifier or blade identifier).

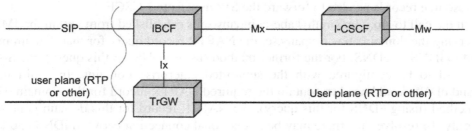

FIGURE 8.12

Network architecture related to ingress routing from another IMS network.

The anchoring of the media in TrGW also has an implicit topology-hiding effect. Without anchoring, the SDP answer provided to the other network would contain the transport address of, for example, a media-plane entity in the operator's network, such as A-SBG or IM-MGW. By media anchoring in TrGW, the SDP answer will contain the transport address of the TrGW. During SIP session establishment, it is preferable to negotiate equal codec(s) between users, so as to prevent the necessity for media transcoding at TrGW (or another entity).

Alternatively, IBCF allows the user plane to be established end to end, without traversing the TrGW. In such an example, the user plane may *theoretically* not traverse the operator's network at all. That would be the case when the destination subscriber is currently "IMS roaming", i.e. registered through a P-CSCF in an IMS network other than the Home IMS network (see Figure 8.13).

Another group of interworking is formed by calls arriving from the circuit-switched (CS) network, such as the Public Switched Telephony Network (PSTN) or Public Land Mobile Network (PLMN) – see Figure 8.14.

Operators often restrict interconnection with the IMS network to PSTN or PLMN. Often, the PSTN or PLMN to which the IMS network interconnects is the PSTN or PLMN of

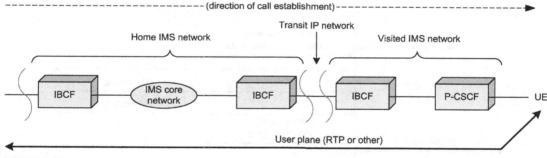

FIGURE 8.13

Transparent end-to-end routing of media.

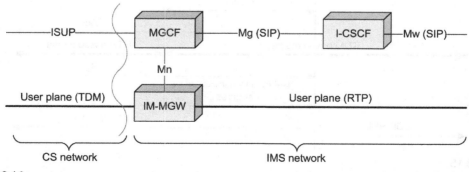

FIGURE 8.14

Network architecture related to ingress routing from a CS network.

the same operator. Even when the operators use both a PSTN and a PLMN, interworking with IMS may be confined to PSTN. In that manner, the operator has fewer interworking agreements to arrange. When a call from the operator's IMS network needs to be routed to a network from another operator, existing PSTN interconnect agreements are then used. Interworking between the IMS and CS networks is done through the media gateway control function (MGCF) and IP multimedia–media gateway (IM-MGW).

MGCF caters for protocol conversion between the control-plane protocol used in the CS network, typically ISUP, and the control-plane protocol used in the IMS network, SIP. Instead of ISUP, the CS network may use BICC or SIP-I. ISUP and SIP are different paradigms, as far as protocol messages are concerned. ISUP and SIP also use different state models. ISUP messages need to be mapped to corresponding SIP request or response messages and vice versa. Not all capability that is defined on ISUP can be mapped to SIP, and vice versa.

Figure 8.15 shows a typical message sequence for a call arriving from a CS domain.

3GPP TS 29.163 specifies the exact mapping between ISUP messages (and parameters) and SIP messages (including headers, parameters, SDP). Within the IMS network, there is no

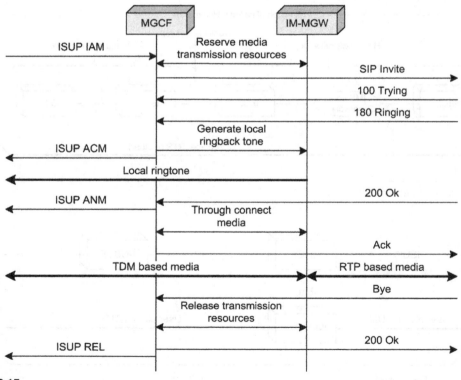

FIGURE 8.15

Message sequence for ingress routing from a CS domain.

ringback tone generated in the user plane; the SIP terminal establishing a call is responsible for its own ringtone generation to the calling party. When MGCF receives 180 Ringing from the remote party, it instructs the IM-MGW to generate local ringback tone into the CS network.

When MGCF receives ISUP IAM, it creates a SIP INVITE and sends the INVITE to a "SIP server". MGCFs may have the capability to configure *SIP profiles*. MGCF parses the received ISUP IAM through number analysis (e.g. A-number analysis and B-number analysis), resulting in the selection of a SIP profile. The SIP profile includes a set of parameters that are used for the SIP signaling from MGCF into the IMS network, such as:

- SIP server address; default will be I-CSCF, for terminating SIP session handling
- Transport protocol (UDP, TCP)
- URI format (tel: URI versus sip: URI) for R-URI and for certain SIP headers.

When ISUP in the CS network is being transported over the SS7 network, the ISUP signaling must first be converted to ISUP over IP. ISUP over IP is done in accordance with 3GPP TS 29.202. Refer to Figure 8.16.

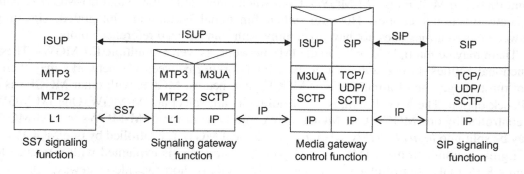

FIGURE 8.16

ISUP SS7-based signaling adaptation to ISUP IP-based signaling.

Source: 3GPP™ TS 29.163 v8.13.0, figure 2. © 2010 3GPP™ and TRs are the property of ARIB, ATIS, CCSA, ETSI, TTA and TTC who jointly own the copyright in them. They are subject to further modifications and are therefore provided to you " as is" for information purposes only. Further use is strictly prohibited.

The conversion between ISUP over SS7 and ISUP over IP is done in a signaling gateway (SGw). SGw represents a functional entity. Practically, the SGw is often integrated in MGCF.

Besides conversion between ISUP and SIP, there is also conversion between TDM-based media and RTP-based media. It is a common goal that there is no transcoding required for the media. For example, the call in the CS domain may be established with a circuit for 64 kb/s PCM A-law encoding (G.711). The same coding may be used in the packet-switched network. Whereas the media in a CS network is transported through a TDM channel, in E1 links (or STM links) the same media is transported in an IMS network through RTP encapsulation. The RTP media stream will have an RTCP channel associated with it.

For the CS–IMS interworking case, the media is always traversing the IMS network (via IM-MGW). For the IMS–IMS interworking case, on the other hand, IBCF may decide not to anchor the media in a TrGW. The reason media anchoring is necessary in media gateway CS–IMS interworking is because both control plane and user plane use different protocols in the respective networks.

MGCF may be deployed as a stand-alone node, in which case it is referred to as a media gateway controller (MGC). Alternatively, MGCF may be integrated in mobile softswitch (MSS) or in telephony softswitch (TSS) – see Figure 8.17.

Also, the IM-MGW may be integrated in circuit-switched mobile MGW (CS-MGW), as used in the mobile network. These integrations of functional entities allow for optimized network deployment, with fewer nodes and fewer interfaces. The MSS that includes MGCF may be a visited MSC, i.e. MSC with connection to radio access network (RAN). The interface between MSS with integrated MGCF and combined CS-MGW + IM-MGW is a combination of the Mc reference point (between the MSC server and CS-MGW) and Mn reference point (between MGCF and IM-MGW). Even when integrated, MGCF and IM-MGW remain separate functional entities. However, when functional entities are integrated, signaling between these entities does not have to comply with standardized reference points.

There may be multiple MGCFs in the IMS network, as well as multiple IM-MGWs. These functional entities do not contain user data. Interworking between CS networks may therefore run through one of a multitude of MGCFs, e.g. nearest MGCF, with other MGCFs as a fallback option. The MGCF will, when handling a call, select an IM-MGW. One IM-MGW is controlled by one MGCF. One MGCF may control one or more IM-MGWs. An IM-MGW may be split into *logical MGWs*, whereby each logical MGW is controlled by one MGCF.

Egress routing, from an IMS network to other networks, is explained with reference to Figure 8.18. Only control-plane entities are shown. ViG is short for video gateway.

When S-CSCF handles originating SIP session establishment, one of the steps to be taken by S-CSCF is to determine the destination serving network. The destination serving network

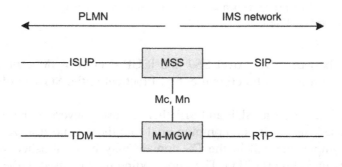

FIGURE 8.17

Integration of IMS border gateway in PLMN nodes.

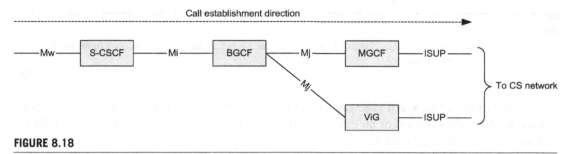

FIGURE 8.18

Egress routing through BGCF (break-out).

is derived from the R-URI. If a call is established to a phone number, then the S-CSCF may attempt to obtain a SIP URI associated with the phone number. The SIP URI would then be used for egress routing. Obtaining the SIP URI from the phone number is done through ENUM query (see section 8.7 for more details on ENUM). If, however, the ENUM query by the S-CSCF does not render a SIP URI, then the S-CSCF will initiate "break-out". The rationale of the break-out is that the S-CSCF is a SIP server and can route only on SIP URIs, not on phone numbers. The break-out implies that the call will be routed to another network that is capable of routing on phone numbers. An example of such a network is PSTN or PLMN. This other network may be the network that is determined to be the serving network for the destination subscriber. Alternatively, the network to which the break-out occurs is not determined to be the serving network for the destination subscriber, but will act as an intermediate network, routing the call further to the appropriate serving network.

For break-out, the S-CSCF forwards the SIP INVITE (or other request message) to a break-out gateway control function (BGCF). The BGCF is a SIP proxy that forwards the request further, to a break-out gateway. The BGCF may be deployed on a designated host or may be integrated into S-CSCF. BGCF will select the (most) appropriate break-out gateway for this request, which may be based on one or more of the following criteria:

- Originating party (calling party)
- Destination party (called party)
- Destination network
- Number portability information

Note

Indication of destination network or number portability of this call may have been received by S-CSCF from ENUM and included as a parameter in the R-URI.

- Media session type – SDP offer (call type; e.g. voice or voice + video)
- Preferred long-distance carrier.

Note

Preferred long-distance carrier could form part of the calling subscriber subscription profile or may be provided by an application server.

The selection of break-out gateway by the BGCF may be based on internal configuration or may be done with the aid of an external database query. The exact behavior of the BGCF, with respect to selecting break-out gateway, is vendor specific.

A common use of the BGCF is the selection of an MGCF, for routing the call into the CS domain, such as PLMN. Within the CS domain, the call would then be routed further to the appropriate destination, based on techniques like B-number analysis, number portability, and intelligent networks. Routing the break-out call through an MGCF implies that the media of this call will be anchored in an IM-MGW, controlled by the MGCF. Break-out gateway selection based on call type may be done to apply distinctive routing for voice calls and video calls. Whereas voice calls may be routed to a voice-only MGCF, video calls would be routed to a video-capable MGCF, also referred to as video gateway (ViG).

For specific interworking cases, the BGCF may select an IBCF for break-out. This method may be applied when the BGCF has determined that a SIP-based network-to-network interworking agreement exists with the destination network. In this case, the destination subscriber may be an IMS subscriber, but the operators have not shared ENUM information. This case is illustrated in Figure 8.19. The break-out from BGCF to IBCF is subject to certain conditions, such as:

1. Request URI is a SIP URI.
2. The domain name part of the R-URI points to the network serving the called party; this would not have to be the case when the R-URI is a phone number with an enterprise-specific domain name.

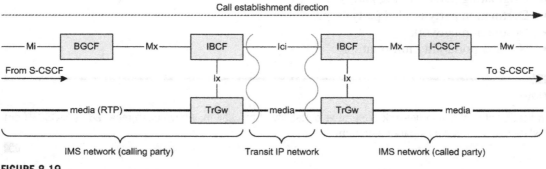

FIGURE 8.19

Break-out to other IMS network.

The default action by the IBCF in the receiving IMS network is to forward the INVITE request (or other SIP request) to I-CSCF for further processing, including HSS query and forwarding to S-CSCF.

There may be multiple IBCFs in an IMS network, as well as multiple TrGWs. Neither of these maintains a subscriber state between sessions. Per break-out/break-in case, one of a group of IBCFs may be selected, typically through DNS-based load sharing. DNS-based load sharing implies that DNS returns multiple host names in response to an SRV query. The DNS client uses one of these hosts for IP routing. The multiple host names received from DNS may have a priority or a weight factor associated with them.

8.5 REGISTRATION

In this section, we describe the registration process in detail. The goal of registration was briefly mentioned in Chapter 6, namely to create and maintain binding in a SIP registrar. Such binding allows the establishment of IP communication sessions to the subscriber using a public user identity. The binding is used to forward the IP communication sessions to the subscriber's current IP address. Before looking into the registration procedure, we consider the subscriber data structure in HSS. An IMS subscriber is provisioned in HSS with one or more of each of the following two identification items:

- **IMS public user identity (IMPU).** IMPU is used to identify an IMS subscriber over the public communication network, e.g. to establish a communication session with that subscriber. It is also used as identification of the originator of a communication session. The IMPU conforms to the structure of Universal Resource Identifier (URI). Examples include:
  - sip:john.smith@abc-company.se
  - tel:+31163279911
  - sip:+31163279911@my-company.nl
  - sip:204004123456789@ims.mnc004.mcc204.3gppnetwork.org
- **IMS private user identity (IMPI).** IMPI is used together with the IMPU to identify the service profile of the subscriber and to determine what authentication should be used for the registration. IMPI has the format of Network Access Identifier (NAI) as defined in IETF RFC 4282 (it has the form of a URI, excluding the schema). Examples of IMPI include:
  - john.smith@abc-company.se
  - +31163279911@my-company.nl
  - 204004123456789@ims.mnc004.mcc204.3gppnetwork.org

The IMPU *sip:204004123456789@ims.mnc004.mcc204.3gppnetwork.org* and the IMPI *204004123456789@ims.mnc004.mcc204.3gppnetwork.org* are temporary IMPU and IMPI respectively. Temporary IMPU and temporary IMPI may be used when IMS registration takes place from a mobile phone (GPRS, UMTS) that is not equipped with an IMS subscriber identification module (ISIM). 3GPP TS 23.003 describes how temporary IMPU and temporary IMPI are constructed, based on the International Mobile Subscriber Identity (IMSI).

IMPU and IMPI form part of the subscriber profile in HSS (see Figure 8.20). In the example in Figure 8.20, IMPU #2 is associated with both IMPI #1 and IMPI #2. IMPUs within one implicit registration set (IRS) may have the same service profile or may have different service profiles. The concept of IRS is explained in a later section.

When an IMS subscriber registers, in other words creates binding in S-CSCF, the IMS subscriber provides his/her IMPU (or one of his/her IMPUs) to the IMS network and optionally IMPI. By providing IMPU to the IMS network during registration, the subscriber indicates *which* IMPU or group of IMPUs should be registered.

Let's first look at the registration process, step by step. We will first consider the *initial registration*. An explanation is given with reference to Figure 8.21.

Registration includes, among others, the following two procedures:

- **Authorization.** The HSS determines whether the subscriber trying to register is a subscriber to this network.
- **Authentication.** Authentication is performed jointly by the HSS and S-CSCF. It is verified that the person or application attempting to register is entitled to use this subscription (has the right credentials).

A step-by-step description of the registration process follows.

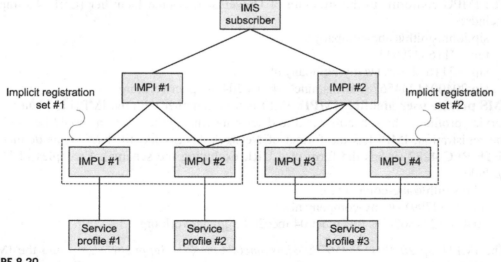

FIGURE 8.20

Relation between IMPU and IMPI in HSS subscriber profile (example).

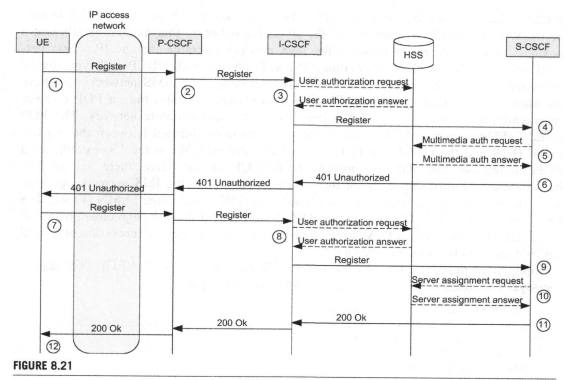

FIGURE 8.21

IMS registration message sequence.

Step 1. SIP REGISTER from UE to P-CSCF

Before the terminal can initiate registration, it has to obtain IP connectivity. The IP connectivity may be obtained through corporate LAN, residential WLAN, 3G mobile access, etc. The IP connectivity should allow for communication over (public) IP infrastructure, so the IMS terminal (user equipment, UE) can communicate with the public IMS network. IMS registration would now be initiated by sending a registration request "towards" the IMS network. However, the subscriber should first acquire the following information:

- Home IMS realm: this is the domain name of the IMS network that this subscriber belongs to – for example, ims-operator.se
- IMPU: as described
- IMPI: as described
- P-CSCF address: this is the address (IP address or domain name) of the P-CSCF through which the registration should be done.

The user of the IMS terminal may configure these information elements. Alternatively, the IMPU, IMPI, and Home IMS realm may be stored on ISIM. The

P-CSCF address may be obtained with DHCP or may be obtained from the mobile network. Since DHCP is a service of the IP access network, this method entails that a business relationship exists between the IMS network operator and the IP access network operator. The method of obtaining P-CSCF address with DHCP will typically be restricted to subscribers of the associated IMS network(s), i.e. IMS networks that have an administrative agreement with this IP access network. Establishing a PDP context, meanwhile, may obtain a P-CSCF address from the user's mobile network. The PDP context (PDPc) represents the data connection that is established between the 3G terminal (UE) and the GGSN (or PDG in the LTE network). When the UE establishes a PDPc, it requests the GGSN to provide the P-CSCF address. Here, there will need to be a business relationship between the mobile operator and the IMS network operator. It would normally be restricted to cases where a GGSN in the home PLMN is used. If a GGSN in the visited PLMN is used for PDPc establishment, then the agreement based on which GGSN in the visited PLMN may be used may also include the provision of a local P-CSCF address from the visited GGSN.

Equipped with the required information, the UE can construct a SIP REGISTER request message. The REGISTER message may be as follows (example):

```
REGISTER sip:ims-operator.se SIP/2.0
Via: SIP/2.0/UDP 153.88.57.74:19392;branch=z9hG4bK-d8754z-b5a9b86e100957a2-1
Route: sip:p-cscf.ims-operator.se;lr
Max-Forwards: 70
Contact: <sip:john.smith@153.88.57.74:19392;transport=udp>
To: "John Smith"<sip:john.smith@abc-company.se>
From: "John Smith"<sip:john.smith@abc-company.se>;tag=c297646c
Call-ID: NmZmMjE4ZmQzY2VhZDFjZGRlZGRkMjRiZmRjMTZmZWM.
CSeq: 1 REGISTER
Expires: 3600
Allow: INVITE, ACK, CANCEL, OPTIONS, BYE, REFER, NOTIFY, MESSAGE, SUBSCRIBE, INFO
User-Agent: X-Lite 4 Release 4.0 stamp 58832
Content-Length: 0
```

The *destination* of the REGISTER request is the Home realm (sip:ims-operator.se). One of the S-CSCFs in that domain will be the receiver of this REGISTER message. The URI in the To header indicates *for whom* the registration request is meant. This is the IMPU through which John Smith wishes to register. The From header indicates *by whom* the REGISTER message is sent. It is possible to send a REGISTER message *on behalf of* another entity, in which case the From header and To header would differ.

The Route header contains the P-CSCF address that was configured, obtained from the access network or obtained from the ISIM. The contact header contains the IP address for this SIP phone, as well as the port number. The port number may have been obtained from the operating system (if the SIP phone is PC soft client). The UE may now send the

REGISTER message. As described in Chapter 7, the routing of the message is based on the topmost Route header. If the Route header contains a P-CSCF address in the form of a domain name (as opposed to IP address), the UE will apply DNS name resolving and may receive multiple P-CSCF host names. These host names may have equal priority or different priorities. The UE will select one P-CSCF for registration. If the selected P-CSCF is determined to be not operational (e.g. an IP signaling failure is detected), then the UE may attempt to register through another P-CSCF.

If this is the initial registration for this terminal, then the terminal has not yet established a security association with the P-CSCF; the subscriber has not yet been authenticated. This may imply, for example, that the REGISTER message is sent to the default port for SIP signaling, 5060 (for UDP, TCP, and SCTP).

Step 2. SIP REGISTER from P-CSCF to I-CSCF

The P-CSCF receiving the REGISTER request message determines that the request message is destined (as indicated in the R-URI) for a domain (realm) that forms part of the administration for this P-CSCF. As described, the P-CSCF used for registration may be a P-CSCF in an IMS network other than the Home IMS network. In this case, the P-CSCF could be handling registration for subscribers from a multitude of IMS networks.

The P-CSCF does not forward the REGISTER request message directly to S-CSCF for the subscriber. Instead, it forwards the REGISTER request message to I-CSCF, leaving the selection of S-CSCF to the I-CSCF. The P-CSCF may use a set of preconfigured I-CSCF host names for forwarding the REGISTER request message. Alternatively, the P-CSCF may use DNS to obtain an inbound SIP server for the home realm of this REGISTER request. The latter would be the case when the P-CSCF is located in an IMS domain other than the Home IMS domain. In that case, the REGISTER request will be routed via border proxy nodes, IBCF, on the border of both the visited IMS network and the home IMS network. External DNS may steer the REGISTER request via these border nodes. Within the Home IMS network, the REGISTER message may be processed by any in a group of I-CSCFs.

The P-CSCF includes a Path header in the REGISTER message. The Path header contains the address (host name) of this P-CSCF. It indicates to the S-CSCF that initial request messages destined for this subscriber should be routed via this P-CSCF. For example:

```
Path: sip:term@p-cscf2.ims-operator.se;lr
```

The address contained in the Path header will be an address (host name) associated with this specific P-CSCF. When SIP REGISTER is sent from P-CSCF to S-CSCF via one or more IBCFs, these IBCFs do not have to be included in the Path header. The routing of initial SIP requests via IBCF(s) would instead be done through DNS or through a network configuration.

A *term* user part may be included in the SIP URI in the Path header. When the terminating INVITE request destined for a user arrives at the P-CSCF of that user, the P-CSCF can determine from this *term* user part that this INVITE request represents a terminating INVITE request.

Since the P-CSCF represents the *visited IMS network*, the P-CSCF also includes an identification of this visited IMS network in the REGISTER request. This identification is contained in the *P-Visited-Network-ID* header. A designated SIP header constitutes the P-Visited-Network-ID. The HSS in the Home IMS network may determine from this header whether an IMS roaming agreement exists with that visited IMS network.

Step 3. User Authorization

The I-CSCF that receives the REGISTER message contacts the HSS, allowing the HSS to authorize the user. If the user is not known to the HSS, then the HSS rejects the registration request. The I-CSCF will in this case send an unsuccessful final response on the REGISTER request to the user; the registration attempt has failed.

If there is currently an S-CSCF assigned to the subscriber, then the HSS informs the I-CSCF about the address of this S-CSCF, allowing the I-CSCF to forward the REGISTER request message to that S-CSCF. If there is currently no S-CSCF assigned to the subscriber, then the HSS provides *S-CSCF capabilities* to the I-CSCF. The S-CSCF capabilities indicate to the I-CSCF what capabilities the S-CSCF to be selected for this subscriber should support. Individual S-CSCF capabilities may be tagged as *mandatory capabilities* and *optional capabilities*. The I-CSCF uses an internal mapping table to select an S-CSCF supporting the indicated capabilities. The I-CSCF may further use load-sharing rules for allocating subscribers according to a set of weighting factors to the available S-CSCFs, or apply other rules for distributing subscribers over the available S-CSCFs. The I-CSCF and S-CSCF always reside in the same IMS network, so the I-CSCF can be configured with addresses (e.g. domain names) of available S-CSCFs. When one S-CSCF is currently not operational, the I-CSCF may have (temporarily) marked that S-CSCF as *not operational* and hence not select that S-CSCF for this registration.

The communication between I-CSCF and HSS is based on Diameter. Diameter is a protocol used for authentication, authorization, and accounting in IP networks. It is specified in IETF RFC 3588 ("Diameter base protocol"). The Cx reference point in IMS is used between I-CSCF and HSS, as well as between S-CSCF and HSS and is based on the Diameter base protocol. 3GPP TS 29.228 and 3GPP TS 29.229 describe the Diameter messages used for this reference point. The Diameter messages used for the subscriber authorization, between I-CSCF and HSS, are:

- **User authorization request (UAR)** – this is the message sent from I-CSCF to HSS.
- **User authorization answer (UAA)** – this is the message sent from HSS to I-CSCF.

Refer to 3GPP TS 29.228 and 3GPP TS 29.229 for the various information elements that may be carried in these messages.

Step 4. SIP REGISTER from I-CSCF to S-CSCF

Routing from I-CSCF to S-CSCF is based on the address (domain name) received from the HSS or on the address selected by I-CSCF (previous step). This address is placed as a Route header in the REGISTER request.

Step 5. Multimedia Authentication Request

When the S-CSCF receives the REGISTER request, it ascertains whether it has a subscriber record for this subscriber. If this registration request is an initial registration, the S-CSCF does not have a subscriber record. Strictly speaking, the REGISTER request would contain sufficient information to complete the registration: contact address and public user identity. However, we first want to perform authentication, verifying that the user is an authenticated user of this specific subscription. To this end, the S-CSCF will contact the HSS to request that it provide authentication instructions. The S-CSCF uses the Diameter protocol hereto (Cx reference point). The Diameter messages used are:

- **Multimedia authentication request (MAR)** – MAR is used from S-CSCF to the HSS, to request the HSS to provide authentication instructions.
- **Multimedia authentication answer (MAA)** – MAA is used from the HSS to SCSCF, containing the authentication instructions.

The type of authentication may differ by device for one subscriber. When one subscriber uses different SIP devices, a different IMPU and IMPI may be used for the registration of these devices. For example, a SIP phone that is equipped with ISIM will use authentication based on authentication credentials stored in the ISIM card, whilst registration through a soft SIP phone may be done with a user name and password. The combination of IMPU and IMPI, as reported in the REGISTER request, is used by the HSS to select the required authentication method.

The above-mentioned credentials are shared between subscriber record in HSS and subscriber. The credentials may have the form of a password stored in HSS and memorized by the subscriber. Alternatively, they may consist of secret information stored on an ISIM card. The HSS uses the credentials (password, secret information) to generate a *challenge* (see Figure 8.22).

The challenge, also referred to as "nonce", is generated in HSS and is conveyed to S-CSCF, which conveys it to the UE, via the P-CSCF. It has the form of a character string. The secret information is not transported between the UE and the IMS network. The challenge is used for a single authentication attempt only, i.e. the HSS generates a new challenge every time authentication takes place. The UE applies the same response-generating algorithm as the HSS. The S-CSCF will inform the UE about the algorithm used for authentication. The response generated by the UE, denoted "Response'" in the figure, is equal to the response generated by the HSS only when the UE uses the same secret information as used in HSS.

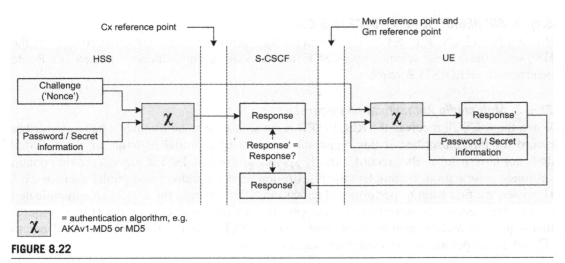

FIGURE 8.22

Challenge – response mechanism for subscriber authentication.

Step 6. Provoking Authentication Response from SIP UE

When the S-CSCF has received the challenge and the response from the HSS, it sends a final response *401 Unauthorized* to the UE, including the challenge. The S-CSCF stores the response (i.e. the outcome of authentication algorithm execution) received from the HSS. The sending of 401 Unauthorized to the UE terminates the REGISTER transaction. The 401 Unauthorized traverses both the I-CSCF and the P-CSCF on its way to the UE.

The S-CSCF has at this point (optionally), as part of the multimedia authentication procedure, also received information from the HSS for the establishment of a security association between the UE and P-CSCF. This security association is also based on shared information between the UE and HSS. The S-CSCF forwards this information to the P-CSCF in the above-mentioned 401 Unauthorized, together with the challenge.

The P-CSCF forwards the challenge to the UE. From this point onward, communication between the P-CSCF and the UE takes place through a protected channel, called a security association. This entails, among other things, the P-CSCF informing the UE that it (the UE) shall address the P-CSCF via a port associated with the security association. Hence, while the first REGISTER message may be sent to the default port, subsequent messages are sent to a different port. The security association is regularly refreshed during the registration lifetime of the UE. 3GPP TS 33.203 provides further background on security association in IMS.

Step 7. Authenticated SIP REGISTER from UE

The UE will, in response to receiving the challenge, generate a response using the authentication algorithm indicated in the 401 Unauthorized, using the received challenge and using the locally stored or locally obtained secret information. This secret information may, for example, be user input (password). The UE then sends a *second* SIP REGISTER request, containing the generated response. This second SIP REGISTER request is sent over the established security association to the P-CSCF.

The P-CSCF will again forward the REGISTER request message to I-CSCF, allowing the I-CSCF to select the appropriate S-CSCF. The I-CSCF for this second REGISTER request may be a different I-CSCF than the I-CSCF used for the first REGISTER request.

Step 8. User Authorization Between I-CSCF and HSS

The I-CSCF will again apply a Diameter user authorization request to the HSS, as for the first registration. This time, the I-CSCF will receive the S-CSCF address from the HSS, instead of S-CSCF capabilities. This is because the registration and authorization process has already started in a particular S-CSCF, so this second REGISTER message shall end up at the same S-CSCF.

Step 9. S-CSCF Processing SIP REGISTER Request

When the S-CSCF receives the REGISTER request message, it will associate this REGISTER message with the process that was started with the first REGISTER message. Presuming that the UE applied the correct secret information, i.e. an authenticated user sends the REGISTER message, the S-CSCF will determine that the response generated by the UE (Response'; see Figure 8.22) is equal to the response previously received from the HSS. The S-CSCF can now accept the registration and store information (received in the REGISTER request) such as:

- Contact binding (subscriber IMPU and contact address)
- Terminal capability
- Path (P-CSCF address that was included by P-CSCF in the REGISTER message).

Step 10. Server Assignment

REGISTER is a SIP method in the formal sense of the word, so a successful REGISTER transaction request should be closed with a final response, such as 200 OK. Before the S-CSCF confirms the registration by sending 200 OK, it will obtain a user profile from the HSS. The user profile is a set of information elements that are needed in S-CSCF for handling SIP sessions to and from this subscriber, traversing this S-CSCF. Also, the HSS will be informed about the completion of the registration, so it will know that the subscriber is now registered in this particular S-CSCF. Here, the S-CSCF applies a Diameter server assignment request (SAR) to the HSS. SAR is, as its name implies, a request to the HSS to assign this S-CSCF ("server") to the subscriber. From now on, as long as the subscriber is registered in this S-CSCF, all originating and terminating SIP requests for this subscriber will be sent through this S-CSCF.

The HSS assigns this S-CSCF to the subscriber and marks the subscriber as 'registered'. The HSS returns Diameter server assignment answer (SAA) to S-CSCF, containing one or more of the following information elements (this list is not exhaustive):

- **Phone-context.** This element may be used for normalizing phone numbers during the original SIP session establishment; strictly speaking, when a SIP session is established toward a phone number not in the global E.164 format, the session request will already include a phone-context; an S-CSCF may use this phone-context that is part of the user profile in cases where phone-context is missing from the request.

- **Associated IMPUs.** The associated IMPUs constitute a set of IMPUs that form an implicit registration set.
- **Initial filter criteria (IFC).** The IFC are used for service invocation during SIP session establishment.
- **Charging data function (CDF) address(es).** The CDF address(es) are used for the dispatching of charging records.

Each of these elements will be described separately in subsequent sections.

Step 11. 200 OK to UE

When the S-CSCF has obtained a user profile from the HSS, the registration procedure is complete and the UE may now receive confirmation, in the form of 200 OK. This 200 OK, which traverses both the I-CSCF and the P-CSCF, contains the following information elements, in the form of SIP headers (the list is not exhaustive):

- **Service-Route.** This SIP header contains the S-CSCF address. It will be stored by UE and will be used by UE for directing outgoing initial SIP request messages. Using the Service-Route by UE has the effect that SIP sessions and stand-alone SIP transactions are established through this S-CSCF. The Service-Route is defined in IETF RFC 3608.
- The Service-Route is also stored in P-CSCF. P-CSCF may verify and ensure that the Service-Route is included in outgoing initial SIP request messages (originated by a SIP terminal). Practically, P-CSCF may hide the Service-Route for the end-user, since Service-Route constitutes a network-internal address, which is normally not publicly disclosed.
- **P-associated-URI.** This SIP header contains a list of one or more IMPUs. These IMPUs reside in the same implicit registration set (IRS) in the HSS. By registering one IMPU, all IMPUs become registered.
- **P-charging-function-addresses.** This SIP header contains a set of addresses of nodes (entities) that may be used for online charging or offline charging. This header is stored in P-CSCF and is included in outgoing SIP session establishment requests. It is not forwarded to the UE.

Step 12. 200 OK Received by UE

When the UE receives the 200 OK, the registration is complete from a terminal point of view. The terminal stores information received in the 200 OK, such as Service-Route and P-associated-URIs. The terminal is now ready to establish and to receive SIP sessions and stand-alone SIP transactions.

8.5.1 Registration Relationships

Figure 8.23 depicts the relationships that are established resulting from the provisioning and registration of the subscriber (terminal) in the IMS network. Table 8.5 describes these relationships.

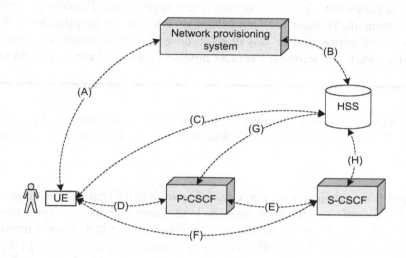

FIGURE 8.23

Relationships in IMS network resulting from registration in IMS.

Table 8.5 Inter-Entity Relationships in IMS Network

(A)	SIP terminal is configured with information relating to the user (e.g. IMPU) and with network address information (e.g. Home IMS domain). In addition, the IMS subscriber has obtained authentication credentials, such as password.
(B)	HSS has received subscriber data from provisioning system.
(C)	HSS and UE share secret information. In addition, UE has obtained from HSS a list of available IMPUs that are associated with this terminal.
(D)	UE and P-CSCF have established an "access relationship", comprising, among other things, security association.
(E)	S-CSCF has stored P-CSCF address (Path) and P-CSCF has stored S-CSCF address (Service-Route). Communication between UE and S-CSCF will be routed via this P-CSCF.
(F)	UE and S-CSCF have a registration relationship. The S-CSCF is registrar for UE. S-CSCF and UE have exchanged addresses.
(G)	P-CSCF has obtained subscriber information from HSS, required for handling outgoing SIP sessions, such as a set of IMPUs associated with the terminal and a set of addresses of the charging data function (CDF) to be used for the charging of SIP sessions from this UE. Charging of a SIP session may take place from various entities; P-CSCF is one of these.
(H)	HSS maintains the address of the S-CSCF of this subscriber. It uses this address for steering initial SIP requests (SIP session or stand-alone SIP transaction) to this S-CSCF.

8.5.2 **Periodic Re-Registration and De-Registration**

The registration that is applied by the SIP terminal needs to be "refreshed" periodically, i.e. the terminal must perform "periodic re-registration". Without periodic re-registration, the binding in S-CSCF will expire. The S-CSCF will in this case discard the contact binding: the terminal will no longer be contactable (receive SIP requests) and will not be able to establish SIP sessions until it has performed a new registration. Periodic re-registration is a protection mechanism. Without such a mechanism, when a SIP terminal loses IP connectivity (e.g. network disconnection or loss of radio coverage) or is terminated improperly (e.g. power disconnection), that terminal's contact binding in S-CSCF would remain indefinitely.

Note

Periodic re-registration in IMS may be compared with *periodic location update* in GSM. If a GSM terminal does not perform a periodic location update at designated intervals, the registration of that GSM phone in the network expires.

Short registration expiry time leads to frequent REGISTER transactions, which may constitute a large portion of the total network load. Long registration expiry time obviously reduces that burden on the network, but has the disadvantage that when a terminal has lost network connection, the S-CSCF will contain a non-functioning binding for a longer duration.

During registration, terminal and IMS core network "negotiate" a periodic registration time, also referred to as "expiry" time. When a terminal initiates registration, it provides a contact address to the IMS network, including an expiry value. For example:

```
REGISTER sip:my-ims-network.se SIP/2.0
Via: SIP/2.0/UDP 153.88.47.176:34634;branch=z9hG4bK-d8754z
Max-Forwards: 70
Contact: <sip:john.smith@153.88.47.176:34634>
To: "John Smith"<sip:john.smith@my-company.ims-operator.se>
From: "John Smith"<sip:john.smith@my-company.ims-operator.se>;tag=da870fc4
Call-ID: Y2M5ZjMxNDNhNzYwZDNiMjU1YWRhYmI0OGIyMjg4Mzk.
CSeq: 1 REGISTER
Expires: 3600
Allow: INVITE, ACK, CANCEL, OPTIONS, BYE, REFER, NOTIFY, MESSAGE
User-Agent: X-Lite 4 release 4.0 stamp 58832
Content-Length: 0
```

The SIP header **Expires: 3600** indicates that the terminal wishes to register with an expiry value of 3600 s. The IMS network itself applies a (configurable) range of registration expiry values. The proposed value of 3600 s may therefore lie within the range or outside the range. The following situations are distinguished:

- **Proposed registration time value is within the range.** S-CSCF accepts the registration expiry. The expiry value is copied back in the 200 OK that is sent from S-CSCF to the terminal in response to the REGISTER request.
- **Proposed registration time value exceeds maximum value.** S-CSCF sets the registration time value to the maximum acceptable (for S-CSCF) value and accepts the registration. The (adapted) expiry value is included in the 200 OK.
- **Proposed registration time value is less than minimum value.** The S-CSCF rejects the registration request by sending a final response *432 Interval Too Brief* to the terminal. This final response includes the minimum value that is acceptable for the S-CSCF. The UE should now send a new registration, with an expiry time not less than the indicated minimum value.

The SIP terminal will typically apply periodic re-registration at half the negotiated expiry time. Hence, if 3600 s was negotiated, then the terminal will re-register every 30 minutes. The expiry value (3600 s) may also be included as URI parameter in the Contact header, as follows:

```
Contact: <sip:john.smith@153.88.47.176:34634>; expires=3600
```

The re-registration SIP message flow is similar, but not identical, to the SIP message flow for initial registration. The SIP REGISTER request message is sent to the P-CSCF that was discovered during initial registration. If a security association is established with the P-CSCF, then the SIP REGISTER request message is sent over that security association. When the REGISTER message is forwarded to the I-CSCF and the I-CSCF has contacted the HSS, as for initial registration, then the HSS provides the address of the assigned S-CSCF to I-CSCF. Hence, registration is directed to the same S-CSCF. The S-CSCF determines that the subscriber, as indicated in the To header in the REGISTER message, is registered in this S-CSCF and that the S-CSCF has a user profile. Hence, S-CSCF will not have to apply the Diameter *server assignment* procedure with the HSS. S-CSCF will not apply authentication for the re-registration. S-CSCF responds with 200 OK, which will complete the re-registration.

One may wonder why the periodic re-registration is sent via I-CSCF and not directly to the S-CSCF, using the Service-Route and bypassing I-CSCF? The reason is that registration may have to be shifted to another S-CSCF. When the S-CSCF where the subscriber had registered has become non-operational, then the I-CSCF may steer the next registration from that terminal to another S-CSCF. The other S-CSCF will then obtain a user profile from the HSS, using Diameter signaling used for initial registration.

Note

The I-CSCF may detect that the S-CSCF is not operational when it (the I-CSCF) does not receive a response on forwarding the REGISTER to the S-CSCF or when it receives an ICMP error message, indicating that the addressed host is not contactable.

There is a subtle aspect regarding re-registration in the S-CSCF. The S-CSCF has to determine whether the registration from this subscriber relates to *re-registration* (from a terminal that had registered already) or to *initial registration* (from another terminal). Here, the S-CSCF analyzes the contact address offered in the registration request. When the contact address offered in the registration request corresponds to a contact address for which the subscriber currently has a binding, the registration constitutes a re-registration. Otherwise, the registration constitutes initial registration from another terminal (or another SIP client on the same terminal/host). The subscriber will then have two bindings registered in the S-CSCF.

When a SIP terminal is switched off "in the proper manner", the SIP terminal will deregister from the network. The de-registration uses the same SIP messaging sequence as does the re-registration. The REGISTER message will, this time, contain an expiry value of 0, indicating that the binding shall be removed. The 200 OK in response to the REGISTER request with expiry value 0 includes no contact address(es), which is an indication that there is no binding for this terminal, i.e. the terminal is de-registered. When the S-CSCF has removed the binding, it instructs the HSS to remove the S-CSCF address from that subscriber's record. The P-CSCF will, as a result of the de-registration procedure, also discard its subscriber data.

If there are multiple contact addresses registered in the S-CSCF for this subscriber, then de-registration of one contact address will not affect the other binding(s) for that subscriber. As long as there is at least one binding present in the S-CSCF for this subscriber, that S-CSCF address will be retained in the HSS.

8.5.3 Implicit Registration Set

As mentioned in previous sections, an IMS subscriber may have multiple public identities (IMPUs), such as:

```
sip:john.smith@my-company.ims.se
sip:+46107152111@my-company.ims.se
sip:+46705453658@my-company.ims.se
```

This subscriber would be contactable under any of these identities. The subscriber's public identities may be associated with all of the subscriber's registered terminals (contact addresses). Alternatively, individual public user identities may be associated with a subset of the terminals. For example, a subscriber has two terminals: a desktop-bound SIP phone and a cellular mobile SIP phone. The subscriber has three public user identities, as in the above example. Figure 8.24 depicts two situations where these IMPUs are associated with the subscriber's terminals in different ways.

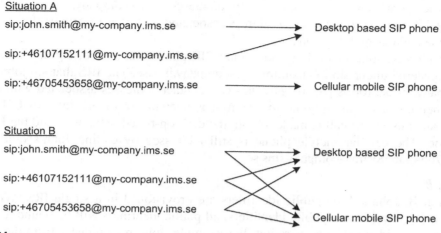

FIGURE 8.24

Mapping from IMPU to terminal.

The mapping from IMPU to terminal depends on the way in which the IMPUs are provisioned in the HSS and depends on the manner in which the respective terminals register in the IMS network.

IMPUs of an IMS subscriber may be provisioned in HSS in an *implicit registration set* (IRS). When two or more IMPUs of a subscriber are given in an IRS, then, when from one of these IMPUs a contact binding is deposited in S-CSCF, all IMPUs in that IRS will be bound to that contact address. This is referred to as *implicit registration*. For example, when sip:john.smith@my-company.ims.se   and   sip: +46107152111@my-company.ims.se   are included in an IRS and John Smith registers from a SIP terminal and identifies himself with sip:john.smith@my-company.ims.se, both IMPUs will be registered. In this case, the HSS provides the complete IRS to S-CSCF during the registration procedure.

Let's have a look at the two examples given above.

Situation A

Sip:john.smith@my-company.ims.se   and   sip: +46107152111@my-company.ims.se   are included in an IRS. John registers from his desktop-based SIP phone and identifies himself with sip:john.smith@my-company.ims.se. As a result of this registration, sip:+46107152111@ my-company.ims.se will also be registered. Both IMPUs will be bound to the contact address provided by the desktop-based SIP phone. Sip: +46705453658@my-company.ims.se is provisioned outside an IRS. John registers from his cellular mobile phone and identifies himself with sip: +46705453658@my-company.ims.se. This IMPU will be bound to the contact address provided by that phone.

When John is addressed on sip:john.smith@my-company.ims.se or on sip: +46107152111@my-company.ims.se, the S-CSCF forwards the SIP request to the contact address associated with these identifiers; the call will be made to the desktop-based SIP phone. When John is addressed on sip: +46705453658@my-company.ims.se, S-CSCF forwards the SIP request to the contact address associated with that identifier; the call will be made to the cellular mobile phone.

In situation A, John could have additional SIP phones registered, e.g. a PC-based soft phone, registered under sip:john.smith@my-company.ims.se. The IRS that sip:john.smith@my-company.ims.se forms part of, now has two contact addresses associated with it. When John is then contacted at sip:john.smith@my-company.ims.se or on sip: +46107152111@my-company.ims.se, the call is made to both the desktop-based SIP phone and the PC-based soft phone. The cellular mobile phone is still addressed by calling John's other IMPU, sip: +46705453658@my-company.ims.se.

Situation B

In situation B, John's three public identifiers are provisioned in a single IRS. When John registers using any of these public identifiers, all public identifiers will be bound to the contact address provided in the registration. For example, John may register from the desktop-based SIP phone and from the cellular mobile SIP phone. When John is contacted on any of his identifiers, the call is made to both phones.

Two enhancements to SIP are closely related to the potential for a subscriber to have multiple IMPUs provisioned. These are P-called-party-ID and P-preferred-identity. Both will be described briefly.

A brief explanation is required here on the IMPUs used in the above examples, containing a phone number. The IMPUs that are included in the P-associated-URI are IMPUs that form part of a subscriber's profile in the HSS. The IMPUs in the HSS do not contain a URI parameter. Specifically, the user=phone parameter is not provisioned in the HSS. So, some of the IMPUs in the P-associated-URI contain phone numbers, but there is no user=phone associated with these URIs. The user=phone parameter is explained in a later section.

8.5.3.1 P-called-party-ID

Depending on subscriber provisioning in the HSS, one contact address binding in S-CSCF may have multiple IMPUs associated with it. Therefore, when an IMS subscriber receives a call on his/her SIP phone, the call may have been established to one of a group of IMPUs. When the S-CSCF forwards the INVITE request to the subscriber's terminal, it applies *retargeting*, i.e. it replaces the Request URI (R-URI) in the INVITE request by the registered contact address of the subscriber's terminal. So, the INVITE arriving at the subscriber's terminal would not reveal *how* (s)he was called.

To overcome that dilemma, the P-called-party-ID SIP header should be used. The S-CSCF will, when applying the above-mentioned retargeting, include the P-called-party-ID SIP header in the INVITE request. The P-called-party-ID will be set to the R-URI as valid just prior to the retargeting. That R-URI will contain the IMPU on which the subscriber was called. The subscriber now knows, when receiving the INVITE request, whether (s)he was called on his/her business number or on his/her private number (for the example case where the subscriber has these two numbers provisioned in the HSS in the same IRS). For example:

```
P-called-party-ID: sip:john.smith@my-company.ims.se
```

An application in the terminal may show this called number on the terminal's display or take other action, like playing an adapted ringtone. Such action would be terminal specific. The P-called-party-ID has a distinctly different meaning than the To header. The To header bears no intrinsic information about the routing of the SIP session and how the destination subscriber was identified. This stems from the fact that the initiator of the SIP session may already have applied a specific, non-routing-specific, value for the To header; in addition, an application server may alter the To header. The P-called-party-ID, on the other hand, is set by the S-CSCF and has a direct relation to the routing of the SIP session and how the destination subscriber was identified.

8.5.3.2 P-associated-URI and P-preferred-identity

When a subscriber registers from a SIP terminal, the subscriber identifies him/herself by providing his/her identity in the To header in the REGISTER message. This identity will be one of his/her IMPUs. As discussed, the IMPU under which the subscriber registers may form part of an IRS. So, not only will the provided IMPU be registered, but also the other IMPUs in this IRS. The set of IMPUs that becomes registered resulting from this registration is reported in the 200 OK in response to the REGISTER request. For example:

```
SIP/2.0 200 OK
P-associated-URI:  sip:john.smith@my-company.ims.se,
                   sip:146107152111@my-company.ims.se,
                   sip:146705453658@my-company.ims.se
```

The terminal now knows, after the registration procedure, that these IMPUs are available as legitimate subscriber identification. They may be used during establishment of the SIP session or stand-alone transaction toward the subscriber and during establishment of the SIP session or stand-alone transaction by the subscriber. In the former case, the subscriber could receive any one of these identities in the P-called-party-ID header. For the latter case, the subscriber may indicate to the network *how* (s)he would like to identify him/herself in the forward direction. In other words, when John Smith is registered with IMPUs   sip:john.smith@my-company.ims.se,   sip:+46107152111@my-company.ims.se and sip:+46705453658@my-company.ims.se, then John may, when establishing a call, use any of these IMPUs to identify himself to the destination party. John has received a list of identities associated with him, during registration. He can choose any one of these as legitimate identification. John can do so by including the P-preferred-identity header in the SIP request. For example:

```
INVITE sip:wendy.jones@abc-company.se SIP/2.0
P-preferred-identity: "John Smith" <sip:146107152111@my-company.ims.se>
```

The P-CSCF handling this INVITE request uses the P-preferred-identity header to set the P-asserted-identity header. The P-asserted-identity header constitutes a network-asserted identification of the calling party (when P-asserted-identity is present in a response message, it constitutes a network-asserted identification of the destination party). The P-CSCF has verified that the URI contained in the P-preferred-identity is a valid IMPU for this subscriber (see Figure 8.25).

If the INVITE does not contain the P-preferred-identity, the P-CSCF uses the topmost (default) IMPU to set the P-asserted-identity. The display name that may be included in the P-preferred-identity is free format text. The P-CSCF will not apply any semantic validation on that text, but will copy it into the P-asserted-identity.

We notice in this example that the P-preferred-identity URI contains a phone number, but there is no user=phone parameter included. This results from the fact that the URIs contained in the P-associated-URI do not include a phone number, as explained above.

8.5.4 Third-party registration

This section contains a brief introduction of the concept of application server (AS). The concept and mechanism of application servers in IMS is described in detail in section 8.8. This section describes only the registration aspect of the AS. The method of registration as such is described in an earlier section. Figure 8.26 shows the IMS registration procedure, where the registration procedure is enhanced with *third-party registration*. Figure 8.26 illustrates a situation where no authentication is applied by the S-CSCF.

An AS in the IMS network may need to be aware of the registration state of a subscriber. This may be required for its internal subscriber database, for example, to mark the subscriber as *registered*. For the designated call scenarios, application service logic, residing in SIP-AS, that is controlling the call, can adapt its logic processing to the registration state of the subscriber.

When the S-CSCF has completed the registration by the subscriber, it will have obtained a subscriber profile from the HSS. The subscriber profile contains, among other things,

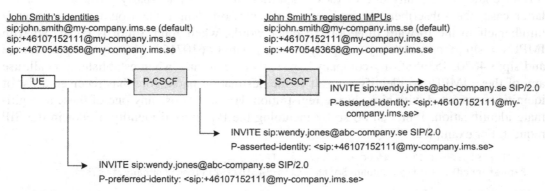

FIGURE 8.25

Use of P-preferred-identity.

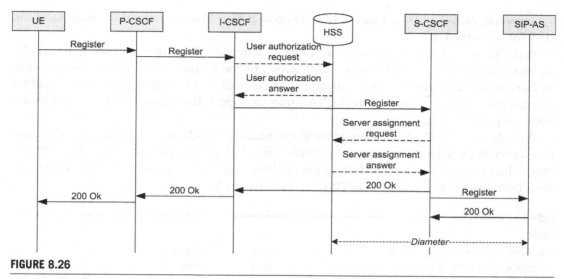

FIGURE 8.26

Third-party registration.

service trigger criteria for invoking IMS application service logic when a SIP session is established. The subscriber profile may also contain service trigger criteria for notifying IMS application service logic directly resulting from this registration. This is depicted by the REGISTER, 200 OK in Figure 8.26. The REGISTER and 200 OK between S-CSCF and SIP-AS form a normal SIP transaction. This SIP transaction is, *from a pure transaction point of view*, not related to the REGISTER transaction between terminal and S-CSCF. Instead, the S-CSCF acts as UAC for initiating the REGISTER transaction. Only a subset of the information elements contained in REGISTER sent from terminal to S-CSCF is included in the REGISTER sent from S-CSCF to SIP-AS. For example, contact address (of the SIP terminal) is not included.

The address of the SIP-AS to which the REGISTER request message should be sent is contained in the aforementioned service trigger criteria for notifying IMS application service logic. Said service trigger criteria may contain multiple SIP-AS addresses. In this case, S-CSCF will initiate an (identical) REGISTER transaction to each SIP-AS.

The response from SIP-AS on this REGISTER request, which should normally be 200 OK, has no effect on the actual registration. In other words, SIP-AS cannot, for example, *disallow* the registration.

The application service logic may, when receiving the third-party registration, apply Diameter signaling with the HSS to obtain a subscriber application service profile. In this manner, the SIP-AS uses the HSS for permanent subscriber data storage. Subscriber data are then kept in the SIP-AS only when the subscriber is registered. The reference point between SIP-AS and HSS is known as the Sh reference point. If there are multiple HSSs in the network, then the SIP-AS would have to contact SLF first, to find out the address of the HSS for the served subscriber.

Periodic re-registration, subscriber-initiated de-registration, and network-initiated de-registration also lead to third-party notifications to the respective SIP-ASs. When a subscriber de-registers, third-party notification indicates explicitly that the subscriber (public user identity) is de-registered. The SIP-AS may then push the subscriber data again to the HSS, for permanent storage.

As indicated, the third-party registration mechanism gives limited information only about the subscriber's registration state, especially considering multiple terminals. The SIP-AS may, when receiving initial third-party registration, subscribe to *registration event notifications* from the S-CSCF. Refer to IETF RFC 3680 for further details.

Note

The term "third-party registration" is sometimes also used as a term for registration of a user identity by another entity. This other entity must in such a case have relevant information available to perform the registration, such as public user identity, contact address, and home realm. This form of registration can be applied only within a trusted environment, as it would not be possible for the *other* entity to establish a security association *on behalf of* the entity to be registered.

8.5.5 Application-initiated registration

A special category of registration is "application-initiated registration", also known as third-party registration. This form of registration entails that the registration of a SIP entity is performed by an entity other than the involved entity itself (see Figure 8.27).

Third-party registration is defined in SIP, but is not formally included in the IMS standard. Hence, support of this mechanism is an operator option. The application server (AS) is

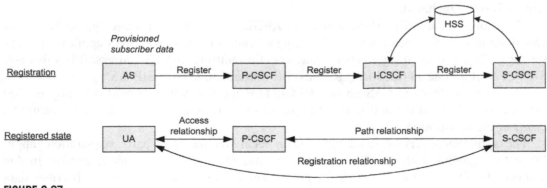

FIGURE 8.27

Application-initiated registration.

provided with subscriber data, such as IMPU and contact address. The AS is also configured with network data, such as home realm and P-CSCF address. The AS may now initiate a REGISTER transaction as would normally be applied by a UE. Regular interaction with the HSS by I-CSCF and S-CSCF applies. The REGISTER message contains in the To header the IMPU of the subscriber to be registered. For example:

```
REGISTER sip:some-ims-network.se SIP/2.0
From: sip:registration-server.some-ims-network.se; tag=asdfqwrtdsflksd
To: sip:john.smith@my-company.se
```

Hence, the registration is performed *on behalf* of another entity, namely the UE. As a result of this form of registration, a binding is placed in S-CSCF, providing a link between the IMPU and the contact address of the UE. The UE is now ready to establish and receive SIP sessions or stand-alone SIP transactions.

The AS will have to apply regular re-registrations, as normal. The AS might have obtained the subscriber data by other means than actually being provided with it, e.g. through traffic activity. If there is no functional connection between the AS and the UE, then the information that is returned in the 200 OK will not be available to the UE. One example is the Service-Route. This would imply that the UE would not be able to use the Service-Route as Route header when sending a SIP INVITE request. In addition, the UE would have to be configured to use the same P-CSCF address for sending an INVITE request as used by the AS during registration. This form of AS-initiated registration therefore works in specific network scenarios only, e.g. an IMS network with a single P-CSCF and a single S-CSCF. To overcome this dilemma, the AS-initiated registration may be sent directly to I-CSCF instead of through a P-CSCF. As a result, there will not be a P-CSCF in the registration relationship between UE and S-CSCF. When the UE establishes a SIP session, it would send the INVITE to an I-CSCF, with the Route header including the "orig" parameter. The I-CSCF uses the services from the HSS to obtain the address of the S-CSCF where the calling subscriber, who is sending the INVITE request, is registered.

In general, application-initiated registration is applied in special cases, not for regular SIP terminals. For example, the application server needs to be provided with a contact address for the terminal; alternatively, the application server would need to be able to dynamically obtain a contact address. It would further not be possible to establish a security relationship between the UE and the P-CSCF (if P-CSCF were used), since a security association is normally established between specific sets of address and port number. The AS and UE will normally have different address and port number.

A practical situation where application-initiated registration may be applied is the case where the UE is constituted by a machine with which trusted SIP signaling can take place. This "machine" may, in turn, be a gateway node that communicates with another communication network, for example a Public Switched Telephone Network (PSTN) or the Public Land Mobile Network (PLMN). Application-initiated registration hence becomes a means of linking non-SIP terminals into the IMS domain.

8.6 SESSION ESTABLISHMENT

Now that we have studied how IMS subscribers register in the IMS network with one or more terminals and with one or more public identifiers (IMPUs), we will look at SIP session establishment (based on an INVITE transaction). We will consider an end-to-end situation,

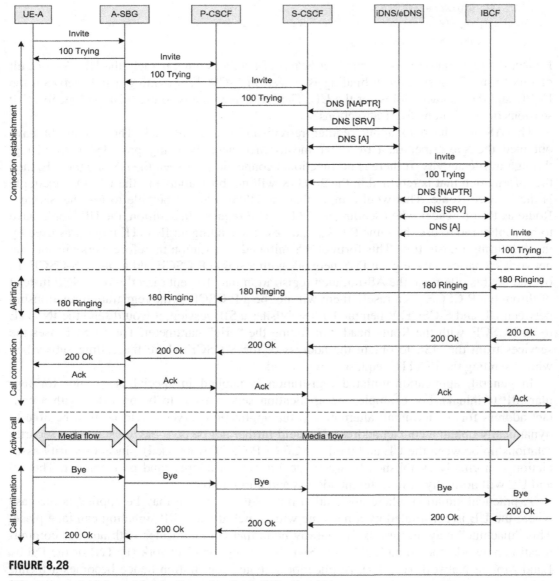

FIGURE 8.28

SIP session establishment (originating subscriber's IMS network).

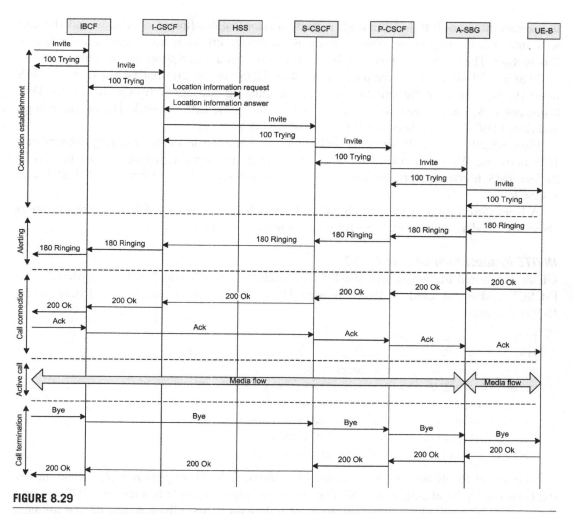

FIGURE 8.29

SIP session establishment (terminating subscriber's IMS network).

with the originating subscriber and destination subscriber belonging to different IMS networks. Explanation is given with reference to Figures 8.28 and 8.29.

The sequence shown in Figures 8.28 and 8.29 applies to fixed access IMS. The sequence is also applicable to IMS used in mobile networks, with the following difference. Mobile IMS includes *preconditions*. Preconditions entail that when a SIP session is established, calling and called parties negotiate an SDP. When SDP is negotiated, calling and called parties ensure that sufficient bandwidth is available over the access network. When the required bandwidth is reserved (i.e. *precondition* fulfilled) for the calling and called parties, the respective party will inform the other party of the fact that precondition is fulfilled. When the precondition is fulfilled for both parties, session establishment continues and the call will enter the alerting stage. The mechanism of preconditions is not considered further in this section. The reader is referred to IETF RFC 3312 for a description of this feature.

Figure 8.28 shows SIP signaling (as well as DNS) in the originating subscriber's IMS network. We consider the situation that the originating subscriber resides in his/her IMS home network, i.e. is registered via a P-CSCF in the Home IMS network. Hence, there is no additional IBCF located between P-CSCF and S-CSCF.

Figure 8.29 shows SIP signaling and Diameter signaling in the terminating subscriber's IMS network. We again consider the situation where the terminating subscriber resides in his/her IMS home network, so there is no additional IBCF located between S-CSCF and P-CSCF.

A brief description of the SIP signaling flow follows below. The SIP message examples show SIP headers that are relevant for the explanation at hand.

INVITE Request from UE-A to A-SBG
UE-A has stored the address of the P-CSCF address through which it had registered. The P-CSCF address is used as Route header. UE-A also includes the Service-Route in the INVITE request:

```
INVITE sip:wendy.jones@company-y.se SIP/2.0
Via: SIP/2.0/UDP 164.48.60.241:55034;branch=z9hG4bK-d87543-
Route: <sip:p-cscsf1.ims-network-x.se;lr>
Route: <sip:orig@s-cscsf1.ims-network-x.se;lr>
From: "John Smith"<sip:john.smith@company-x.se>;tag=64232014
To: "Wendy Jones"<sip:wendy.jones@company-y.se>
Call-ID: 12rjhafvsnadf8wyt5njsdffgmsdfngowe5iu
Contact: <sip:john.smith@153.88.53.230:44816>
```

The use of Route header(s) for sending the initial INVITE request requires some further explanation. In the example, the INVITE request contains a Route header for a P-CSCF and for a S-CSCF, both addresses in the form of a domain name. UE-A routes on the topmost Route header, so, UE-A would have to apply a DNS query to resolve the P-CSCF domain. The following DNS queries would be needed:

- NAPTR-record query, to obtain a list of supported services for this domain.
- SRV-record query, to obtain the host name for the host that supports SIP over UDP or TCP, depending on which transport protocol the SIP phone has been configured to use.
- A-record query, to obtain the IPv4 address (or AAAA-record query if IPv6 address needs to be obtained).

There is typically an access session border gateway (A-SBG) at the edge of the IMS network, between end-user and IMS core network. So, the INVITE request must be sent via the A-SBG. The DNS query for host name (SRV) may, to that end, return the host name of the A-SBG as opposed to the host name of the P-CSCF. The A query then results in the IPv4 address of A-SBG. The INVITE request may then be sent to the A-SBG. Meanwhile, the topmost Route header still contains the domain name of the P-CSCF, p-cscf1.ims-network-x.se in this example.

The DNS query performed by the UE is not shown in the figure. The UE uses its DNS access, which applies iterative DNS resolution in order to obtain the required information from the DNS server responsible for p-cscsf1.ims-network-x.se. The DNS client resident in the host on which UE-A runs (e.g. a desktop-based PC) may have the required DNS answers still in cache.

Instead of using a Route header, a SIP terminal may also be configured to always route to a designated domain or IP address. That domain or IP address may belong to a designated A-SBG and that A-SBG may have an integrated P-CSCF. In this case, the INVITE request may be sent without a Route header for the P-CSCF.

It is further possible that the P-CSCF (whether or not integrated in A-SBG) did not forward the Service-Route in the 200 OK during registration. In that case, the UE did not get a Service-Route, so it will not set a Route header for the S-CSCF. The P-CSCF must then take care to add a Route header to the INVITE request, the Route header being set to the Service-Route applicable for that subscriber.

The Contact header contained in the INVITE request will typically be identical to the contact address that was previously deposited in the S-CSCF. The UE *could* use a different port number for the Contact header in the INVITE request than was used for the Contact header in the (protected) REGISTER request.

The INVITE request will also contain an SDP offer. The media session details contained in the SDP offer will depend, among other things, on the capability of the UE, attached devices (e.g. camera), and user configuration (preferred audio and video codecs).

As soon as the UE has sent the INVITE request, it will be prepared to receive media in accordance with the offered SDP. This implies that it has activated, per offered media line, the appropriate RTP receiver on the port indicated in the SDP offer. When a particular media line in the SDP offer contains more than one codec, the UE will be prepared to receive a media stream on the respective port in accordance with any of the offered codecs!

The 100 Trying response from A-SBG to UE-A would have the following structure:

```
SIP/2.0 100 Trying
Via: SIP/2.0/UDP 164.48.60.241:55034;branch=z9hG4bK-d87543-
From: "John Smith"<sip:john.smith@company-x.se>;tag=64232014
To: "Wendy Jones"<sip:wendy.jones@company-y.se>
Call-ID: 12rjhafvsnadf8wyt5njsdffgmsdfngowe5iu
```

The 100 Trying provisional response is generated by the server transaction (ST) in the A-SBG and not by the A-SBG application. For this reason there is no To tag in the 100 Trying. Chapter 7 explains the concept of server transaction and client transaction (CT).

Processing the INVITE Request at A-SBG

Assuming that the INVITE request was sent by UE-A to the P-CSCF based on a Route header, the A-SBG would, in turn, use that Route header to forward the INVITE request to the P-CSCF. An A-SBG will typically act as back-to-back user agent (B2BUA). In addition, it will apply topology hiding.

Processing the INVITE Request at P-CSCF

One of the tasks of the P-CSCF is to set the network-asserted identity of the calling subscriber, the P-asserted-identity. Here, the P-CSCF has to match the incoming request with a registered subscriber record. If the INVITE request arrives at the P-CSCF over the security association, then this security association forms a pointer to the subscriber record. Otherwise, if no security association is used, the P-CSCF may use the URI contained in the From header and the address in the Contact header as indication of calling subscriber.

```
INVITE sip:wendy.jones@company-y.se SIP/2.0
Via: SIP/2.0/UDP 192.168.34.55;branch=z9hG4bK-oiweykjlshadf
Via: SIP/2.0/UDP ...
Route: <sip:orig@s-cscsf1.ims-network-x.se;lr>
From: "John Smith"<sip:john.smith@company-x.se>;tag=667232
To: "Wendy Jones"<sip:wendy.jones@company-y.se>
Call-ID: klasddjhffjksdahfo934785sdkldjjfh
P-asserted-identity: <sip:john.smith@company-x.se>
Contact: <sip:john.smith@153.88.53.230:44816>
P-charging-vector: icid-value=0de22e1f0266bc409a6da71ce544c3
P-Charging-Function-Addresses: ccf=192.168.34.240; ccf=192.168.34.241
```

The set of Via headers present in the INVITE sent from the P-CSCF to the S-CSCF depends on the behavior of the A-SBG. The A-SBG may, for example, apply topology hiding. In that case, there will be a Via header present from the A-SBG, but not from UE-A. Likewise, when topology hiding applies, the Contact header will contain a contact address relating to the A-SBG and not the contact address provided by UE-A.

The P-CSCF will also add a P-charging-vector header and, optionally, a P-Charging-Function-Addresses header.

Processing the INVITE Request at S-CSCF

The S-CSCF will, when receiving the INVITE request, determine from the topmost Route header that this message is destined for this node. This Route header contains the Service-Route that this S-CSCF had previously sent to the SIP terminal, during registration. There is a subtle detail about this Route header (Route: <sip:orig@s-cscsf1.ims-network-x.se;lr>), namely the *orig* user part. The S-CSCF may receive initial SIP requests related to the establishment of the *originating* SIP session and initial SIP requests relating to the establishment of the *terminating* SIP session. The S-CSCF behaves differently in these two cases. The

S-CSCF uses the Route header, used to forward the initial SIP request to the S-CSCF, to distinguish between originating SIP request and terminating SIP request.

- **Originating SIP request.** The URI contained in the Route header is equal to the Service-Route provided by that S-CSCF to the terminal, during registration.
- **Terminating SIP request.** The URI contained in the Route header is received by I-CSCF from the HSS. The S-CSCF address in the HSS is, in turn, received from S-CSCF during registration (server assignment procedure (Diameter)). The S-CSCF address stored in the HSS is not specifically associated with the originating or terminating procedure and does not contain an orig indication. When S-CSCF receives an initial SIP request with topmost Route header not containing orig as user part, the request is determined to be a terminating request.

Note

Instead of using "orig" as user part in the Service-Route, the S-CSCF could also include a URI parameter indicating originating behavior. This parameter may be a constructed parameter, containing various subscriber information elements.

Having determined that the INVITE request relates to the establishment of the initial SIP session, the S-CSCF has to associate the request with a served user, i.e. the user who is sending the request. As described earlier, this is indicated in the P-asserted-identity header. The S-CSCF tries to find a "match" between the P-asserted-identity in the INVITE request and a stored user profile. Assuming that the S-CSCF contains a user profile for sip:john.smith@company-x.se, the S-CSCF will utilize that user profile for further processing. Otherwise (although that should normally not occur), the S-CSCF would return final response 404 Not Found.

Note

When a SIP terminal receives 404 Not Found, it is ambiguous whether this final response is caused by (i) the *originating subscriber* not being registered in the S-CSCF or (ii) the *terminating subscriber* not being served in the target IMS network (i.e. no matching subscriber record present in the HSS in the target IMS network).

The S-CSCF will, as a first step, augment the P-asserted-identity header. More specifically, the S-CSCF may add a *second* P-asserted-identity to the INVITE request. The P-asserted-identity that is present in the INVITE request may form part of an implicit registration set (IRS); an example is given in Table 8.6.

Table 8.6	
Implicit Registration Set #1	**Implicit Registration Set #2**
sip:john.smith@company-x.se	sip:+46705453658@company-x.se
tel:+46107152111	tel:+46705453658
<potentially more IMPU(s)>	<potentially more IMPU(s)>

Both sip:john.smith@company-x.se and sip:+46107152111@my-company.ims.se constitute public identifier for John Smith. The S-CSCF behaves as follows:

- When the P-asserted-identity contains a SIP URI, S-CSCF adds the topmost Tel URI from the IRS, if present, to the P-asserted-identity.
- When the P-asserted-identity contains a Tel URI, S-CSCF adds the topmost SIP URI from the IRS, if present, to the P-asserted-identity.

So, the P-asserted-identity will, in our example, be constructed as follows:

```
P-asserted-identity: <sip:john.smith@company-x.se>
P-asserted-identity: <tel:+46107152111>
```

The calling subscriber is, from here onward, identified with both a SIP URI and a Tel URI. The rationale is that when the signaling branches out to a non-IMS (and non-SIP) network such as PSTN, the Tel URI P-asserted-identity may be used to set the calling party number in the ISUP initial address message. It is common that subscribers are always provisioned in the HSS with both a SIP URI and a phone number, with the corresponding SIP URI and Tel URI provisioned in an IRS.

Special note should be taken of the following. The INVITE request message may contain two P-asserted-identity headers. The sequence of these two headers is not fixed. As a result, it is not possible to determine from the INVITE message *which* of the two P-asserted-identity headers was set by the P-CSCF and *which* one was added by the S-CSCF.

Note

It is not possible to assign an IMPU to more than one IRS. When a subscriber has multiple IRSs, then it is not possible to assign a phone number (Tel URI) to these multiple IRSs. These IRSs would then need to contain different phone numbers. Practically, the different IRSs may be associated with different terminals of the subscribers and, hence, the different terminals would have different phone numbers associated with them.

An application server acting as back-to-back user agent would be able to alter the P-asserted-identity, though.

The next step from the S-CSCF is the invocation of one or more *services*, based on the subscriber's user profile. These *services* are formed by applications, *service logic*, executed on designated application servers. The user profile may differ per IRS and may differ per IMPU within an IRS. John may have different service invocation associated with sip:john.smith@company-x.se than with sip:+46705453658@company-x.se

A further step of S-CSCF is to apply *outbound routing*. The SIP session established by John Smith is targeted toward sip:wendy.jones@company-y.se. So, the S-CSCF has to forward the INVITE request to a network that serves that target address. We take a closer look at the R-URI, sip:wendy.jones@company-y.se. The addressed user is *wendy.jones*, who is a user of domain *company-y.se*. Domain company-y.se is an administrative domain, in this case an enterprise. The S-CSCF has to obtain details of the network that serves company-y.se. The S-CSCF applies the DNS resolving procedures as already described:

1. Obtain a list of IP services available for the domain company-y.se. Here, S-CSCF will request the NAPTR record associated with company-y.se from DNS.
2. Obtain host name(s) of a SIP server for that domain (offering SIP over UDP or SIP over TCP). Here, S-CSCF will request the SRV record associated with the required SIP service from DNS.
3. Obtain IP address(es) of the host (or IP address(es) of one of the hosts). Here, S-CSCF will request an A-record for the respective host from DNS.

Assuming that enterprise **company-y.se** is served by IMS network **ims-network-y.se**, the S-CSCF would obtain the host name and the IP address(es) of an inbound SIP server of that IMS network (ims-network-y.se) from the DNS. This information facilitates the S-CSCF forwarding the INVITE request to the target IMS network.

A few notes on DNS provisions for this basic SIP session establishment case are as follows:

1. The domain name **company-y.se** needs to be provided in DNS, and the NAPTR record associated with that domain needs to be available (through generic DNS tree structure) to the S-CSCF of the calling party, **ims-network-x.se**.
2. It is further assumed the same (IMS) network serves all users of the administrative domain **company-y.se**. This is because the process for determining the destination IMS network is based on the domain name and does not take the user part into consideration.

Processing the INVITE Request at Outbound IBCF

When the S-CSCF determines, by checking the internal configuration, that the destination IMS network is a different IMS network than the network this S-CSCF belongs to, then the S-CSCF would typically forward the INVITE request to a network boundary node, an IBCF. Forwarding the INVITE request to the inbound SIP server of the destination network would then be done by the IBCF. Figures 8.28 and 8.29 both show an IBCF. There may be an IBCF at the boundary of the originating IMS network, where the INVITE request leaves that IMS network. There may also be an IBCF at the boundary of the destination IMS network, where the INVITE request enters that IMS network. Hence, the SRV record received by the *outbound SIP server* of the originating IMS network would be an SRV record associated with an IBCF of the destination IMS network and not an SRV record associated with an I-CSCF of the destination IMS network.

The outbound IBCF may apply various SIP header policies, e.g. remove designated SIP header(s). This may depend on the relation with the destination IMS network, e.g. whether adequate trust relation is in place. For example, when the INVITE request contains a **privacy** indication associated with the P-asserted-identity indicating that the P-asserted-identity should not be revealed to the called party, but an adequate trust relation between the originating IMS network and destination IMS network is not in place, then the IBCF may remove the P-asserted-identity (and the privacy indicator). In addition, the IBCF may apply topology hiding.

Let's not forget the user plane! The IBCF may force the user plane (carrying the media) to be routed through a transition gateway (TrGW). This may be done for two reasons:

1. The operator may want to apply control over the media plane, such as topology hiding (sensitive information contained in some RTP/RTCP messages).
2. The operator ensures that the control plane and media plane between the two networks are routed via the same IP interconnect infrastructure.

Forcing the media plane to be routed via TrGW requires that the IBCF selects a TrGW for this SIP session. The IBCF then adapts the SDP offer in the INVITE to contain the media termination address associated with the selected TrGW. Likewise, the SDP answer that will eventually be sent back to the calling party will also be modified by IBCF to contain the termination address associated with the selected TrGW (see Figure 8.30).

When the IBCF is ready to forward the INVITE request to the destination IMS network, it will behave as described for the S-CSCF. Namely, it will apply DNS resolution in order to obtain the host name and IP address of the inbound SIP server of the target IMS network.

The routing of the INVITE from the originating IMS network to the terminating IMS network is a topic of ongoing discussion between (IMS) operators. The designated IP transmission network (IPX), for example, may be used for this purpose. There is also further standardization ongoing in 3GPP for the IMS interconnection.

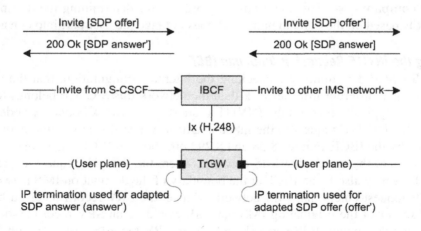

FIGURE 8.30

TrGW used at network boundary. (SDP offer' is adapted SDP offer; SDP answer' is adapted SDP answer).

Processing the INVITE Request at Inbound IBCF

The IBCF at the destination IMS network, acting as inbound SIP server, has a similar role to the IBCF at the origination IMS network, acting as outbound SIP server. It may force the user plane through TrGW, it may apply topology hiding, it may alter designated SIP headers, etc. A further task of the IBCF, when acting as inbound SIP server, is to forward the INVITE request to an I-CSCF. For handling a terminating initial SIP session request, forwarding to I-CSCF is default behavior. However, an incoming INVITE request may also have to be forwarded to another designated device, such as a media server or an application server; this may depend on the R-URI in the INVITE request. For our current explanation, we will consider only the situation where the terminating INVITE request is forwarded to I-CSCF. IBCF has a number of I-CSCF addresses configured and selects one of the available I-CSCFs or uses DNS-based load sharing.

As an implementation option, the IBCF and I-CSCF may be integrated in one node.

Processing the INVITE Request at I-CSCF

The main task of the I-CSCF is to forward the INVITE request to the S-CSCF that is serving the addressed subscriber. The addressed subscriber is identified in the R-URI in the INVITE message. The I-CSCF contacts the HSS by sending a Diameter message location information request (LIR). If the destination subscriber, in our example sip:wendy.jones@company-y.se, currently has an S-CSCF assigned, then the HSS returns the address of that S-CSCF. There are two cases where the subscriber has an S-CSCF assigned:

1. The subscriber is marked in the HSS as "registered". This state implies that the subscriber is registered in an S-CSCF with one ore more contact bindings.
2. The subscriber is marked in the HSS as "unregistered". This state implies that the subscriber's user profile has been sent to S-CSCF, but there is no contact binding available in that S-CSCF.

In case 1, the HSS sends the S-CSCF address to the I-CSCF in a Diameter location information answer (LIA). The I-CSCF can in that case forward the INVITE to the received S-CSCF address, using the S-CSCF address as the topmost Route header. In case 2, the HSS also sends the S-CSCF address to the I-CSCF and I-CSCF will also forward the INVITE to that S-CSCF. The S-CSCF will, in the latter case, apply terminating_unregistered call behavior.

If the destination subscriber does not currently have an S-CSCF assigned, then the HSS will not send an S-CSCF address to the I-CSCF. Instead, the HSS sends *S-CSCF capabilities*. The I-CSCF uses the set of S-CSCF capabilities to select an S-CSCF that can handle the terminating call for the subscriber. Selecting the S-CSCF by the I-CSCF for this case is done in the same manner as S-CSCF selection by I-CSCF during registration.

We will, for now, consider the situation where the destination subscriber is registered, so the I-CSCF has received an S-CSCF address and forwards INVITE to that S-CSCF. The I-CSCF will typically not apply record routing. So, the INVITE request and responses traverse the I-CSCF, but other transactions for this SIP session do not.

Processing the INVITE Request at S-CSCF

When S-CSCF receives the INVITE request, it determines from the topmost Route header that the INVITE is meant for this S-CSCF and it determines that the INVITE relates to a terminating initial request (since there is no orig indication in the Route header). So, the next step for S-CSCF is to find a match for the INVITE. Since we're talking about a *terminating call* here, the *served* subscriber is identified by the R-URI. So, S-CSCF will search its internal records of user profiles to find a record containing sip:wendy.jones@company-y.se. Assuming that Wendy Jones is registered in this S-CSCF, S-CSCF will find the corresponding user profile.

We recall that, for an originating call, when the IMPU through which the served subscriber is identified is determined to belong to an IRS, the S-CSCF may add a second P-asserted-identity. This functionality is not formally applicable for a terminating call. As an implementation option, the S-CSCF may apply R-URI replacement, prior to invoking IMS service(s). The S-CSCF may check whether the R-URI is contained in an IRS. If that is the case, then the S-CSCF may replace the R-URI by the topmost SIP URI in that IRS. In this manner, IMS service invocation for a terminating call for a particular subscriber will always be done with the R-URI (in the INVITE sent up to SIP-AS) containing the same URI, namely the primary SIP URI for this subscriber, regardless of the URI on which the subscriber is called. This may facilitate service logic processing in SIP-AS, specifically the step of matching the R-URI with the subscriber record.

In this description, we are not yet including the invocation of SIP applications from the S-CSCF. That will be described in a later section.

The next step for the S-CSCF is to build a *target set*. Thus far, the call was directed to the destination subscriber's IMPU (or one of the subscriber's multiple IMPUs). Now that the call has arrived at that subscriber's S-CSCF, the subscriber's registrar, the call can be retargeted to the contact address of that subscriber. In line with the principle of SIP, the S-CSCF replaces the R-URI in the INVITE by the contact address associated with that URI. So,

```
INVITE sip:wendy.jones@company-y.se SIP/2.0
```

is replaced by

```
INVITE sip:153.88.47.176:34634; transport=tcp SIP/2.0.
```

In this example, sip:153.88.47.176:34634 is the contact address that Wendy Jones had, during registration, bound to her IMPU. The URI parameter transport=tcp was also provided by Wendy Jones during registration, as part of the contact address. It indicates that Wendy Jones prefers to be contacted through TCP for initial SIP requests. Without transport indication in the contact address, a network default transport is applied, e.g. UDP.

Building a target set by S-CSCF may result in multiple INVITE requests being generated by the S-CSCF. This is the case when Wendy Jones had registered through more than one terminal and two or more of these registered terminals had registered explicitly or implicitly (through implicit registration) with the called IMPU (sip:wendy.jones@company-y.se). The

INVITE request will in this case be *forked* to these multiple contact addresses. This implies that multiple, identical INVITE request messages are generated by the S-CSCF, each sent to one particular contact address.

The next step taken by the S-CSCF is setting the P-called-party-ID. The subscriber was called on a particular IMPU. That IMPU, contained in the R-URI, was replaced by the contact address. In order to preserve in the INVITE the information about the IMPU on which the subscriber was called, the P-called-party-ID header is included in the INVITE message. The P-called-party-ID is filled with the URI contained in the R-URI, just prior to target set creation. The resulting INVITE request message will therefore be:

```
INVITE sip:153.88.47.176:34634; transport=tcp SIP/2.0
P-called-party-ID: <sip:wendy.jones@company-y.se>
```

It should be remembered that the P-called-party-ID contains the URI as available to the S-CSCF just prior to target set creation. This URI may not always be a true reflection of how the destination subscriber was called. The URI used to address the destination subscriber may have been modified already. For example:

- The destination subscriber was addressed by the calling party using a phone number; the phone number was replaced by a SIP URI, resulting from ENUM query by the calling party's S-CSCF.
- I-CSCF in the destination subscriber's IMS network has modified R-URI from SIP URI format to Tel URI format.
- S-CSCF has replaced R-URI by the primary SIP URI from the same IRS.

Service designers (end-to-end solution designers) will realize that these identity alterations should in principle alter only the *format* of an identity, but not the contents thereof. This may be done through careful network configuration (such as providing ENUM and defining IRSs).

A further step taken by the S-CSCF is to add a Route header to the INVITE request. The Route header should contain a copy of the URI that was contained in the *Path* header that was included by the P-CSCF in the REGISTER request and that the S-CSCF had stored for this binding. In this manner, the INVITE request destined for the called subscriber's SIP terminal will be routed through the P-CSCF. For example:

```
INVITE sip:153.88.47.176:34634; transport=tcp SIP/2.0
Route: sip:term@p-cscf2.ims-operator.se;lr
```

Here, please take note of the "term" user part in the Route header URI. This user part was also present in the Path header in the REGISTER request. The S-CSCF has stored the entire Path header, including URI user part and including URI parameters. The reason for the P-CSCF including the "term" user part is as follows: by receiving the INVITE with a Route header including "term", the P-CSCF knows that this INVITE relates to a terminating request message. Without the "term" user part, the INVITE request would be interpreted as an originating request message. A further aspect differentiating originating and terminating initial request messages is the security association and port number:

- An originating request message arriving at a P-CSCF arrives through security association that was established with the SIP terminal (*if* a security association was established); the request message arrives at the port number assigned to that security association.
- A terminating request message arriving at P-CSCF arrives at the default port for SIP. This is because the terminating request for P-CSCF originates from S-CSCF in the same network, residing in a secure zone.

When the REGISTER request for the called party (at the time of registration) is sent through one or more IBCFs, between the P-CSCF and the S-CSCF, the (one or more) IBCFs do not have to add a Path header in the REGISTER request. The reason for this is that the routing of the initial request message from the S-CSCF to the P-CSCF and to the terminal may traverse one or more IBCFs based on DNS configuration or based on S-CSCF configuration; no designated Path header is needed for this.

The S-CSCF is now ready to dispatch the INVITE request. There is a Route header present, so the INVITE will be sent to the URI contained in that Route header, i.e. to the P-CSCF.

Processing the INVITE Request at P-CSCF

The P-CSCF determines from the Route header that this INVITE request is meant for this P-CSCF and will remove that Route header. Next, the P-CSCF determines the served subscriber. This is derived from the R-URI of the INVITE, containing the contact address for this subscriber. The last hop for SIP signaling can now be taken. The INVITE is forwarded to the SIP terminal based on the contact address. It is common for the P-CSCF to apply topology hiding, so the called party's terminal will not receive network information contained in system headers like Via, Record-Route, and Contact.

An INVITE sent to UE-B may run through an access SBG (A-SBG). The A-SBG will force the media to be routed through it as well. Here, the A-SBG adapts SDP offer and answer, in the same manner as it is done by the IBCF.

Processing the INVITE Request at UE-B

Finally, the UE-B receives the initial INVITE request. The UE, e.g. a soft SIP phone, receives the INVITE on the IP address, transport protocol, and port number that it had provided during registration. As discussed, this IP address may be an IP address that it had received from the P-CSCF, as part of the procedure of establishing a security association. The UE will have a listener installed, listening for SIP over UDP or SIP over TCP (whichever is applicable) on the port number previously provided to the S-CSCF. So, when the INVITE request arrives and assuming the UE is operational and has adequate listener installed, the UE takes the role of user agent server (UAS) and starts a new SIP session service logic. The UE has determined from the absence of a tag in the To header that this

INVITE is an initial INVITE request. The server transaction (ST) process of this UAS instance will return a 100 Trying provisional response, in the same manner as all intermediate SIP proxies have generated a 100 Trying. The call may now be offered to the end-user. Various information elements (headers or parameters) from the INVITE request may be used to notify the called party about this call, such as:

- **P-asserted-identity (PAI).** This header may be used as calling party identity; the PAI may contain a SIP URI, a phone number, or both. Either one or both may be shown on the phone's display. The PAI may contain a display name, which may also be shown to the called user.
- **From.** If no PAI is present, then the URI contained in the From header may be used for A-party number or name display. The From header contains a SIP URI or a phone number, but not both. The From header is not asserted by the network. The shown A-party number may, for example, be a service-provided number.
- **To.** The URI contained in the To header may be shown to the user. The To header may be used as informal addressing of the called party. Both the From header and the To header may contain a display name.
- **Subject.** The calling party, or an intermediate SIP application, may add a Subject header to the INVITE, to reveal the subject of the phone call, e.g. Subject: phone conference.
- **SDP offer.** The SDP offer may include one ore more media offers, such as voice and video. When the SDP offer includes both voice and video, the SIP phone may prompt the user to accept the call as a voice call or video call (including voice component and video component). The phone may, based on configuration, restrict the call to voice call.

So, the phone is now in alerting phase and the called party may answer the call or decline the call. To indicate to the calling party that the phone has entered the alerting phase, a 180 Ringing provisional response is generated. The 180 Ringing provisional response traverses the same set of SIP proxies as the INVITE request message (refer to Chapter 7 for the routing of the response message).

The 180 Ringing will typically not yet contain an SDP answer, for the following reasons:

- An SDP answer has to be sent *reliably*; if, besides the 180 Ringing, there is no need for further SIP signaling between INVITE and 200 OK, then the 200 OK is the first response that is sent reliably. So, the SDP answer is included in the 200 OK.
- The user might influence the SDP answer, e.g. indicate whether to answer as voice call or as video call.

From here onward, the SIP signaling between called party, UE-B, and calling party, UE-A, follows the signaling principles described in Chapter 7. If the called party answers, a 200 OK final response is sent, including SDP answer. The INVITE transaction is complete and is followed by an ACK transaction. The SIP session between UE-A and UE-B is established, whereby the SIP session runs via a number of intermediate SIP proxies. A media session between UE-A and UE-B is established, whereby the media session traverses a number of media proxies. This is illustrated in Figure 8.31.

As long as the SIP session between UE-A and UE-B remains established, the corresponding media session may remain active. When the SIP session is terminated, e.g. by UE-A or UE-B initiating a BYE transaction, the respective SIP entities will tear down the media session as well.

8.6.1 Media Gating

The media session is under direct control of the associated SIP session. The media of that media session may start to flow at designated points in the SIP session establishment. This is shown in Figure 8.32. Figure 8.32 and the following description are fixed network oriented. For mobile IMS deployment, preconditions are used. In this case, the through connection of media is dependent on the fulfilment of the applicable preconditions. This is not further illustrated or described in this section. 3GPP TS 23.228 provides further background and description on preconditions.

Figure 8.32 shows only those entities in the SIP signaling path that have an influence on the media transfer. SIP proxies, e.g. S-CSCF, are not shown. Non-SIP entities such as HSS, ENUM, and DNS do not influence (control) media transfer either.

The actions taken by the respective SIP entities, with respect to media gating (i.e. allowing media to pass through), are given in Table 8.7.

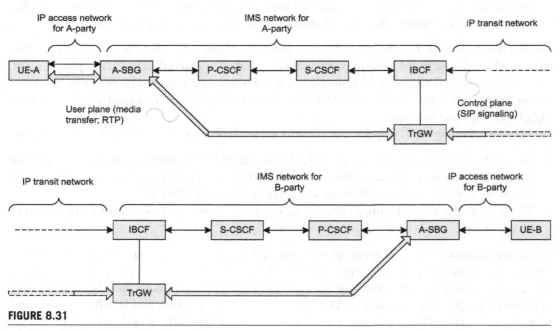

FIGURE 8.31

Established SIP session and media session.

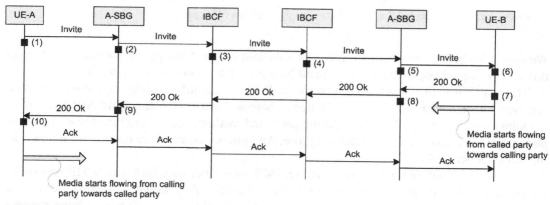

FIGURE 8.32

Media gating for regular call.

8.7 USING PHONE NUMBERS

In the description so far, we have focused mainly on the use of a SIP URI for identifying the destination subscriber for establishing a SIP session or stand-alone transaction and for identifying the originating subscriber. For example:

```
INVITE sip:wendy.jones@company-y.se SIP/2.0
```

Table 8.7 Actions Taken by SIP Entities With Respect to Media Gating

(1)	When UE-A has initiated the SIP session establishment, it is prepared for receiving media in accordance with the SDP offer that it has included in the INVITE. The rationale for being prepared for media reception at this point is the fact that media from the remote party may arrive prior to the 200 OK. In addition, *early media* may be received.
(2)	The A-SBG reserves resources for media transfer in both directions. At this point of the SIP session establishment, A-SBG allows media to flow in an upstream direction, i.e. to the calling party.
(3), (4)	IBCF reserves resources, in TrGW, for media transfer in both directions. IBCF instructs IBCF to allow media to flow in both directions. The rationale is that IBCF leaves it to the respective A-SBGs to apply gating policy and to decide when media transfer may take place.
(5)	The A-SBG for the called party behaves in a similar manner as A-SBG for the calling party. It reserves resources for media transfer in both directions and allows media to flow to the calling party.
(6)	UE-B receives an INVITE request and determines that the SDP offer is acceptable. At this point, it will not expect any media from the calling party yet. Neither will it transfer media to the calling party.
(7)	When B-party answers the call, the SDP answer is provided to the calling party. UE-B prepares its user-plane resources for media reception and media transmission, in accordance with the SDP answer. UE-B will start media transmission.
(8), (9)	As the 200 OK traverses the A-SBG of B-party and the A-SBG of A-party, the respective A-SBGs allow media transfer in both directions.
(10)	UE-A receives the SDP answer and allows media transfer in the forward direction. The ACK transaction does not have a further effect on the media gating in A-SBG.

```
From: sip:john.smith@company-x.se; tag=lkuykluasddflkas
To: sip:wendy.jones@company-y.se
```

We have also briefly mentioned a few times the use of good old phone numbers in IMS. In this section we will describe in more detail how phone numbers are used in IMS.

The capability to use phone numbers in IMS, to identify calling or called subscribers, was introduced *retrospectively*. Hence, whereas the use of meaningful SIP URIs (e.g. sip:john.smith@...) was deemed to be the preferred method, it was concluded that the use of phone numbers was too widespread to ignore. And phone numbers are needed for interworking with CS networks.

We will first look at the way in which an IMS subscriber may establish a SIP session or initiate a stand-alone transaction toward a phone number. A designated schema, defined in IETF RFC 3966, is defined for phone numbers, *tel:*. It allows for defining a phone number as an *address* and uses that address to identify the destination subscriber. The rationale is that IMS subscribers should be able to address another subscriber, e.g. for establishing a call, in the manner they are used to. A number of examples are:

```
Tel:+31163279911
Tel:0163279911; phone-context=+31
Tel:905-206-6500; phone-context=+1
```

The "+" character indicates that the number string represents a global number string. The "-" characters are "visual separators"; they aid in visual representation of a number and have no meaning for URI analysis and call establishment. The Tel URI may be used in call establishment, e.g.:

```
INVITE tel:+31163279911 SIP/2.0
```

This INVITE request message, when initiated by a SIP terminal, follows the regular routing through the P-CSCF and S-CSCF. The S-CSCF then has to process the R-URI, containing the Tel URI, to determine the destination network. Here, two main tasks are defined for the S-CSCF: number normalization and ENUM query. Figure 8.33 shows the position of these two procedures in the S-CSCF. Both will be described in subsequent sections.

8.7.1 Number Normalization

The S-CSCF receives an originating initial INVITE request. It matches the served subscriber, indicated in P-asserted-identity, with a user profile locally available in the S-CSCF. The R-URI contains a phone number. In order to further process this phone number, it has to be normalized into international format, if this is not already the case. Number normalization is done with the use of *phone-context*, a parameter that accompanies a phone number when normalization is needed. So, the number normalization process uses the R-URI and the phone-context as input. The phone-context may have the form of a number string, for

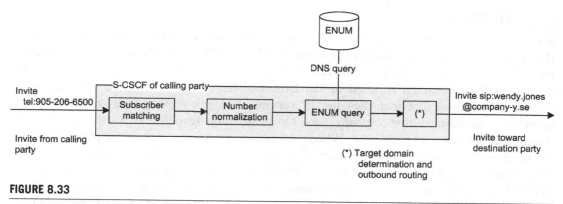

FIGURE 8.33

Number normalization and ENUM query.

example +1. A subscriber from North America would typically have + 1 as phone-context. The phone-context may be made available to the number normalization process in the following ways:

- Subscribed – the user profile, received from the HSS, contains a phone-context.
- Provided by UE – the R-URI in the INVITE request has a phone-context added to it, e.g. INVITE tel:905-206-6500; phone-context=+1 SIP/2.0.
- Provided by A-SBG – the A-SBG may add the phone-context, if this is missing.
- Default – in the absence of provided or subscribed phone-context, the S-CSCF may use a default phone-context. Under normal configuration and operational conditions, this should not occur.

In our example, the INVITE request to tel:905-206-6500 with phone-context +1 would typically be normalized to +19052066500. Remember that the "-" characters are visual separators and do not form part of the phone number. The phone-context is not simply a prefix for the number. Rather, the phone-context is used as *input selector* for the number normalization process. The actual modification of the number depends on the configuration of the number normalization tables in the S-CSCF.

Note

Number normalization in the S-CSCF may be compared with B-number analysis in switching nodes such as the mobile switching services center (MSC) in the GSM network. Whereas number normalization in IMS uses phone-context as input parameter (besides the actual number), B-number analysis in switching nodes uses B-number origin (BO). BO in MSC may, for example, be associated with the IMSI range that the calling subscriber belongs to.

The use of phone-context is described further in IETF RFC 3966.

8.7.2 ENUM query

So, the S-CSCF has now ended up with a number in international format, +19052066500. It must now find out what destination network the INVITE should be sent to. This is done through *ENUM query*. The rationale of ENUM query is as follows. Addressing within the context of the IP domain is based on the Universal Resource Identifier (URI), whereby a *domain* is required that identifies the IP services network serving that resource. When addressing sip:john.smith@company-x.us, the destination resource is a user (john.smith) within the domain company-x.us. When addressing a resource through a phone number, we need to obtain a *user* and a *domain*, for reasons given above. Hence, when addressing John Smith through his phone number, +19052066500, we would need to get a corresponding URI, associated with the same John Smith, in the form user@domain. For this purpose, the ENUM database is defined. ENUM is short for E.164 number. The numbering plan for which the number to URI mapping is defined in IMS is the E.164 numbering plan as defined by ITU-T in recommendation E.164.

So, ENUM would contain various mappings, e.g.:

+19052066500   john.smith@company-x.us
+19052066525   wendy.jones@company-y.us

The E.164 numbering plan has a hierarchical structure. When considering +19052066500, this may be dissected into:

1   Country code (North America)
905   Area code (Ontario, Canada)
206 6500   Subscriber number.

This hierarchical number structure has led to the concept of mapping the E.164 numbering plan into the DNS domain tree. The rationale of defining E.164 numbers as a *domain* is that operators may use their DNS infrastructure for storing and sharing E.164 number to SIP URI mapping information. Also, DNS allows for a distributed database. A particular mapping from E.164 number to SIP URI may then be defined once, in a designated network, and then other operators may have access to it, provided that an adequate agreement is in place.

So, how do we define E.164 phone numbers in a domain tree structure? When considering a number like 19052066500, we notice that when analyzing the number left to right, digits have *decreasing* significance. One may consider the "1" to be more significant ("higher up in the tree") than the digits "905". When, on the other hand, analyzing a domain such as sales.company-x.ims.operator-x.se, one will notice that, on analyzing the domain fields left to right, the fields have *increasing* significance. The field "se" has a higher position in the global domain tree than "operator-x". This reversed order of significance, when comparing E.164 numbers with IP domains, has led to the following convention. To represent an E.164, international format phone number as a domain, we reverse the digit string and we consider all digits as a domain field. This leads to the following representations:

E.164 number	Represented as domain in ENUM
+19052066500	0.0.5.6.6.0.2.5.0.9.1
+19052066525	5.2.5.6.6.0.2.5.0.9.1

These domains, 0.0.5.6.6.0.2.5.0.9.1 etc., could then be administered in DNS and would be accessible by DNS clients, within applicable constraints that may be imposed for DNS queries. Note that the "+" character is not explicitly included as domain field. This is because the mapping is defined for *international* format E.164 numbers, so the E.164 number domains contained in ENUM are by definition international E.164 numbers. There is still one aspect missing. The "1" is not a top-level domain. Hence, 0.0.5.6.6.0.2.5.0.9.1 and other E.164 phone numbers that are represented in this manner cannot be administered in DNS. Rather than internationally defining digits as top-level domain, the E.164 number domains are placed underneath another designated domain, namely e164.arpa. The top-level domain arpa refers to the Advanced Research Projects Agency (ARPA[2]), the creator of the ARPANET, which is the predecessor of the current Internet. The e164 domain indicates that what's underneath (in the tree) represents E.164 numbers. So, the representations in ENUM will be:

E.164 number	Represented as domain in ENUM
+19052066500	0.0.5.6.6.0.2.5.0.9.1.e164.arpa
+19052066525	5.2.5.6.6.0.2.5.0.9.1.e164.arpa

Further details are found in IETF RFC 3761.

When a S-CSCF is handling originating SIP session establishment towards an E.164 number, it constructs an E.164 number domain out of the normalized phone number, as described above. It then wants to obtain a SIP URI relating to this phone number, from ENUM. For this purpose, the DNS NAPTR query is used. A NAPTR query may result in a rewrite rule, based on regular expression. Effectively, a SIP URI replaces the E.164 number domain. So, the next requirement is that for each E.164 number domain, a NAPTR resource record is defined in ENUM. For example:

E.164 number domain	SIP URI
0.0.5.6.6.0.2.5.0.9.1.e164.arpa	sip:john.smith@company-x.us
5.2.5.6.6.0.2.5.0.9.1.e164.arpa	sip:wendy.jones@company-y.us
0.5.5.6.6.0.2.5.0.9.1.e164.arpa	sip:andrew.johnson@ims.operator.us
8.5.5.6.6.0.2.5.0.9.1.e164.arpa	sip:+19052066558@ ims.operator.us; user=phone

When the S-CSCF has carried out the NAPTR query to ENUM and has obtained a SIP URI, it may establish the call or transaction as normal. So, the S-CSCF would, for example, apply a DNS query to obtain the host name and IP address of an inbound SIP server associated with domain company-x.us.

---

[2] The name ARPA was later changed into Defense Advanced Research Projects Agency (DARPA).

When the S-CSCF carries out a NAPTR query to ENUM, but ENUM does not have a record for that number domain, the S-CSCF receives an error response. The S-CSCF will then apply break-out. The S-CSCF routes the INVITE to a break-out gateway control function (BGCF), which then (typically) selects an MGCF for breaking out the call to the CS domain.

Whereas ENUM intrinsically has a DNS-like, distributed data structure, sharing of ENUM data between operators is not yet common practice. One reason for this is the fact that many IMS operators still use PSTN or PLMN for interconnection. Hence, when establishing a call to a phone number representing an IMS subscriber, the call may still have to take a break-out to a CS network. Regular number-based routing is then applied to route the call to the destination network, which may be an IMS network. The routing to the destination network may include methods like number portability and IN services (e.g. CAMEL service).

A further issue with ENUM is that it is not transparent *who* should be responsible for administering the resource records for certain E.164 number domains. Consider, for example, 8.0.9.3.1.6.1.5.6.1.3.e164.arpa. The domain part 1.3.e164.arpa identifies the Netherlands (country code=31). 6.1.3.e164.arpa identifies a *mobile number* within the Netherlands. But for the domains below that, it cannot be stated which operator owns that number domain, due to number portability. A similar issue exists with number portability. Operators within one country often keep a copy of the entire NP database for that country, including numbers that are not served by that operator.

For these and other reasons, IMS operators currently often have in their ENUM database only E.164 number domains administered of their own subscribers.

8.7.3 Public ENUM versus Carrier ENUM

The E.164 number domain format discussed so far, with domain under e164.arpa, is known as Public ENUM. The idea behind Public ENUM is that the subscriber of a particular E.164 number should also be the owner of the IP service that is associated with E.164 number in ENUM. And that owner should then be able to administer his/her resource record in ENUM. In that manner, the subscriber could define, for example, an email address against the E.164 number or an Instant Message service. These IP services may be suitable when Public ENUM is queried by a (VO)IP network that has the capability to use any of these received IP service addresses for establishing the corresponding communication method.

Carrier ENUM is more directed towards (IMS) VoIP operators. Carrier ENUM constitutes ENUM infrastructure that is owned, administered, operated, and accessible by (IMS) VoIP operators. The operators can then ensure that the resource record contained in ENUM will relate to an IMS service. Carrier ENUM, also known as Infrastructure ENUM, is described in IETF RFC 5526.

A third category of ENUM is Private ENUM. Private ENUM entails that the ENUM database is used in a confined area, e.g. within one operator's network. The domains administered in this ENUM do not necessarily have to have e164.arpa as top-level domain.

8.7.4 **Phone Number Representation Through SIP URI**

Besides using a Tel URI for representing a phone number, a SIP URI may also be used. The use of SIP URI instead of Tel URI may be needed in various cases. Two examples of SIP URI containing a phone number are:

```
sip:+31163279911@ericsson.ims.nl;user=phone
sip:+31163279911@ims-operator.nl;user=phone
```

When an IMS subscriber establishes a call to a phone number and uses a SIP URI to contain that phone number, the S-CSCF determines from (i) the user part (which contains number digits) and (ii) the user=phone parameter that this URI represents a phone number. The S-CSCF will then continue SIP session establishment based on the phone number, including ENUM query.

Note

SIP phones that establish a call to a phone number often omit the user=phone parameter. S-CSCFs are often capable, through configuration, to consider a SIP URI that contains phone digits as user part, as a phone number even when user=phone is absent from that SIP URI.

The DNS resource record that is provided in ENUM for a particular E.164 number domain may also be a SIP URI containing a phone number. For example:

E.164 number domain	SIP URI
0.0.5.6.6.0.2.5.0.9.1.e164.arpa	sip:+19052066500@company-x.us;user=phone

The S-CSCF would in this case continue SIP session establishment towards sip:+19052066500@company-x.us;user=phone, for which it would query DNS for the inbound SIP server for the domain company-x.us. Within the IMS network that is serving users of the domain company-x.us, this subscriber would be provisioned in the HSS with that phone number as IMPU, so that person is contactable under that number. One reason for providing a SIP URI phone number in ENUM even when that subscriber has a SIP URI not containing a phone number is as follows. When a call is established for a subscriber based on a phone number and that phone number would be replaced by a non-phone number SIP URI by ENUM, then the fact that this subscriber was called on his/her phone number is no longer visible in the signaling. Such information may, however, be important for charging purposes.

There is a subtle behavior in the I-CSCF for handling calls to subscribers identified with a phone number contained in SIP URI. For this example, where the call is established to sip:+19052066500@company-x.us;user=phone, the I-CSCF would convert the sip:+19052066500@company-x.us;user=phone into tel:+19052066500. Further call handling is then based on the R-URI containing tel:+19052066500 (see Figure 8.34).

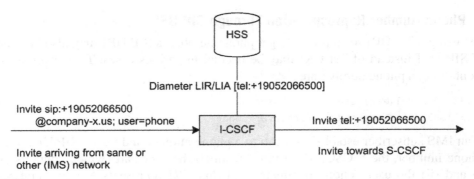

FIGURE 8.34

Number reformatting by I-CSCF.

One specific reason for this number reformatting by the I-CSCF is as follows. The call destined for sip:+19052066500@ims-operator.us; user=phone may have come from a non-IMS network, in which case the call was converted by an MGCF from ISUP signaling to SIP signaling. The MGCF would have received a called party number in ISUP IAM and then creates a SIP INVITE request message containing that called party number as R-URI. The MGCF may be configured to construct an R-URI in sip: URI format or tel: URI format. In the former case, the called party number is used as user part for the SIP URI (+19052066500), but the domain is based on MGCF-internal configuration, e.g. ims-operator.us. The resulting URI would therefore contain a network-generic domain (ims-operator.us) and not an enterprise-specific domain (company-x.us). If the called subscriber is an enterprise subscriber, then that subscriber may not have any URI with this network-generic domain. So, an HSS query for sip:+19052066500@ims-operator.us would fail. That subscriber will, by convention, have a corresponding Tel URI provisioned in HSS. So, by converting the SIP URI into Tel URI by the I-CSCF, the HSS query for this enterprise subscriber will succeed.

8.8 APPLICATION SERVERS IN IMS

This section introduces the concept of application servers in an IMS network. The application server is connected through a number of reference points to the IMS core network. It is through service logic deployment in a SIP application server that an operator may control the multimedia sessions in the IMS network, and other IMS services.

8.8.1 Introduction and Concept

Application servers have a specific and prominent role in IMS. IMS differs from a CS network in this regard. In CS networks, calls may be established without the necessity for an application server (service control point, SCP) to gain control over the call. This is because the CS network itself is fully geared for offering a *telephony service*. In addition, the CS network contains a large set of supplementary services. In the IMS network, on the other hand, the core network elements like the CSCF and HSS allow mainly for *SIP connectivity*, i.e. the ability to establish a communication session (media session) between two end-points in the IMS network, or between a user in the IMS network and a user in another (non-IMS) network. The SIP terminal may assert a limited set of services, such as:

• Call forwarding (SIP terminal may retarget an INVITE request message)
• Call barring (SIP terminal may administer a list of barred calling parties)
• Call transfer (REFER method may be used for that purpose)
• Return of a 3xx-class final response code, to deflect a call to another destination.

Similarly to CS networks, it is desirable for operators to offer services (user experience) that go beyond the basic capability of the IMS network. Also, as will be shown in this section, the IMS network is very conducive for the development and deployment of powerful value-added service (VAS). A VAS is implemented in the IMS network through application servers (AS), also referred to as SIP application servers (SIP-ASs). To start off with, there is an interesting similarity in some of the basic principles of SCP in a CS network and AS in a IMS network. This is shown in Figure 8.35.

As we go through this section, we will see a few more similarities between the two networks, as far as VAS is concerned.

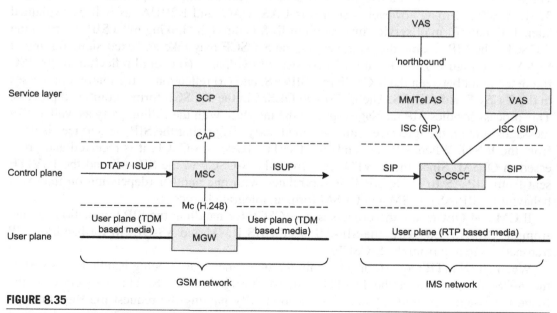

FIGURE 8.35

Comparing SCP in GSM network with SIP-AS in IMS network.

Service assertion in IMS is essentially a matter of manipulating the SIP session that is established between two parties in the IMS network. A SIP application server (AS) is capable of applying actions like changing the destination of a SIP request message ("retargeting"), adding, removing or modifying certain headers, and modifying the SDP offer. A SIP-AS may also generate a SIP transaction within a SIP session, generate SIP sessions, or terminate SIP sessions. The exact capability that is available for a SIP-AS depends on the *role* that that SIP-AS takes. Commercial telecommunications application platforms support these various roles for the SIP-AS.

Whereas the SIP-AS, strictly speaking, acts on the control plane, i.e. on the SIP signaling, it indirectly controls the user plane as well. The characteristics of the user plane, as well as the termination points of the user plane, are described in the SDP. Hence, by modifying the SDP, the SIP-AS has control over the user plane.

When comparing GSM with IMS, it is important to realize that in IMS the *basic telephony services* and *supplementary services* are realized through a combination of the IMS core network (CSCF in particular) and Multimedia Telephony (MMTel) Application Server (AS). Value-added services may be asserted in IMS through a northbound interface in MMTel AS or directly through the ISC reference point. The northbound reference point from MMTel AS, which may be implemented by Parlay X, is described further in Chapter 11.

8.8.2 The ISC Reference Point

The SIP-AS is defined as a SIP proxy in accordance with the SIP specification IETF RFC 3261. A SIP-AS may, however, also act as UAS, UAC, and B2BUA, as will be explained later. The main form of service invocation in IMS is through "linking in" a SIP-AS from the S-CSCF. The SIP session that is traversing the S-CSCF may take an "excursion" through a SIP-AS (see Figure 8.36). Figure 8.36 consists of two parts: (i) overall reflection of the ISC reference point between the S-CSCF and SIP-AS, and (ii) reflection of the internal processes in the S-CSCF and SIP-AS. The ILSM and OLSM in the S-CSCF form a combined process. This process handles SIP messages sent on the interface with the calling party as well as SIP messages sent on the interface with the called party. Triggering the SIP-AS services is done from the ILSM. When SIP-AS sends the INVITE back to S-CSCF, it is received and processed in OLSM. Since the INVITE sent from the S-CSCF to the SIP-AS and the INVITE sent from SIP-AS to the S-CSCF are correlated with one another (depending on the exact behavior of SIP-AS), ILSM and OLSM form an integrated process.

ILCM and OLCM are the processes in S-CSCF for handling the SIP transactions to and from the SIP-AS. Likewise, the SIP-AS has the AS-ILCM and AS-OLCM for handling SIP transactions to and from the S-CSCF.

When an INVITE request, or other initial request message, is being handled by S-CSCF, the S-CSCF may deflect the INVITE request through a SIP-AS. This deflection of the request message constitutes "service invocation". By passing the request message through SIP-AS, the SIP-AS gains control over the request message and may apply required actions

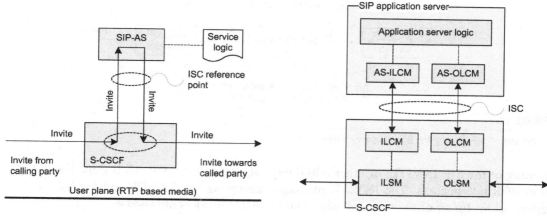

FIGURE 8.36

ISC reference point between S-CSCF and SIP-AS. Key: AS, application server; AS-ILCM, AS incoming leg control model; AS-OLCM, AS outgoing leg control model; ILCM, incoming leg control model; ILSM, incoming leg state model; OLCM, outgoing leg control model; OLSM, outgoing leg state model.

on it. This invocation of a SIP service is based on service trigger information of the served subscriber. This service trigger information is known as initial filter criteria (IFC). The IFC is provisioned in HSS, as part of the user profile, and is transferred to S-CSCF during registration. The IFC consists of a set of trigger definitions, indicating for SIP methods that a service should be triggered. The user profile may contain IFC for different call cases:

- Originating – this IFC will be evaluated by S-CSCF when handling an originating request
- Terminating – this IFC will be evaluated by S-CSCF when handling a terminating request
- Originating_unregistered – this IFC will be evaluated by S-CSCF when handling an originating request for a subscriber who is not registered (→ unregistered)
- Terminating_unregistered – this IFC will be evaluated by S-CSCF when handling a terminating request for a subscriber who is not registered (→ unregistered)
- Call_diversion – this IFC will be evaluated by S-CSCF when handling a retargeted call.

The address of the application server is included in the respective IFC. Let's now have a look in more detail at how the service invocation occurs and how the INVITE may be sent from a SIP-AS back to the S-CSCF. The reference point between the S-CSCF and the SIP-AS, the IMS service control interface (ISC), uses SIP as its protocol. The signaling between S-CSCF and SIP-AS is standard SIP. The SIP-AS may be linked in as a SIP proxy, having the effect that a SIP session may transparently traverse the S-CSCF, then the SIP-AS, and then the S-CSCF again (see Figure 8.37).

The "S-CSCF process" in the figure represents a process in the S-CSCF. The third block, "S-CSCF process", represents the same process as the first block of the same name. This

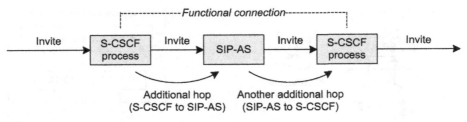

FIGURE 8.37

Transparent SIP routing for SIP-AS service invocation.

routing of the SIP INVITE, or other initial request, through the SIP-AS is done through the user of Route headers. The INVITE messages traversing the S-CSCF and the SIP-AS are given below, for an originating call case. Only selected headers are shown.

```
INVITE from P-CSCF to S-CSCF
INVITE sip:wendy.jones@ims-network-y.se SIP/2.0
Route: sip:orig@s-cscf1.ims-network-x.se;lr
```

The Route header for the INVITE message arriving at the S-CSCF is the Service-Route that the S-CSCF provided to the UE during registration.

```
INVITE from S-CSCF to SIP-AS
INVITE sip:wendy.jones@ims-network-y.se SIP/2.0
Route: sip:service-1@application-server1.ims-network-x.se;lr; session=originating
Route: sip:i33i76412378964178236@s-cscf1.ims-network-x.se:25000;lr
```

The topmost Route header in the INVITE message sent to the SIP-AS is copied from the IFC that is applicable for this call case. It contains the address (domain name) of the application server and (optionally) a user part. The AS may use the optional user part for service logic selection (there may be various services deployed in the same SIP-AS). The parameter session=originating indicates to the SIP-AS that this service invocation relates to the establishment of the originating session. Routing this INVITE request message to the SIP-AS takes place in accordance with standard SIP routing, i.e. the S-CSCF will apply DNS name resolving on the domain of the topmost Route header.

The second Route header is added by the S-CSCF. This Route header is not contained in the IFC, but is constructed by the S-CSCF itself. It contains the address (domain) of this S-CSCF, allowing the SIP-AS to route the INVITE message back to the S-CSCF, after the SIP-AS had done the required inspection and modification on that Route header. The user part contained in the URI in this example Route header, i33i76412378964178236, is known as an original dialog identifier. It is a process identifier, pointing to the process in the S-CSCF that is handling this IFC evaluation. Hence, when S-CSCF receives this INVITE request, it determines from the presence of the original dialog identifier that this INVITE should be forwarded to the respective IFC evaluation process. The S-CSCF may then continue SIP session establishment. This example also includes a port number in the second Route header. In this manner, the S-CSCF receives an initial request messages over the ISC reference point on a designated port number. Hence, when a request message arrives at port 2500, the S-CSCF designates that it is related to the ISC reference point.

The structure of the original dialog identifier is S-CSCF vendor specific. The structure of this element does not affect the SIP-AS. The SIP-AS retains the second Route header in the INVITE message; it does not analyze the original dialog identifier. An S-CSCF may embed various process parameters into the original dialog identifier.

The second Route header may contain an IP address instead of a domain name (host name):

```
INVITE from SIP-AS to S-CSCF
INVITE sip:wendy.jones@ims-network-y.se SIP/2.0
Route: sip:i33i76412378964178236@234.16.38.11:25000;lr
```

As we can see, the SIP-AS has removed the topmost Route header from the INVITE message it had received from the S-CSCF, as would be expected. What remains is the second Route header, which now becomes the topmost Route header. The SIP-AS routes the INVITE message based on this Route header, i.e. the INVITE message is routed back to the S-CSCF.

```
INVITE from S-CSCF to destination
INVITE sip:wendy.jones@ims-network-y.se SIP/2.0
Route: sip:icscf.ims-network-y.se;lr
```

The S-CSCF has received the INVITE back from SIP-AS and continues IFC evaluation. If there are no further trigger criteria in the IFC, then the S-CSCF continues SIP session establishment, based on the R-URI in the INVITE.

Figure 8.38 shows the position of IFC evaluation in the S-CSCF in relation to other processes in the S-CSCF, for the originating SIP session handling.

Number normalization and ENUM query apply only when the R-URI, after IFC evaluation, contains a phone number.

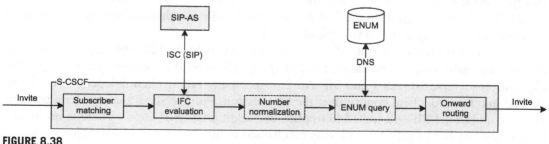

FIGURE 8.38

IFC evaluation in relation to other processes in S-CSCF.

The SIP-AS, acting as a SIP proxy, will remain part of the INVITE transaction signaling, as is normal for a SIP proxy. If the SIP-AS wants to also receive SIP messages after the SIP session has reached the established state (active call), the SIP-AS has to apply record routing or has to act as a B2BUA.

One relevant aspect of service invocation in IMS is the following. As described, a SIP-AS gains control over a SIP session by having the SIP signaling chain traverse the SIP-AS. This is done through transparent proxy routing or B2BUA. This implies that the S-CSCF will, when required, forward an initial request message, such as INVITE, to SIP-AS. The protocol used for session establishment, namely SIP over the Mw reference point, is the same as the protocol used for service invocation, namely SIP over the ISC reference point. As a result, the S-CSCF does not have to perform any protocol mapping. All SIP messages, headers, and parameters, as far as applicable, may be transparently forwarded from the S-CSCF to the SIP-AS and vice versa. This includes standardized or proprietary headers and parameters not recognized by the S-CSCF. This is different when compared with service invocation in CS networks. For CS networks, switching nodes apply a designated protocol for the service invocation, the Intelligent Network (IN) application part (INAP) or the CAMEL application part (CAP). INAP/CAP are different protocols than the protocol used for call establishment, the ISDN user part (ISUP). The service-switching function (SSF) in the switching node has to apply mapping between ISUP and INAP/CAP. This mapping relates to messages and parameters, but also to the state models. This difference in service invocation constitutes an advantage for IMS, compared to IN. New SIP headers and parameters will become available to SIP-AS without impacting S-CSCF.

8.8.3 Service Chaining

An age-old problem is the issue of invoking multiple services. In CS networks, there is limited capability to invoke multiple IN services. CAMEL specifies the possibility to invoke *subscribed services*, *subscribed dialled services*, and *serving network dialled services*. For INAP-based service invocation, vendors typically apply proprietary methods for invoking multiple services in one originating or terminating call.

For IMS, the IFC-based service triggering method allows for transparent invocation of multiple services in succession. When the S-CSCF has invoked one service and has received the INVITE request back from the SIP-AS, it continues IFC evaluation, which may result in a second service being invoked (see Figure 8.39).

Block (1) represents the task of subscriber matching; block (2) represents the tasks of number normalization (if needed), ENUM query (if needed), and onward routing. The application servers, SIP-AS 1 and SIP-AS 2, respectively are invoked independently of one another and have identical control over the SIP session. From a SIP signaling chain point of view, an additional SIP-AS linked in implies two additional "SIP hops". There is no formal limit to the number of SIP-ASs that may be chained in this manner. The invocation of two or perhaps three SIP-ASs from S-CSCF is normal for a SIP call – for example, one SIP-AS for call control application and another SIP-AS for a designated number screening service.

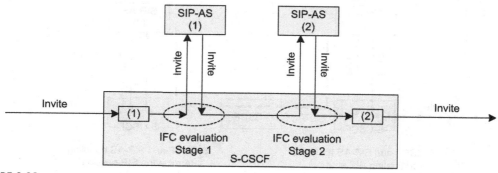

FIGURE 8.39

Service chaining in IMS.

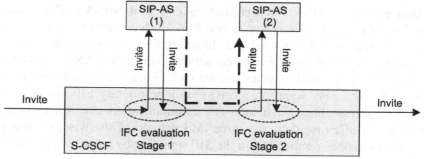

FIGURE 8.40

Interaction between SIP-ASs for one call half.

Although the services are invoked independently of one another and may not be aware of one another's existence, they can influence one another. For example, service logic in SIP-AS 1 may alter certain SIP headers. The SIP signaling continues with the modified SIP headers and so SIP-AS 2 receives the modified SIP header. If SIP-AS 2 modifies a SIP header or changes the destination of a SIP request, then SIP-AS 1 may not be aware.

There exists the possibility, though, for the SIP-ASs invoked for one call to interact with one another. Two examples are given below:

- **Interaction between SIP-ASs for one call half.** Refer to Figure 8.40. The dashed line represents the transfer of a proprietary SIP header. The nature of SIP allows an entity, such as SIP-AS, to add a proprietary SIP header into a request message. This SIP header is transferred transparently in the forward direction. In this example, the proprietary header would be transferred to S-CSCF and from S-CSCF to SIP-AS 2. This header may, for example, contain information that would be needed by SIP-AS 2 to apply charging. SIP-AS 2 would remove the header from the INVITE request.

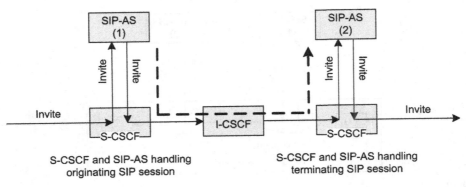

FIGURE 8.41

Interaction between SIP-AS of originating call and SIP-AS of terminating call.

- **Interaction between SIP-AS of originating call and SIP-AS of terminating call.** (Refer to Figure 8.41.) In this example, one SIP-AS, handling an originating call, wants to convey information to another SIP-AS, handling the terminating part of the same call. This example may relate to an enterprise service, where SIP-AS 1 wants to convey to SIP-AS 2 that the call qualifies as an "on net call", i.e. a call between users of the same enterprise. SIP-AS 2 may want, on receiving this proprietary SIP header, to apply special service logic handling. Such inter-AS communication may typically apply when the respective S-CSCFs reside in the same IMS network. Otherwise, the proprietary SIP header would have been removed from the SIP signaling by a border gateway, IBCF.

8.8.4 SIP-AS as proxy, B2BUA, UAC, or UAS

We have shown in previous sections how a SIP-AS may be linked into the SIP signaling chain by an S-CSCF. The SIP-AS may then remain in the SIP chain, as regular SIP proxy. The SIP-AS may, however, assume different roles. Different roles for the SIP-AS give the SIP-AS different capability, but also have an effect on the core network, on the way in which the S-CSCF will continue the SIP session establishment. The four roles that a SIP-AS may take are: proxy, back-to-back user agent (B2BUA), user agent client (UAC), and user agent server (UAS). We will describe each of these briefly. It will become clear that the ISC reference point has some distinctively different characteristics than the Mw reference point (between CSCFs)!

8.8.4.1 Proxy

The SIP-AS forwards the INVITE request message (or other request message) back to the S-CSCF. The dialog that is established from the S-CSCF to the SIP-AS is the same as the dialog that is established from the SIP-AS to the S-CSCF (see Figure 8.42).

There may be multiple SIP dialogs established through the SIP-AS. This may result from SIP forking by the S-CSCF, for example. These multiple dialogs traverse the SIP-AS transparently.

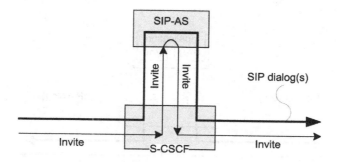

FIGURE 8.42

SIP-AS acting as SIP proxy.

The capability for the SIP-AS, when acting as a SIP proxy, is limited. It may modify the R-URI and it may modify non-system SIP headers. When the establishment of the SIP session results in an unsuccessful final response, it may retarget ("forwarding proxy"), by sending a new INVITE request to a different destination than the previous one (unsuccessful INVITE).

When the SIP-AS applies retargeting the following will be considered:

- When retargeting is applied during termination of call handling and SIP-AS has not yet responded to S-CSCF, the S-CSCF will stop terminating call handling when it receives the retargeted INVITE request. The S-CSCF will create a diverted (retargeted) call, directed to the modified R-URI.
- When retargeting is applied for an originating call by the SIP-AS in response to receiving an unsuccessful final response, the S-CSCF will not evaluate the remaining IFC again. This is because the remaining IFC was already evaluated when the S-CSCF had received the first INVITE back from SIP-AS.

A SIP-AS acting as SIP proxy can be further subdivided into the following categories:

1. Non-record routing proxy. The SIP-AS will be part of the INVITE transaction. Subsequent SIP signaling will not pass the SIP-AS. This method may be applied when the SIP-AS needs the capability to influence session establishment, e.g. number translation, but has no interest in further controlling the call.
2. Record routing proxy. The SIP-AS will be in the entire SIP session signaling flow. This gives the SIP-AS the ability to modify SIP headers, as deemed necessary, for the entire call. One example is as follows. If the SIP-AS wants to modify the URI contained in the From header or in the To header, that modification should be done consistently throughout the call, i.e. in all SIP messages. The reason for this is that SIP entities that were designed in accordance with RFC 2543 may use the From URI or To URI for dialog recognition. Hence, modifying the From URI in the INVITE request message only may lead to a dialog problem for a later SIP message.

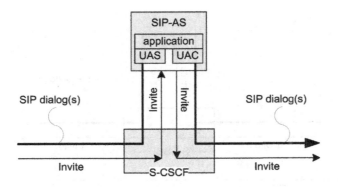

FIGURE 8.43

SIP-AS acting as back-to-back user agent.

8.8.4.2 Back-to-back user agent

By acting as a Back-to-Back User Agent (B2BUA), the SIP-AS gains more control over the call (see Figure 8.43). The INVITE request message arriving at the SIP-AS is terminated in a user agent server (UAS) process. For the INVITE that the SIP-AS sends back to the S-CSCF, the SIP-AS starts a UAC process. An application in the SIP-AS correlates these two processes and conveys SIP messages between the UAC process and the UAS process. There are now separate SIP dialogs upstream of the SIP-AS and downstream of the SIP-AS. Acting as B2BUA is more complex for the SIP-AS (and, hence, more costly to implement and test). This is because the SIP-AS has to map various system headers (e.g. Call-ID, Contact, Record-Route) and has to map the dialog tags (From tag, To tag). However, acting as a B2BUA gives the SIP-AS more capability. Specifically, the SIP-AS has independent control over the call legs. This gives the SIP-AS the following capabilities (not exhaustive):

- **AS-controlling parallel alerting.** Instead of letting the S-CSCF fork the INVITE to multiple registered contact addresses, the AS sends multiple INVITE messages to S-CSCF, one for each device to which the call shall be offered. This method requires that the respective devices have registered with IMPUs that do not reside in the same IRS.
- **Termination of a call at any moment.** For example, due to charging restrictions (credit depletion).
- **Creation of follow-on calls.** When the remote party ends the active call, the SIP-AS can create a new INVITE request message to a new destination.
- **Generation of mid-call announcements.** This requires connecting the calling or called party to an announcement device during the established SIP session.
- **Creation of conference calls.** This requires the establishment of one or more additional outgoing SIP sessions, as well as connecting the user plane of the participants of the call to a conference bridge.

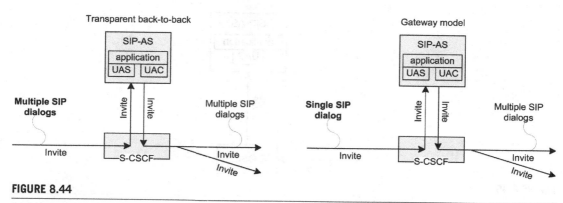

FIGURE 8.44

SIP-AS applying the gateway model.

One very prominent use of the SIP-AS acting as B2BUA is the "SIP gateway model" (see Figure 8.44). The left SIP-AS in Figure 8.44 illustrates a scenario where the SIP-AS maps each dialog that is created downstream to a corresponding dialog upstream. The multiple downstream dialogs may also have resulted from the SIP-AS itself creating multiple SIP sessions, i.e. SIP-AS sending multiple INVITE messages (AS-controlled forking). For each of these dialogs, there may be independent SDP offer–answer sequences; the SIP-AS conveys the SIP messages transparently, whilst acting as a B2BUA. The calling party, which may be a SIP terminal, a media gateway controller or other entity, will in this case be capable of receiving multiple SIP dialogs.

The right SIP-AS in Figure 8.44 applies the gateway model. There are multiple dialogs downstream. However, the SIP-AS maps these multiple downstream dialogs to a single upstream dialog. The SIP-AS may be compelled to apply this method when the calling SIP entity, e.g. a gateway, does not support multiple early dialogs.

The gateway model is more complex than the transparent B2BUA. For the gateway model, SIP-AS may have to map the SDP answer received through one dialog on to an UPDATE request message to the calling party. This is specifically required when SIP-AS had already forwarded one SDP answer to the calling party. A second SDP answer received by SIP-AS would have to be mapped to a new "SDP offer" by the SIP-AS, conveyed in an UPDATE transaction! Not all use-cases can be supported through the gateway model, including for example preconditions for the mobile IMS.

8.8.4.3 User Agent Server

When a SIP-AS receives a SIP INVITE request (or other request), it may act as UAS in the strict sense of the word. That is, the SIP-AS starts a UAS process and responds to the calling party through this UAS process, without creating a corresponding UAC process (see Figure 8.45).

The UAS provides a final response to the initiating party, which then closes the INVITE transaction. Typically, the final response will be an unsuccessful final response, since there

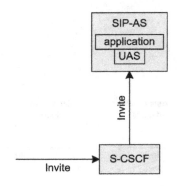

FIGURE 8.45

SIP-AS acting as UAS.

is no SIP session to be established. A SIP-AS may apply the UAS model when it is applying incoming call barring, for example. The SIP-AS is being invoked from the S-CSCF for a terminating call and SIP-AS determines that the call may not be established. The SIP-AS may in that case return a 486 Busy Here response:

```
SIP/2.0 486 Busy here
Retry-after: 1200 (Out of office)
```

The 486 Busy Here response is followed by an ACK from the calling entity up to the SIP-AS. The Retry-After header provides additional information to the calling party. Service logic in the SIP-AS may determine (calculate) that, from this moment, the call will be available in 1200s (20 minutes). The "(Out of office)" text string is free-format text that may be compiled by the service logic. This example may apply for a personal-assistant service with coupling with a person's Outlook agenda.

The fact that a SIP-AS is acting as a UAS, as opposed to a proxy or B2BUA, is not explicitly known or detected by the S-CSCF. The S-CSCF sends the INVITE request up to the SIP-AS, over the ISC reference point, and is prepared to receive an INVITE back containing the original dialog identifier. Instead, S-CSCF receives a *486 Busy Here* response, without having received an INVITE back. This terminates the INVITE transaction for the S-CSCF, after the ACK is processed.

8.8.4.4 User Agent Client

A SIP-AS acting as user agent client (UAC), finally, implies that the SIP-AS is generating a call not related to ongoing call handling. When SIP-AS is acting as a B2BUA, it already includes a UAC instance.

8.8.5 Public Services

A prominent set of services in IMS is formed by public services. Public services represent services that are generally accessible by subscribers of the IMS network or from other networks. Public services constitute, functionally, terminating services, residing in a SIP application server (SIP-AS). Subscribers who want to make use of these services do not need a subscription for it. Public services in IMS are identified with a SIP URI or a Tel URI, in accordance with the respective IETF standards (RFC 3261 for SIP URI and RFC 3966 for Tel URI). A URI that represents a public service is known as public service identity (PSI).

Note

A Public *service* identity (PSI) is not to be confused with public *user* identity (PUI): the term PUI was initially used to identify a user.

Two forms of PSI are distinguished, *distinct PSI* and *wildcard PSI*:

- **Distinct PSI.** This category of PSI relates to a service for which a distinct identity exists for access to that service. For example:
  sip:helpdesk@service-company.nl
  tel:+31163279000
- **Wildcard PSI.** This category of PSI relates to a service that may be accessed by using a public identity that falls within a regular expression. A few wildcard PSI examples are given, with italics below denoting specific PSIs that would match that particular wildcard PSI definition:

  sip:freephone-!.*!@service-company.nl
  sip:freephone-service@service-company.nl
  sip:freephone-sales@service-company.nl
  sip:freephone-support@service-company.nl
  tel:+31800!.*!
  tel:+318001234
  tel:+31800567811
  sip:+31900![0-9]{4}!@ims.operator.nl
  sip:+319001234@ims.operator.nl
  sip:+319009876@ims.operator.nl

A (wildcard) PSI represents a service that is hosted by a *particular* IMS operator. The PSIs that are given above, for example, may represent services that are hosted by different IMS operators. When a SIP session is initiated toward a PSI, the routing of the SIP signaling to the IMS network hosting that PSI is done via standard SIP routing methodology. This implies, among other things, that when a SIP session is established to a PSI from the IMS domain and the PSI is addressed as Tel URI, then ENUM is needed to obtain the corresponding SIP URI. This is a special form of ENUM query, namely *wildcard ENUM*. For example, if a call is established to tel:+318001234, then the ENUM query will be made on the following domain: 4.3.2.1.0.0.8.1.3.e164.arpa. The following SIP URI may be returned: sip:+318001234@ims-operator-x.nl; user=phone. ENUM would, for this wildcard PSI,

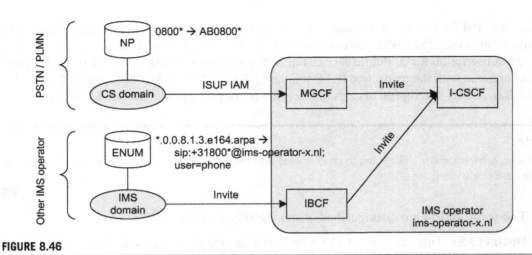

FIGURE 8.46

Routing of SIP INVITE to IMS network operating a PSI.

still have a single entry only, namely 0.0.8.1.3.e164.arpa. Adequate replacement rules are then provided to get the ENUM client to construct the required PSI URI. Call routing then continues to sip:+318001234@ims-operator-x.nl;user=phone, for which the first step by S-CSCF is determining the address of a SIP server of the IMS domain ims-operator-x.nl.

Break-in and break-out, as described, also apply for calls to (wildcard) PSI services. Figure 8.46 represents call routing to a wildcard PSI service hosted in a particular domain. In this figure, a call is established to +318001234, which is a specific number of the range +31800*. That range represents a wildcard PSI hosted by IMS operator-x.nl. Figure 8.46 shows control plane only.

When the call is established in a CS domain, the routing of the call to the IMS domain may be based on number block assignment or on number portability (NP). An NP database may provide a network routing number, e.g. "AB" in the example. When the call is established in the IMS domain, the routing of the call is based on the ENUM query. (There could also be break-out to the CS domain, followed by routing to the IMS domain and then break-in to the IMS domain.)

When the call was established in, or routed through, the CS domain, MGCF will convert from ISUP initial address message (IAM) to SIP INVITE request message. The R-URI in the INVITE contains the called party number from ISUP IAM, +318001234. R-URI may be a Tel URI or a SIP URI, depending on MGCF configuration.

When the call was established in the IMS network and there is no break-out to the CS domain, then the SIP INVITE request message arrives at IBCF. The R-URI in the INVITE request message contains the specific PSI (+318001234) in SIP URI format, sip:+318001234@ims-operator-x.nl; user=phone.

Once the INVITE request message destined for +318001234 has arrived at the IMS network, the MGCF or the IBCF sends the INVITE request message to an I-CSCF of this operator.

Note

It is assumed that the MGCF sends INVITE to the I-CSFC. The MGCF may be configured to select another entity to receive an incoming INVITE, e.g. based on the called party number in ISUP IAM.

If the call was established in the same IMS network as the network hosting the PSI, then the call will not be routed through the IBCF.

The I-CSCF now receives the INVITE containing the specific PSI as destination number. We distinguish three methods to invoke the PSI service:

1. Subdomain-based PSI triggering
2. Direct PSI triggering
3. Indirect PSI triggering.

We will explain each of these routing methods briefly.

8.8.5.1 Subdomain-based routing

Subdomain-based routing entails that the I-CSCF determines, based on the domain of the PSI, that the INVITE should be sent to a designated application server (see Figure 8.47). The reference point between the I-CSCF and the IP-AS is called the Ma reference point. Subdomain-based PSI routing is applicable, as the name implies, for cases where the PSI is reflected in the domain, rather than in the user part. For the PSI used in the example, sip:+318001234@ims-operator-x.nl; user=phone, subdomain-based PSI routing would not be applicable. A PSI like sip:heldesk.company-x.nl would be more logical to use for this method. With subdomain-based PSI routing, the mapping of PSI to SIP-AS is done in I-CSCF, through network configuration.

The I-CSCF can now forward the INVITE to the SIP-AS. The address of the SIP-AS is configured in the I-CSCF for this particular PSI. The PSI may still have a user part

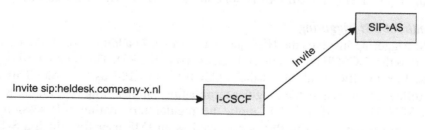

FIGURE 8.47

Subdomain-based PSI routing.

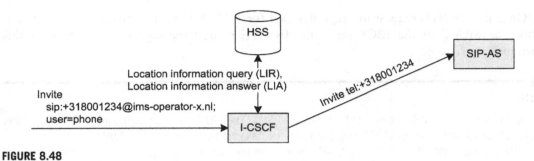

FIGURE 8.48

Direct PSI triggering.

associated with it; the user part may be used by the SIP-AS to select the required service or to adapt the processing of the selected service logic.

The SIP-AS will detect from the INVITE that it is not received over the ISC reference point. For example, the INVITE contains a single Route header; there is no second Route header with original dialog identifier. The SIP-AS must apply service logic behavior that is specific for this form of service invocation.

8.8.5.2 Direct PSI triggering

Direct PSI triggering involves an HSS query as normal (see Figure 8.48). In our example, the R-URI in the INVITE arriving at I-CSCF contains a phone number (the user part contains digits, preceded by a "+", and the user=phone parameter is present). The I-CSCF therefore converts the SIP URI to Tel URI. This is standard behavior, not specific for PSI. The I-CSCF uses the Tel URI for the HSS query, Diameter location information request (LIR). The HSS applies *subscriber matching*. In this example, tel:+318001234 matches the wildcard PSI tel:+31800!.*!. The HSS has a SIP-AS address configured for this wildcard PSI and returns that address to the I-CSCF in Diameter location information answer (LIA). The I-CSCF, in turn, behaves in the same manner as if the address received from the HSS were an S-CSCF address. The I-CSCF forwards the INVITE to the SIP-AS.

Again, the SIP-AS detects from the INVITE that it has not been received over the ISC reference point and applies the appropriate service logic behavior. With direct PSI triggering, the mapping of PSI to the SIP-AS is done in the HSS, through configuration.

8.8.5.3 Indirect PSI triggering

Indirect PSI triggering involves the HSS query as normal. In addition, the INVITE for the PSI is forwarded to the S-CSCF that is currently assigned to the PSI. The S-CSCF triggers the PSI service based on the IFC available in the S-CSCF for this PSI user profile. This method of triggering follows, to a great extent, regular terminating call handling (see Figure 8.49).

The I-CSCF, HSS, and S-CSCF apply the regular terminating SIP session establishment procedure for the PSI. The PSI is provided as an IMS user for which a S-CSCF may be assigned. The first time a SIP session is established for a PSI for which indirect PSI

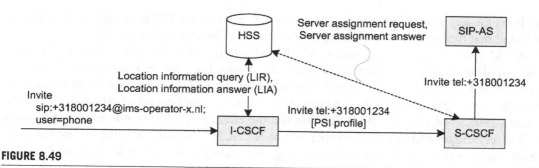

FIGURE 8.49

Indirect PSI triggering.

triggering applies, the INVITE will be forwarded from the I-CSCF to the S-CSCF as for a regular terminating INVITE request. The I-CSCF will select an S-CSCF that conforms to the S-CSCF capabilities indicated by the HSS. The S-CSCF does not have a user profile for this PSI, so it will request a profile from the HSS, using Diameter server assignment request/answer messages. This S-CSCF becomes assigned to this PSI and the registration state of the PSI changes to unregistered. Presuming that the PSI user profile contains IFC, the S-CSCF invokes the PSI service. This time, the PSI service triggering is done through the ISC reference point, including a second Route header with original dialog identifier. The session case for this form of triggering will be terminating_unregistered.

The next time that a SIP session is established for this PSI, the HSS provides the address of the assigned S-CSCF to the I-CSCF and the I-CSCF forwards the INVITE to that S-CSCF. The S-CSCF then determines that it has a user profile for this PSI, so it can invoke the required service without contacting the HSS to obtain a user profile. All in all, indirect PSI triggering is identical to terminating unregistered service triggering.

Although the PSI service is now invoked via the ISC reference point, the PSI service must still apply specific handling. It cannot just forward the INVITE back to the S-CSCF, based on the second Route header, since the PSI is not *registered* in S-CSCF; there is no contact address for the PSI.

A special SIP header is defined to facilitate the wildcard PSI matching in the S-CSCF. When the HSS receives the Diameter LIR for a wildcard PSI, it allows the SIP session request to be forwarded from the I-CSCF to the S-CSCF that is currently assigned to this wildcard PSI. So, when the S-CSCF receives a terminating INVITE for this wildcard PSI, it *could* do the matching, i.e. match tel:+318001234, as received in the R-URI in the INVITE, with tel:+31800!.*!, which is the wildcard PSI stored in the S-CSCF. To prevent that the S-CSCF has to support the wildcard matching procedure (for which regular expressions have to be applied), the P-profile-key SIP header is introduced. The HSS has applied the matching procedure and has found the wildcard PSI for this call (tel:+31800!.*!). The HSS includes the wildcard PSI in the Diameter LIA to the I-CSCF and the I-CSCF includes the wildcard PSI in the INVITE request to the S-CSCF. The latter will be done in the P-profile-key parameter. The S-CSCF can now retrieve the PSI user profile, if available, from internal memory, without having to apply the regular expression check.

A PSI service may be accessed regularly, especially when the PSI service is a service like a (public) helpdesk, information desk, etc. Every time the PSI service is called, the S-CSCF will refresh the unregistered state time for the PSI. If this timer is set to 10 minutes, then as long as calls for this PSI arrive with a maximum interval of 10 minutes, the PSI remains associated with this S-CSCF. This has the effect that a PSI that is triggered through the indirect trigger method tends to stick to one S-CSCF. "Breaking the PSI up" into smaller number ranges may mitigate this shortcoming; for example, instead of having one wildcard PSI tel:+31800!.*!, the following wildcard PSIs may be defined:

Tel:+318000!.*!
Tel:+318001!.*!
Tel:+318002!.*!
Etc.

These PSIs may end up being spread over a multitude of S-CSCFs. Also, the average duration between successive calls to this PSI will increase, increasing the chance that a PSI is de-registered from an S-CSCF, after which an S-CSCF (another S-CSCF) may be assigned to that PSI again.

The subdomain-based PSI triggering method and the direct triggering method access the PSI service directly from the I-CSCF. In that case, the PSI service is not tied to one S-CSCF (it is not tied to any S-CSCF), which may lead to better scaling and load distribution.

Figure 8.50 shows the architecture for a typical use-case for PSI service triggering. The PSI service triggering in this example is done through the indirect triggering method. In the example in Figure 8.50, the PSI service may be a helpdesk number, for example. The service applies a round robin method to establish a call to an available helpdesk agent. Here, the SIP-AS applies retargeting. One aspect of terminating service logic handling is as follows.

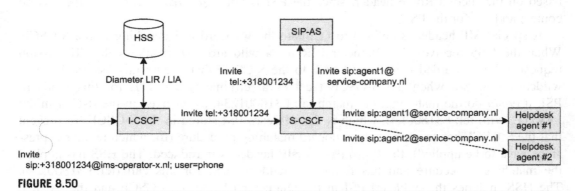

FIGURE 8.50

Architecture for PSI service with indirect triggering.

When the service logic changes the R-URI it has received in the incoming INVITE and sends the INVITE back to the S-CSCF over the ISC reference point, it applies *retargeting*. The S-CSCF will in that case stop IFC evaluation and start originating call establishment. The originating call establishment in this case is done on behalf of the served subscriber, which in this case is the PSI service. The PSI service profile is available in the S-CSCF; the registration state of the PSI service in the S-CSCF is unregistered. The S-CSCF will now establish the call to the indicated target, e.g. a helpdesk agent. The establishment of the retargeted call includes number normalization (if applicable), ENUM query (if applicable), and onward routing. If the S-CSCF supports *originating unregistered service triggering*, the S-CSCF may at this point apply IFC evaluation for the retargeted call.

Although the subdomain-based PSI triggering and direct PSI triggering methods are efficient ways of triggering a PSI service, they have an impact on service logic behavior. For a PSI service that is triggered through subdomain-based PSI triggering or through direct PSI triggering, there is no user profile in an S-CSCF. A PSI service may want to retarget the incoming call, e.g. to connect the calling party to an announcement device or to an agent. Such resource connection or call establishment would have to be done through an S-CSCF. The S-CSCF is needed for onward routing and for break-out to the CS domain, if applicable. However, since there is no S-CSCF assigned to this PSI, to which S-CSCF can the PSI service send the INVITE request? The PSI service was not triggered through the ISC reference point, so it does not have a relation to a S-CSCF.

The PSI service may in this case have to establish the retargeted call through an I-CSCF with an indication of the *originating* call (**orig** parameter in the Route header). I-CSCF can, through HSS query, forward the call to an S-CSCF, which it has selected based on S-CSCF capabilities received from the HSS. The selected S-CSCF determines, when receiving the INVITE, that it has no user profile and then applies a Diameter server assignment request (SAR) to retrieve the user profile, so it can establish the SIP session. For this purpose the S-CSCF would have to support originating unregistered service triggering. The next time the PSI service establishes a retargeted call, the S-CSCF may still be assigned to this PSI. The I-CSCF would receive that S-CSCF's address from the HSS and forward the INVITE to that S-CSCF. That S-CSCF then does not have to retrieve the user profile from the HSS, since it has the profile already. The SIP INVITE sent from the SIP-AS to the I-CSCF would have to contain the P-served-user-identity, containing the wildcard PSI (e.g. tel:+31800!.*!) and not a specific PSI (e.g. tel:+318001234). This is because, in the latter case, the HSS would have to perform subscriber matching with regular expression analysis (when receiving a Diameter location information request, LIR, from I-CSCF) and the same for the S-CSCF, when receiving the INVITE from I-CSCF.

Hence, whereas the *triggering* of a PSI service may be done efficiently through subdomain-based triggering or direct triggering, the service logic execution may still require that an S-CSCF be assigned to the PSI.

A distinct PSI may overlap with a wildcard PSI. For example, we define the wildcard PSI tel:+31800!.*!. We also define the distinct PSI tel:+318001234. When a call is established to tel:+318001234, this would match in the HSS with the distinct PSI tel:+318001234 and not with the wildcard PSI tel:+31800!.*!.

8.8.6 Service-initiated Session Establishment

The UAC model may be used by the SIP-AS for creating a new call, also known as *call out of the blue* (see Figure 8.51). Figure 8.51 shows two cases where the SIP-AS generates a new call. The generation of a SIP session from the SIP-AS toward a remote party will not be sufficient for setting up a "call". To set up a call, the SIP-AS will have to establish two SIP sessions to exchange user plane details (SDP offer/answer) with the respective remote parties. But let's first consider one single SIP session generated by the SIP-AS.

The SIP session that is established by the SIP-AS will be an originating SIP session. The services of an S-CSCF for handling this originating SIP session are needed for:

- Originating service invocation. An originating service may, for example, apply charging for the subscriber on whose behalf the call is established; this charging service may work independently of the service logic from which the call is established.
- Number normalization, if applicable.
- ENUM query, if applicable.
- Outbound routing, to the remote party. This may be a destination in a IMS network or in another network, in which case break-out applies.

For these purposes, the SIP session will be established through the S-CSCF or, more specifically, through the S-CSCF where the served subscriber is registered. The served subscriber is, in this regard, the subscriber whose user profile will be used for the handling of the SIP session in the S-CSCF. This subscriber may or may not be registered in an S-CSCF at this moment. The SIP-AS may not have this knowledge. The SIP-AS may use the Sh reference point, which is defined between a SIP-AS and the HSS, to obtain the address of the

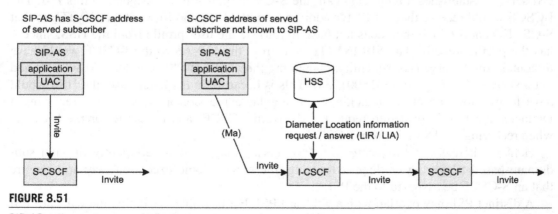

FIGURE 8.51

SIP-AS acting as user agent client.

S-CSCF currently assigned to the served subscriber. This served subscriber may, for example, be the subscriber who is using a click-to-call service on the Internet. If the SIP-AS has obtained the S-CSCF address for the served subscriber, the SIP-AS may send the INVITE to that S-CSCF, via the ISC reference point. Otherwise, the SIP-AS may send the INVITE via the I-CSCF, as will be explained.

Sending the INVITE via the I-CSCF instead of via ISC has some advantages:

- SIP-AS does not have to determine whether the served subscriber is registered and in which S-CSCF.
- When an S-CSCF uses different port numbers for the ISC reference point and for the Mw reference point, SIP-AS would have to take measure to obtain that port number. The S-CSCF address stored in the HSS is the S-CSCF that is provided to I-CSCF, in response to Diameter message location information request (LIR). That address is then used to forward a request message to S-CSCF over the Mw reference point.

When SIP-AS sends an INVITE request to S-CSCF over the ISC reference point, then S-CSCF takes the following steps:

- Determine that the INVITE request constitutes an originating INVITE from SIP-AS. The absence of an original dialog identifier is one indication that this INVITE is an originating INVITE.
- Determine the served subscriber. S-CSCF may look at the P-asserted-identity header or the P-served-user header in the INVITE request.
- Apply profile matching. S-CSCF uses the identity of the served subscriber to obtain the user profile from its internal memory.

When these tasks have been performed, the S-CSCF continues to establish the originating SIP session, including (i) service invocation, depending on the presence and contents of IFC in the user profile, (ii) number normalization, if needed, (iii) ENUM query, if needed, and (iv) outbound routing.

When the SIP-AS sends the INVITE request to the I-CSCF, the procedure will be as follows. The SIP-AS is in this case configured with the address, typically in the form of a domain name, of the I-CSCF in this operator's network.

Note

When this SIP-AS does not reside in the IMS network, e.g. because it is operated by a third party, the sending of an INVITE to the I-CSCF would have to be done via border gateway(s), such as the IBCF. That situation is not explored further.

The I-CSCF does not hold subscriber data; the INVITE from the SIP-AS may be sent to one of a multitude of I-CSCFs. The I-CSCF domain name configured in the SIP-AS may therefore resolve, through DNS query, into a multitude of host names and corresponding IP addresses. The SIP-AS may apply load sharing or round robin over these I-CSCFs, which is normal behavior.

When the I-CSCF receives an INVITE request, its default behavior is to apply an HSS query based on the destination of the request, i.e. based on the R-URI in the INVITE. For a terminating call, the R-URI identifies the served subscriber. If, however, the INVITE request arriving at the I-CSCF relates to an originating call, the I-CSCF will forward the INVITE to the S-CSCF related to the *calling subscriber* as opposed to the *called subscriber*. For this reason, the I-CSCF will in this case base the HSS query on P-asserted-identity. A designated parameter in the Route header in the INVITE is used to indicate to the I-CSCF that it should apply this originating INVITE handling:

```
INVITE sip:+31163279911@company-y.nl SIP/2.0
Route: sip:icscf.ims-network.nl; orig
P-asserted-identity: sip:john.smith@company-x.nl
```

The I-CSCF now applies the HSS query (location information request, LIR) including sip:john.smith@company-x.nl in the Diameter message to the HSS. The S-CSCF address that the I-CSCF receives from the HSS is used as Route header to forward the INVITE to the S-CSCF:

```
INVITE sip:+31163279911@company-y.nl SIP/2.0
Route: sip:scscf-1.ims-network.nl; orig
P-asserted-identity: sip:john.smith@company-x.nl
```

Refer to Figure 8.52. The orig parameter is needed in the INVITE from the I-CSCF to the S-CSCF as well. Without it, the S-CSCF would apply terminating SIP session behavior. The S-CSCF may now apply the SIP session handling as required.

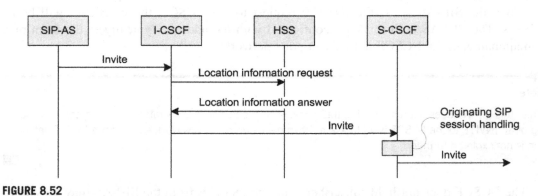

FIGURE 8.52

Originating HSS query by I-CSCF.

Figure 8.53 provides a signal sequence diagram showing session establishment to both parties for the service-initiated session establishment. In the example in Figure 8.53, the SIP-AS establishes two SIP sessions, one to "A-party" and another one to "B-party" (these terms are in quotation marks since, for this call case, they are not A-party and B-party in the traditional sense), and the SIP-AS sends an "empty INVITE" to "A-party", entailing that the INVITE does not include an SDP offer. The rationale is that SIP-AS has, at this point, no knowledge of the user plane that will be established between the respective parties. The empty INVITE prompts the "A-party" to offer an INVITE. Assuming that 180 Ringing from the "A-party" is not sent reliably (i.e. 100rel and PRACK are not used), the SDP offer will be provided in the 200 OK from the "A-party". The SIP-AS uses the SDP offer, if acceptable to the SIP-AS, to construct an INVITE to the "B-party". The SDP answer received from the "B-party" is then included in the ACK to the "A-party". "A-party" and "B-party" have now been facilitated to exchange SDP, where the "A-party" was given the lead in the SDP negotiation.

One issue associated with this method is that there will be a relatively long (from a transaction handling point of view) duration between the receipt of the 200 OK from "A-party"

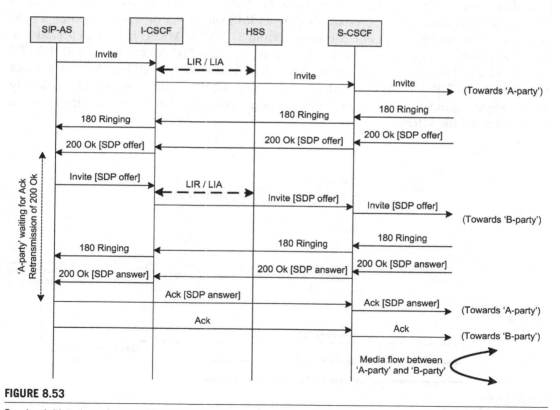

FIGURE 8.53

Service-initiated session establishment with direct media transfer.

and the sending of the ACK to the "A-party". There may therefore be a number of retransmissions of the 200 OK from the "A-party". The SIP-AS may therefore (have to) cancel the call establishment if the "B-party" does not answer within the maximum retransmission time of the 200 OK.

8.8.7 User Interaction

This section gives a brief overview of user interaction in IMS. Specifically, it describes how a SIP-AS can connect an IMS subscriber to special resource equipment and how the SIP-AS can control that resource equipment to play announcements to a user and apply other forms of user interaction. Figure 8.54 shows the architecture applicable to user interaction.

Two main entities are defined for user interaction:

- Media resource function controller (MRFC)
- Media resource function processor (MRFP).

The MRFC is an entity on the control plane. A SIP-AS may establish a SIP session toward an MRFC. This SIP-AS may establish this SIP session during service logic processing, e.g. when controlling a call established by an IMS subscriber or when executing a public service. The establishment of a SIP session from SIP-AS to MRFC is done in accordance with standard ISC reference point rules. For example, when SIP-AS receives a SIP INVITE request for a call to a remote destination, SIP-AS may retarget this call to MRFC. The exact behavior of the SIP-AS, with respect to establishing a SIP session with MRFC, is highly service logic dependent. Normal SDP offer/answer rules apply for the SIP session between the SIP-AS and the MRFC.

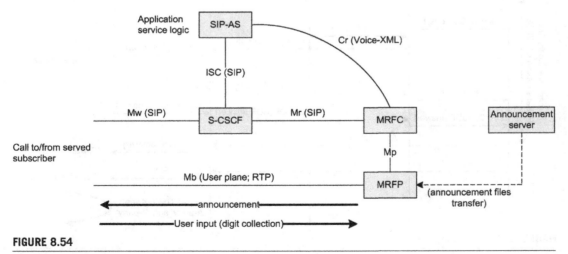

FIGURE 8.54

Architecture relating to user interaction.

The MRFP is the media handling device. An MRFP acts under control of an MRFC. One MRFC may control multiple MRFPs. One MRFP is, however, controlled by just one MRFC. When the SIP-AS has established a SIP session with an MRFC, it instructs the MRFC to apply media handling. The MRFC here selects an MRFP. The MRFP is responsible for allocating IP address and port number for the RTP termination for the user plane; that information is provided to MRFC and is used in the SDP answer. The media handling that the MRFP is instructed to apply may be the streaming of an announcement to the served subscriber or to collect user input (DTMF digits) from the served subscriber and to report the collected input to the MRFC, which can then provide that information to the service logic in SIP-AS.

The address of the MRFC is configured in SIP-AS, typically as domain name. If there are multiple MRFCs deployed in the network, then the controlling SIP-AS may select one of them or DNS-based round robin may be applied. The latter method would be applied to select between MRFCs of equal capability.

The MRFC and MRFP may be integrated into a single media server, referred to as the Media Resource Function (MRF). In practical deployment, the MRFC may be integrated in a SIP-AS. In this case, the SIP-AS has a direct media control interface with one or more MRFPs.

When there is a multitude of MRFPs deployed in the network, e.g. for reasons of load sharing and geographic redundancy, then an operator may administer the media source files, the actual announcements, in a centralized announcement server. When an MRFP is instructed to stream a particular announcement, it retrieves the announcement from the announcement server and keeps it in cache.

For conveying the announcement or user interaction instruction from SIP-AS to MRFC, various methods may be applied:

1. Announcement playing instruction is conveyed in the SIP session signaling from SIP-AS to MRFC. This method may be applied for single announcements. For example:

```
INVITE sip:annc@mrfc.ims.operator.se;
  play = file://announcementserver.ims.operator.se/prepaid-announcements/
    announcement1.wav; early=yes SIP/2.0
```

The user part of the R-URI, "annc", tells the MRFC that announcement service is required. The URI parameter "play=..." identifies the required media file. The "early=yes" URI parameter indicates that MRFC should apply media streaming through *early dialog*. IETF RFC 4240 states that the method of providing announcement through early media is deprecated. However, this method may still be supported in some MRFCs.

2. MRFC establishes a user interaction control relationship with the SIP-AS. This relationship may be formed by an HTTP session through which Voice XML scripts are sent from SIP-AS to MRFC and answers are sent from MRFC to SIP-AS. This method is applied for complex script execution. For example:

```
INVITE sip:dialog@mrfc.ims.operator.se;
  voicexml=http://vpn-node1.ims.operator.se/voice-xml-scripts/voice-xml-
    script_25.vxml<session-id>
```

When the MRFC receives this INVITE, it will construct an HTTP GET request message to obtain a Voice XML script. The voicexml parameter in the INVITE request contains the address of the node where the service execution is taking place. This address is augmented with a session identifier. The session identifier is transparently included in the destination of the HTTP GET request message and is used by the application server to associate this HTTP GET request with the corresponding service logic processing instance.

Figure 8.55 shows an example sequence diagram for user interaction that is taking place during establishment of an originating call.

The SDP that is offered in the INVITE from the SIP-AS to the MRFC is identical to the SDP offer from the calling party. The rationale is that there will be a regular SDP offer–answer sequence taking place between the calling party and the MRFC. The MRFC establishes a SIP session in response to the INVITE, i.e. it responds with a 200 OK. The SIP-AS does not want to convey this 200 OK, including SDP answer, to the calling party, since that would constitute a call answer. Hence, the SIP-AS conveys the SDP answer in a provisional response to the calling party (SIP-AS is acting as a B2BUA). By using a *reliable provisional response*, the SIP-AS provokes a provisional acknowledgement (PRACK) from the UE, upon which the SIP-AS will send the ACK to the MRFC. The UE now has a dialog established relating to the user interaction. Media streaming (announcement) may now take place in association with this dialog.

When the playing of the announcement is complete, the SIP session between the SIP-AS and MRFC is terminated. The SIP-AS may then establish the SIP session toward the remote destination. The dialog that was established for the user interaction will not be used further. When the remote destination provides a provisional or final response, the SIP-AS will convey that response to the calling party through a dialog *other* than the dialog that was used for conveying the SDP answer from the MRFC.

8.8.7.1 Voice XML

When Voice XML is used for user interaction, there will be a sequence of HTTP requests and HTTP responses between the MRFC and the SIP-AS. This sequence is used to convey instructions from the SIP-AS to the MRFC, for the playing of an announcement, or for the collecting of user input. Figure 8.56 gives an example sequence between the SIP-AS and the MRFC. SIP signaling between the SIP-AS and MRFC traverses the S-CSCF. That is not shown in the figure.

The first HTTP Get from the MRFC to the SIP-AS is used to obtain Voice XML script. The Voice XML script may be tailored to this specific service logic procession instance. That is, the announcement and the variable components in the announcement may be specific for this call from this subscriber. The response from the SIP-AS contains the Voice XML script. The script contains instructions for the MRFC on announcement(s) to play. When the MRFC has received the response from SIP-AS, it can complete the SIP session

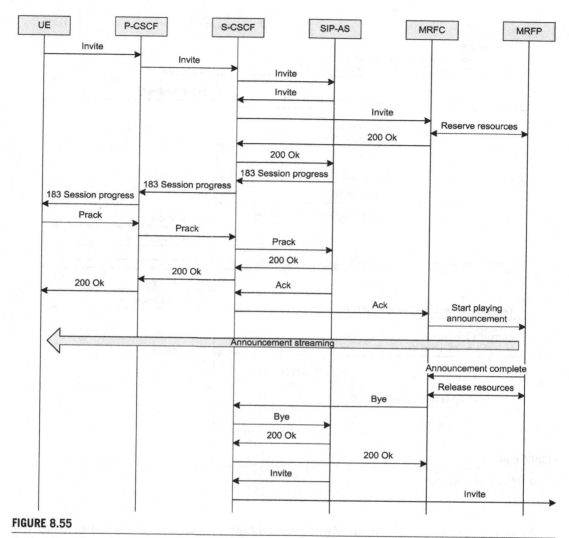

FIGURE 8.55

Announcement playing at call establishment.

establishment, i.e. send 200 OK back to SIP-AS. The World Wide Web Consortium (W3C) standardizes the Voice XML language. The Voice XML script received by the MRFC at this point should also contain an instruction on how to behave when the current script execution is completed. The completion of playing the announcement(s) may constitute the completion of the script execution. Typically, the script indicates that MRFC will contact the SIP-AS again, to obtain the next Voice XML script to execute. That next Voice XML script may include an indication that the SIP session should be terminated.

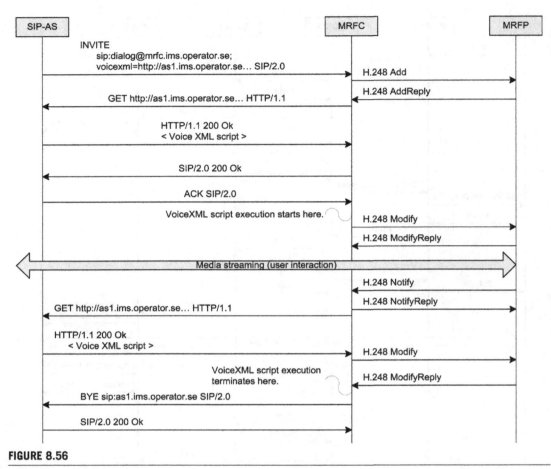

FIGURE 8.56

Voice XML script execution.

As proprietary enhancement in MRFC, various mappings may be defined between the SIP INVITE request arriving at the MRFC and the HTTP Get request sent from the MRFC to the SIP-AS. For example, various SIP headers may be included as proprietary parameters in the Get request. These parameters may, in turn, be used by the SIP-AS for internally routing the GET request (process instance selection) or for script execution, for example.

8.8.8 Unregistered Service Invocation

In order to handle a SIP request in S-CSCF, such as INVITE or MESSAGE, a user profile is required for the served subscriber. When a subscriber is registered as a user in the IMS network, the INVITE and other request messages from or to that subscriber are directed to the

S-CSCF currently assigned to that subscriber. There are cases, however, where a subscriber is not registered in an S-CSCF, but we still want to apply service handling for a call (or message) to or from that subscriber. Two examples are:

- Terminating a call to a subscriber whose phone is switched off. This subscriber is not registered in an S-CSCF and is therefore not contactable; the call will, for example, be diverted to voicemail.
- Service-initiated session establishment on behalf of an unregistered subscriber.

8.8.8.1 Terminating Unregistered Session Handling

When the I-CSCF receives an initial INVITE request for a subscriber, it applies an HSS query (Diameter location information request, LIR) to obtain that subscriber's S-CSCF address (returned in Diameter location information answer, LIA). Figure 8.57 illustrates the situation where the destination subscriber turns out not to have an S-CSCF currently assigned.

The I-CSCF receives, in Diameter LIA, *S-CSCF capabilities* from the HSS instead of the S-CSCF address. The I-CSCF then selects an S-CSCF that meets the required capabilities, as indicated in these S-CSCF capabilities. This S-CSCF selection procedure is the same as during initial registration, where the I-CSCF must also decide which S-CSCF should be assigned to a user. The S-CSCF determines, when receiving the INVITE request, that it does not have a user profile for the served subscriber; since this is a terminating call, the served subscriber is represented in the R-URI in the INVITE request line. Hence, the S-CSCF applies Diameter procedure server assignment request/answer (SAR/SAA). With this

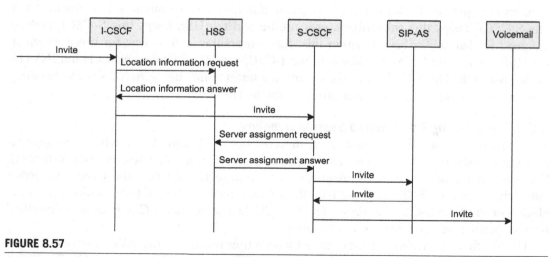

FIGURE 8.57

Terminating unregistered session handling.

Diameter procedure, S-CSCF obtains the user profile of the subscriber, so it can apply terminating SIP session handling. By receiving the user profile from the HSS, the status of the subscriber changes from *not registered* to *unregistered*. The subscriber is still not *registered*; the S-CSCF does not have a contact address for the subscriber. So, it is not possible to forward the INVITE to a terminal. The rationale of this procedure is that the user profile received from the HSS contains IFC and the IFC contains a trigger definition for *terminating_unregistered* session. The S-CSCF will hence use that trigger definition in the IFC to forward the INVITE request message to an application server.

The application service logic should not attempt to establish the call *in the normal manner*, i.e. it should not send the INVITE unmodified back to the S-CSCF, since that would lead to call establishment failure, due to no contact address being present. Instead, the service logic should apply action like retargeting the call to voicemail, as shown in Figure 8.57, or to the media server (for playing an announcement).

The Route header that is used for the service invocation contains an indication that this service invocation relates to a terminating unregistered SIP session. For example:

```
INVITE sip:john.smith@company-x.se SIP/2.0
Route: sip:service-x@scscf1.ims.operator.se;lr; session=terminating_unregistered
```

It is not uncommon that a SIP application does not apply differentiated handling for session=terminating_unregistered compared to session=terminating. In other words, the application ignores the fact that the subscriber is unregistered. The effect is that the application attempts to establish the call, resulting in an unsuccessful final response generated in the S-CSCF toward the SIP-AS, upon which the SIP-AS may still decide to retarget the call to voicemail or take other action.

Resulting from the S-CSCF having received the user profile from the HSS, the subscriber remains *unregistered* for a system-configurable duration, e.g. 10 minutes. For that duration, this S-CSCF keeps this subscriber's user profile and the HSS keeps this S-CSCF address assigned to that subscriber. If within this duration another call arrives for this subscriber, the HSS returns the S-CSCF address to the I-CSCF and the I-CSCF forwards the INVITE to that S-CSCF. The S-CSCF can then apply the terminating_unregistered session handling without first having to obtain a user profile from the HSS.

8.8.8.2 Originating Unregistered Session Handling
The explanation of this situation is with reference to Figure 8.58, where an application server establishes a call *on behalf of* a user (service-initiated session establishment). Service-initiated session establishment is described in section 8.8.6. If the served subscriber currently has no S-CSCF assigned, then the I-CSCF receives S-CSCF capabilities from the HSS instead of the S-CSCF address. The I-CSCF then selects an S-CSCF, as also described for terminating unregistered session handling.

The S-CSCF determines that it does not have a user profile for the subscriber that is identified as *served subscriber* in the incoming INVITE request. Hence, the S-CSCF applies

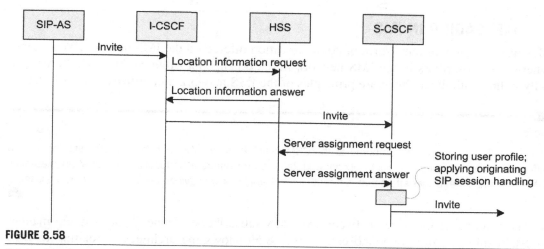

FIGURE 8.58

Originating unregistered service triggering.

Diameter procedure server assignment request/answer (SAR/SAA). The S-CSCF obtains the user profile of the subscriber, so it can apply originating SIP session handling. This method is known as *originating unregistered* session handling. It may include service triggering, based on IFC in the user profile.

Resulting from the S-CSCF receiving the user profile from the HSS, the subscriber's registration status has changed to *unregistered* (see also terminating unregistered session handling).

The served subscriber for service-initiated SIP session establishment may be identified in the P-asserted-identity in the INVITE that is sent from the SIP-AS. The P-asserted-identity is, however, also used for number presentation to the destination subscriber. The subscriber on whose behalf the call is established may, however, not be the same as the name or number that is to be used as calling identity. For this purpose, the P-served-user header may be used:

```
INVITE sip:+31163279911@company-y.nl SIP/2.0
Route: sip:icscf.ims-network.nl; orig
P-asserted-identity: sip:john.smith@company-x.nl
P-served-user: sip:helpdesk@company-x.nl
```

The identity that is used as *calling party* will be sip:john.smith@company-x.nl. Meanwhile, the S-CSCF handling this originating call will be the S-CSCF that is assigned to sip:helpdesk@company-x.nl. Here, the I-CSCF will, when handling an originating INVITE request, apply the following logic:

- If the INVITE request contains the P-served-user, then apply an HSS query based on the identity contained in the P-served-user header.
- Otherwise use the identity contained in the P-asserted-identity to apply the HSS query.

8.9 MESSAGING IN IMS

Messaging is an additional form of communication offered on the IMS network. Messaging, where two subscribers in the IMS network may exchange *message components*, is realized fully in line with the architecture principles of the IMS network and utilizes standard SIP.

Note

This is a different approach compared to the GSM network, where voice/video calls are established in a different manner, through different network architecture and different protocols, than for short messages (SMS). This has the further effect that different services apply for voice/video calls than for short messages.

The sending of messages between two IMS subscribers is done in an identical manner to establishing a voice or video call. Figure 8.59 shows the architecture relating to IMS messaging.

The User Equipment (UE) of the respective IMS subscribers is registered in P-CSCF and in S-CSCF in IMS, with a contact address as normal. The UE is then ready to establish a SIP session and for stand-alone SIP transactions. The transfer of a message to and from a UE is done through a designated SIP method, MESSAGE. The use of SIP MESSAGE is defined in IETF RFC 3428. 3GPP TS 23.228 specifies how the messaging specified in IETF RFC 3428 is applied in IMS.

Sending a message from one UE to another UE is done by initiating a MESSAGE transaction from the sending party to the destination party. The initiation of this MESSAGE transaction follows the rules that apply for the initiation of an INVITE transaction, including for example (not exhaustive):

- Addressing the destination party (through SIP URI or through Tel URI)
- Use of Route headers to route the MESSAGE request message
- Use of Via headers for routing response message(s)
- Dialog identification (From, To, Call-ID)
- Transaction client state model and transaction server state model
- Inclusion of a message body to convey application information.

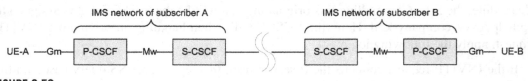

FIGURE 8.59

Architecture for messaging in IMS.

Two ways to transfer a message between two IMS subscribers are:

1. Immediate message
2. Messaging session.

Method 1 entails the use of SIP MESSAGE as transaction. It results in the transfer of a message component, such as text string, to the destination party. When the SIP MESSAGE transaction is complete, the messaging action is complete as well. If the recipient of the message wishes to respond, the recipient applies a separate SIP MESSAGE transaction. The use of MESSAGE between two subscribers may be done outside a call or during a call.

Method 2 entails that a SIP session is established through the INVITE transaction. The initiator of the SIP session invites the destination party to engage in a message transfer session.

We will first look at the instant message method of message communication.

8.9.1 Instant Message

An example of message exchange between two IMS users is shown below. Not all SIP messages involved are shown. The message sequence would be similar to the message sequence used for establishing a SIP session with INVITE, with some differences:

- There is no 100 Trying used. The rationale is that a response on MESSAGE should be "immediate". When the sender does not receive a response, it will retransmit the MESSAGE request.
- There are normally no provisional responses used.
- There is no ACK transaction applied.

8.9.1.1 MESSAGE Transaction Initiated by Originating Party

```
MESSAGE sip:wendy.jones@abc-company.se SIP/2.0
Via: SIP/2.0/UDP 153.88.17.1:42770;branch=z9hG4bK-d8754z-340e1d1f07dee94a-1---
d8754
Max-Forwards: 70
From: "John Smith"<sip:john.smith@abc-company.se>;tag=affabdc3
To: "Wendy Jones"<sip:wendy.jones@abc-company.se>
Call-ID: NjZlMTgyNjNjNmJjOGExOWM1NGRiNDYxZDk5ZGQxNTM.
CSeq: 2 MESSAGE
Allow: INVITE, ACK, CANCEL, OPTIONS, BYE, REFER, NOTIFY, MESSAGE
Content-Type: text/html
User-Agent: X-Lite 4 Release 4.0 stamp 58832
Content-Length: 64
<font face="Arial" size="2">It's 12h00. Let's have lunch.</font>

SIP/2.0 200 OK
Via: SIP/2.0/UDP 153.88.17.1:42770;branch=z9hG4bK-d8754z-340e1d1f07dee94a-1---
d8754
From: "John Smith"<sip:john.smith@abc-company.se>;tag=affabdc3
```

```
To: "Wendy Jones"<sip:wendy.jones@abc-company.se>;tag=11126f16
Call-ID: NjZlMTgyNjNjNmJjOGExOWM1NGRiNDYxZDk5ZGQxNTM.
CSeq: 2 MESSAGE
Content-Length: 0
P-Asserted-Identity: "Wendy Jones"<sip: wendy.jones@abc-company.se>
User-Agent: X-Lite 4 Release 4.0 stamp 58832
```

The *Content-Type: text/html* header in the MESSAGE request indicates that the body in this SIP MESSAGE request contains HTML-based text. The actual body consists of a single, formatted text string ("It's 12h00. Let's have lunch.").

When the party that had sent the message starts typing a further message, this typing action may be revealed to the destination party, informing the destination party that another message is about to follow. For example:

```
MESSAGE sip:wendy.jones@abc-company.se SIP/2.0
Via: SIP/2.0/UDP 153.88.17.1:42770;branch=z9hG4bK-d8754z-51cbd6d54c5444ec-1---
d8754z
Max-Forwards: 70
From: "John Smith"<sip:john.smith@abc-company.se>;tag=affabdc3
To: "Wendy Jones"<sip:wendy.jones@abc-company.se>
Call-ID: NjZlMTgyNjNjNmJjOGExOWM1NGRiNDYxZDk5ZGQxNTM.
CSeq: 3 MESSAGE
Allow: INVITE, ACK, CANCEL, OPTIONS, BYE, REFER, NOTIFY, MESSAGE
Content-Type: application/im-iscomposing+xml
User-Agent: X-Lite 4 Release 4.0 stamp 58832
Content-Length: 312
<?xml version='1.0' encoding='UTF-8'?>
<isComposing xmlns='urn:ietf:params:xml:ns:im-iscomposing'
xmlns:xsi='http://www.w3.org/2001/XMLSchema-instance'>
<state>active</state>
<lastactive>2011-01-21T12:26:46Z</lastactive>
<contenttype>text/html</contenttype>
<refresh>60</refresh>
</isComposing>
```

The *Content-Type: application/im-iscomposing+xml* header indicates that the body part in this MESSAGE request contains an XML script in accordance with IETF RFC 3994. This RFC defines the message exchange sequence that supports users who are engaged in a message sequence to be informed about one another's activity, such as "busy composing a message...".

The MESSAGE transaction may be subject to value-added services, in the same manner as for voice/video calls (see Figure 8.60).

When a user initiates a MESSAGE, the S-CSCF may forward the MESSAGE request to a SIP-AS, based on IFC for that subscriber. The IFC may include an indication that a

MESSAGE transaction, when sent as a stand-alone transaction, will be forwarded to an application server. The Application Server (AS) may take the following roles:

- **Proxy.** The SIP-AS forwards the MESSAGE back to S-CSCF, after it has suitably modified the MESSAGE, e.g. modified the URI in the From header and/or To header or modified the destination of the MESSAGE. A MESSAGE transaction does not establish a SIP session. Hence, the SIP-AS will not act as *record-routing* proxy in this case.

  The S-CSCF now continues routing the MESSAGE to the destination indicated in the R-URI.
- **B2BUA.** The SIP-AS forwards the MESSAGE back to S-CSCF, but sends it back through a different dialog.
- **UAS.** The SIP-AS acts as user agent server by sending a final response to the sending party.

The invocation of a value-added service for MESSAGE is equally applicable for originating MESSAGE handling and terminating message handling.

The above-described MESSAGE handling may be applied outside a (voice/video) call or during a call. In the latter case, the MESSAGE may still be handled independently of the ongoing call. That is, the MESSAGE transaction is not strictly associated with the INVITE transaction. The identities used in the MESSAGE transaction (From, To, P-asserted-identity, Request URI) may differ from the corresponding identities used in the INVITE session. Whereas the INVITE transaction may be sent to tel:+46107152400, the MESSAGE transaction could be sent to sip:wendy.jones@abc-company.se. These two identities may relate to the same user and may arrive at the same terminal.

Figure 8.60 also shows that a SIP-AS may generate an instant message to a SIP terminal. This may, for example, be done within the context of an ongoing call. The choice of sending the SIP MESSAGE by SIP-AS to an I-CSCF or S-CSCF is the same as for voice call establishment by SIP-AS.

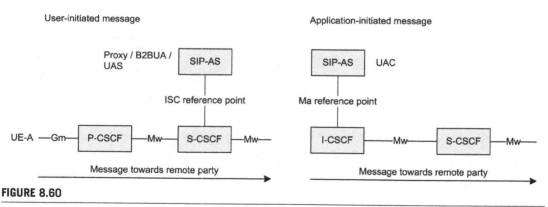

FIGURE 8.60

Value-added service for messaging in IMS.

8.9.2 **Messaging Session**

When a SIP session is established between two users, the MESSAGE transaction may be used as an *in-session* transaction, related to the existing SIP session. The effect of such MESSAGE transfer is that these in-session MESSAGE transactions follow the same path as the INVITE session. Specifically, the MESSAGE transactions would traverse SIP-AS. An operator may want to restrict SIP-AS invocation for voice/video calls.

IETF RFC 4975 describes how two SIP terminals may establish a SIP session for the purpose of message transfer. The message transfer is done according to the Message Session Relay Protocol (MSRP). The SIP session between the two parties is established in accordance with regular SIP routing methodology, including the invocation of one or more value-added service(s). The messages themselves are transferred as payload. The transport details applicable for this message transfer are negotiated in accordance with the SDP offer–answer model. Hence, the message transfer, i.e. the payload, may now take a different path than the SIP messages.

MMTel and Other IMS Enablers

9.1 INTRODUCTION

It may be a bit surprising that the header for this chapter says "enablers" and not "services". The point we are trying to make is that standard services like telephony, presence, and messaging deserve the designation "service" in the sense that they are directly visible and usable to the end-user, as opposed to under-the-hood functionality such as routing and media conversion. However, when a standard service becomes ubiquitous, application creators find ways of incorporating those capabilities into their designs. As soon as the value of this is recognized, the idea occurs that it might be a good thing if the interface between the application and the supporting service is standardized, or at least well known (i.e. a de facto standard). This is the mechanism that turns a service into an enabler, with the emergence of an API as the event that turns it from a proof-of-concept hack to an architectural layer supporting an evolved business model. So, in this chapter we will look at a couple of those services cum enablers, and discuss how you can incorporate them in your design.

9.2 A MORE IN-DEPTH LOOK INTO MMTEL

As you may have gathered from the brief introduction in Chapter 1, the best way to describe MMTel (Multimedia Telephony) is that it is the joining of the tried and true telephony service definition with modern technology. So, it has evolved from both today's telephony and Internet technologies, giving it the following properties:

- **Interoperable services provided by multiple operators.** You can call anyone, not just those that have signed up to the same operator as you.
- **Interoperability with legacy networks.** This is a key property. Being able to call and be called by legacy phones (both fixed and mobile) is essential to build a mass market quickly.
- **Use of telephone numbers.** E.164 numbers (as they are known in the trade) may be legacy, but they are very well known and work in both the PSTN and the MMTel domains.
- **Built with telco characteristics.** Many of the features of IMS and services like MMTel are put into the architecture to secure scalability to very large systems, and to guarantee the levels of availability that the general public expects from the telephony service.
- **Built on well-known Internet protocols.** This means that the highly cost-effective technology that is developed in the IP world is directly usable.

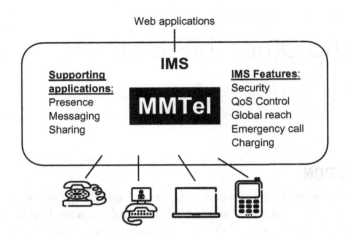

MMTel in context.

Also, MMTel is purposely built to retain and improve key operator values:

- It supports business models evolved from today's
- Open and extendable
- Standardized multi-vendor interfaces, avoiding vendor lock-in and promoting competition
- Offers migration from today's services
- Offers integration with Internet services
- Meets regulatory requirements – first-line telephony services are typically regulated more strictly than "leisure" voice services.

Figure 9.1 summarizes MMTel's role and value proposition: it provides the service logic for connecting real-time, low-latency media streams between users over multiple access forms. It does this globally, with QoS control, with the possibility to handle emergency calls, to charge fairly and accurately, securely, and with the required means to oversee traffic. Additionally, as the upward line hints, through northbound APIs, it is possible to let external applications affect call handling, but more about that later.

9.3 BASIC MMTEL ARCHITECTURE

In Figure 9.2 a typical MMTel deployment is shown. This is the basis for the VoLTE (voice over LTE) standard: the way of delivering voice over next-generation mobile networks.

Control signaling goes to the CSCF as usual, and is then routed (using iFC triggering criteria downloaded from the HSS at registration) over ISC to the MMTel application server. If

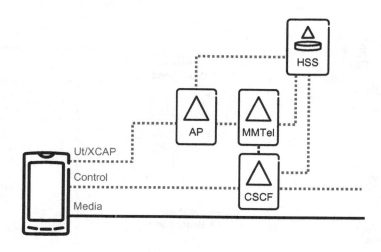

FIGURE 9.2

MMTel in e.g. LTE.

the session is aimed at another IMS terminal, signaling is forwarded back through the core network (possibly forwarded on to another network); if the session needs to be routed to a CS network it goes to the right in the figure, as described in other chapters.

However, there is a new feature here that we should note: the Ut/XCAP connection. It is basically configuration information delivered from the client terminal over HTTP using XCAP (XML Configuration Access Protocol), setting up things like call diversion, black-list/white-list, etc., i.e. what is usually called supplementary services in telecommunications. In legacy networks, this is handled through methods like *21# commands (or, to be exact, the equivalent network protocol messages); the MMTel method leverages much more modern and extensible mechanisms. We will discuss XCAP and its use later, when we describe IMS Presence in Section 9.7.

9.4 GOING DEEPER AND WIDER

Figure 9.3 expands on the previous theme, showing some of the potential deployment scenarios. This is by no means a complete picture; the point is to show that IMS and MMTel were designed to handle a very wide range of device and access technologies. However, the good news for developers of add-on services is that they don't really have to worry: all this is taken care of inside the system. It can be treated like the black box described in Chapter 3.

Now, what can actually be delivered over these networks?

- **Voice** in various forms. Using the inherent negotiation mechanisms, the end-points (and the networks) create the best combination of capabilities that can deliver what the user wants (and has paid for). This includes fully conversational voice as well as "half

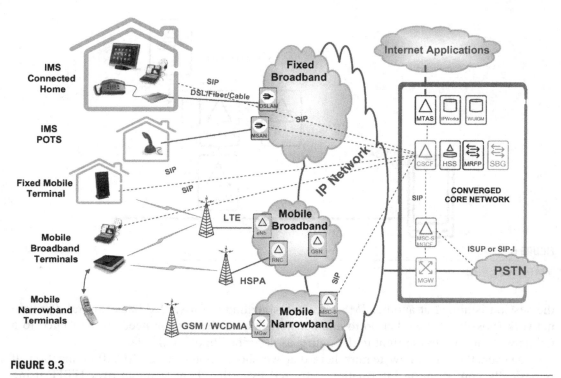

FIGURE 9.3

MMTel connecting everything.

duplex" (i.e. push-to-talk). Also, QoS parameters like bit rate and codec types are agreed upon, so that features like high-definition audio and stereo can be incorporated. Typical codecs in the mobile scenario would incorporate the newest developments like AMR, both narrowband and wideband. Note that the term "wideband AMR" refers to the audio bandwidth (about 7 kHz, a major improvement over PSTN's 3.1 kHz); it isn't coupled to the W in WCDMA, which refers to the signal bandwidth over the air (5 MHz).

- **Video:** one way (see what I see) or two way. Codecs (e.g. H.263, H.264) and bit rates are negotiated as for voice, but the fact that both media streams are controlled by the same machinery also means that latency can be controlled – this is essential to achieve lipsync, for instance.
- **Text.** A bit surprising, maybe, but having a text stream as part of the conversation can be quite handy. For instance, if you take a phone call while in a meeting you may want to respond by typing instead of talking, while still being able to listen to the caller.
- **Add/drop media in any combination.** This is where the in-session modification capabilities of SIP are put to good end-user use. What started as a regular phone call can, for example, be seamlessly translated into a video session and back again.

- **Share media.** Image, video clips, or any file. This is essentially a convenience function but a very important one – sending something to whoever you are talking to is very easy, as the addressing is implicit (no need to start another app and then trying to find your conversation partner again in what may be a different set of addresses kept by that application).

- **Supplementary and value-add services (VPN, white-list/black-list, forwarding, etc.).** This is where the inheritance from classical telephony comes in. You may wonder why it is necessary – surely, with new architectures come new services? True, but one of the ways of deploying IMS and MMTel is for fixed-line replacement (as discussed in Chapter 3), and for the end-user it is very convenient if what worked yesterday still works today. In the extreme case, the end-user device may still be a regular PSTN phone, connected to an RJ-11 port (the hole that fits the square clear plastic plug that tends to be on the end of the wire that connects your phone) on the broadband router at home, so the end-user interaction may actually still be based on *# combinations.

- **Conferencing.** MMTel also defines how to create conferences (preplanned or ad hoc), and how participants are added and removed. Note the power of standards and symmetry – conferencing applies to all media types (but most of them require network resources, of course).

- **Media transport processing.** When devices happen to have incompatible codecs installed, the network can provide conversion services. If the device is a PC (and the users are reasonably technically inclined) these issues are normally handled by downloading codecs, but apart from the inconvenience factor, it is most likely not even possible for the majority of mobile phones.

- **Media handling in general.** MMTel clients and networks handle issues like adaptive jitter buffer, time scaling, RTP payload optimizations, packet loss handling, echo cancellation, and noise reduction. The point here is again the value of higher abstractions – by adding MMTel to your arsenal of application support components, you get a wide range of sophisticated and complex features automatically.

- **Smart supplementary services.** As has been mentioned before, a key feature of MMTel is that it includes in the standard a range of supplementary services based on SIP. For example:
  - Communication handling based on media awareness
  - Forwarding and barring based on media type
  - Add multiple participants with a single SIP INVITE
  - Communication based on presence awareness
  - Forwarding to voicemail based on presence status
  - Multiple access lines
  - Ad-hoc conferencing
  - One number for multiple devices
  - Privacy on-demand
  - Or the service you design!

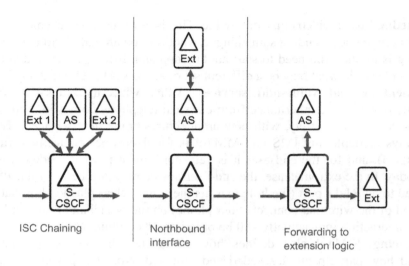

ISC Chaining   Northbound   Forwarding to
                         interface   extension logic

FIGURE 9.4

Options for extending the MMTel control logic.

The next section will discuss how you can accomplish what the last bullet hinted at: providing additions to MMTel.

9.5 ADDING TO MMTEL

Previously, we promised that MMTel is extendable and modifiable; here, we will go through the ways that this can be done.

First, it should be noted that services can be – and usually are – invoked on both the originating and the terminating side. However, as they are generally situated in different networks, the originating and terminating services are executed independently; the only mechanism connecting them is that they can see and modify the SIP messages passing between the end-points.

Secondly (as shown in Figure 9.4), there are three main ways of modifying an existing service like MMTel.

9.5.1 ISC Chaining

This is also known as "the bookend model". In this case, the extra services are invoked before or after the core service (or perhaps both). This is easily done for things like number translation or black-listing, even though in the black-list case it may be wise to invoke it both before and after (as the MMTel AS may have redirected to another number). The drawback is of course that the only information link possible is what can be conveyed through SIP headers. For example, if your app is routed to precede MMTel, you will not be aware of what happened to the call (e.g. a Tel URI may have been translated). Conversely, if you are linked in after, you might never see the INVITE (if the MMTel AS decides to reject it for some reason). If you add both upstream and downstream ASs, you might be tempted to add proprietary headers. This is typically not a good idea, although some equipment vendors use it. Of course, back-end methods like external databases (or shared state, if both AS instances can be guaranteed to execute on the same container) can always be used, but require that iFC routing is properly designed.

Another issue you need to consider is if your application just needs to do something to the first INVITE, or if you want to stay linked in for the duration of the session. The latter case is necessary, for example, in cases where you have allocated physical resources (a media server, say), or you charge for events happening during the session. If so, you need to insert a Record-Route header into the INVITE as it is passed forward (JSR289 provides operations for this).

9.5.2 Northbound Interface

This is an explicit call-out to an external application, invoked as part of the call processing in the MMTel AS node itself, i.e. where in the processing chain this happens and what protocol is used is often vendor dependent, although Parlay X can provide much of the necessary functionality. However, the GSMA OneAPI initiative is aimed at aligning this, so an application developer can have as large a target market as possible.

If your application is deployed in the operator's network, your application will typically have direct access to the NBI. If outside the trust domain, you will typically interact through a service gateway that exposes the equivalent of the NBI, potentially restricted as per the operator policies. REST-based APIs are becoming increasingly popular for this kind of interaction, as they need comparatively little in terms of middleware support on the part of the (external) application server. Because of the asymmetric nature of HTTP (the basis for REST), various tricks can be employed to simulate asynchronous behavior from the network to the application; long polling, COMET, and Bayeux[1] are typical examples.

9.5.3 Forwarding to Extension Logic

Topologically, this is rather similar to the post-connection bookend variety, but there are two major differences. The first is in the invocation: in the bookend case, this happens by iFC evaluation in the S-CSCF; in the forwarding case, the MMTel AS rule execution leads to the SIP signaling being retargeted to a node like an IVR (Interactive Voice Response) MRF. The second major difference is that the receiving AS terminates the session; it is not forwarded directly to another party (the target AS becomes the receiver of the session).

An interesting case is where the terminating AS is a gateway, perhaps providing a web service interface to a non-SIP aware application, inside or outside the operator's trust domain. Then, the SIP session as such is terminated in the gateway, but the logical end-point

---

[1] svn.cometd.com/trunk/bayeux/bayeux.html

for the conversation is on the other side of the web service interface. This covers a number of Parlay X operations; one way of seeing Parlay X is actually as the web service view of communication networks. The GSMA OneAPI, building on OMA work, translates most of the Parlay X operations into REST equivalents, to make them even more accessible to application developers and – perhaps more importantly – to a wide range of third-party service providers.

Of course, with the gateway concept in place, it is also possible (indeed, probably even more common) that the initiative is taken from the external server side – for example, an external application sending an IMS text message to an IMS device.

9.5.4 **Web Interfaces on the Client Side**

Although not shown in Figure 9.4, this can technically be seen as a variant of the extension logic case (an application in the client using web services and/or HTTP/REST to interact with the IMS network). However, an alternate way of looking at it is as a lightweight way of implementing client-side functionality. There are of course limitations, the main one being that typical browser environments classically have not been given access to crucial parts of the device platform, such as codecs. Ongoing development in the device arena and emerging standards like HTML5 may remove most of these limitations, opening up new and interesting ways of capitalizing on IMS capabilities.

9.6 USE-CASE: CALENDAR-BASED ROUTING

As an example, consider this scenario: you want your call routing to depend on what the subscriber has in his or her calendar. A way to implement this would be for the MMTel AS to issue a call direction request to an application server that queries the calendar. Based on whether the user is busy or not, the calendar-routing AS response directs the MMTel AS to forward the call to the currently registered device(s) or to voicemail. The application may also check against special rules set by the user, e.g. if a family member calls and the meeting is not organized by the boss, the call will be forwarded anyway. Also, an item could be added to the user's to-do list to call this person.

Conversely, another use-case may be to set up a conference call as a meeting in the calendar: when the meeting starts, the app server will instruct the MMTel AS (or rather the conferencing part of it) to set up a conference with the called members, and then initiate calls to them.

Note that the calendar needs to know only the SIP URIs (or phone numbers) of the participant, plus the URI of the scheduling AS. All the details of finding the users at the proper devices are handled by MMTel and IMS; the calendar system is never involved in actually handling the communication side of things. Also, given that the scheduling AS can be a part of the operator's trust domain, it can generate the appropriate charging information. What happens to that information is described in Chapter 10.

9.7 IMS PRESENCE

The Presence function in IMS is the result of cooperation between standards bodies, in this case Open Mobile Alliance (OMA), 3GPP (and 3GPP2), and the Internet Engineering Task Force (IETF), in a pattern that is becoming common. Basic protocols are developed in IETF; core capabilities and the architecture supporting them is the job of 3GPP, and then organizations like OMA build end-user applications. Also, as global standards tend to have many options, organizations with here-and-now interests like GSMA carry out profiling, down-selection, and practical interoperability specs (and testing), such as the RCS initiative. ITU TISPAN also refers to 3GPP and OMA for presence and XCAP data management.

Specifically, IMS Presence is thus implemented using the Session Initiation Protocol (SIP) and XML Configuration Access Protocol (XCAP) as defined by IETF and 3GPP, with semantics and logical network elements defined by OMA. The presence information format that is used is Presence Information Data Format (PIDF) with extensions from IETF and OMA.

IMS Presence is, then, a feature that provides similar functionality as its Internet counterparts, but with adaptations to the realities of wireless communications, as well as exploiting the capabilities of the IMS infrastructure.

As Figure 9.5 shows, the presence server collects information from a variety of presence sources – not just the user (as is normally the case in classical presence), but also through

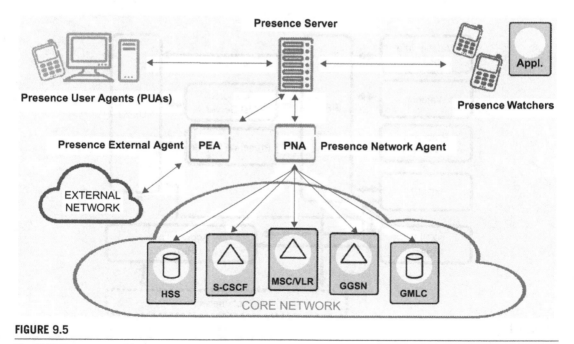

FIGURE 9.5

The IMS Presence architecture.

the concept of presence network agents (both internal and external), where additional information such as location and registration status can be fed in and presented. In classical presence systems, the presence watchers (the consumers of presence information) are essentially human users interested in the status of friends. In IMS Presence, there is additionally the ability to subscribe to watcher information, i.e. finding out who is watching a certain person or, as the case may be, an entity using the presence mechanism to spread interesting information; you can, for instance, watch a weather service and get updates whenever interesting information appears.

A piece of terminology needs to be defined here: the standard does not refer to persons as the entities being described; instead, it talks about a "Presentity" as the thing you publish to (to indicate status change) or subscribe to (to find out if something happens). More about the data model is given later.

9.7.1 Presence as Defined by OMA

Figure 9.6 expands on the simple architecture we described previously. It lists a number of reference points and nodes that you will probably never come close to as an application developer, but it is useful to understand some aspects of what they do, as they affect how a client (and/or an application server) interacts with the presence system.

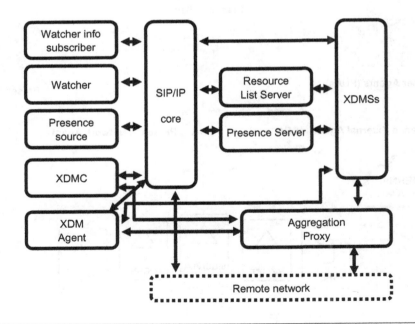

FIGURE 9.6

OMA Presence architecture, simplified.

9.7.1.1 Presence Server

The main functionality of the presence server (PS) is to accept, store, publish, and distribute presence information and watcher information.

9.7.1.2 Resource List Server

The resource list server (RLS) accepts and manages subscriptions to presence lists, enabling a watcher to subscribe to the presence information of multiple Presentities in one subscription transaction. This is an example of how the IMS Presence architecture has included optimizations to avoid unnecessary signaling over the air interface. Network efficiency aside, the main impact of taking classical PC-based implementations and transplanting them unchanged to the mobile environment is typically that battery life is severely impacted. Mechanisms like the RLS alleviate such issues significantly.

9.7.1.3 Aggregation Proxy/Cross-Network Proxy

The aggregation proxy (AP) is the single point of contact for XCAP requests; i.e., operations relating to application specific user and service data (more about this later). The OMA XDM architecture is used not only for presence, but also to store configuration information for MMTel, for example. Thus, one of its tasks is to route XCAP requests to the appropriate XDMS, including load balancing across instances. Another important job of the AP is to perform authentication and authorization of the requestor, as the requests are received over HTTP and not SIP (where the CSCFs in the IMS architecture have that responsibility). Additionally, the AP handles network-to-network interworking, which is the reason for the cross-network proxy(CNP) designation.

9.7.1.4 XDMS

The top level of IMS Presence consists of the following categories of XDMS data, named (somewhat strangely) "application usage" (AU):

- **Presence XDMS**
  - *Presence rules AU*. Authorization rules specific to the presence service enabler.
  - *Persistent presence AU*. Presence information that is persistent over a period of device inactivity. These presence data are therefore sometimes referred to as "hard state" presence as they have no expiry time like normal "soft state" presence.
- **Presence content XDMS.** This AU manages media files such as images/avatars so that the presence source can upload them, including a URI pointing to a certain media file as part of presence information. An authorized watcher uses the URI to obtain the file.
- **RLS XDMS.** Resource list AU manages presence lists specific to the RLS
- **Shared list XDMS**
  - *Resource list AU*. This allows a user to store lists that are common to several applications, such as the presence contact list used by the RLS and authorization list used by the presence server.

- *Group usage list AU*. The group usage list AU is used to store URI lists that could, for instance, be used to store session invitations received from other users or as a "book marking" of what groups a user belongs to.
- **Shared group XDMS.** This promotes sharing of communication group documents across different enablers. A group document may be used as a distribution list by Instant Messaging or by PoC, for example, in a fleet management use-case.
- **Shared policy XDMS.** These documents contain a common policy definition that may be used as a "black and white" list controlling communication by different enablers, such as Instant Messaging.
- **XCAP server capabilities.** This is a single document that may be used when an XDMC would like to determine what extensions, application usages, or namespaces a server supports before making a request.
- **XML documents directory.** An XDMC can use the XCAP directory request to retrieve information about which documents and versions it has stored in respective AUs. It is normally used by a client to check whether it has the same versions of its documents cached as the AU has stored.

9.7.1.5 XDMC
This logical network element is used to manage presence data from an operator point of view. It typically also exposes Java and/or web services interfaces for integration of applications using XCAP to interact with presence. Note, however, that the normal way for an application to request information is of course SIP SUBSCRIBE/NOTIFY.

9.7.2 Interacting with the Presence System
The functional components are shown in Figure 9.7 along with a basic use-case. Before a user can access PGM functions, the user must register on to the IMS network. When the terminal is switched on it sends an SIP REGISTER request to the S-CSCF. The S-CSCF will then as usual perform authentication of the user and store the registration information. As an additional security measure the P-Asserted-Identity SIP header is checked.

Figure 9.7 summarizes most of what goes on between clients and the presence server. The presence server manages all data on behalf of the Presentity (i.e. the person). Watchers subscribe to Presentity data; of course, the device a person uses can be both a source of updates and a watcher of other Presentities' data.

In addition to the Presentity, other trusted presence data sources can update presence information – for example, network status, device location.

9.7.2.1 Subscribing to Presence Data
A watcher subscribes to Presentity information by issuing an SIP SUBSCRIBE signal. Four kinds of SUBSCRIBE operations are used:

- Subscribe with Time-Span – initiates a time-constrained subscription for presence information
- Subscription Poll – one time request for presence data

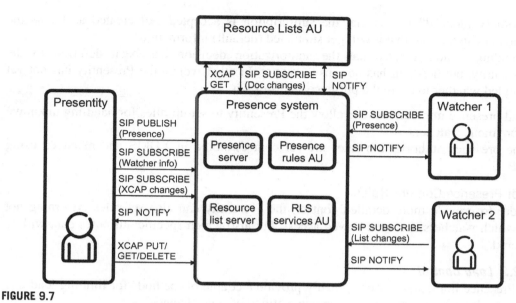

FIGURE 9.7

Basic presence functionality.

- Subscription Refresh – used to refresh the current subscription before the expiry time elapses
- Subscription Terminate – terminates the ongoing subscription.

A watcher can subscribe to presence information related to a single Presentity or a list of Presentities. The latter method results in less load on the network and less battery drain in the client (it is more battery efficient to send a whole transaction as compared with the same amount transmitted in several sessions; this is related to the signaling necessary to change radio link state). Subscriptions to single Presentities are handled by the presence server; subscriptions to lists of users are handled by the resource list server.

The RLS will also keep track of changes in the lists during the time of the presence subscription by subscribing to changes in the relevant XDMS data; this means that the client need not reissue the subscription when a list changes status. Again, this feature saves on signaling over the air.

9.7.2.2 Authorization
Two kinds of authorization rules can be applied: subscription based and content based.

Use of Subscription Authorization Rules
The presence server checks subscription requests against the access lists. The response is "allowed", "blocked", "politely blocked" or "pending":

- "Allowed" means that the watcher is allowed to see this presence information ("blocked" of course means that access is rejected).

- "Politely blocked" means that the subscription is accepted and created as far as the watcher can tell, but that it will get simulated (default) information.
- "Pending" status means that the authorization decision has been deferred to the Presentity, but he/she/it has not answered yet. In other words, the Presentity has not yet decided whether to share the presence status or not.

Typical presence implementations allow the Presentity to set up rules for handling unknown or anonymous watchers.

The presence authorization rules are (of course) stored in XDMS and managed using XCAP.

Use of Presence Content Rules

In order to provide more detailed control, the Presentity can also set rules governing not only which watchers that are allowed access, but also which specific attributes data will be delivered.

9.7.2.3 Load Control

IMS Presence implementations typically provide a couple of methods for limiting load that may affect how often your application gets notified of state changes.

Watcher-Based Filtering

A watcher can select what data it is interested in, by specifying a filter in the body of the SUBSCRIBE request. This is yet another case where IMS Presence tries hard to eliminate unnecessary traffic.

Rate Limitation

Note that IMS Presence implementations can optimize traffic in ways other than explicit filtering. For example, the server can wait a specified time to see if more updates are available for a given watcher; it then delivers all the changes in one burst. A watcher can specify its own timeout value, in order to further reduce the update load (watcher-based throttling). Yet another load control mechanism is the subscription refresh timer; this is sent to the client as part of the response to the SUBSCRIBE signal.

9.7.2.4 Publishing Data

Finally, we come to the operation that actually makes data available to the watchers. A Presentity makes its presence data available using an SIP PUBLISH request. These updates are known as "soft state", which means that they have a limited life span (expiry time). If the data are still valid after the life span, the Presentity must refresh the data some time before the timer expires.

The Etag mechanism (part of HTTP 1.1, see IETF RFC 2616) is used to validate updates ("PUBLISH modify"), ensuring that updates from different devices do not result in inconsistent data.

If the Presentity wishes to explicitly remove data, the "PUBLISH remove" variant is used.

When data are changed, as a result of the initial PUBLISH, after a modify or after a remove (explicit or through timeout), any subscribing watchers (subscribing directly or by subscribing to a list where this Presentity is a member) are notified.

All watchers who have stated an interest in any of the changed presence data, due to a modification/terminate request or a timeout, will be notified as appropriate.

9.7.3 The Presentity Data Model

IMS Presence data describes a number of different facets of communication. The Person element models the information about the Presentity. The layers in Figure 9.8 show the different kinds of information and their relationships:

- **Person related.** At this level, we find data that are not associated with any specific service or device, but that rather relate to the Presentity as a representation of the physical person – for example, "mood" and "willingness to communicate".
- **Service related.** Here we find attributes that are relevant for a given service – for examples, "service specific availability" (logged on using a mobile phone, but nothing else), "service specific willingness" (phone is logged on, but the person would not like to be disturbed using that channel, but for example messaging is OK), and references to the device used to access the service.
- **Device related.** This information is related to the device irrespective of the service used – for example, "device type" and "device location".

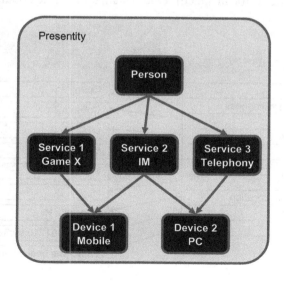

FIGURE 9.8

The Presentity data model.

The basis of the Presentity data model is the IETF PIDF framework (Presence Information Data Format). PIDF is very flexible, including the ability to extend the schema when necessary. Additionally, IMS Presence systems may support schemas like RPID (Rich Presence Information Data), CIPID (Contact Information in Presence Information Data format), Geopriv, and OMA (Open Mobile Alliance) schemas:

- RPID defines a large number of attribute values such as activity, mood, and device.
- CIPID defines extensions for contact information such as home page, display name, and sound.
- Geopriv defines attributes describing the location of a user or device.
- OMA defines additional attributes relating to a user, service, and device.

9.7.3.1 Personal Data

As can be seen in Figure 9.9, this is where we find information that is naturally associated with the person behind the Presentity. The canonical example is of course "mood"; if your mood depends on the device you are talking into you should probably contact your communications provider. Willingness to communicate here is on a general level, irrespective of service or device.

9.7.3.2 Service Data

In this day and age, we communicate in many different ways, and sometimes simultaneously. The service data model shown in Figure 9.10 captures those attributes that are relevant for a specific service (IP address, for instance). One instance of service data is stored per service.

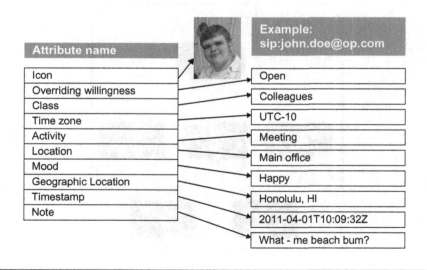

FIGURE 9.9

Presence personal data.

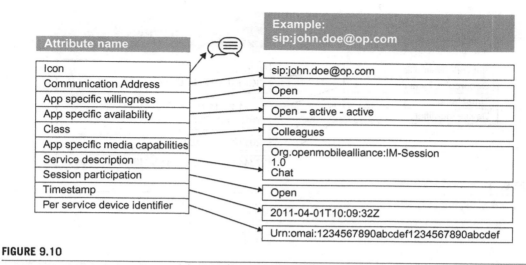

Attribute name		Example: sip:john.doe@op.com
Icon		sip:john.doe@op.com
Communication Address		
App specific willingness		Open
App specific availability		Open – active - active
Class		Colleagues
App specific media capabilities		
Service description		Org.openmobilealliance:IM-Session 1.0
Session participation		Chat
Timestamp		Open
Per service device identifier		2011-04-01T10:09:32Z
		Urn:omai:1234567890abcdef1234567890abcdef

FIGURE 9.10

Presence service attributes.

On this level the user can indicate if he or she wants to be reached by this particular means. Note: it is very unlikely that a user is exposed directly to this level of detail; how much of the inherent flexibility that is exposed depends entirely on how the client software is designed.

9.7.3.3 Device Data
Services – obviously – terminate on devices, so the device level is also needed. However, it would be unwise to embed the information describing the device in service data, as devices typically support various forms of service (as a simple example, a standard GSM phone supports both voice and SMS).

In this data set (see Figure 9.11) we find information about device type and capabilities, for example.

9.7.4 XDM Data Management
In order to manage data stored in the XDM servers of the network, a user issues HTTP/XCAP requests using GET, PUT, and DELETE methods. In this way various document manipulation operations (e.g. retrieving, adding, and deleting document elements or attributes) are executed.

A client (or an application server) creates an XCAP URI that indicates both the application (the application usage ID part) and the addressed Presentity. The element to create, read, update, or delete is specified using a subset of XPath (essentially, the URI parts – the terms between the "/" separators – direct the XDMS to travel down the XML document tree by naming the branches). The actual operation to perform is selected by the standard HTTP methods GET, PUT, and DELETE. For more details, see IETF RFC 4825.

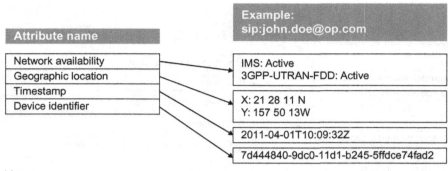

FIGURE 9.11

Presence device attributes.

9.8 FINDING THE RIGHT DEVICES

As this chapter has shown, multiple services can be provided on the same IMS core. As the different services are not necessarily supported on all devices, there is obviously a need to ensure that signals get routed appropriately. In order to do this, IMS uses tools provided by IETF RFC 3840/41. This enables the developer to partly influence the way the network will perform *forking*: i.e., how the network forwards a service invitation to the invitee's devices/contacts (single or multiple, in parallel or in sequence). Take for example a Chess application with the communication architecture we described in section 1.6. The chess commands are coded as plain text exchanged by means of MSRP (IETF RFC 4975) between the two players. The game starts when the inviter selects the intended opponent (the invitee) from her contact book and presses the chess application icon. This causes a chess communication invitation to be sent to the invitee. Now, the developer is very likely to prefer the network to forward the invite only to those of the invitee's devices which supports this particular Chess service. Why? Because otherwise there is a risk of "ghost ringing", i.e. that this invitation reaches an unprepared device which will then try to do the best it can. For instance, it may end up in the invitee's telephony client, where it will not make much sense to the person holding the device. It may very well support MSRP and plain text in general, but will have no clue on how to decipher or encode the Chess service commands it receives.

In such case the developer would need to make sure the feature parameters *explicit* and *require* are both included in the Accept-contact header carrying the chess feature tag sent with the INVITE. This will make the invitee's network fork only to those devices having explicitly registered their Chess capability. A developer must therefore make sure that the device-side application will register the Chess capability, since otherwise no Chess communication is possible with that device.

The chess INVITE would look like this:

```
INVITE tel:+12129876543
Accept-contact:*; +g.3gpp.iari-ref= "urn%3Aurn-7%3A.3gpp-application.ims.iari.
worldsbestdeveloper.chess"; require; explicit
```

This INVITE will reach those devices of the invitee having registered the Chess capability, like this:

```
Contact: sip:u1@h.example.com; +g.3gpp.iari-ref= "urn%3Aurn-7%3A.3gpp-application.
ims.iari.worldsbestdeveloper.chess"; audio; q=0.2
```

Let's now look at a different example on how a developer can use: *explicit* and *require*. Assume a developer has made a "Talk 'n Draw" application – basically a whiteboard communication with voice. The developer may want the user who invokes the application to be able to fallback to voice only in case neither of the invitee devices is whiteboard capable. In such case the developer would need to apply the *require* feature parameter to the audio capability preference but apply the "softer" *explicit* feature parameter to the whiteboard capability. Hence, when the user selects the invitee and presses the 'Talk 'n Draw' application icon, the invitee's network will not fork to any invitee devices that have registered a non-voice capability (audio = "FALSE"), such as a web-camera might register. This is because the "require" feature parameter disqualifies a device with a known non-matching capability.

If there are two remaining devices where one has registered both the audio and the whiteboard capabilities, and the other one has registered audio only, the network will alert the former first due to its better matching score with the invitee preferences. It gets a score of 1.0 as two out of two capabilities are matching, whereas the latter gets a 0.5 score. In the case of no answer from the higher score device, the remaining device will be used, partly since it supports the required audio/voice capability, and partly because the whiteboard capability had an *explicit* preference only, opening up for the network to fork even to devices not having registered the desired capability at all, albeit as a last resort. This way, the developer has ensured a wanted behavior, i.e. that the inviter will at least be able to talk to the invitee in the event none of the invitee devices has the developer's whiteboard application capability.

The whiteboard example INVITE would look like this:

```
INVITE tel:+12129876543
Accept-contact:*; audio; require
Accept-contact:*; +g.3gpp.iari-ref= "urn%3Aurn-7%3A.3gpp-application.ims.iari.
worldsbestdeveloper.whiteboard"; explicit
```

This will reach an invitee device which has registered (SIP REGISTER) a contact like this:

```
Contact: sip:u1@h.example.com;+g.3gpp.iari-ref= "urn%3Aurn-7%3A.3gpp-application.
ims.iari.worldsbestdeveloper.whiteboard"; audio; q=0.2
```

Or like this:

```
Contact: sip:u2@h.example.com; audio; q=0.2
```

But not like this (due to a known non-matching with the required audio capability):

```
Contact: sip:u3@h.example.com; audio=FALSE; video; q=0.2
```

Nevertheless, a developer must be aware that no matter how hard she tries to make the application express the wanted network forking behavior by means of the feature parameters *explicit* and *require*, she can never override the priority provisioned by the invitee utilizing the q-value (see IETF RFC 3261) as shown in examples above. A higher q-value means a higher preference. The way this works is that if the invitee for some reason would have registered a higher q-value, e.g. q = 0.3, with the audio-only device "u2" in the example above, forking takes place to the "u2" device before the "u1" device despite the fact that the latter has a higher score result (1.0 over 0.5 for u2). In other words, the capability matching affects the forking order within devices having the same q-value only. This example shows that the q-value that was registered by device "u2" above, "destroys" the forking from the application perspective since it is obvious that the device with the wrong capabilities can be alerted before the device with the right capabilities. Therefore, a developer of an IMS stack or application should cater for the case where the q-value is not registered in the SIP REGISTER process by the device. This is for the sake of letting an inviter's application reach the invitee device which is the most suitable to handle the application. This is in the best interests of both the inviter and invitee. The q-value would function well in an ideal world where every device of an invitee has exactly the same set of capabilities. But that is unfortunately not the world we currently inhabit. Different devices have different capabilities and that is the reality that a developer must consider.

Also worth knowing about forking is that there are legacy devices out there that do not have the ability to register any capabilities at all, as they were launched before IETF RFC 3840/41 was released. Such devices are "immune" to capability matching processes like the ones outlined herein, in the sense they always get the highest score (1.0). The way this works is that, for example, if the whiteboard invitation above was to an invitee who had three devices registered with the same q-value, the "immune" one will be forked to in parallel with the one with the best capabilities matching the result, i.e. score 1.0. This is because the network can never be sure of what capabilities such "immune" device may have or not have.

To make a long story short, use tags to avoid surprises. And if you find yourself surprised, then that is the time to come back and re-read this section.

9.9 CONCLUSION

In this chapter, we have briefly introduced two of the major services built on IMS: Multimedia Telephony and Presence. Other services exist: Push-to-Talk and VideoShare were mentioned in Chapter 1, and IPTV will be touched upon in Chapter 14. We chose to describe MMTel and Presence for two reasons: first, they show a couple of patterns demonstrating how services are built on top of the IMS core. Secondly, they are good examples of services that you can leverage when you build your own applications.

Charging

10.1 INTRODUCTION

The title of this chapter is probably not the first thing a developer thinks of. But it might very well be the first thing that a sponsor/investor thinks of: How can this product idea generate a return on investment? Chapter 2 discussed at length how the business models of today and tomorrow look (and might look respectively); in this part we will be considering the practical side of getting paid in a bit more detail.

Actually, before going into that subject, let us point out that sometimes the point is not to get paid for the software per se. Much of the past years' open source and freeware development has been driven by an honest desire to do something useful for the community at large, to gain peer respect, for the intellectual challenge, or just simply for fun. If your project falls into any of these categories, you can safely skip this chapter, but if your plan for the future somehow includes quitting your day job, this might be of some interest.

By the way, note that in telecoms it is common to separate charging from billing: charging is the process of gathering enough information about resource usage to figure out who should pay whom how much for what; billing is converting charging information to a document that instructs one party (usually the subscriber) to pay another party (usually the service provider) specifying why and how much. In modern telephony systems (notably mobile systems) there is also a distinction between prepaid and postpaid approaches, but more on that later. Also, it is useful to consider that, in the telco world, operators are peers providing the same services. In typical scenarios, one operator gathers the revenue (bills the customer), but the cost of actually producing the service may be shared between two operators: consider the case when you are calling someone served by another cellular operator. In some cases – notably in the USA – the receiving party pays for his or her traffic, but otherwise the norm is that the calling party pays. Therefore, there has to be some money flowing between operators: settlement is thus a large part of the financial side for the telco companies. Actually, it becomes even more complex if you consider that both the caller and the callee can be roaming, involving even more business transactions behind the scenes. But, as we tend to come back to: you shouldn't have to worry about this as a service developer – your dealings will be with the IMS provider you connect to (or deliver components to); the rest is taken care of for you.

10.2 OBVIOUS AND NOT SO OBVIOUS WAYS OF GETTING PAID

If your background is from inside of the telco industry, you probably think of charging as simply counting call minutes and/or bits transferred, multiplying by the going rate (not quite that easy, as we will see), and sending a bill. The business model is simple: somebody signs a contract, someone else provides connectivity (audio, broadband, etc.), and money changes hands. If, on the other hand, you grew up as a web designer, unless you were one of the lucky few that turned an idea into a runaway success – think eBay – getting paid for your efforts may still be a bit of a mystery. And remember the basic prerequisite for getting paid (and/or becoming famous): your app/service needs to do something that other people will find useful, clever, amusing, or otherwise desirable; if not, there won't be much to charge for!

Let us begin with listing the potential sources of income:

1. The end-user. OK, this is still a major option. This essentially means selling your software, physically or online. It typically requires that you maintain the relation to the subscriber in detail, as you would probably want to use some sort of licensing.
2. Subscriptions. Even though World of Warcraft is sold outright, the interesting part of the business model is still the recurring subscription fees.
3. Pay-per-use. The user pays per something measurable (time, volume, events, virtual items bought, etc.)
4. Advertisers (or more accurately the people running the ad delivery system, borrowing space in your app or web page), as exemplified by Google AdSense and Apple's iAds.
5. Communication service providers may operate your application, either on your behalf (sharing revenue) or by having bought a license from you.
6. Sales of something else. Sometimes, the reason to develop and sell software is to promote the sale of, say, hardware (Apple apparently makes most of their money through hardware sales). Content may be the main source of revenue (a player app that is coupled to coded content).
7. Donations. The shareware concept is still alive and kicking.
8. Donations, but to something else. You might encourage your users to donate to their favorite charity. Agreed, it does not qualify as revenue to your company, but it may still be a very worthwhile reason to write and distribute something.

10.3 MONEY MAKES THE APP GO AROUND

In the following, we will go through these cases in a little more detail.

10.3.1 Selling to the End-user Through a Store

In the past year or so, the app store has emerged as one of the most powerful ways of letting small (and big!) application developers reach their customers. The mechanisms are rather

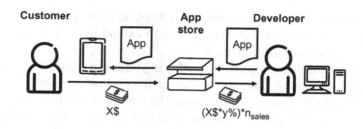

FIGURE 10.1

Selling through an app store.

straightforward: you supply the app, the app store makes it findable, any revenue is split between you and the app store vendor (see Figure 10.1). The examples are of course well known: Apple's AppStore, Android's Market. The telco operators are creating their own alternatives; from a developer's point of view, the WAC (Wholesale Application Community) probably makes the most sense, since the intent is to create an environment where the application can be delivered through any WAC-compliant app store to be run on any WAC-compliant device.

Note that this is about delivering a piece of software to a device and getting paid in return (typically, 70% of the revenue goes to the developer); the app does not have to do anything relating to communications. Examples abound: single-player games, notepad software, media players, etc. The kinds of apps we talk about in this book have an additional property: they use communications services provided by IMS to enhance an experience (extending games to multiplayer scenarios, extending the notepad by immediately publishing on the web, playing network-based content, etc.). Or, indeed, the app itself may be creating an entirely new communication service. But this only makes it a special case of apps in general, so it is still a very valid revenue opportunity. However, the flip side is that you only get paid once, so in order to have a sustainable business model, you need to figure out a number of improvements that will entice your users to upgrade.

If you don't feel like linking your fate and/or reputation to an app store, you can of course organize payment through, for example, PayPal, providing unlock keys in exchange for payment.

While we're on this subject, don't forget the freemium model, i.e. providing a limited version of your app that gives the user some feel for what it can do, but where the really good stuff is only available in a paid version ("pro").

10.3.2 Selling Over and Over Again

This is otherwise known as the subscription model. It is a bit trickier to implement, as you need to verify that the user's subscription is still valid. This works fine if your application has a network side to it (such as a content server): then all your license management can happen on the server side, where it is easier to protect (still not trivial, but at least you have a bit more control). App stores like, for instance, Apple's, allow you to sell from inside your app; this provides a comfortable way for the user to extend his or her subscription without having to leave the context of the app itself.

The subscription model is particularly useful if you are paying for some consumables, cellphone airtime, for instance. Say you have created a multiplayer game, where the players in a team can talk to each other over cellphones (let's assume that this is a game where you are supposed to be out and about – some kind of treasure hunt, for example). The gamers pay you a monthly fee and don't expect to pay for the voice part: this means that you need to pay for the outgoing legs (let's assume for simplicity that we are in a part of the world where the calling party pays), i.e. you will have recurring costs (running the servers, providing voice services); obviously, you will need recurring revenues.

As mentioned above, you can sometimes let the app store manage this for you, or you can choose to do the administration server-side.

10.3.3 Pay-per-use

A variant of the subscription model can be implemented using tighter cooperation (and some sort of contract) with the communications service provider (the operator). If your app drives traffic (phone calls, data, SMS, etc.) that the operator gets paid for, he might be willing to share some of that revenue. Of course, this model does not work in flat-rate scenarios, as more traffic in those cases is only a cost to the operator. However, the operator may still want your app on his network if it attracts the right demographics, but then we are more into the scenario described in section 10.3.5.

If the usage in question is not directly linked to traffic (such as buying virtual paraphernalia in a game), you will typically have to keep track of that yourself (cf. Linden dollars in Second Life, exchangeable for "real" cash). For instance, in Apple's AppStore, you have the possibility to do in-app purchases, but this mostly takes care of the job of getting real money flowing; you will have to keep track of the actual "thing" (content, health points, map sections, etc.) that was purchased.

10.3.4 Advertising

This is the traditional way of generating income on the web, by allowing some space on a web page for the ad broker to fill with what might make sense for this viewer, at this time, in this context. It has the advantage of being a proven business model in the classical browser-running-on-a-PC-connected-to-the-web context. Google's AdSense is, of course, the canonical example here (see Figure 10.2).

However, mobile applications – in particular if running on a phone – typically run under constraints of a small screen, sometimes limited bandwidth, and possibly most importantly subject to power consumption constraints. Or maybe without connectivity at all – think airplane mode. Therefore, for mobile apps there are mobile-adapted solutions such as AdMob (bought by Google in 2010) and iAd (released by Apple in 2010). In either case, the main

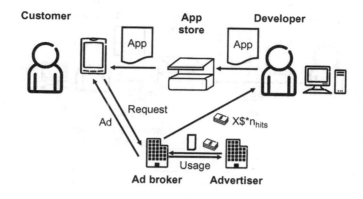

FIGURE 10.2

Advertising.

job done by the middleman is to make it much easier to make the guy with the billboard (you) meet the guy with the poster (the advertiser) and bring a message in front of the customer (your user, potentially also a customer of the advertiser). The middleman can then bring all contacts to the table, including his old web customers, instantly creating a very large base of potential revenue generators. Note, though, that the margins in this business are very low, so unless you have managed to build something massively successful (sorry, not "unless", of course we meant "until") the revenue streams will be moderate.

As a comment, note that of course you can also be an advertiser if you wish, using these methods to market your own app. One advantage is that you are already in front of a member of your target group: a device owner!

10.3.5 Letting Someone else do the Heavy Lifting

One way of getting paid is of course to share revenues with someone else selling your product. The app store concept is a version of this (the app store owner/operator is basically retailing your product), but we should point out that it is of course still very possible to sell your app to, say, an operator who then sells it to his customers, probably under his own brand (i.e. what you provided was what is called a white-label product). Flat-fee or revenue-share-based contracts are possible options (see Figure 10.3). In a sense, this is very close to the traditional model where the initiative is from the other side: an operator ordering the development of a piece of bespoke software from an outside party. However, here we assume that you came up with the idea, and you are now searching for channels to your customers.

This option has interesting side-effects that relate directly to the comms side of your app (i.e. the IMS capabilities). In this case, it is much easier for the operator to relate connectivity usage to the traffic generated by your app: charging (including promotional bundles, zero rating of certain types of traffic, etc.) now becomes an operator-internal issue.

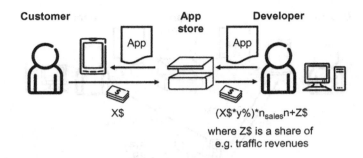

FIGURE 10.3

Revenue sharing.

10.3.6 Sell Something Else

Maybe your main business isn't at all writing software for mobile devices? For us writing this book this is of course a very strange idea, but there are successful examples out there. For instance, Apple seems to be making most of their revenues selling hardware, which is the reason that OS X retails for a lot less than Windows 7. Another example is, if you are running a radio station, your main income would be from ads, i.e. what you really sell is airtime to advertisers. And if you design a smartphone app that provides single-click access to the streamed version of your radio channel, you have just increased the attractiveness of your business. Other examples are the driving school tutoring app, the find-my-nearest-Hup-mobile-dealership app, etc.

10.3.7 Count on your Fellow Man

In other words, provide your app as shareware/donationware. Again, this is a very old and well-established model, essentially just requiring a bank account (or similar, such as PayPal) that your users can send money to if they feel you have provided something worthwhile. The upside for you is that you don't need to keep track of anything (although your accountant – and your local taxman – might want you to), as your software does not need to incorporate any license validation mechanisms. Licensing tend to be rather hard to do in practice, and should be left to the professionals. Unfortunately, they tend to charge professionally for their services.

As we have noted before, the zero-rated version of this – freeware – can be a useful first step in the freemium model.

10.3.8 Benefit in an Entirely Different Dimension

Just to round things off on a very altruistic note, the donations you encourage need not be to you, but could be to your favorite charity. Instead of giving away old clothes and used furniture, why not write the world's best to-do shopping list and suggest that your users donate €10.00 to Save the Children. Of course, this app will use IMS messaging to allow your family members to exchange items, including updating the "done" box across all connected phones so you don't get two of everything – and if mother adds something to the list, father's list will have a call button on that item so he can call immediately and figure out details. A messaging button to request the same info in an IMS text message also avoids the need to leave the app and bring up the built-in message GUI. See, another clever app idea for you!

10.4 THE MECHANICS OF CHARGING

In the remainder of this chapter, we will elaborate upon how charging works for the IMS connectivity your app will be using. Also, we will show you a few ways you can feed charging info from your app into the normal charging and billing system of the operator. Note that, with few exceptions, this means that the server side of your app most likely executes on the operator's network, as you need to be inside their trust domain. Some APIs (e.g. Parlay X or its REST derivatives) allow you to insert charging information also from external servers, but of course this means that you have established enough credentials with the operator to allow such signaling. Note that in many cases, the operator's billing system must also be made aware of your added information so that it can be presented correctly and clearly on the subscriber's bill.

Two distinct modes will be described: online and offline charging. As a user, you have probably come across these variants in the form of prepaid and postpaid contract versions respectively. In GSM, the prepaid model was implemented using Intelligent Network (IN) technology, which had the side-effect of a certain level of coupling between services and payment form: if you wanted to convert from a prepaid to a postpaid account, you sometimes found yourself with a different service set. As time and technology moved on – and as prepaid became very successful, in some markets it was the dominating contract form – it became clear that IMS needed a redesign of the charging architecture, as described in 3GPP specs TS32.240 (Charging principles for 3G networks) and TS32.260 (Charging for IMS). Here, online and offline essentially means the following (borrowing the definition – in italics – from 32.240):

- **Online charging.** *Charging mechanism where charging information can affect, in real-time, the service rendered and therefore a direct interaction of the charging mechanism with bearer/session/service control is required.* "Direct interaction" means, for example, that the charging system notes that there is no money left in the account; hence a message is generated telling the user how to top up the account. I.e. the charging system must be capable of returning a response with a delay that does not add noticeably to the call setup time.
- **Offline charging.** *Charging mechanism where charging information **does not** affect, in real time, the service rendered.* Essentially, this means that Charging Data Records can be collected and processed in batch.

As can be seen, the terms prepaid and postpaid are not used. Indeed, combinations are possible and useful: a postpaid account with a credit limit is in technical terms a prepaid account, but with a non-zero credit limit.

Note

Although CDR (Charging Detail Record) information is primarily collected for the purpose of billing, it is also very useful for capacity and trend analysis, cost allocation, and auditing. This explains why CDRs are generated and collected also in online charging scenarios, even though no actual bills are produced. Thus, a by-product of the charging system is a huge bank of information about users and their behavior.

Below, we will now work through the basics of how these two methods work. But remember that if your app isn't running inside the operator infrastructure, it is unlikely that you will see many of the details in practice. It may still be useful to understand some of what happens behind the scenes, in order to have a complete grasp of how your users' experience will play out.

10.4.1 Offline Charging

As can be seen in Figure 10.4, many nodes in the IMS architecture, at both the core and application levels, are capable of producing charging information. This information is

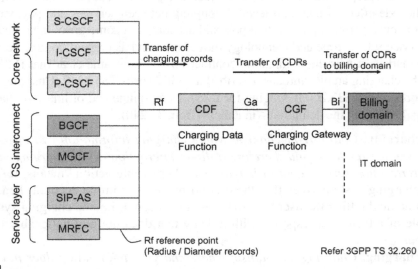

FIGURE 10.4

IMS offline charging.

delivered over the Rf reference point using the Diameter protocol, as specified by IETF (RFC 3588). Diameter is extensible by design, so IETF RFC 3588 essentially specifies the basic mechanisms. In this case, the payload is described in IETF RFC 4006: Diameter Credit Control Application. This use of an IETF protocol (or maybe protocol family tree is the appropriate term) is very typical of how 3GPP and IETF have cooperated fruitfully during the last few of years: IETF designing the protocols, 3GPP placing them in a complete architecture. SIP is of course the prime example, but many others exist.

Nodes and reference points are as follows:

- Charging data function (CDF) – collects charging records from different nodes.
  - The address of the CDF to use by an entity is obtained from P-charging-vector (conveyed in SIP signaling) or is determined from the node configuration.
  - For "IMS roaming" subscribers, CDF in the visited IMS domain may be used.
- Charging gateway function (CGF) – combines and formats charging records (CDRs), packaging them into charging files for the billing system. (Note: prior to 3GPP Rel-6, CDF and CGF were combined in the charging control function, CCF).
- Rf – transfer of Radius/Diameter charging records.
- Ga – transfer of charging records (CDRs) to CGF.
- Bi – transfer of charging files to the billing system.

Use-case Example

In this case (see Figure 10.5), user A is setting up an IMS voice call to user B, with MMTel AS (Multimedia Telephony application server, shown as MTAS) being the invoked application server. The MMTel AS sends an accounting request (ACR) for a number of reasons: when the session has been established, during the session due to interim timer expiry, due to media change, and when the session terminates. The CDF responds with the appropriate accounting answer (ACA) message.

10.4.2 Online Charging

In general, some of the same principles apply (see Figure 10.6), such as the CGF being the intermediary between the runtime system (including OCS, the Online Charging System as defined in the standard) and the back office (as was mentioned before, "billing domain" here is something of a misnomer, as no bills are being produced – instead, these CDRs are mainly used for analytical purposes). Also, Diameter is the basis for Ro (as well as Rf). The main difference is that, in the online case, the requesting node needs to wait for the response in order to be able to proceed with the signaling (or not, as the case may be).

The response (credit control answer, CCA) includes not only OK or not OK, but also information regarding what the user can afford to do (how long can the call be, how much data can be transferred, etc.). This is allocated in chunks; when (or slightly before) this chunk has been consumed, the AS reissues a CCR (credit control request) in order to find

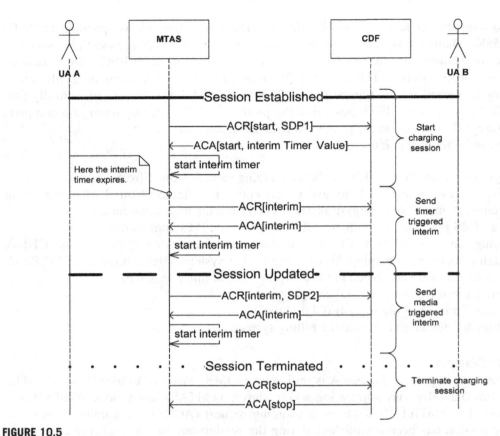

FIGURE 10.5

Offline charging signal flow.

out if the call can continue. In the meantime, the user might have been doing other things that has used up credits, or topped up the account. Similarly, any time something happens to the session (media added or removed, for instance) the CCR/CCA procedure is executed.

When the session is finally terminated, any outstanding credits are returned to the OCS. This means that the intervals between CCRs can be large enough to hold the network load to a reasonable level, while maintaining fairness towards the user.

Use-case Example

Again, an MMTel call is being set up (see Figure 10.7). However, here the MMTel application server (MTAS) sends a credit control request (CCR), expecting a credit control answer to be fed into the call handling logic.

In addition, the originating MMTel AS can play an announcement to the originating subscriber at session establishment if the credit is low, very low, or empty before the INVITE

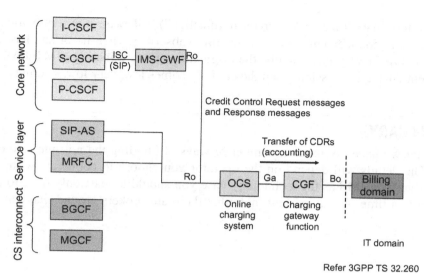

FIGURE 10.6

IMS online charging.

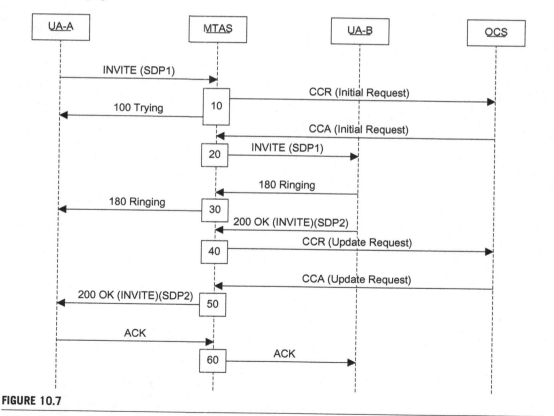

FIGURE 10.7

Online charging signal flow.

is sent to the terminating network ("credit notification"). Likewise, the terminating MMTel AS can play an announcement to the terminating subscriber when he or she has answered if the credit is low or very low. Finally, the originating and terminating MMTel AS plays an announcement during the session when the credit becomes low, very low, or empty.

10.5 SUMMARY

In this chapter we have discussed a number of ways of building the revenue side of a business case. The details will depend on the operator you choose to cooperate with; the main message is that with IMS as the technology base you should at least only have to deal with one of them at a time. An added hint: note that there are brokers around that can help you with this.

Interworking with Legacy Networks

11.1 INTRODUCTION

This chapter provides the user with "the bigger picture", by placing the IMS network in a larger context. In Chapters 7 and 8, we introduced a number of interconnection points between the IMS network and other networks. For example, the media gateway control function (MGCF) and IP multimedia media gateway (IM-MGW). These entities allow connection of the IMS network to circuit-switched (CS) networks, e.g. the Public Switched Telephony Network (PSTN) or the Public Land Mobile Network (PLMN). In this chapter, we will consider the impact of network interconnection (connecting IMS network to CS network) on the availability of services in the IMS network.

A further topic we will look into is the application of Intelligent Network (IN)-based value-added services (VAS) for IMS subscribers. Within the context of the IMS network, IN services are often referred to as *legacy services*. This topic of applying legacy services in the IMS network has led to many initiatives and has resulted in many products like service brokers. This chapter will put these various products in the right context.

11.2 THE BIGGER PICTURE – CONNECTING IMS TO THE OUTSIDE WORLD

Figure 11.1 shows where IMS fits in the bigger picture. It depicts an IMS network, with terminals connected to it, and connection to other networks. The diagram shows also the position of a SIP application server (SIP-AS) in the IMS network. The SIP-AS may be controlling a call that is established from a SIP terminal (user equipment, UE) destined for a subscriber outside the IMS network. The SIP-AS may also be controlling a call that originates from outside the IMS network, such as a public service call or a call destined for a subscriber of the IMS network.

The communication between the IMS network and the other networks runs via designated gateways and through a transit transmission network. Interconnection between the IMS network and PSTN or PLMN may, alternatively, be realized without a transit network. The interconnection gateway that allows for interconnection between PSTN/PLMN and IMS may be integrated in PSTN/PLMN (e.g. in an MSC of the PLMN).

The gateway between IMS and PSTN/PLMN, which will be described in more detail in subsequent sections, offers interconnection for, primarily, call establishment. More

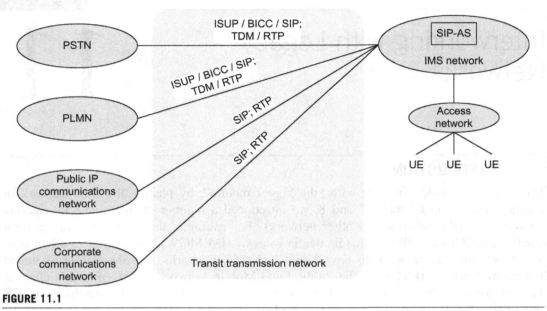

FIGURE 11.1

Interconnecting IMS network to other networks.

specifically, it allows a call that spans two or more networks to be established. For example, a call established in PLMN, destined for a subscriber in IMS; or a call established in IMS, destined for a subscriber in PSTN. Not all telephony services in IMS have corresponding services in non-IMS networks. Services in IMS are typically controlled through SIP signaling. Hence, when certain (value-added) services are applied in the IMS network, these services may not be available to a calling party from another, non-IMS, network. This aspect should be borne in mind by application developers in IMS.

An example of a call that spans more than two networks is the following. A PLMN subscriber (GSM subscriber) establishes a call to an IMS subscriber. The call is routed from the GSM network to the IMS network. This routing may be based on B-number analysis or on number portability information. The call traverses the CS–IMS network boundary. The IMS subscriber, in turn, is contactable on multiple terminals: a SIP terminal and a GSM terminal. The SIP terminal of the called subscriber is currently switched off. The call is offered to the called subscriber's GSM phone. Hence, the call traverses the IMS–CS network boundary once again.

For interconnection between IMS and non-IMS networks, protocol mapping has to be applied. Table 11.1 shows the protocol mapping for interworking between IMS and GSM. Table 11.1 is not exhaustive; there may be protocols other than RTP in the user plane.

Diameter is not subject to protocol conversion at the network boundary. So far as Diameter is used for IMS, it is used only inside the IMS network.[1] If a switch in the CS

---

[1] Diameter is used for various IP networks. Here, we are considering IMS only.

Table 11.1 Protocol Mapping Between IMS and GSM

Protocol in IMS Network	Protocol in CS Network	
SIP	ISUP	ISDN user part; call control protocol used in circuit-switched (CS) networks
	BICC	Bearer independent call control; successor of ISUP
	SIP-I	SIP tailored for use in CS networks
RTP, RTCP	TDM	Time division multiplex; traditional media transportation methodology in CS networks
	ATM	Media transfer over asynchronous transfer mode
	RTP, RTCP	When BICC or SIP-I is used, media transport in a CS network may be done with RTP. When BICC is used for control plane, no RTCP is used for the user plane. For RTP in a CS network, *framing* is used on the Nb reference point (Nb reference point between two CS-MGWs)

network contains MGCF, then that MGCF may have an interface for generating Diameter charging records, e.g. Rf reference point. One may argue that the MGCF component in the switch is functionally a part of the IMS network.

We will describe in the next sections some of the intricacies of protocol mapping, and call handling in general, at MGCF and IM-MGW. Video interworking is described in a separate section, due to the additional complexity associated with interworking between video in a CS network and video in an IP network.

11.3 INTERWORKING THROUGH MGCF AND IM-MGW

11.3.1 General

Figure 11.2 shows example interworking for a call for which the control plane traverses MGCF. It is emphasized that *MGCF* represents a functional entity. It may be embodied as a stand-alone node or may be integrated in another node, such as mobile softswitch (MSS).[2] Likewise, the IM-MGW may be a stand-alone node or may be integrated in another node. For control-plane interworking in the MGCF, the following interworking cases are defined:

ISUP (over IP or TDM) $\leftarrow \rightarrow$ SIP
BICC (over IP or ATM) $\leftarrow \rightarrow$ SIP
SIP-I (over IP) $\leftarrow \rightarrow$ SIP

MGCF facilitates interworking between ISUP and SIP, whereby the ISUP runs over IP infrastructure. More specifically, ISUP is applied as *MTP3 user* on top of M3UA. MTP3

---

[2] Mobile softswitch refers to the MSC server, containing control-plane entity only.

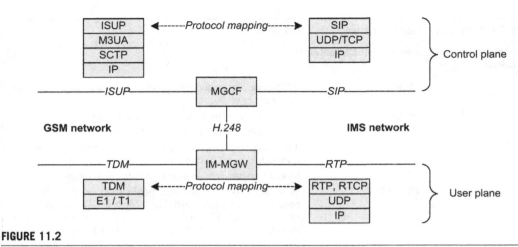

FIGURE 11.2

Protocol mapping at CS–IMS boundary.

refers to the topmost protocol layer of the message transfer part (MTP), which is one of the layers in the OSI protocol stack for the SS7 network. "MTP user" is a protocol on top of MTP3. MTP3 user adaptation (M3UA) is a protocol adapter function. It allows conventional SS7 protocols such as ISUP and SCCP to be transported over IP infrastructure, using SCTP as transport protocol. If the ISUP in the CS network runs over TDM infrastructure (SS7), then a signaling gateway (SGW) is required. The SGW converts between ISUP over TDM and ISUP over M3UA/SCTP/IP. An SGW may be deployed as a stand-alone entity or may be integrated in another node, e.g. the IM-MGW or MGCF. In the latter case, SGW integrated in MGCF, we have TDM-based ISUP running up to MGCF; conversion from TDM-based ISUP to M3UA-based ISUP then occurs inside the MGCF.

For media transfer, TDM-based media is converted to RTP. TDM-based media is transported through a time slot in an E1 (European standard) or T1 (American standard) connection. The media conveyed in the TDM time slot in a CS network is encapsulated into RTP messages in the IMS network and vice versa. The "protocol mapping" in this case consists of *unpacking* and *repacking*, depending on whether the same codecs are used on the CS side of the call and on the IMS side of the call. If the same codecs are used, e.g. both G.711 A-law, then the actual media is not modified; there is no transcoding taking place in that case. Hence, if the TDM-based media represents PCM-encoded speech in accordance with G.711, then the RTP messages will convey the same G.711 PCM-encoded speech. For RTCP, there is no mapping taking place; there is no equivalent to RTCP in a CS network. RTCP messages are generated by the media entities in the IMS network, based on the characteristics of the RTP media transfer. If different codecs are used in the CS and PS domains, then transcoding is needed in IM-MGW (see Table 11.2 for a non-exhaustive overview of transcoding cases). Table 11.2 shows the G.711 codec; G.711 is typically used in wireline networks (PSTN).

Table 11.2 Transcoding in IM-MGW (non-exhaustive)

Codec Used in CS Domain	Codec Used in PS domain
G.711 A-law	G.711 A-law
	G.711 μ-law
	G.723.1
	G.726-24
	G.726-32
	G.729
	telephone-event

The conversion between G.711 A-law and telephone-event, for example for the transfer of a DTMF digit, constitutes a different form of transcoding than the other cases. A group of consecutive G.711 A-law PCM samples may represent a DTMF digit. This is then reflected in the RTP stream as telephone-event.

If the media transmission in the CS network is based on IP or ATM, e.g. RTP with Nb framing, then no TDM–RTP protocol conversion is needed. The IM-MGW has to apply reframing for the transfer of RTP between the CS and IMS networks.

11.3.2 Protocol Mapping

We will now concentrate on the ISUP–SIP mapping in MGCF. Figure 11.3 shows an example call case for call connection from a CS network to the IMS network. When the IMS network represents *mobile IMS*, then we would also expect a 183 Session Progress provisional response, a PRACK transaction, and an UPDATE transaction. These additional response messages and transactions would be used for resource reservation in the access network. This is not shown in Figure 11.3.

When MGCF receives an ISUP initial address message (IAM), it selects an IM-MGW for this call and reserves media transmission resources in that IM-MGW. Specifically, the IM-MGW allocates termination for a TDM circuit and allocates termination for RTP media transmission. The RTP media resources that are to be reserved will be in accordance with the required call type. For example, different resources will be needed for a voice call than for a video call. When the ISUP IAM relates to a video call, then MGCF will select an IM-MGW that is capable of video interworking.

The INVITE that is constructed by MGCF contains SIP headers that contain information copied from ISUP IAM. Various mappings are applied (see Table 11.3).

Called party number, calling party number, and additional calling party number in ISUP are represented as *phone numbers*. Hence, their respective corresponding representation in SIP will have the form of a SIP URI with user = phone or a Tel: URI, depending on MGCF configuration. The phone number in the Request URI in the INVITE request will be normalized to international format. If one of these identifications is represented as SIP URI

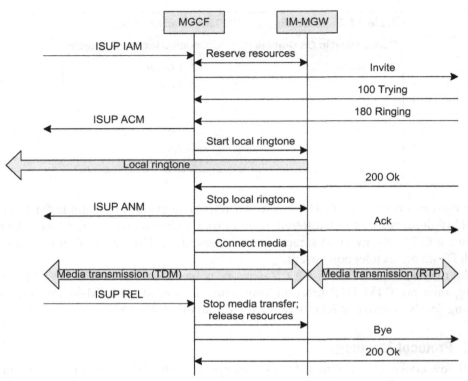

FIGURE 11.3

Call establishment and call release from CS domain to IMS domain. ACM, address complete message; ANM, answer message; IAM, initial address message; REL, release.

Table 11.3 ISUP to SIP Mapping

Information element in ISUP IAM	Header in SIP INVITE
Called party number	Request URI
	To
Calling party number (CgPN)	P-asserted-identity
	From (if no additional calling party number present, A-CgPN, in ISUP IAM)
	Privacy
Additional calling party number	From
Hop counter	Max-Forwards
3GPP TS 29.163 specifies rules for the mapping of CgPN and A-CgPN depending on their respective values and availability in ISUP IAM.	

(with user=phone), then the MGCF uses a configured domain, which will typically correspond to the *realm* of the IMS operator. For example, a call arrives at the MGCF, destined for +31651613900. The realm of the IMS network is ims-operator.nl. The MGCF generates the following INVITE request message:

```
INVITE sip:+31651613900@ims-operator.nl; user=phone SIP/2.0.
```

As an implementation option, the MGCF may use ENUM to obtain the domain associated with the destination phone number; the use of ENUM by MGCF is, however, not standardized.

When the call in the CS network was subject to one or more call forwardings, this may be indicated in the ISUP IAM through the information elements Original called number, Redirecting party, and Forwarding information. These information elements are mapped to the History-Info header in the INVITE request. The syntax and semantics of that header are described in IETF RFC 4244.

Various other information elements from ISUP can be mapped to corresponding SIP headers. Support of these headers depends on, for example, support of supplementary service interworking by the MGCF. 3GPP TS 29.163 provides an extensive overview of parameter mapping between ISUP and SIP.

The SDP offer in the INVITE is derived from the type of call in the CS network. The type of CS call is reflected in the ISUP information elements Transmission medium requirements (TMR), User service information (USI), and High layer compatibility (HLC). Mapping tables in MGCF determine the SDP offer based on these parameters. If the MGCF offers additional codecs, compared to the codec that applies in the CS domain for this call, then this may result in the necessity of transcoding in the IM-MGW.

When MGCF receives 180 Ringing, it will instruct the IM-MGW to generate a *local ringtone* through the CS media connection to the calling party. The rationale of the local ringtone is that in a CS network the ringback tone is generated by the remote switch, using the media channel established for this call, whereas in the IMS network ringback tone is generated by the terminal itself. For this reason, the MGCF has to generate the ringback tone in this call case. It is, however, also possible that the MGCF receives an 18x provisional response indicating that early media is available from the remote party. This indication may have the form of the P-early-media header (see IETF RFC 5009) with value "Sendonly". In that case, the MGCF instructs the IM-MGW to allow media to pass through backward. A typical use-case example is the playing of a call setup announcement. If MGCF does not receive an indication about early media, the through connecting ("gating") of early media depends on MGCF configuration. For example, MGCF may allow backward media transfer directly after having sent an INVITE request.

When 200 OK is received, MGCF instructs IM-MGW to allow media to pass in both directions.

The reception of ISUP REL by MGCF leads to termination of the SIP session (BYE, 200 OK). MGCF will at this point release the resources in IM-MGW. Alternatively, the remote

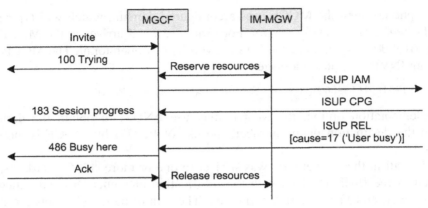

FIGURE 11.4

IMS to SIP call establishment – call failure.

party releases the call, resulting in MGCF receiving a BYE request. MGCF then sends an ISUP REL to the calling party.

Designated ISUP–SIP message mapping is also defined for call establishment failure. Call establishment failure may originate from the calling or called party:

- **Calling party abandon.** This leads to ISUP REL to be received by MGCF, whilst MGCF was waiting for the final response on the INVITE transaction. The MGCF will then cancel the INVITE transaction, in the manner described in Chapter 7 (CANCEL method).
- **Call connection in the forward direction fails.** This may be due to a variety of reasons, including, but not limited to, called party busy, subscriber not available, redirection, invalid destination, and network failure. MGCF receives an unsuccessful final response (3xx–6xx) in that case. MGCF applies ACK to the remote end and sends ISUP REL to the calling party.
(The 3xx response is a final response and would normally not lead to MGCF generating another SIP INVITE request.)

MGCF allows for call establishment from the CS domain to the IMS domain, and from the IMS domain to the CS domain. Figure 11.4 illustrates call establishment from IMS to a destination in the CS domain, e.g. PLMN. In this example, the call establishment fails and a SIP-AS acting on behalf of the calling party wants to take action based on the reason for call establishment failure.

The ISUP release is mapped to an unsuccessful final response on the INVITE transaction. The ISUP cause in the release message is used to select the final response. In the example in Figure 11.4, ISUP cause 17 ("User Busy") results in 486 Busy being sent to the calling party. ISUP cause codes are defined in ITU-T Q.850. The mapping between ISUP cause value and INVITE final response is defined in 3GPP TS 29.163. Table 11.4 shows a subset of this mapping.

Table 11.4 ISUP Cause Value to INVITE Final Response Mapping (not exhaustive)

ISUP Cause	Final Response
17 (user busy)	486 Busy Here
18 (no user responding)	480 Temporarily Unavailable
19 (no answer from the user)	480 Temporarily Unavailable
20 (subscriber absent)	480 Temporarily Unavailable
21 (call rejected), Location = 000 (User)	603 Decline
21 (call rejected), Location has other value than 000 (User)	480 Temporarily Unavailable
22 (number changed)	410 Gone
24 (call rejected due to ACR supplementary service)	433 Anonymity Disallowed
25 (exchange routing error)	480 Temporarily Unavailable
27 (destination out of order)	502 Bad Gateway
28 invalid number format (address incomplete)	484 Address Incomplete
29 (facility rejected)	500 Server Internal Error
31 (normal unspecified)	480 Temporarily Unavailable

A problem that would occur for a SIP-AS acting on behalf of the calling or called party is that a multitude of release causes in ISUP map to the same final response. For example, when considering the subset of ISUP cause values given above, ISUP cause values 18, 19, 20, 21, 25, and 31 map to 480 Temporarily Unavailable. Hence, information is lost that may be needed for service logic processing in the SIP-AS. This will result in this particular functionality not being offered to the calling or called party.

To overcome this dilemma, the MGCF may copy ISUP cause information into the final response (see also IETF RFC 3326). A *Reason* header may be included in the final response. For example:

```
SIP/2.0 480 Temporarily unavailable
Reason: Q.850; cause=18; text="no user responding"
```

The issue of cause value is equally applicable when an unsuccessful final response of 4xx-, 5xx-, or 6cc-class, relating to an INVITE transaction, has to be mapped to an ISUP REL. A SIP-AS that is controlling the call in the IMS network may include a Reason header in the final response, e.g. when call establishment has failed. The Reason header is then used by MGCF to set the cause value in ISUP REL.

11.3.3 MGCF SIP Signaling Capability

When a call is established from the CS domain into the IMS domain, the supported SIP capability of the MGCF may be limited. An application server operating in the IMS network controlling this call needs to have the flexibility to cater for the lack of support of certain capability. Likewise, a call being controlled by an application server in IMS may break out downstream to the CS network. The MGCF for that break-out may also impose certain restrictions on SIP signaling. A SIP-AS will therefore be prepared in the case where certain capability is supported by the MGCF and in the case where that same capability is not supported by the MGCF. This section highlights a few of these aspects that service designers should be aware of.

11.3.3.1 Multiple Early Dialogs

The MGCF on the upstream side may not support multiple early dialogs. In this case, the SIP-AS would be compelled to map between (i) multiple dialogs in the downstream direction and (ii) a single dialog in the upstream direction. These multiple dialogs in the downstream direction may be the result of:

- Action taken by the AS, such as AS-based forking or call forwarding on busy; in these cases, different SIP sessions are established by the SIP-AS, either simultaneously (e.g. parallel alerting) or sequentially (e.g. call forwarding on busy).
- S-CSCF-based forking. The AS in this case is not in control of multiple dialog establishment; however, the AS will receive multiple provisional responses relating to different dialogs, hence the AS must still map these multiple dialogs to a single upstream dialog (if multiple SIP-ASs are invoked from the S-CSCF, with ISC chaining, then only one SIP-AS would have to apply this mapping of dialogs).

There is no designated SIP header that may be included in the INVITE request (from MGCF) indicating whether the MGCF supports multiple early dialogs (and if yes, how many). Hence, a SIP-AS will be suitably configured, in accordance with network capability. That is, the SIP-AS will be configured in accordance with the capability of the MGCF of the IMS network in which the SIP-AS operates.

The method of converting, by SIP-AS, between multiple downstream early SIP dialogs and a single upstream early SIP dialog is commonly referred to as the "gateway model".

11.3.3.2 Reliable Provisional Response

When a call is established from the CS domain into the IMS domain, MGCF generates an INVITE request and indicates its supported SIP capability. More specifically, the MGCF indicates the supported SIP methods and the supported SIP extensions. For example:

```
INVITE tel:+31163279900 SIP/2.0
Allow: INVITE, ACK, CANCEL, BYE, UPDATE, PRACK
Supported: 100rel
```

The indication that 100rel (reliable provisional response) is supported means that the MGCF has the capability to initiate a PRACK transaction, when requested through a provisional response. A service in the IMS network may connect the call to an MRF, for playing a call establishment announcement. Whereas the MRFC may establish the SIP session with 200 OK, the SIP-AS controlling this call may not want to send a 200 OK to the MGCF, since that would result in the call being answered towards the calling party and charging for the calling party being commenced. Therefore, the 200 OK – ACK between MRFC and SIP-AS would be mapped to 183 Session Progress – PRACK between SIP-AS and MGCF (see Figure 11.5). Intermediate proxies between MGCF and SIP-AS and between SIP-AS and MRFC, such as CSCF, are not shown. MRF in the figure represents MRFC and MRFP.

The MRF starts media streaming as soon as it has received acknowledgement from the calling party, in this case MGCF, that it has received the SDP answer.

If, however, MGCF does not support the reliable provisional response (albeit not likely), then SIP-AS cannot provoke a provisional acknowledgement from MGCF. In addition, SIP-AS cannot forward the SDP answer from MRF to MGCF, since the SDP answer shall be transported in a reliable (provisional) response. In this scenario, SIP-AS may send the ACK to MRF without forwarding the SDP answer to MGCF. MRF then starts streaming media without MGCF having received the SDP answer. Remote source filtering for early media would then not be possible at IM-MGW. Hence, applying remote source filtering for early media requires support for reliable provisional response.

Note

Remote source filtering entails that a SIP entity accepts remote media only from the remote IP address and port number that was provided in the SDP answer applicable for this SIP dialog.

Whereas SIP-AS can at this stage not send the SDP answer in the backward direction, it may still send a 183 Session Progress to MGCF including the P-early-media header, informing the MGCF that it will permit backward media to flow through IM-MGW. This will work only when MGCF allows for gating media before having received the SDP answer.

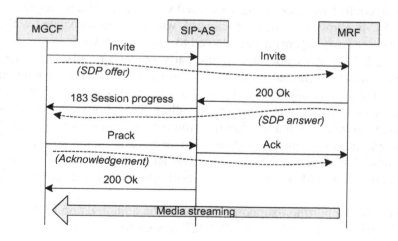

FIGURE 11.5

SDP offer–answer mapping at SIP-AS, with reliable provisional response.

11.3.3.3 Redirection (3xx)

According to standard 3GPP TS 29.163, the MGCF does not, by default, support the 3xx-class of final responses (redirection). MGCF in this case will send an ACK message in response to receiving the 3xx response and it will release the call in the CS network, i.e. send an ISUP REL. The remote end that had sent the 3xx response is not informed about the non-execution of the request indirectly conveyed through the 3xx response.

The support or non-support of 3xx by MGCF is not explicitly indicated in the INVITE (an MGCF may have implemented support for 3xx). A SIP-AS shall take care with forwarding a 3xx final response in the upstream direction, as the 3xx may not be supported upstream. If SIP-AS does not have sufficient (implicit) indication that 3xx is supported upstream, a SIP-AS should rather apply the redirection itself, i.e. generate the redirected INVITE itself.

11.3.3.4 Early Media Authorization

When a SIP-AS is controlling a call for a service number, e.g. a helpdesk, it may apply service logic that includes the playing of an announcement to the calling party prior to connecting the calling party to an available agent. The playing of the announcement to the calling party may be done as early media. This may be done in order not to penalize the calling party (playing the announcement through an established and active SIP session would result in the calling party having to pay for the call). Whilst the connection to the agent is being established, however, early media transfer from the agent (or PBX that the agent is an extension of) to the calling party may not be desirable. Early media from the agent might interfere with the application service logic, for example when call hunting applies. The application service logic may play an announcement whilst call hunting is taking place.

The P-early-media header may be used by the SIP-AS to control for *individual* early SIP dialogs on whether media should be allowed to be transferred to the calling party. Hence, the SIP-AS can control whether media associated with the SIP dialog related to the connection with MRFC is allowed to stream backward, whilst media associated with a SIP dialog relating to a remote party (agent) is not allowed to stream backward.

If, however, MGCF does not support the early media authorization, then the SIP-AS would not have this fine-grained control over the various (possible) media streams to the calling party. MGCF may apply a configuration setting to determine under which circumstances to allow early media streaming to the calling party. For example, when MGCF has received an SDP answer in a dialog relating to the pre-call connection announcement, it may gate media streaming in the upstream direction (specifically media associated with this dialog). When MGCF at a later stage receives a 180 Ringing provisional response in a dialog relating to a connection with a remote party (agent), the MGCF may revert to local ringtone generation.

Fortunately, the support/non-support of early media authorization by MGCF is indicated through the P-early-media header in the INVITE request sent by MGCF:

```
P-early-media: supported
```

Hence, service logic in SIP-AS can apply appropriate behavior when early media authorization is not supported in MGCF.

11.3.3.5 New SIP Methods

Various SIP extensions are defined in the form of additional methods. These methods are defined in separate RFCs. These methods extend the capability of standard SIP, as specified in IETF RFC 3261. The supported SIP methods are indicated in the Allow header in the INVITE request or in the 200 OK response. For example:

```
Allow: INVITE, ACK, CANCEL, BYE, UPDATE, OPTIONS
```

We will briefly look at a few methods and consider the impact when not supported in MGCF.

- REFER (IETF RFC 3515). REFER is used to transfer a call to another destination (cf. explicit call transfer, ECT, in the GSM network). MGCF typically does not support the REFER method. MGCF would respond with 403 Forbidden.

  A SIP-AS that offers an advanced call control service may process a REFER request, when this was received from a served subscriber, and apply appropriate service logic execution for achieving the requested call transfer.
- UPDATE (IETF RFC 3311). UPDATE is used for updating dialog characteristics during session establishment or during an established session. For example, SDP characteristics may be changed with the UPDATE method. The UPDATE method may also be used during a call for signaling call hold to the remote party. MGCFs commonly support the UPDATE method, at least for updating the SDP.

  When the MGCF does not support UPDATE, then the SIP-AS will not generate or forward an UPDATE request to the MGCF. This may hamper functionality such as resource reservation or two-stage SDP negotiation.

11.3.3.6 Number of Accepted Codecs

When an MGCF establishes a SIP session, in response to receiving an ISUP IAM, it may offer an SDP with a media line containing more than one (speech) codec, e.g. G.711 A-law and G.722 (we are not considering the *telephony events* (IETF RFC 4733) that may be offered in the SDP, since that does not really constitute a "codec"). By virtue of offering these multiple codecs, the IM-MGW selected by the MGCF should be prepared for receiving media encoded in accordance with either one of these codecs, i.e. the IM-MGW will be able to receive RTP media streams containing G.711 A-law data as well as RTP media streams containing G.722 data, but not at the same time!

The remote party, in turn, may accept both codecs and generate a 200 OK final response (when the called party answers) and provide an SDP answer with both codecs contained in the accepted media line. The IM-MGW may, resulting from this SDP answer (or even before), receive G.711 A-law media or G.722 media.

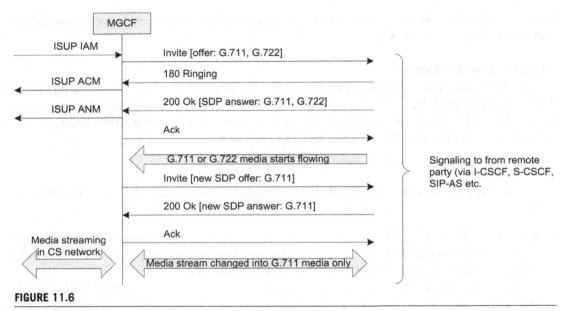

FIGURE 11.6

"SDP downgrade" by MGCF.

In this situation, the MGCF would want to "downgrade" the media session to a single codec. Figure 11.6 shows the signaling sequence for this situation.

MGCF in this example did not get a chance to negotiate a single SDP before answer. The first SDP answer was provided in 200 OK. MGCF cannot prevent media from starting to flow from the remote party at this point. The downgrading of the SDP to a single codec is done through a re-INVITE (or through an UPDATE) in the established SIP session. The exact behavior of the MGCF in this case will be implementation specific.

One issue associated with signal sequence as in the above example is the following. The MGCF, acting as UAC for the INVITE transaction, initiates a subsequent INVITE transaction (re-INVITE) after the ACK message. The ACK message is *unconfirmed*; the UAC does not get an indication that the ACK is received by the remote party, acting as UAS for the INVITE transaction. The re-INVITE request message *might* therefore arrive prior to the ACK message. IETF RFC 5407 provides examples of how SIP user agents may deal with such race conditions.

11.3.4 User-plane Interworking

Figure 11.7 shows user-plane interworking at IM-MGW. The example on the left of the figure depicts a call case whereby the same codec is used for the user plane in the CS network as for the user plane in the IMS network. The IM-MGW will have to apply repacking to the media streams: in the CS network, the G.711 samples are transported in chunks, with a bit

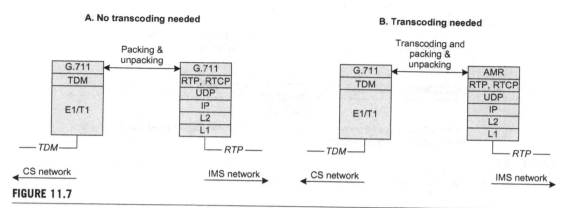

FIGURE 11.7

User-plane interworking at IM-MGW.

rate of 64 kb/s; in the IMS network, the G.711 samples are packed in RTP messages. The example on the right of the figure depicts a call case where transcoding is needed. Generally, transcoding at IM-MGW is avoided, but this may not always be possible. Transcoding consumes additional processing and memory resources, compared to transcoder-free user-plane interworking.

11.3.4.1 DTMF Transfer

For DTMF (dual-tone multifrequency) transfer, special user-plane interworking is defined. Typical usage of DTMF encompasses control of the interactive voice response system (IVR), such as voicemail, help desk, and telephone banking. This use of DTMF may span the CS and IMS networks. DTMF tone is carried through the user plane[3] as in-band speech. The DTMF tone will pass the IM-MGW when a call passes the CS–IMS boundary.

Within the IMS network, DTMF tones may be transferred as designated RTP messages (see Figure 11.8). If an IM-MGW wants to transfer DTMF tones and other telepony events as designated RTP messages, this capability will be negotiated in the SDP offer–answer exchange during media session establishment.

IETF RFC 4733 defines a method for conveying DTMF tones in RTP media streams by means of designated RTP messages. For each DTMF tone, a designated code is defined to be used in the RTP message. The DTMF code contained in this RTP message is known as a *telephony event*. The RTP header reveals that the RTP message contains a telephony event. Besides DTMF tones, other telephony events are defined in IETF RFC 4733 as well. IM-MGW may convert between in-band transfer of DTMF (in the CS network) and RTP message-based transfer of DTMF (in the IMS network).

---

[3] When bearer independent call control (BICC) is used in the CS network, DTMF tone may be carried through the control plane, with designated BICC messages. This is not further elaborated upon in this section.

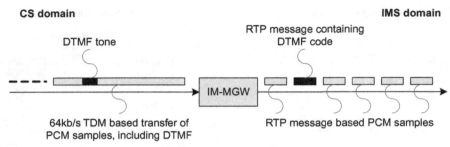

FIGURE 11.8

DTMF transfer through IM-MGW.

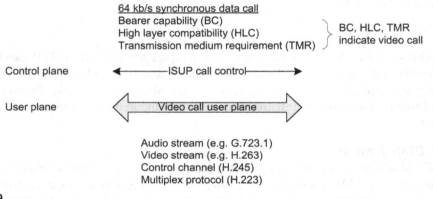

FIGURE 11.9

Video call in a circuit-switched network.

11.4 VIDEO INTERWORKING

Video interworking between CS and IMS networks requires video-capable MGCF and IM-MGW. The combination of video-capable MGCF and video-capable IM-MGW is often referred to as video gateway. The main reason for the need for special MGCF and IM-MGW is the fact that, in a CS network, video codec and video coding characteristics negotiation takes place in the user plane as opposed to the control plane. This is shown in Figure 11.9.

The ISUP signaling in the circuit-switched network indicates that a call is a video call, with 64 kb/s synchronous data transfer with multimedia (audio + video) content and control channels. When the call is answered, bidirectional media transfer over the user plane starts and the respective call parties negotiate video codec characteristics.

For video call interworking at a video gateway, we identify the following two issues:

1. The data streams carried over the user plane in the CS network will undergo different treatment in IM-MGW; some data streams are forwarded into the IMS network over RTP

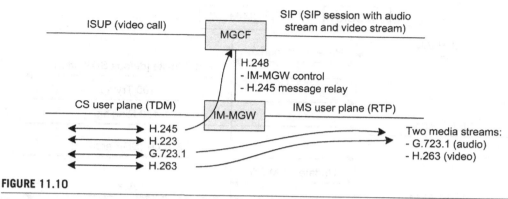

FIGURE 11.10

Video call interworking at video gateway.

(user plane); other data streams are forwarded to MGCF for use in SIP session establishment (control plane).

2. The SDP offer to be included by MGCF in SIP INVITE is based on the configuration in MGCF; SDP may be updated after call establishment.

Further explanation of this is given with reference to Figure 11.10. The architecture shown in Figure 11.10 relates to a specific call case, ISUP to SIP video call establishment. Also, there are different ways to establish the call between the CS and IMS networks. In the example in Figure 11.10, the MGCF determines from the bearer capability in the ISUP initial address message (IAM) that a video call is established with H.223 multiplex protocol and H.245 multimedia control protocol in the user plane. Since the call is not yet established, the H.245 codec negotiation and terminal capability exchange, between the calling party and IM-MGW, has not yet taken place. Hence, IM-MGW has not been able to inform MGCF about the terminal capability. MGCF may, as one implementation option, apply a default voice codec and video codec, e.g. G.723.1 + H.263 or AMR + H.263. These codecs are then reflected in the media lines in the SDP offer. When the call is established (200 OK received, with SDP answer; ISUP ANM sent), the H.245 negotiation between calling party and IM-MGW will commence. As a result of the H.245 negotiation, the MGCF may send a new SDP offer (see Figure 11.11).

The re-INVITE that the MGCF sends to the remote party should contain an SDP offer containing a video codec and a voice codec that are supported both on the CS side and on the IMS side. Otherwise, transcoding of media streams would be required at the IM-MGW.

The sequence shown in Figure 11.11 is one possible interworking method for video calls. An alternative method of interworking entails the generation of an ISUP answer by MGCF before establishing the call in the IMS network. By doing so, the MGCF completes the codec negotiation in the CS network and can then generate a SIP INVITE with an SDP offer in accordance with video codec capabilities in the CS domain. The video capabilities in the CS domain can then no longer be changed for that call.

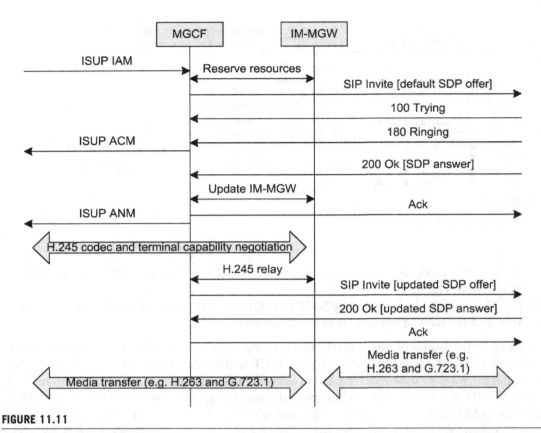

FIGURE 11.11

Signaling for video call interworking.

For the IMS core network, the effect of video interworking is that there may be additional SDP offer–answer messages exchanged (as seen in Figure 11.11). For video call handling in the IMS network, it is important that when break-out to the CS network takes place, a capable gateway is selected for the break-out. For example, when BGCF is involved in break-out to the CS domain, the BGCF may base its selection of MGCF on the offered media type (m-lines in the SDP). When, in this scenario, the destination subscriber in the CS network accepts the call as a CS-only call, the call (control plane and user plane) still traverses the video gateway.

11.5 SUPPLEMENTARY SERVICE INTERWORKING

This section deals with interworking between CS and IMS domains for supplementary services. This interworking is particularly important as operators want to preserve the user

Table 11.5 IMS Supplementary Services versus GSM Supplementary Services

IMS Supplementary Service		GSM Supplementary Service	
Service	3GPP TS	Service	3GPP TS
Originating identity presentation (OIP)	24.607	Calling line presentation (CLIP)	23.081
Originating identity presentation restriction (OIR)	24.607	Calling line presentation restriction (CLIR)	23.081
Terminating identity presentation (TIP)	24.608	Connected line presentation (COLP)	23.081
Terminating identity presentation restriction (TIR)	24.608	Connected line presentation restriction (COLR)	23.081
Communication diversion (CDIV)	24.604	Call forwarding (CF)	23.082
Communication hold (HOLD)	24.610	Call hold (CH)	23.083
Communication barring (CB)	24.611	Call barring (CB)	23.088
Explicit communication transfer (ECT)	24.629	Explicit call transfer (ECT)	23.091

experience for end-users when establishing or receiving calls from non-SIP terminals, e.g. mobile terminals (GSM phones). At the same time, this brings us to an interesting observation. When considering interworking between CS and IMS networks, we have two sets of supplementary services, listed in Table 11.5. This overview is not exhaustive. The reader is referred to 3GPP TS 24.173 for a complete overview of IMS supplementary services and to 3GPP TS 22.004 for a complete overview of GSM supplementary services. We also have supplementary services in the (ISDN-based) PSTN. However, in our explanation, we focus on the services in the mobile network (GSM).

Note

The supplementary services in IMS stem from the "PSTN/ISDN simulation services" as specified by ETSI Tispan.

For communication with the IMS network, the operator will offer the functionality as specified for the IMS supplementary services. When, however, a GSM terminal is used for establishing a call in the IMS network, the operator may prefer to retain the GSM supplementary services end-user experience, for the subscriber calling from/called on a GSM phone. This is shown in Figure 11.12.

IMS supplementary services and GSM supplementary services are, to a great extent, executed through SIP signaling (with the aid of the MMTel application server (AS), a dedicated SIP-AS for multimedia telephony) and ISUP signaling (with the aid of MSC, GMSC, HLR) respectively. We will look at some of the supplementary services and study the signaling at MGCF relating to these services. MMTel is described in Chapter 9.

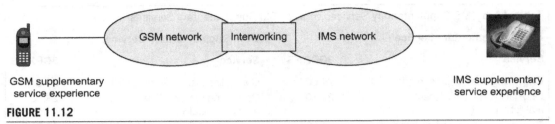

GSM supplementary service experience

IMS supplementary service experience

FIGURE 11.12

End-user experiences for supplementary services.

Table 11.6 Calling Line Presentation	
CS Network (ISUP Information Element)	**IMS Network (SIP Header)**
Calling party number	P-asserted-identity
Additional calling party number	From
Address presentation indicator	Privacy

11.5.1 Calling Line Presentation and Calling Line Presentation Restriction

Table 11.6 shows how calling line presentation is applied in the CS and IMS domains. The mapping in MGCF follows the rows in the table. Whereas the ISUP IAM does not need to have an additional calling party number (A-CgPN), a SIP INVITE always has a From header. Mapping for the case that not all parameters/SIP headers are present in the respective signaling is provided in 3GPP TS 29.163. For example, when no A-CgPN is present in ISUP IAM, MGCF sets the URI in the From header identical to P-asserted-identity. MGCF is normally configured to generate P-asserted-identity and From header as SIP URI with user=phone or as Tel URI. When SIP URI is used, the domain will be set to an operator-generic domain, e.g.

```
P-asserted-identity: <sip:+31169242200@ims-operator.nl;user=phone>
```

An IMS application server (the MMTel application server or another application server) may influence the number presentation by modifying the URI in the From header. For example:

```
From: <sip:+31169242200@ims-operator.nl;user=phone>;tag=568178145
```

may be modified into

```
From: <tel:2200>;tag=568178145
```

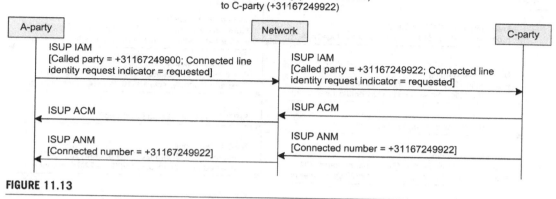

FIGURE 11.13

Connected line presentation in a CS network.

Tel:2200 is not an international format number. However, it is used for display purposes to party B and not for routing. When party B returns the call (call back to presented party A number), the call would be set up (reflected in R-URI) to tel:2200 or to sip:2200@ims-operator.nl;user=phone. An IMS application server acting on behalf of that calling party would translate the R-URI into *sip:+31169242200@ims-operator.nl; user=phone*.

As follows from the above elaboration, for the purpose of interworking between CS and IMS, an IMS subscriber should always have at least one phone number provided as IMPU in the HSS.

When the identity of a subscriber is restricted (i.e. identity shall not be shown to the called party), a SIP-AS may add a Privacy header, e.g. based on service subscription profile or based on enterprise policy. The P-asserted-identity remains available, since it may be required for network routing, charging, and identification purposes. When applying privacy for the calling party, the SIP-AS will also make the URI in the From header anonymous, to sip:anonymous@anonymous.invalid. This *anonymous URI* is standardized and defined in 3GPP TS 23.003.

11.5.2 Connected Line Presentation and Connected Line Presentation Restriction

When a call is established to a remote party, the calling subscriber may receive an indication about the *connected party*. The connected party may differ from the *called party*. Figure 11.13 shows an example for the CS network. In this example, the calling party establishes a call to +31167249900. The ISUP IAM contains a request for connected line identity. The call is forwarded to +31167249922. The ISUP answer message (ANM) carries the directory number of the connected party. The MSC of the calling party (for the case where the calling party is a GSM subscriber) uses the connected number information element in ISUP ANM to set the connected number in the DTAP connect message to the calling party. In this

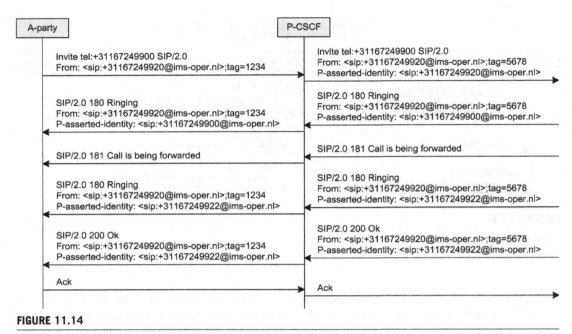

FIGURE 11.14

Terminating identity presentation in IMS.

manner, the calling party is informed about the *eventual* connected party, even when there are multiple call forwardings taking place during call establishment.

Note

ISUP allows for transport of an *additional connected number* in the answer message. Whereas the *connected number* should be a network asserted identification of the connected party, the *additional connected number* may be a service-defined number, to be used for connected line display. In the GSM network, however, it is not possible for a service control function (SCF) to provide or modify the additional connected number. For ISUP–SIP interconnection, there is, however, the possibility that an additional connected number might be set by the MGCF. Refer to 3GPP TS 29.163 for details.

In IMS, the sequence for terminating identity presentation is shown in Figure 11.14. As follows from the figure, P-asserted-identity in SIP is used for subscriber identification in both the forward and backward directions:

- In an INVITE *request* message, the P-asserted-identity identifies the calling party; it may be used for originating identity presentation.
- In an INVITE *response* message, the P-asserted-identity, if present, identifies the party that is responding to the INVITE request.

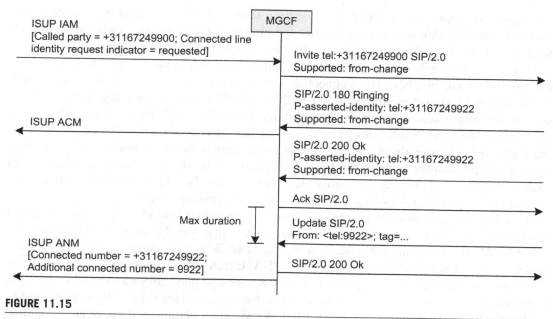

FIGURE 11.15

Terminating identity presentation interworking at MGCF.

This implies that the initiator of the INVITE transaction may receive the identity of the forwarding party, as well as the identity of the eventual connected party.

___

Note

Usage of P-asserted-identity in provisional responses, for display purposes (something like "provisional destination party"), is not formally specified.

___

For both the P-asserted-identity in the forward direction (INVITE request) and the P-asserted-identity in the backward direction (INVITE response), a Privacy header may apply, meaning that the identity of that party will not be revealed to the destination party or calling party, as appropriate.

Figure 11.15 shows interworking at MGCF for the connected number/identity presentation. The figure shows an example of how the identity of the connected subscriber in IMS can be conveyed to the calling party in the CS domain. It also shows how a service in the IMS network can provide an additional connected line identity, for display purposes. One straightforward approach is that P-asserted-identity in final response on the INVITE transaction is copied to the ISUP answer message (ANM). In the example in Figure 11.15, the provisional response on the INVITE transaction also contains connected line identity (carried in P-asserted-identity). However, since the final response, 200 OK, contains P-asserted-identity, it is the latest P-asserted-identity that is reported to the calling party.

The question now is how to convey additional connected line identity to the calling party, e.g. an enterprise number. The connected party's *network-asserted* identity may be tel: +31167249922. However, we would like to display 9922 to the calling party, without affecting the network-asserted connected line identity. In the *forward direction*, an IMS service may utilize the From header to affect the calling line identity shown to the called party. In the *backward direction*, there is no possibility to manipulate the From header or To header to control identity display because the From header and To header are closely associated with dialog identification. Hence, the entity that initiated the INVITE transaction will receive consistent From URI and To URI in the provisional and final responses.

To allow for provision of a *service-controlled* connected line identity, over and above the *network-asserted* connected line identity, MGCF may briefly withhold the sending of ISUP ANM after receiving 200 OK. This facilitates an IMS application to provide a service-specific connected line identity through an UPDATE transaction. Specifically, the UPDATE transaction in the immediate response to the ACK message may carry a modified From URI, compared to the From URI used in the INVITE transaction. This From URI is then used to set the additional connected number in ISUP ANM. The operator will apply an acceptable maximum delay in waiting for an UPDATE. After that maximum delay, the MGCF will proceed as normal, i.e. construct ISUP ANM based on the information received in provisional response(s), if any, and final response.

To allow for provision of a modified From URI, the MGCF has to signal this capability in the INVITE request (Supported: from-change). Further details can be found in IETF RFC 4916 (Connected identity in SIP) and in 3GPP TS 29.163 (ISUP–SIP mapping).

11.5.3 Call Hold and Resume

Call hold generally manifests itself in two ways when applied in the CS or IMS network: (i) transfer of media is temporarily suspended and (ii) signaling is exchanged between the *holding party* (the party applying the call hold) and the *held party* (the remote party who is put on hold). The signaling exchange between holding party and held party is, for example, for display purposes, i.e. providing visual indication that the call is placed on hold. In addition, a remote exchange may provide *call hold tones*, providing additional, audible indication to the remote party that the call is placed on hold. Call hold is, hence, an end-to-end service, involving both parties in the call. We will see in this section how this end-to-end functionality is facilitated when crossing the network domain border.

Figure 11.16 illustrates the signaling plane and media plane for an active call, crossing the CS–IMS boundary twice. This situation may occur when a call is established from a CS domain to an IMS subscriber, where the IMS subscriber forwards the call to a destination in the CS domain.

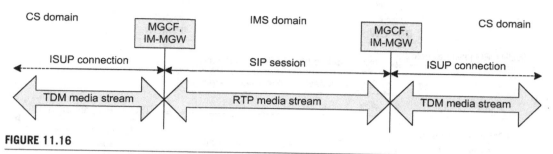

FIGURE 11.16

Active call across CS and IMS domains.

Call hold, i.e. suspending the transfer of speech between two parties, may be initiated by one of the involved parties or may be initiated by a service in the network. Applying call hold by a service in the network may also form part of a more comprehensive service, such as mid-call announcement.

Figure 11.17 shows the signal sequence for call hold initiated by one of the parties in the CS domain.

Call hold and call resume are conveyed in the CS network through an ISUP call progress (CPG) message, with indication "remote hold" or "remote retrieval". In the IMS network, the communication hold is signaled through an SDP update, which may be done with an INVITE transaction or an UPDATE transaction. The updated SDP offer is reflected in the stream attribute "sendonly" (a=sendonly). This has the effect that the remote party will apply *receive only* (stream attribute "recvonly"; a=recvonly) in the answer to the updated SDP offer. The holding party will, at the same time, stop transmitting media.

In a CS network, call hold and resume may function even without the associated ISUP signaling. In this case, media transfer is suspended by the local party (applying the hold), but the remote party will not receive the visual indication that (s)he is put on hold; neither will the remote party receive call hold tones. In IMS, on the other hand, update of the SDP is needed, to inform the remote party that RTP data transfer will be suspended. This is needed to prevent *RTP absence timeout* (a SIP phone may terminate the SIP session when no RTP data arrive for a configurable duration, e.g. 30 s). RTCP message transfer will, however, continue during the held state. This is to ensure that respective end-points may continue monitoring the operational condition of the media transfer connection (a kind of "heartbeat" message on user plane). In addition, a lower media transfer bandwidth may be negotiated with the remote end for the duration of the communication hold. This may be especially relevant for the part of a SIP session conveyed over wireless connection, where bandwidth is a scarce resource. Sufficient bandwidth will be retained for the RTCP message exchange.

The call in the IMS network may be under control of one or more IMS applications, e.g. an IMS application acting on behalf of the called IMS subscriber. The SDP renegotiation signaling will typically traverse the IMS application transparently. However, an application may, as an implementation option, offer service such as *music on hold* (e.g. as part of a set of Centrex

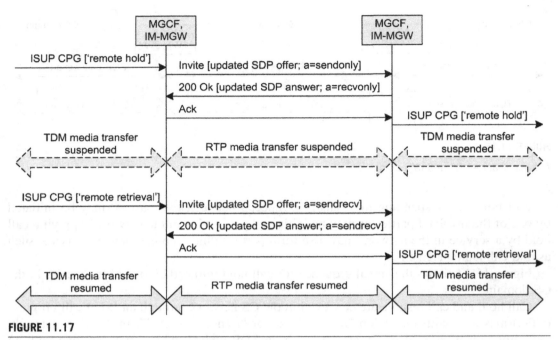

FIGURE 11.17

Call hold and resume spanning CS and IMS networks.

services). The application will in that case not forward the SDP offer with a=sendonly to the remote end. Instead, the application negotiates a new SDP with that remote party to connect that remote party to a media source, for the streaming of music for the duration of the hold by the local party. It is possible to suspend the media streams individually or to suspend all media streams.

11.5.4 Call Forwarding

Call forwarding information in the CS network may be mapped to corresponding call forwarding information in the IMS network. The call-forwarding-related information conveyed in ISUP includes the following:

- **Original called number** – this information element indicates the directory number to which this call was originally established.
- **Redirecting number** – this information element contains the identification of the subscriber on whose behalf the call forwarding is applied.
- **Redirection information** – this information element contains augmenting information, such as forwarding reason and forwarding counter.

In the IMS network, forwarding information is reflected in the History-Info SIP header in a SIP INVITE request message. Refer to IETF RFC 4244. An example is the following:

```
INVITE tel:+468001234 SIP/2.0
P-asserted-id: <sip:alice.jones@my-company.org>
History-Info: <sip:john.smith@my-company.org>;index=1,
       <sip:+46107132600@ims.operator.se;user=phone>;index=1.1,
       <tel:+468001234>;index=1.1.1
```

The first entry in the History-Info represents the original target. The subsequent entries in the History-Info represent the retargeted calls. In the above example, a call is established to sip:john.smith@my-company.org. A proxy or application acting on behalf of John Smith retargets the call to sip:+46107132600@ims.operator.se;user=phone. A proxy or application acting on behalf of +46107132600 retargets the call to tel:+468001234.

If this information is to be converted to ISUP, then we would get the following mapping:

- Original called number – sip:john.smith@my-company.org; this is not possible (see below).
- Redirecting number – +46107132600.
- Redirection information – the forwarding counter would be set to 2.

We notice a dilemma here. The syntax of the History-Info allows for a single URI to be used to identify the destination of each call leg (unlike P-asserted-identity, which may contain both a Tel URI and a SIP URI). If a destination is identified with a SIP URI not containing a number, then that destination cannot be mapped to an ISUP parameter such as original called number. For interworking with CS networks, the use of SIP URI in History-Info therefore has its dangers. For further mapping details between ISUP and SIP for History-Info, refer to 3GPP TS 29.163.

11.6 APPLYING LEGACY VAS IN THE IMS NETWORK

In this section, we will provide a closer look how legacy value-added service (VAS) may be applied in an IMS context. Legacy VAS refers to Intelligent Network (IN)-based services, as developed for CS networks. In fact, the topic of applying legacy VAS in the IMS network ought to be seen in a broader light, as will be explained next. The evolution from CS networks to IMS networks results in network constellations where a mix of old and new network technology has to work side by side. This old and new working side by side applies to both core network (IMS core network and CS core network) and service layer (SIP application server and service control point). We will start by making a brief comparison between these two networks and their respective paradigms. We will then continue by describing how the two may be integrated.

11.6.1 The Starting Point: VAS in the CS Network and VAS in the IMS Network

Let's have a quick look at the position and role of the service layer in the CS network and in the IMS network. We will look at the GSM network as an example CS network.

11.6.1.1 CS network

Two main categories of services that are offered in the GSM core network are *basic services* and *supplementary services*. The group of basic services encompasses basic telephony, video calling, emergency calling, messaging, etc. The group of supplementary services offers additional services that *augment* the basic services. A basic service like basic telephony may be subject to a supplementary service such as call forwarding, call hold, and explicit call transfer. Both basic services and supplementary services are categorized as *subscribed* services, meaning that the subscriber needs a subscription to the service in order to use it. Basic services and supplementary services are supported through capability in core network nodes such as MSC, Gateway MSC, and HLR.

To further enhance the level of service for a subscriber, an operator may apply *value-added services* (VAS). VAS is offered through Intelligent Networks (IN). For the mobile network (GSM, UMTS), VAS is based on the protocols and procedures defined in the CAMEL standard (Customized Applications Mobile network Enhanced Logic).[4] VAS may be applied for services like prepaid calling, number translation, virtual private network, and incoming/outgoing call screening.

Figure 11.18 shows the position of VAS in the CS network (GSM network in this example). VAS is equally applicable, in comparable manner, to the PSTN, but that is not described further here. Figure 11.18 provides a limited view of the GSM core network and the associated service layer. Additional reference points and functional entities are defined; these are not reflected in the figure. The service control point (SCP) is the node where IN service logic resides and is executed. The SCP has a functional connection with the visited MSC (VMSC) as well as with the gateway MSC (GMSC). VMSC and visitor location register (VLR) are normally integrated and form one functional entity.

When calls are established, value-added services may be applied on the call by establishing a control connection between MSC or GMSC and SCP. For different call cases, different control relationships apply. Figure 11.19 shows how this control connection between MSC or GMSC and SCP is realized.

The call control between MSC and SCP is realized through two functional components:

- **gsm Service switching function (gsmSSF).** The MSC may, when establishing a call for a subscriber, invoke an instance of a gsmSSF ("instantiate a gsmSSF"). The gsmSSF allows a control relationship with the gsmSCF to be established.
- **gsm Service control function (gsmSCF).** The SCP instantiates a gsmSCF instance when the service logic is invoked. The service logic can apply control over the call through this control relationship.

The gsmSSF maintains a Basic call state model (BCSM) instance for the call. As the call proceeds through the various stages of call establishment, state transitions occur in the

---

[4] For a comprehensive overview of CAMEL, see Noldus (2006).

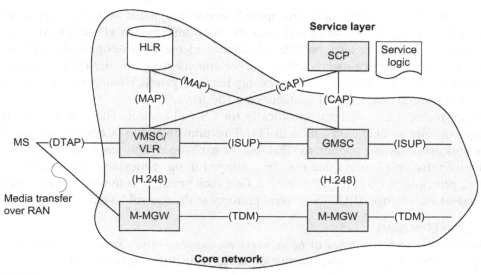

FIGURE 11.18

GSM core network versus service layer. CAP, CAMEL application part; DTAP, direct transfer application part; GMSC, gateway MSC; H.248, media gateway control protocol; HLR, home location register; ISUP, ISDN user part; MAP, mobile application part; M-MGW, mobile media gateway; MS, mobile station; MSC, mobile services switching center; RAN, Radio Access Network; SCP, service control point; TDM, time division multiplex; VLR, visited location register; VMSC, visited MSC.

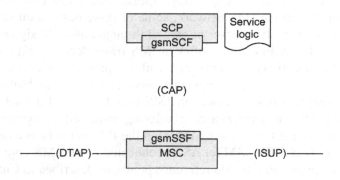

FIGURE 11.19

Call control between SCF and MSC/SSF.

BCSM. State transitions may be reported to gsmSCF, allowing gsmSCF to maintain a mirror image of the gsmSSF BCSM instance.

Control-plane signaling uses DTAP and ISUP. DTAP is used between GSM terminal (mobile station, MS) and MSC; ISUP is used between MSC and other traffic nodes in the

GSM network. The gsmSSF applies mapping between designated information elements carried on DTAP or ISUP messages and corresponding information element CAP messages. The service logic may, in turn, provide information elements in designated CAP messages that will be used to replace specific information elements in a particular ISUP message.

When the call is terminated, the relationship between gsmSCF and gsmSSF is terminated as well; the gsmSSF and gsmSCF instances are "destroyed".

The IN protocol is developed specifically for CS call control. The call state model, the messages (known as operations; refer to ITU-T recommendations X.880, X.881, and X.882 for the concept of remote operations) that may be exchanged between gsmSSF and gsmSCF, and the information elements that may be conveyed in these messages are all grafted on to the basic principle of CS call establishment. One such principle is the use of ISUP (or bearer independent call control, BICC) as control protocol in the control layer.

11.6.1.2 IMS network

For the IMS network, the roles of *basic services*, *supplementary services*, and *value-added services* and their respective implementation are slightly different from GSM.

- **Basic connectivity.** The IMS core network offers communication session establishment capability. An IMS subscriber may establish a SIP session with another IMS subscriber, using the techniques described in Chapters 7 and 8. The SIP session may be used for media transfer of various types, such as speech, video, messaging or specific media application. The IMS network further supports the use of SIP capability for forking a call by S-CSCF (acting as forking proxy; e.g. parallel alerting), diverting a call (through 3xx response or through forwarding proxy), transferring a call (using REFER), etc. We will see later on that, in the IMS network, some of these basic connectivity services are offered through an application server instead of through basic SIP signaling.
- **Multimedia Telephony.** In order to offer in IMS *true* telephony experience, the above-described basic connectivity is augmented with supplementary services. The combination of basic connectivity and supplementary services is known as Multimedia Telephony (MMTel). The supplementary services in IMS therefore fulfill a dual role: (i) offering basic telephony for the subscribers and (ii) offering multimedia communication.

  One main component for offering MMTel in the IMS network is a designated MMTel application server (AS). The MMTel AS is connected to the IMS core network through the ISC reference point (and the Ma reference point), as described in Chapter 8.

  Although the MMTel AS interconnects with the CSCF though ISC and Ma reference points, MMTel AS will not be considered to be a value-added services AS. Instead, MMTel AS is associated with basic telephony and should hence rather be considered to form part of the IMS core network.
- **Value-added services (VAS).** VAS in IMS allows for providing an enhanced service level, through services that go beyond the standardized service set. Examples include call park, call pickup, VPN, group call, and number translation. VAS in IMS is realized mainly through the ISC reference point. VAS resides in a SIP-AS.

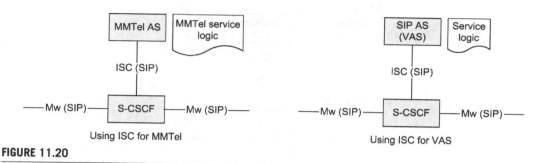

FIGURE 11.20

Position of SIP-AS in the IMS network.

Figure 11.20 shows the position of MMTel AS and SIP-AS in the IMS network.

When an operator wants to augment the Multimedia Telephony with VAS, two or more SIP-ASs need to be "connected to S-CSCF", as in invoking two (or more) SIP-ASs for SIP session establishment, based on initial filter criteria (IFC) in the subscriber profile. The operator will take great care that MMTel services and VAS are triggered in the right order – generally, for originating SIP session establishment, first MMTel AS and then VAS; for terminating SIP session establishment, first VAS, then MMTel AS. It is not excluded that VAS would be integrated with the MMTel AS, leading to a single service invocation from S-CSCF. That would be an implementation option.

Multimedia Telephony services as well as value-added services are designed specifically to interwork with the IMS network. The transaction model, the SIP message routing principles, the support for SIP headers, and the different SIP-AS roles (proxy, B2BUA, UAC, UAS) are all grafted on to the principles of SIP and IMS.

11.6.2 The Challenge: Safeguarding Legacy VAS Investment

A particular and recurring challenge for operators, when introducing IMS, is: How can we safeguard the investment made in legacy VAS? Operators have, over the years, invested large sums in IN services, considering both CAPEX and OPEX. Some of these IN services, such as prepaid, VPN, call assist services, are serving millions of subscribers of an operator. This section describes techniques that may be applied for deploying legacy VAS, where applicable, in the IMS network.

The first step is that we classify IN services into categories. This helps in deciding which IN services should be deployed in the IMS network and which IN services should be rewritten (adapted) to support native VAS protocols in IMS (ISC reference point). Needless to say, this categorization is a *guideline*.

A first category of legacy VAS is labeled *CS/GSM-specific VAS*. These services have no direct use in IMS. They may relate to, for example, routing in a CS network (least cost routing, optimal routing). Some of these services do have a comparable service in IMS, but then that service in IMS would benefit from being developed for the IMS-specific method of

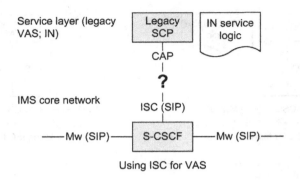

Using ISC for VAS

FIGURE 11.21

How to "connect" legacy VAS to S-CSCF.

invoking VAS and asserting SIP session control. One example of such a service may be carrier preselect.

We identify a second category of legacy VAS, comprising services with similar use-cases in IMS but for these services we do not expect further evolution towards multimedia, i.e. they are intrinsically voice-centric. These services may be reused in IMS, which we will describe below. Examples include access screening or cashless calling.

The third category that we identify consists of services that have clear applicability in IMS and for which evolution towards multimedia is foreseen. This category also contains mass-market services for which the sheer subscriber volume justifies (partial) redevelopment of that service in IMS. This category includes freephone, premium rate, centrex, and ringback tone.

When we look again at Figure 11.20, showing VAS in the IMS network, the challenge we are faced with is illustrated in Figure 11.21. Referring to the earlier explanation that legacy VAS in a CS network is a means to *augment* the basic telephony and supplementary services, when applying this legacy VAS in IMS, it will have the same role of *augmenting* the basic telephony and supplementary services. This implies that the legacy VAS is applied in the IMS network in combination with MMTel AS. This is shown in Figure 11.22.

But let's first see how the legacy VAS may be applied in the IMS network through the ISC reference point. In order to appreciate the impact of this requirement, we have to refine the picture a little, going further into what is labeled "legacy SCP" in Figure 11.21 (see Figure 11.23). What Figure 11.23 tells us is that what we conveniently call "SCP" in architecture diagrams is constituted by a set of functional entities and nodes that together provide the value-added service for a call. The following functional entities are distinguished:

- **Front-end service logic.** This is service logic designed and built specifically for interworking with a CS network, e.g. through CAPv2. This service logic may be deployed on general-purpose hardware (+ operating system and middleware) or on designated telecom servers.

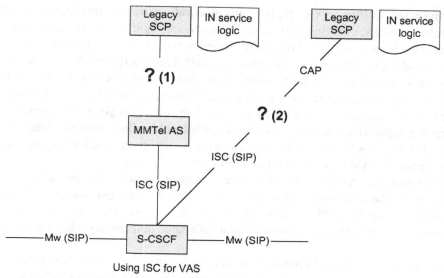

FIGURE 11.22

Combining legacy VAS in IMS with basic telephony and supplementary services.

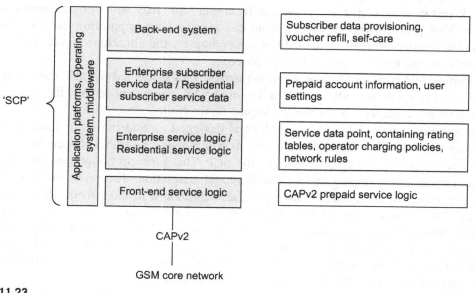

FIGURE 11.23

Dissecting the "SCP".

- **Enterprise subscriber service logic/residential subscriber service logic.** This logic contains rules, policies, charging tables, etc. This service logic, one may argue, is the *real* IN service. This logic is accessed by the front-end service logic and provides instructions to the front-end service logic for controlling the call. The enterprise service logic/residential subscriber service logic may be unaware of the call control protocol (CAPv2 in this case). This logic may reside in the same node as the front-end service logic or may be deployed on separate server(s), to prevent single point of failure and for geographic load sharing.
- **Enterprise subscriber service data/residential subscriber service data.** These data comprise, for example, subscriber-specific information that is needed for the execution of the enterprise service logic/residential subscriber service logic.
- **Back-end system.** This comprises nodes and functional entities: that provide enterprise subscriber service data/residential subscriber service data; that process charging records (billing, accounting, statistics, O&M); and that facilitate user self-care, and voucher upgrade (for prepaid service). Clearly, this component consists of a large number of nodes and subsystems.

When going from front-end service logic to a back-end system, we will see that the respective investment in these components increases. Whereas an SCP with CAPv2 service logic may represent (relative) small investment, the investments in a back-end system may be huge!

It is emphasized once again that the above-described functional decomposition will differ per service; some services, for example, have front-end service logic and subscriber service data combined in SCP and do not have an intermediate layer.

Hence, the goal of legacy VAS reuse in IMS, where deemed applicable, will be to reuse the back-end system, whilst accepting that front-end logic may benefit from porting to IMS intrinsic VAS protocol (i.e. develop a front-end service logic supporting the ISC reference point). The particular solution will differ depending on the situation. Figure 11.24 shows two solutions that may be applied for this purpose.

A further goal is to ensure that when the legacy VAS is applied in IMS, there will be no or minimal impact on the billing system. Existing VAS, when applied in IMS, may generate charging records that have no or limited impact on the billing system.

The solution on the left in Figure 11.24 is designed around the IP multimedia service switching function (IM-SSF). IM-SSF acts as a protocol converter between SIP (over the ISC reference point) and CAPv2 towards the SCP. This protocol conversion includes message and parameter mapping and call state model adaptation. The rationale of IM-SSF is that it is transparent for the SCP whether the service logic is invoked from GSM network or from the IMS network, via IM-SSF. Hence, there should "in principle" be little impact on the SCP. This method may be used when the SCP is a closed system and there is no possibility to enhance the SCP with SIP front-end service logic. For further details in IM-SSF, refer to 3GPP TS 23.278 and 3GPP TS 29.278. IM-SSF is formally specified for CAPv3 services, but operators may deploy IM-SSF equally for CAPv2 services.

IM-SSF solution will not give complete transparency. Not all parameters normally found in CAP Initial DP (IDP) operation are present in the CAP IDP when IM-SSF is used.

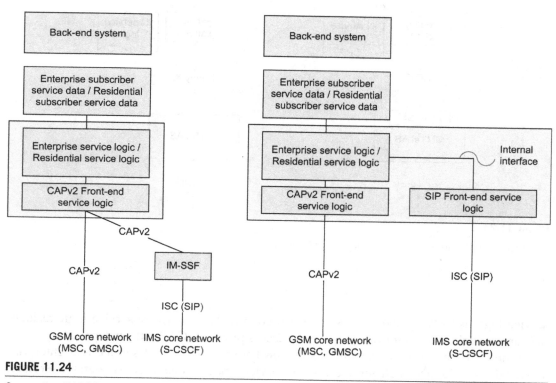

FIGURE 11.24

Connecting IN SCP to the IMS core network.

Likewise, not all CAMEL capability offered by CAPv2 or CAPv3, when used directly with the GSM core network, is available through IM-SSF. Whereas IM-SSF is a standardized entity, an IM-SSF will always have to be adapted to the specific legacy IN service(s) that are to be invoked through IM-SSF. Therefore, flexibility of service development on IM-SSF is crucial for practical deployment. In addition, the (legacy) service itself may be affected, due to the fact that the CAP IDP will not contain exactly the same parameters as when the service is invoked directly from the GSM network. At least some functionality of the service may not be available when triggered through IM-SSF.

The solution on the right in Figure 11.24 entails the enhancement of the SCP. SIP front-end service logic is implemented. The SIP front-end service logic acts as SIP-AS. IMS service triggering from S-CSCF may now be done directly to the application server. The SCP has CAPv2 front-end service logic and SIP front-end service logic. Both logics interact with the enterprise service logic/residential service logic. The rationale of this approach is that it is more efficient (in terms of signaling), has less complexity (fewer network nodes), and offers more session control capability. The latter results from the fact that the SIP front-end

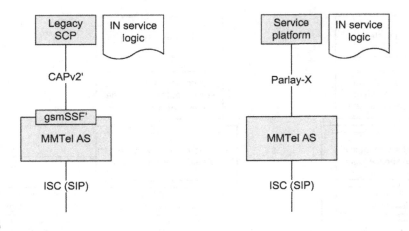

FIGURE 11.25

Northbound control interface for the MMTel application server.

service logic is fully prepared for SIP session control. The enterprise service logic/residential service logic should, for this purpose, be access protocol (CAP, SIP) independent.

It should be understood that the distinction between residential service and enterprise service does not apply to all services. Some services may be classified as operator services, such as freephone and premium rate.

The architecture shown in Figure 11.22 requires some further explanation. The legacy VAS may be applied in the IMS network through "northbound interface" from MMTel AS. The concept of "northbound interface" from an application server is well-known technology. It allows the application server to enhance its processing with service logic from an external service node. Two example implementations of this concept are shown in Figure 11.25.

The left side of Figure 11.25 shows a CAPv2 interface from MMTel AS. The MMTel AS includes a gsmSSF instance. At designated points in the service logic processing in MMTel AS, signaling towards the legacy SCP may be applied. That signaling is done through the gsmSSF instance. The gsmSSF in MMTel is not a genuine gsmSSF; it does not have a connection to an MSC. Likewise, the CAPv2 interface between this gsmSSF and the legacy SCP does not have all the capability as found in the CAPv2 interface between MSC and SCP. Certain parameters will not be present in the CAPv2 protocol, since they are GSM networks specific. So, the notation gsmSSF' and CAPv2' is used in the figure.

The right side of Figure 11.25 shows a Parlay X interface between MMTel and an external service platform. Parlay X is specified in 3GPP TS 29.198 and 3GPP TS 29.199. The use of Parlay X for a northbound interface from the application server is common methodology. Parlay X may, for example, be applied for augmenting the MMTel feature set with newly built functionality on a service platform.

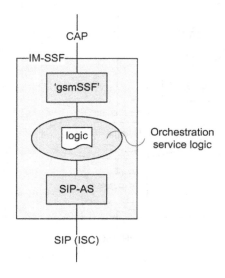

FIGURE 11.26

Orchestration service logic.

11.6.3 **Service Capability Interaction Manager**

Flexibility of service development on IM-SSF was briefly mentioned in the previous section. 3GPP TS 23.278 specifies detailed rules for mapping between SIP messages (requests and responses) and CAP operations. But, still, it is foreseen that an IM-SSF will have to be tailored or developed for a specific deployment, considering the specific legacy IN service(s) that have to be invoked. Hence, rather than having IMS-SSF act as "protocol converter", IM-SSF will be built around orchestration service logic (see Figure 11.26).

The orchestration service logic in IM-SSF is invoked by the receipt of an initial INVITE request. The orchestration service logic then instantiates a gsmSSF process and invokes the required IN service. The figure shows gsmSSF in quotes, since this process is not a real gsmSSF; it does not have a connection to MSC. Rather, it emulates gsmSSF behavior to the legacy service. Mapping between SIP headers and CAP parameters may be determined by the orchestration service logic. It is through this orchestration service logic that operators may adapt the IM-SSF to their specific needs.

The concept of orchestration service logic may be applied in a wider context. Rather than invoking a single CAMEL service, multiple services may be invoked. This is shown in Figure 11.27.

The orchestration service logic may invoke a multitude of services. This constellation is commonly known as the service capability interaction manager (SCIM). IMS standards depict SCIM as an entity that may invoke multiple IMS services, leaving the internal working of this entity to implementors. The spirit of a concept like SCIM is that it should provide flexibility to operators for developing orchestration service logic for invoking a multitude of

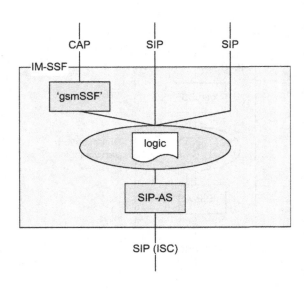

FIGURE 11.27

Service capability interaction manager.

services of different kinds. Any interaction between the respective services will be handled by the orchestration service logic. For example, when the orchestration service logic invokes first a CAMEL service, then a modified destination number received in a CAP Connect operation may be copied into the Request URI (R-URI) in the INVITE request to an IMS application.

Besides invoking external services, SCIM may also locally (on the node itself) execute service logic. That local service logic then forms part of the orchestration service logic. Once again, a flexible service creation environment is crucial for a node offering SCIM functionality. The orchestration logic may have a fair degree of complexity, in order to integrate the external services in a manner that is tailored to these specific services.

Further information regarding strategies for deploying legacy VAS in IMS network can be found in the Ericsson white paper "Next-generation intelligent networks: migrating to IMS", which can be downloaded from www.ericsson.com.

Rich Communication Suite

CHAPTER

12

12.1 INTRODUCTION

This chapter discusses the Rich Communication Suite, or RCS. Chapter 5 covered IMS's role within capital goods software and how to develop applications on a slightly larger scale for companies, in particular companies functioning as system integrators within the global economy. In this chapter, we now focus our attention on consumer goods software, i.e. software that is used by individual consumers on their handsets. Here, we discuss how to use the profiles developed by GSMA in the Rich Communications Suite (RCS) to create applications for end-user consumers. This can be anything from a small app downloadable to a mobile device, or a more large-scale service that allows multiple end-users to communicate and coordinate with one another in new ways.

The first section of the chapter covers the basics of RCS, how it has developed, and what the suite provides. The second part of the chapter discusses what a developer can actually do with RCS that they cannot easily achieve through other means.

As we discussed in Chapter 2, the role of the individual in economic life is increasing in importance due to the expansion and use of digital technologies, in particular the mobile broadband platform. This is perhaps the most commonly recognized and understood area of the consumer goods software market of the mobile communications industry. There is currently a huge focus on apps that are downloadable to a smartphone and for which developers receive either direct payment or are able to use advertising to generate revenue for their efforts. While there is an immense amount of focus on such models right now, consumer goods software in the mobile industry is still relatively immature[1] and will continue to evolve over the next few years. While app stores are currently very popular, there is no company in the world with the financial resources to support every mobile platform in existence. As a result, this will force most "vendors, operators and end-users away from an app store model that will only get more fragmented". While the different app stores are extremely relevant today, in order to develop financially successful applications in an evolving market, a key aspect for a developer to understand is how the changing role of the individual in economic life can be used to capture value within a relatively fluid industrial structure in the

---

[1] A useful comparison is the development of the consumer goods software market in the computing industry during the early 1980s, where many different companies emerged and were replaced rapidly. As the consumer goods software market matured, so did the delivery models associated with it.

communications industries. Simply put: How can a developer capture value and how does RCS help them implement innovative applications in response to these economic changes?

12.2 THE BASICS OF RCS

12.2.1 What is RCS?

Rich Communication Suite is an initiative run by the GSMA. It brings together the main actors from the telecommunications industry from network vendors, operators, handset manufacturers, and device client developers for the creation of an "interoperable, convergent, access-technology-independent rich communication experience to end-users". At the time of writing, there are nearly 100 companies participating in the RCS initiative within the GSMA.

We should clarify that RCS is not a group that defines new standards – they simply bring together already defined services into profiles based on global standards (e.g. IMS), presence, and content sharing, and define requirements for service features. RCS also defines some architectural aspects with regard to the best manner in which to open RCS for combination with Internet and web services. In particular, during 2011, the OMA will be releasing a series of new APIs exposing RCS functionality to the wider development community.

The RCS specifications have been developed to facilitate the introduction of commercial, IMS-based communication services for 3G, LTE, CDMA, and fixed networks. As will be discussed in the following sections, functionality within RCS includes Network Address Book, Enhanced Call through image and video sharing, and also Rich Messaging for both mobile and PC/broadband clients.

12.2.2 Why RCS?

A developer may reasonably ask why there is a need for something like RCS. Surely it is enough for the IMS platform primitives to be exposed as Open APIs at a truly granular level, giving them access to the raw platform, rather than the services that operators, network vendors, and handset manufacturers have decided are most convenient or useful for them to use.

The necessity for RCS lies with two main points:

1. The necessity for global interoperability between socially driven services.
2. Subscriber-centric social networking, rather than platform-centric social networking.

12.2.2.1 Global Interoperability

As discussed in Chapter 2, global interoperability provided by the IMS specifications enables developers to take advantage of the significant economies of scale created by a global standard. RCS extends these economies of scale to services and the individuals that use them. RCS provides a truly globally interoperable standard for social services that is formed *around the subscriber*, rather than around the social platform in and of itself. For example,

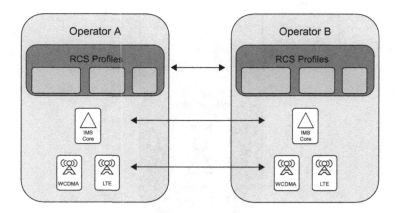

FIGURE 12.1

RCS provides global interoperability.

Facebook provides an excellent social networking site,[2] but users are locked into this platform. RCS provides a globally interoperable standard that allows an end-user to share social information across any handset or fixed device, any operator network, and towards any social network. Through relying on already established interoperable standards defined within 3GPP and OMA, the RCS provides functionality across any device the end-user wishes to use, as illustrated in Figure 12.1.

12.2.2.2 Subscriber-Centric Social Networking

Perhaps the most powerful aspect of RCS is that it does not prescribe a platform as the central function in an end-user's communication sphere as Facebook, LinkedIn, or other social networks do. Rather, it focuses on end-users and how they communicate with one another through different social interactions and technologies, including IMS, MMS/SMS, etc. This is realized technically through the use of the MSISDN to establish services, irrespective of which device the end-user is currently using. This places the subscriber, or the *individual*, directly at the center of the RCS building blocks. This is illustrated conceptually in Figure 12.2.

This is perhaps one of the key strengths of RCS within the emerging global economy – rather than prescribe a platform, it forms the communication building blocks for any platform a developer may wish to build *with the subscriber at the center of that platform*. RCS is therefore significantly more flexible than any social platform that exists today. As we will discuss in the second section of the chapter, this becomes a fundamental and

---

[2] Notwithstanding the many complaints leveled at Facebook, particularly in terms of privacy, no one can deny the success of its platform.

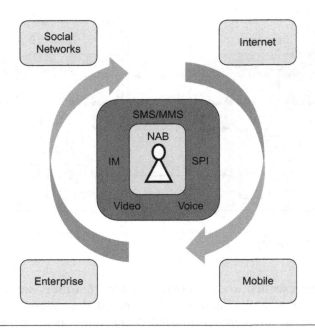

FIGURE 12.2

The subscriber-centric world.

extremely important aspect for developers to understand and be able to manipulate as the communications industries continue to evolve.

12.3 OVERVIEW OF RCS RELEASE FUNCTIONALITY

RCS work commenced in May 2007. Since 2007, there have been several releases of RCS, each with increasing functionality. This section briefly describes the contents of each release and gives an outline of the architecture of each profile. RCS Release 4 is currently under development at the time of writing and the discussion around that release should be taken as preliminary; interested readers are highly recommended to check the GSMA website for further releases.

Table 12.1 provides an overview of the content of each RCS release.

Readers should note that it is beyond the scope of this chapter to discuss in great detail the OMA specifications, and we have already discussed OMA presence in great detail in section 9.7.1 in Chapter 9. Here, we focus on the creation of the RCS profiles that gather together this functionality in a manner that is useful for developers. Any reader wishing to gain further insight in the evolution of OMA enablers mentioned here is directed to the excellent reference book, *The Open Mobile Alliance: Delivering service enablers for next-generation applications*, by Michael Brenner and Musa Unmehopa. Alternatively, visit the OMA website: www.openmobilealliance.org.

Table 12.1 Overview of RCS Functionality

Release	Contents	Release Date

RCS Release 1	• Content sharing (GSMA) for handsets with CS and PS.
	  • Video
	  • Image
	• Chat (OMA IM SIMPLE 1.0)
	• File Transfer (OMA)
	• Presence (OMA)
	• Enhanced Address Book (EAB)
	• Enhanced Messaging
		Q1 2007

RCS Release 2	• Network Address Book (NAB) • Provisioning and configuration of RCS devices **Broadband Access (BA)** • VoIP (3GPP MMTel) • Video Share over multi-devices *(watch video on PC whilst talking on the mobile)* • SMS (3GPP). Send only • Service control **Multi-device (mobile + PC)** • Enhanced presence • Enhanced chat	Q4 2008
RCS Release 3	• Participant list in group chat invitations • Presence enhancements • Geo-location • Nicknamed buddy-invite • Favorite link by label text • Capability blocking control ("Who can I invite?")	Q2 2009

RCS Release 3	• Network Value-Added Services
	  • Enrichment of content and chat with media processing
	• Content sharing enhancements
	  • Voice call not required
	  • Deferred sharing for users of legacy handsets
	Broadband Access
	• SMS send and receive
	• MMS to/from PC
	• Broadband access device may be used as a primary device
		Q4 2009

RCS Release 4	• LTE • Network APIs for interaction between RCS and social networks	TBA, APIs to be developed by OMA[a]

[a]Refer to http://www.gsmworld.com/our-work/mobile_lifestyle/rcs/ to check for the latest updates.

12.4 **RCS RELEASE 1**

RCS Release 1, completed and released from an end-user perspective in 2008, established three main components for the RCS service:

• Enriched Call, specifically Image Share and Video Share
• Enhanced Messaging

- Enriched Phone Book
  (GSMA, accessed 2011).

RCS Release 1 focused on profiling a subset of the IMS service specifications, specifically for mobile 3G devices using "hybrid" radio, i.e. those handsets that ran voice services over traditional circuit-switched technology, but used packet-switched technology to connect to content services such as video and images. As a result, these handsets used the 3GPP User Network Interface (UNI)[3] specification for connecting to mobile networks.

12.4.1 Enriched Call

An enriched call is one that allows end-users to share media content during an ongoing conversation, either pre-recorded or "live" media. Release 1 provided end-users with the ability to share images, video, and other sorts of files with one another. This section provides a brief overview of each of these from the end-user and architectural perspective as a preparation for discussions in the later part of the chapter.

It should be noted that for RCS Release 1, content sharing is only possible when a CS call is ongoing.

12.4.1.1 Related Standardization Efforts

As mentioned, RCS makes use of underlying IMS functionality so that the complexity is hidden from the end-user and the developer. Two aspects are worth mentioning briefly: the use of the CSI feature tag; and radio capability exchange.

12.4.1.2 CSI Feature Tag

Enriched call is built using underlying functionality in the IMS, developed within 3GPP, called combinational services, or "CSI". This standardization work item was created to provide the "technical realization for the combination of circuit-switched calls and IMS sessions when using them simultaneously between the same two users" (3GPP TS 24.279 Release 9).

Essentially, the set of specifications associated with CSI defines how the core network will handle how to combine CS and IMS services simultaneously. It is not necessary to go into great detail here about CSI as the RCS profiles allow developers to use this functionality without needing to understand it in any great detail. Suffice to say, when a mobile device sends an SIP INVITE for enriched call, be it for image or video, one feature tag is always included, namely + g.3gpp.cs-voice. This feature tag indicates that a mobile device is capable of transmitting voice via a circuit-switched call while combining it with an IM session.

---

[3] UNI is a relatively old-fashioned term that describes an open interface definition, marking the delineation of responsibility between a terminal and the network providers. Handset manufacturers use these specifications to implement radio and core network connectivity to ensure that their handsets are interoperable with mobile operator networks worldwide. A UNI is therefore one of the fundamental parts of providing globally interoperable handsets. All handsets that connect to any mobile operator network, irrespective of their OS, will comply with these specifications.

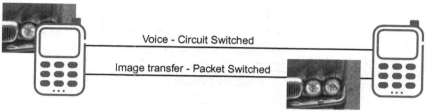

FIGURE 12.3

RCS Image Share – user perspective.

When this feature is found alone in the INVITE, the default service is Video Share. For other services, e.g. Image Share, an additional feature tag augments the INVITE.

12.4.1.3 A Note on Radio Capability Exchange

As mentioned, Release 1 of RCS is for terminals that support both CS and PS services. In order to do this, however, the terminal must be in a radio network cell that supports the simultaneous use of both CS and PS. For all of the descriptions that follow, it is assumed that the terminal supports simultaneous CS and PS usage. This information regarding this capability is *only* exchanged during the establishment of the CS call and is not part of the RCS specifications at all. We mention it here for information purposes only.

12.4.1.4 Image Share – End-user Perspective

Let's take a look at how Image Share works from the end-user's perspective. Using Image Share, end-users may send images to one another during an ongoing phone call. They may select a picture that is stored in their mobile devices gallery, or they make take a "live" picture while they are talking to the other person. This is illustrated in Figure 12.3, where end-user A wishes to share a picture of his or her car during a call to end-user B. End-user A takes the photo and clicks send. End-user B will receive a prompt on his or her mobile screen asking whether they wish to accept the image or not. Alternatively, this may be preconfigured within end-user's B device – to either automatically accept or reject images on a per-user basis (see Figure 12.3).

Figure 12.4, meanwhile, illustrates the call sequence of Image Share, showing the establishment of the CS-Voice session, the capability exchange as discussed in section 12.4.1.3, the SIP INVITE, and an expanded view of the MSRP session used in Image Share.

Prior to sending an image, end-user A, using Mobile_A, calls end-user B, who is using Mobile_B. A voice call is established according to 3GPP TS 24.008 for CS voice.

After this, the capability exchange procedure is performed and the image share icon is highlighted on both end-user's handsets, indicating that they are able to use Video Share.

An Image Share session is established between the two terminals using SIP signaling. At this point, end-user A is able to send an image to end-user B.

1. Mobile_A sends an MSRP message with the image embedded in it. This may be sent as MSRP "chunks", in which case end-user B will receive an indication that an image has been received when all the chunks have arrived at their device.

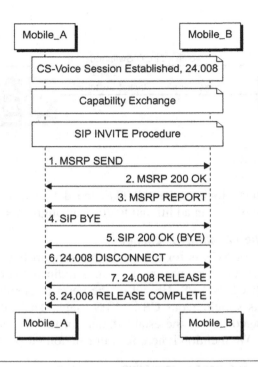

FIGURE 12.4

RCS Image Share – message sequence diagram.

2. On receipt of the image, Mobile_B sends an MSRP 200 OK to indicate successful transmission.

3. An MSRP report is sent by Mobile_B to Mobile_A. At this point, the SIP session is *automatically* torn down.

4. Mobile_A sends an SIP BYE.

5. Mobile_B sends an SIP 200 OK (BYE). End-user A decides to terminate the phone call.

6. Mobile_A sends a DISCONNECT in accordance with 3GPP TS 24.008.

7. Mobile_B sends RELEASE in accordance with 3GPP TS 24.008.

8. Mobile_A sends a RELEASE COMPLETE in accordance with 3GPP TS 24.008.

12.4.1.5 Image Share – Architectural Perspective

Image sharing is described in the GSMA IR.79 document. Figure 12.5 illustrates the network architecture of Image Share. For simplicity's sake, the CS and PS bearers are only indicated on the left-hand side of the diagram.

End-user A establishes a phone call to end-user B via the traditional circuit-switched network. Once this call is established, a device that has RCS Image Share capability will send an SIP OPTIONS request including the +g.3gpp.cs-voice feature tag in the Accept-contact header. If the other end-user's terminal supports Image Share, it will respond during the

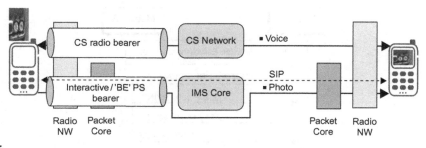

FIGURE 12.5

RCS Image Share architecture.

capability exchange with a 200 OK that contains the correct 3GPP IARI feature tag +g.3gpp .iari-ref=urn:urn-7:3gpp-application.ims.iari.gsma-is in the Contact header. Once the capability exchange has been successfully completed, the Image Share icon will be displayed on the end-user's mobile devices.

Alternatively, end-user B's terminal may respond with an indication that it supports both Image Share and Video Share, which is described in section 12.4.1.6. In such a scenario, the 200 OK would include both +g.3gpp.iari-ref=urn:urn-7:3gpp-application.ims.iari.gsma-is and +g.3gpp.cs-voice. In this case, both the Rich Call Image and Rich Call Video icons would be displayed on the end-users' mobile devices.

An image sharing session is started by the initiating end-user's device sending an SIP INVITE request containing the SIP TEL URL of the CS call. This SIP INVITE carries two feature tags in the Accept-Contact header, as follows:

- The CSI feature tag: +g.3gpp.cs-voice
- The IARI feature tag for GSMA IR.79: +g.3gpp.iari-ref="urn:urn-7:3gpp-application .ims.iari.gsma-is".

Please note that the colon in the URN has to be replaced by "%3A" when included in an SIP message – for example, +g.3gpp.iari-ref="urn%3Aurn-7%3A3gpp-application.ims.iari .gsma-is".

Once end-user B accepts to receive the image, his or her mobile device sends a 200 OK. At this point, an MSRP session is established in accordance with IETF RFC 4975 and the image is transferred as described in section 12.4.1.4. As discussed in previous chapters, the image transmission is directly between the two end-users and does not flow via the SIP-AS. Figure 12.6 shows the message sequence diagram for this use-case.

Once the image transfer is complete, the SIP session is automatically terminated. Images may be shared as many times as desired during an ongoing CS call. Each time a new image is sent, a new SIP session and new MSRP session are established. It should be noted that the sending user is the one who is charged for the transmission of the images. Figure 12.6 illustrates the SIP session establishment in accordance with IMS principles and IR.79.

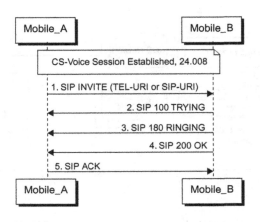

FIGURE 12.6

SIP session establishment during a CS voice call.

Prior to SIP session establishment, end-user A, using Mobile_A, calls end-user B using Mobile_B. A CS-voice session is established in accordance with 3GPP TS 24.008.

1. Mobile_A sends an SIP INVITE with a request header containing the Tel URI or SIP URI of Mobile_B.
2. Mobile_A receives an SIP 100 Trying from the IMS core network, indicating that the IMS core is attempting to establish an SIP session with Mobile_B.
3. Mobile_A receives an SIP 180 Ringing from the IMS core, indicating that Mobile_B has received the SIP INVITE.
4. End-user B accepts the session and Mobile_B sends an SIP 200 OK.
5. Mobile_A sends an SIP ACK.

The general rule is that one image is sent per session, but an image share session can be set up many times during the lifetime of the call.

12.4.1.6 Video Share – End-user's Perspective

Video Share allows end-users to send videos to one another during an ongoing phone call. Using mobile devices that are video capable, they may send a live video stream or a video they already have stored on their device to another end-user while they are talking. This is illustrated conceptually in Figure 12.7, where end-user A wishes to share a video of a waterfall during a call to end-user B. End-user A selects the video icon on his or her mobile device and requests to send video. End-user B will receive a prompt on his or her mobile screen asking whether they wish to accept the video stream or not. Alternatively, this may be preconfigured within end-user B's device – to either automatically accept or reject images on a per-user basis. Figure 12.7 depicts the end-user perspective for Video Share.

It should be noted that Video Share is different to a pure 3G video call. First, it is based on IP technology, RTP, rather in the way CS as 3G video calls are. Secondly it is

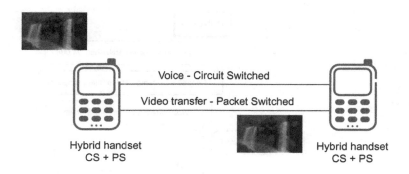

FIGURE 12.7

RCS Video Share – user perspective.

unidirectional and can be initiated or stopped any time during the call without terminating the phone call (Orange, 2009). A pure 3G video call is designed to allow end-users to see each other during an ongoing call – Video Share is designed to allow end-users to share experiences with one another while talking.

The signaling diagram in Figure 12.8 illustrates the high-level end-user view of Video Share, focusing on the establishment and running of the RTP session for Video Share.

Prior to establishing the video stream, end-user A, using Mobile_A, calls end-user B, who is using Mobile_B. A voice call is established according to 3GPP TS 24.008 for CS voice.

After this, the capability exchange procedure is performed and the Video Share icon is highlighted on both end-users' handsets, indicating that they are able to use Video Share. A Video Share session is established between the two terminals using SIP signaling.

At this point, end-user A is able to start sending video to end-user B.

1. Mobile_A sends a request to Mobile_B to establish an RTP session.
2. Mobile_B uses RTCP receiver reports (RR) in order to inform Mobile_A of the quality of service being experienced in the transfer of the video.
3. Mobile_A sends RTCP sender reports (SR), which contain an absolute timestamp, and which Mobile_B uses to synchronize incoming RTP messages. End-user A has finished sending video to end-user B and therefore terminates the video stream using the correct icon on their device.
4. Mobile_A sends an RTCP (BYE) indicating that the video stream has ended. Mobile_A automatically terminates the SIP session that had been established for Video Share.
5. Mobile_A sends an SIP BYE.
6. Mobile_B sends an SIP 200 OK. After the SIP session is terminated, the CS-Voice session remains intact. End-user A decides to terminate the phone call.
7. Mobile_A sends a DISCONNECT message in accordance with 3GPP TS 24.008.
8. Mobile_B sends a RELEASE message in accordance with 3GPP TS 24.008.
9. Mobile_A responds with a 3GPP TS 24.008 RELEASE COMPLETE.

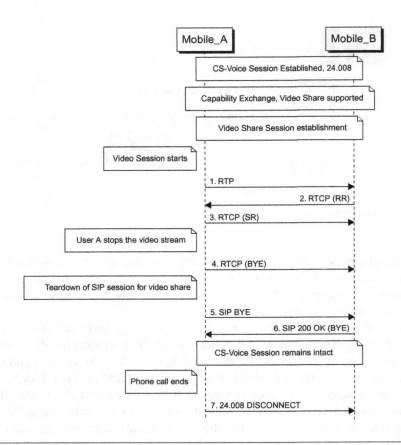

FIGURE 12.8

Signaling diagram for Video Share.

12.4.1.7 Video Share – Architectural Perspective

Video Share is described in GSMA document IR.74. Figure 12.9 illustrates the architecture of Video Share. CS and PS bearers are again shown only on the left-hand side of the diagram.

End-user A establishes a phone call to end-user B via the traditional circuit-switched network. Once this call is established, a device that has RCS Video Share capability will send an SIP OPTIONS request including the +g.3gpp.cs-voice feature tag in the Accept-contact header. If the other end-user's terminal supports Video Share, it will respond during the capability exchange with a 200 OK that contains the feature tag +g.3gpp.cs-voice. Once the capability exchange has been successfully completed, the Video Share icon will be displayed on the end-users' mobile devices.

Alternatively, end-user B's terminal may respond with an indication that it supports both Image Share and Video Share. In such a scenario, the 200 OK would include both +g.3gpp .iari-ref="urn:urn-7:3gpp-application.ims.iari.gsma-is" and +g.3gpp.cs-voice in the Contact

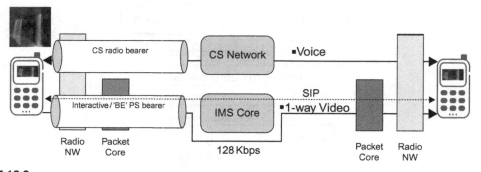

FIGURE 12.9

RCS Video Share architecture.

---

header. In this case, both the Image Share and Video Share icons would be displayed on the end-users' mobile devices.

The initiating end-user's device sends an SIP INVITE request containing the SIP Tel URL of the CS call in order to start a video-sharing session. This SIP INVITE carries a feature tag in the Accept-Contact header, CSI feature tag: +g.3gpp.cs-voice.

For Video Share, H.263-2000 profile 0 level 45 is mandated, which provides about 128 kbps transmission of video.

Once end-user B accepts to receive the video, his or her mobile device sends a 200 OK. At this point, an RTP/UDP session is established, as discussed in Chapter 7, and the video begins streaming. As discussed in previous chapters, the video transmission occurs directly between the two mobile devices and does not flow via any SIP-AS servers.

When an end-user wishes to stop receiving or sending video, they select this on their mobile device. At this point, the session is terminated, but the voice call continues via the circuit-switched network.

Again, it is the originating end-user who will be charged for the transmission of the video stream.

Figure 12.10 illustrates the establishment of the SIP session for Video Share.

Prior to SIP session establishment, end-user A, using Mobile_A, calls end-user B, using Mobile_B. A CS-voice session is established in accordance with 3GPP TS 24.008.

1. Mobile_A sends an SIP INVITE with a request header containing the Tel URI or SIP URI of Mobile_B.
2. Mobile_A receives an SIP 100 Trying from the IMS core network, indicating that the IMS core is attempting to establish an SIP session with Mobile_B.
3. Mobile_A receives an SIP 180 Ringing from the IMS core, indicating that Mobile_B has received the SIP INVITE.
4. End-user B accepts the session and Mobile_B sends an SIP 200 OK.
5. Mobile_A sends an SIP ACK.

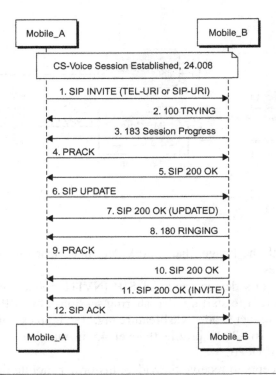

FIGURE 12.10

Establishment of the SIP session for Video Share.

12.4.2 Enhanced Messaging

Enhanced Messaging allows end-users to have a "new" messaging experience based on SMS/MMS capabilities, and also provides File Transfer and Chat facilities. Enhanced Messaging provides a unified interface for an end-user's communication needs – all options that are available on his or her mobile device in a so-called "conversational view". This means that all SMS/MMS, IM, voice, and video calls are presented in a "threaded" communication history (Orange, 2009).

The following subsections describe the File Transfer and Chat profiles.

12.4.2.1 Enhanced Messaging – File Transfer

A File Transfer session allows end-users to share files with one another via an MSRP message and relies on the underlying OMA IM specifications. A simplified illustration of the architecture is given in Figure 12.11. For the sake of simplicity, the originating and terminating instances of the IM-AS and IMS core networks are shown as one box. This complexity is hidden from developers and is therefore not shown in detail in this chapter.

End-user A selects another end-user, B, to establish a chat session with. User A's mobile device sends an SIP INVITE request with the SIP Tel URI of end-user B. In the

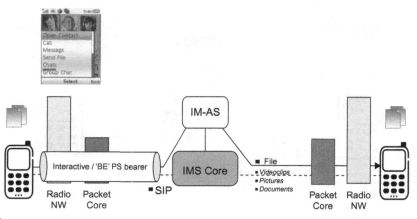

FIGURE 12.11

Enriched Messaging – File Transfer.

Accept-Contact header, the feature tag +g.oma.sip-im is indicated. This feature tag indicates that the terminal wishes to establish an OMA IM session with the other terminal. The SDP direction attribute is set to "sendonly". If end-user B accepts to receive the file, his or her mobile device sends a 200 OK, and the session is established. Files are transferred via an en-bloc mode MSRP message.

Upon transfer of the file, the session is terminated automatically. In contrast to the transfer of images in Image Share, messages flow via the IM SIP-AS.

End-user A, i.e. the party that selects to send the file, will be charged for the transmission of the file by his or her mobile operator, depending on operator policy.

12.4.2.2 Enriched Messaging – Chat

Enriched Messaging allows end-users to have an interactive chat session by simply selecting the other party from their Enhanced Address Book and choosing "Chat". RCS Chat is based on the OMA IM session-mode specification. While it is possible to establish ad-hoc chat sessions, only the initiating end-user can add extra parties to the chat session. RCS Chat is illustrated in Figure 12.12.

Figure 12.13 shows a simplified architecture. Again, the originating and terminating IM-AS and IMS core are illustrated as one box each, rather than displaying the full complexity within one diagram.

End-user A selects the person that they wish to establish a chat session with. His or her mobile device then sends an SIP INVITE request with end-user B's phone number as the "target address", the SIP Tel URL or SIP URL. The SIP INVITE also includes the OMA IM feature tag in the Accept-Contact header, +g.oma.sip-im. In addition, the SIP INVITE will also include a subject header that contains the first message of the chat session, e.g. "Hi, do you have a minute?"

FIGURE 12.12

RCS Chat – user perspective.

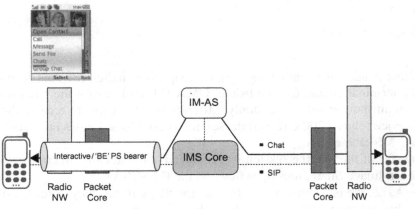

FIGURE 12.13

RCS Chat – architecture perspective.

If end-user B wishes to chat with end-user A, they merely start typing a response and a 200 OK is sent to the originating mobile device and the session is established. In accordance with OMA IM, each chat message is sent via an MSRP message. Every MSRP message is sent via an IM application server on the core network.

It is possible to send a small image of a preconfigured size limit within the Chat session. If the picture is too large, the mobile device will instead establish a separate File Transfer session in parallel to the Chat session.

Within RCS Release 4, RCS Chat is being updated to follow OMA CPM and to allow end-users who have only SMS capability on their mobile devices to also participate in RCS Chat sessions.

All the chat messages pass the IM application server for charging purposes. If any images are too large and a separate File Transfer session is established, this is charged according to RCS File Transfer, described in the previous section.

12.4.3 Enriched Phone Book

Enriched Phone Book allows end-users to be connected and receive updates about their favorite contacts through a concept called Social Presence Information. In order to receive information about one another's updates, end-users must first establish a social relationship with one another. Social Presence Information is based on the profiling of OMA Presence 1.0 and 2.0 specifications. An end-user may publish:

- Free text
- Web link
- Profile photo
- "Hyper" availability.

Through this, an end-user may publish a profile and will receive updates when his or her contacts update theirs (Orange, 2009). An end-user may also synchronize his or her legacy vCard with the network address book.

The architectural implementation of the RCS Social Presence Information is illustrated in Figure 12.14.

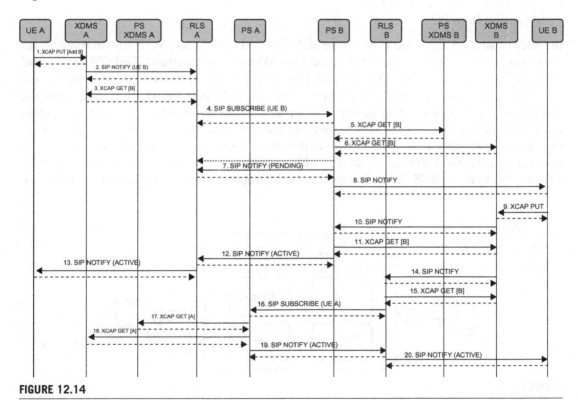

FIGURE 12.14

Sequence diagram for RCS Social Presence Information.

12.5 RCS RELEASE 2

RCS Release 2 was released in June 2009. In addition to the RCS Release 1 functionality, Release 2 introduced several new features, which are described in the following subsections.

12.5.1 Broadband Access

Release 2 was largely about extending RCS to broadband accesses, i.e. all-IP, as well as the mobile domain. An example of such a broadband access device is a PC in someone's home – see Figure 12.15.

RCS Release 2 therefore extended the following profiles, described in the Release 1 section, to the broadband domain:

- Presence
- Chat
- File Transfer
- Image Share and Video Share
- Voice over IP based on MMTel
- SMS send from PC.

In order to support Enriched Call Image and Enriched Call Video, VoIP was introduced for broadband access clients in the form of the IMS MMTel specification (3GPP TS 24.173 and 3GPP TS 26.114), but also GSMA IR.92 (VoLTE). MMTel is described in Chapter 9 and we will not therefore review it again here. From an RCS perspective, this meant that the feature tag +g.3gpp.icsi-ref:="urn:urn-7:3gpp-service.ims.icsi.mmtel" was now acceptable in the Accept-Contact header. This feature tag indicates that the voice session will be handled using VoIP, rather than the circuit-switched network as it was for the mobile devices described in the RCS Release 1 section, although it is of course still possible to use RCS 1.0 specifications as RCS 2.0 is designed to be interoperable and compatible with RCS 1.0.

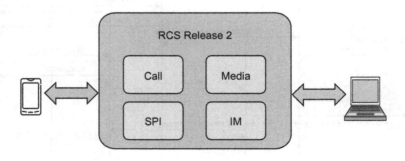

FIGURE 12.15

RCS Release 2.

Please note, however, that both Image share and Video share are treated as separate SIP sessions even when a call is established via a SIP session based on MMTel.

12.5.2 Multi-Device Environment

RCS Release 2 also introduced the concept of a multi-device environment. This allowed end-users to have multiple devices under the same number (MSISDN), where he/she could use services from any of the terminals (see Figure 12.16). When a phone call or session is established towards an end-user, all devices will ring simultaneously. Once the end-user accepts or rejects the call, the other devices stop ringing. If the user has accepted the call, the session is established with the device that they have answered on.

12.5.3 Enriched Call – Multi-Device

RCS Release 2 introduced a new concept, allowing for Enriched Call functionality to be split between the broadband access device and the mobile terminal. This means that for Image Share and Video Share, an end-user is able to receive the voice call on his or her mobile terminal, while the image or video is displayed on his or her PC or other broadband access device. This relies on the mobile softswitching solution, as illustrated in Figure 12.17.

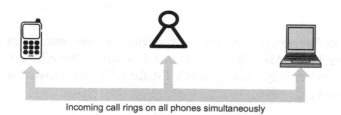

Incoming call rings on all phones simultaneously

FIGURE 12.16

Multiple devices.

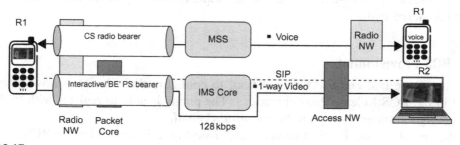

FIGURE 12.17

Enriched Call – multi-device.

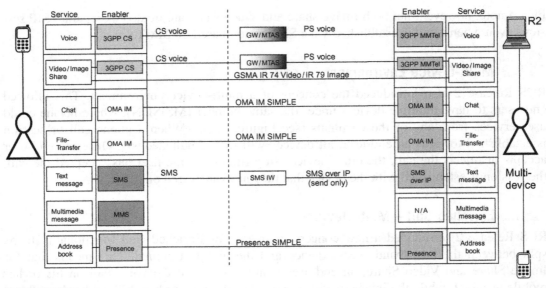

FIGURE 12.18

Interworking for the multi-device environment and broadband access.

The full range of interworking for the multi-device environment and broadband access is illustrated in Figure 12.18. Also illustrated in this diagram are the relevant specifications that the GSMA RCS services rely on, e.g. OMA IM for RCS file transfer, or SIMPLE for the network address book.

12.5.4 Network Address Book

The Network Address Book was the same as in RCS Release 1 as described in section 12.4, but provided network storage and access to the address from any of the devices that the end-user had registered with.

12.5.5 RCS Provisioning

In order to ensure that end-users do not need to configure their devices, provisioning was included in the RCS Release 2 specifications. The phone is able to register with RCS without any configuration necessary on the end-user's behalf.

It is the operators' and handset makers' responsibility to install and make RCS provisioning available to end-users. It is therefore beyond the scope of this book, other than to say that RCS is designed to work "out of the box" for end-users in the same way as SMS or

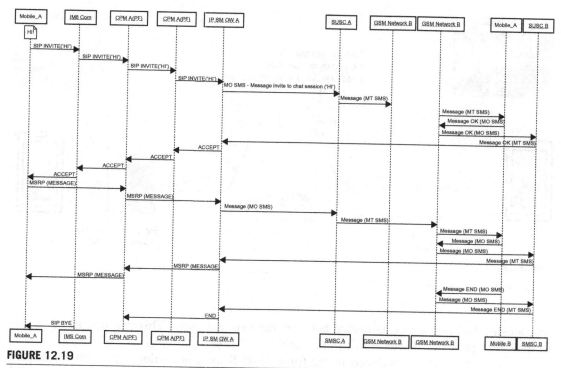

FIGURE 12.19

RCS Chat interworking towards SMS.

voice services. The RCS functionality is available on the handset and does not need any configuration by end-users.

12.6 RCS RELEASE 3

RCS Release 3 introduced several new functions to the RCS specifications, including the use of a fixed broadband access device such as a PC as the primary device, rather than the mobile device needing to be mobile. In addition, Image Share and Video Share were updated to allow content sharing between devices that do not require the terminals to be involved in an ongoing phone call. RCS 3.0 also allows end-users on legacy terminals to still receive the content from Image Share or Video Share sessions by having the delivery of the content deferred until later. Fixed broadband devices are also able to now send and receive SMS and MMS, rather than just receive them, as illustrated in Figure 12.19.

Presence was also updated to include geo-location information and the ability to see the capabilities of all the contacts within the address book, even if a "social presence"

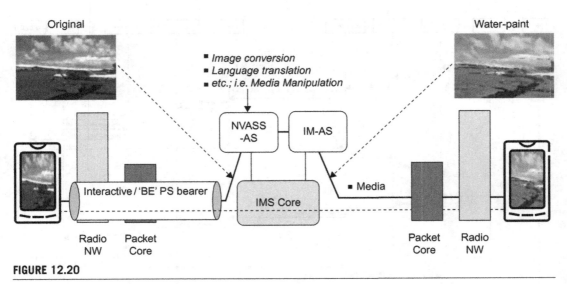

FIGURE 12.20

RCS Release 3 – using an NVAS.

relationship had not been established between the two end-users. This was described as part of the OMA presence discussions in Chapter 9.

A new service was introduced in the form of RCS used in conjunction with a Network Value-Added Service (NVAS) that enables the enrichment of content during ongoing RCS sessions with media processing. Examples are language translation or image manipulation. This is achieved through hooking an NVAS-AS into the IMS sessions, as illustrated in Figure 12.20.

12.7 RCS RELEASE 4

This section is a preliminary description of ongoing work within GSMA RCS specifications. Readers are strongly recommended to check the RCS website for a full and complete definition of the APIs developed within the framework.

The RCS Release 4 specification work handles the development of API profiles that OMA will develop for release. These APIs will allow interaction between elements of RCS and other services, for example social networking sites such as Facebook or enterprise applications that allow end-users within a company to coordinate with one another more effectively. This may be viewed as the creation of an RCS community, but as we will discuss in subsequent sections, this is not strictly speaking true: RCS allows for much greater flexibility than many current social networking sites. This flexibility, in conjunction with global interoperability, allows RCS developers to harness new aspects of the communications industries as and when they emerge (see Figure 12.21).

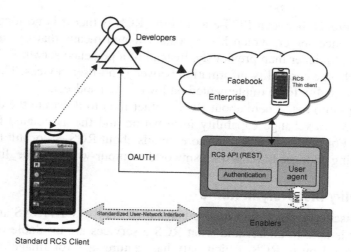

FIGURE 12.21

Conceptual overview of RCS Release 4 APIs.

RCS Release 4 has adopted LTE ("4G") in order to improve the user experience of all the RCS services of Releases 1–3 thanks to the higher speed offered by the higher band-width LTE provides over the radio interface. For example, if two users are located under LTE access, Video Share can utilize up to 768 kbps video by using H.264 baseline profile level 1.3.

RCS Release 4 has also endorsed the voice-over-IP solution for LTE, also known as VoLTE (GSMA IR.92), for its all-IP broadband access connected devices, e.g. a PC, in order to secure good interoperability in PS-based voice communication with VoLTE mobile devices.

12.8 RCS-e

During February 2011, several operators[4] active in RCS specifications launched a new initiative named RCS-e,[5] which is a carefully selected subset of the RCS profiles based on those with highest consumer demand. RCS-e is designed to lower the entry barrier for RCS by allowing operators (and developers) a method to use RCS functionality without needing to implement the full RCS profiles.

These operators have taken a joint approach to OEM handset manufacture, ensuring that the RCS-e profiles will be available and "just there" for end-users to access.

---

[4] T-Mobile, Deutsche telecom, Orange, Telecom Italia, Telefonica, Vodafone, Bharti Airtel, Telenor.
[5] For more information and specifications, see www.gsma.org.

The main difference between RCS-e and "full" RCS is that it is no longer necessary to implement a presence server, which keeps costs low and means that operators are able to roll out RCS much quicker than previously. Instead of a presence server, RCS-e utilizes SIP OPTIONS to exchange capability information between end-user devices. This approach also allows RCS functionality to be implemented on lower-end handsets.

We will not go into detail here about RCS-e other than to illustrate the difference in use of SIP OPTIONS to exchange capability information and the associated feature tags that RCS-e supports. For those interested in more details about RCS-e, consult the GSMA website at the following address: http://www.gsmworld.com/our-work/mobile_lifestyle/rcs/.

12.8.1 Capability Discovery in RCS-e

The capability discovery process in RCS-e is handled via SIP OPTIONS and is the process by which a user is able to understand what RCS-e services are available on another user's terminal. In comparison to RCS, which only has feature tags for Image and Video Share, there are several more feature tags that are possible with RCS-e. These feature tags are sent in the Accept-Contact header of the SIP OPTIONS message by the requesting terminal. The responding terminal sends the response in the Contact header of the 200 OK message. Table 12.2 lists the feature tags that are possible.

The final three feature tags in the table are worth explaining and relate to RCS-e clients that supplement this functionality with RCS 1.0 and 2.0 standards functionality. Those that are able to support "social profile information via presence" use the "+g.3gpp.iari-ref="urn:urn-7:3gpp-application.ims.iari.rcse.sp" feature tag. Those clients that are able to use the presence-based discovery mechanism use a new feature tag, "+g.3gpp.iari-ref="urn:urn-7:3gpp-application.ims.iari.rcse.dp".

A final feature tag is possible that allows operators to define their own services. This is defined using the feature tag "+g.3gpp.iari-ref="urn:urn-7:3gpp-application.ims.iari.rcse.<operatorID>.<servicename>".

Figure 12.22 shows the capabilities exchange process for RCS-e.

1. End-user A, using mobile device A, sends an SIP REGISTER request to the IMS core.
2. The IMS core responds with an SIP 200 OK to the SIP REGISTER. At this point, Mobile_A restarts its registration timer.
3. Mobile_A sends an SIP PUBLISH message with its capabilities to the IMS core.
4. The IMS core responds with an SIP 200 OK.
5. When mobile user A decides that they wish to exchange terminal capabilities with mobile user B, their device sends an SIP OPTIONS request to the IMS core.
6. The IMS core routes the SIP OPTIONS request to mobile user B.
7. Mobile_B responds to Mobile_A with an SIP 200 OK message, containing the capabilities of the Mobile_B terminal.

If the terminating user has multiple devices, the terminating network applies a SIP-AS that sends multiple OPTIONS AS to each device, and aggregates the capabilities into one OPTIONS response back to the user that sent the OPTIONS request.

Table 12.2 RCS-e Service and Feature Tag Value

RCS-e Service	Feature Tag
Image Share	+g.3gpp.iari-ref%3Aurn-7%3A3gpp-application.ims.iari.gsma-is
Video Share	+g.3gpp.cs-voice
IM Chat	+g.3gpp.iari-ref="urn:urn-7:3gpp-application.ims.iari.rcse.im *only* identifies the IM service
File Transfer	+g.3gpp.iari-ref="urn:urn-7:3gpp-application.ims.iari.rcse.ft uniquely identifies the File Transfer service
Social Profile Information using Presence	+g.3gpp.iari-ref="urn:urn-7:3gpp-application.ims.iari.rcse.sp
Capability discovery using Presence	+g.3gpp.iari-ref="urn:urn-7:3gpp-application.ims.iari.rcse.dp
Tailored operator service	+g.3gpp.iari-ref="urn:urn-7:3gpp-application.ims.iari. rcse.<operatorId>.<service name>

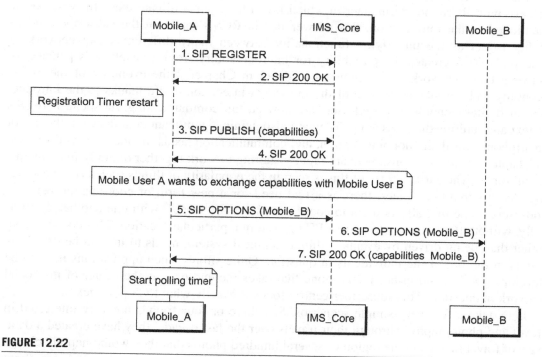

FIGURE 12.22

Capabilities exchange process.

12.9 USING RCS APPLICATIONS TO CAPTURE VALUE

The previous section outlined the basics of the RCS suite of profiles for communication between end-users. Much of what was described above may look reasonably simple to many developers – apart from the global interoperability, much of the functionality is available from existing web or Internet-based services. Presence services exist in many different forms and flavors, many social networks exist, and there are certainly many VoIP solutions available to end-users already today. Indeed, many use-cases presented for RCS can at first sight seem to be rather boring: most examples are of a person watching a football match and wanting to share it with their friends. There are equally as many use-cases that describe shopping scenarios, where someone wishes to ask advice about whether to buy a particular pair of shoes or not and the user starts a phone call and sends an image to their friends at the other end of the call while discussing the shoes. In our era of tech-savvy populous and mobile Internet devices that serve up a multitude of information and entertainment in a variety of different formats, these are really not stand-out examples that will have end-users flocking to services.

What, then, does the RCS suite provide for developers beyond these simplistic exchanges of images, video, presence, and text? The answer to this, of course, lies in what you can achieve with RCS and the manner in which it can help to coordinate these for you. As we mentioned in the introduction, this answer lies in RCS's focus on the subscriber and its ability to act as the underlying functionality between many different social networks – essentially RCS creates the possibility for interoperability from a developer's perspective between these networks. We discussed in detail in Chapter 2 the evolution of the global economy and the increasing role of the individual in economic life thanks to the introduction of digital technologies. End-users are now au fait communicating via social networks or text and multimedia messaging. They are also gaining significant confidence in the use of smartphones for their shopping, search, and communication needs. In addition, as discussed in Chapter 2, end-user consumers are rapidly starting to work together digitally in many different forms. The earliest form of this is seen in the popularity of things such as crowdsourcing. As this trend continues, there is a real need for a new form of coordination between end-users – one that allows them to coordinate flows of "work"[6] with one another dynamically, without the necessity to be at a PC or even in a particular location. This coordination, rather than being driven by events within a technical system, needs to instead be driven by events in the life of the individuals in question. Quite simply, such applications need to be developed with a subscriber focus – one that takes the end-user as the center of the social network in question. This subscriber-centric focus is exactly what RCS provides.

Let's take a few very simple scenarios. Say three or four friends are very interested in travel and photography. Through their travels over the last decade, they have created a database of travel images in the region of several hundred photos that they would happily charge a small fee for usage of. In addition, they have gathered experience and knowledge about how to resolve problems when traveling in dangerous regions and they have established a

---

[6] Or, indeed, play. For the purpose of this chapter, we use "work" in the loosest definition of the word, i.e. there are a series of events between end-user consumers that need to be coordinated between one another.

contact network with other travelers who share their interests. Together, they have created a new website that they wish to develop and grow a user base for; each of the friends has a particular area of the website that they are responsible for. One takes care of the editing, another takes care of the image database and the use of images on the website, while another handles the contract negotiations. Each of these aspects requires coordination between the individuals in question, in particular the editing of text and associated images. Naturally, however, these three friends are constantly traveling either for work or for pleasure, so there is rarely an opportunity for all of them to meet specifically to manage the website. Phone conferences are difficult and unwieldy places to discuss placement and usage of images. Without installing an expensive enterprise system, there is little ability for them to coordinate contracts, invoices, and customer relationship management. While there are many systems that allow people to work remotely with websites, there are very few that are able to handle the coordination of *workflows* between individuals.

Another scenario concerns the creation of a local community newspaper for a small village in, say, rural Wales. This is created by a group of dedicated volunteers who collect local news stories as they hear them, take requests from new parents to put pictures of their babies in, and prepare a list of local events people may wish to attend. Much of this is done by hand and as a result is extremely time-consuming. The volunteers would very much like some kind of digital means to coordinate with one another, capture images, take notes of events and stories. Critically, however, these need to be sent to one another for review and inclusion in a repository of some kind that could then be used to produce the final version of the newspaper for printing and distribution at a computer in the local cafe. Many of the volunteers work in remote areas and do not have ready access to either a computer or a reliable broadband connection.[7] All of them, however, have mobile devices for their work. If it were possible to capture images, add them to text and send them on to the next volunteer for approval and eventually for typesetting this would speed up the production of the newspaper. It would also allow for the inclusion of last-minute events that perhaps might otherwise have to be omitted due to time limitations.

Alternatively, several people may wish to view a sporting event together and coordinate their video streams and photographs together in order to create some kind of multimedia broadcast that would be viewed live by people who had not been able to attend the event in person. Through using the mechanisms provided by RCS, a developer could help coordinate these flows of information between the different end-users.[8]

Each of these scenarios may be viewed as a *supply chain* of digital information flowing between individuals. In order to efficiently manage an individual's time, these supply

---

[7]Yes, such countries do exist. The UK is a great example of a "developed" nation that as yet does not have full broadband coverage across the nation. In addition, many lesser developed nations have already "leapfrogged" fixed broadband and gone directly to mobile.

[8]Melding video streams together from separate handsets is not handled by RCS, but it could be used in conjunction with a media platform that handled such technologies and be used to form a social network around such a platform.

chains need to be handled in the best digital manner possible. In the same way as enterprise systems increased the efficiency of supply chains of corporations during the late 1990s and early 2000s,[9] digital technologies could easily be applied to help individual end-users streamline their daily processes in order to reduce the time spent fiddling around with mobile devices, PCs, and other digital means of communication.

There is a major difference between using digital technologies to coordinate flows of work between individuals and the coordination of work within an enterprise, however. Enterprise systems, in particular for supply chain management, are all centrally controlled and top-down driven. Most companies establish their supply chain processes prior to introducing an enterprise management system. They then "recreate" these processes digitally through some means, often through hiring expensive business analysts who then implement it in BPEL or similar. Tasks are then shunted around such systems between the relevant employees for their input and approval. In order to achieve this, the boundaries of the firm must be strictly established; a company must know who its suppliers are and how much bargaining power it has during negotiations regarding price.

The individuals described in our scenarios above, however, are not colleagues in the strictest sense of the word. They are a loose collaboration of different people, joined together by a common social interest. The drive to coordinate work is not because they wish to push the price of commodities down, but to reduce effort spent on producing something that they enjoy. Through improving workflow coordination between individuals in these examples, they are instead freeing up time for more important (non-digital!) social activities – they are able to enjoy the final product, rather than focusing on producing it. The digital technology becomes a fundamental enabler of their life experience, rather than merely a tool that they must constantly reconfigure for different parts of different processes.

One of the key factors in each of the scenarios above is that they are about managing a distributed workload between a series of people that are not otherwise connected via traditional frameworks or structures. The structures are truly social structures of loosely connected people. In this sense, they are not structures that can be captured within any platform framework – they are dynamic and reformed over time as volunteers need to leave the newspaper due to the illness of a loved one, or one of the friends suddenly decides to spend several months traveling to the North Pole.

Another important issue is that the individuals involved also do not define their role within these groups in accordance with the device they are using. Employees of a large company, even if they using a laptop at home, know that they are in some sense "at work" when they access corporate systems via a VPN. In contrast, an individual participating in a social endeavor with a few friends will still view him or herself as part of that endeavor whether they are using a mobile phone or a PC in the local library. The experience is not related to

---

[9] And continue to do so today.

the platform but instead how a group of individuals relate and coordinate with one another around events within that group. Each of these events triggers a session when digital communication is required. The overlying management of such sessions is where the developer can innovate, create, and capture real value for end-users, be it for "work" purposes as described in the scenarios above or for more fun scenarios that may be about different types of media such as video or other content.

This is exactly what RCS enables – a social network based around the individual end-user, irrespective of device and the basis to create social networks driven by events, rather than by a platform.

How would one of these scenarios actually be implemented using RCS, however? Let's take the example of using RCS to organize and coordinate communication between a group of friends at a sporting event who wish to broadcast their output on the web so that other people may watch and comment at the same time as they are broadcasting.

Let's say we have three main people watching a surfing competition on a beach in Sydney. One person has a slightly older feature phone and an RCS 1.0 client that supports image sharing and social sharing. Another friend has a smartphone with RCS 2.0 profiles that supports video sharing. Both phones have voice services via VoLTE or CS. Meanwhile a friend who is a keen surfer and has recently moved to a new city wants to watch the surfing competition on a PC and listen to his friends' discussing it like they used to, rather than just watch it on TV. Using RCS functionality, they establish a small social group between one another for the purposes of the surfing competition. In order to create an outside broadcast of the event, this group of friends needs to utilize some form of broadcasting platform, which is able to blend together images, videos, and the voice-based commentary/reporting from the different people. In order to do this, they use an NVAS on RCS, which allows them to interact with the broadcasting platform in a social manner. The operator is able to offer this NVAS broadcasting service to many different groups of end-users, allowing for the creation of a new form of distributed journalism between small groups of individuals.

RCS used in conjunction with the broadcasting NVAS helps the friends to coordinate the flow of interaction between them, allowing them to switch between one another's videos, images, and voice commentary. The friend sitting on the PC is also able to interact with his friends and provide feedback on the surfing competition and his friends' reporting abilities (see Figure 12.23).

Using RCS functionality, three end-users on very different terminals, with different client software on both mobile and fixed devices, are able to coordinate and interact with one another in a social manner. Due to the ability of RCS to hook in complex NVAS services such as the broadcasting platform, the social capabilities of RCS can be used to help coordinate extremely complex use-cases in the digital era. This is something that no other social network platform can offer – and of course we have already discussed how RCS allows a developer to interact with other social networks.

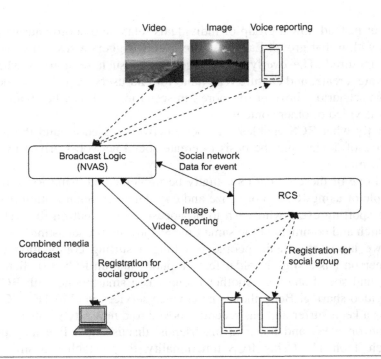

FIGURE 12.23

RCS used in conjunction with broadcasting.

12.10 CONCLUSIONS

RCS gives a developer the possibility to create an application once and run it on any operator's network and on any device without needing to reconfigure the clients. Provisioning and maintenance is instead provided by the RCS standard and is handled automatically by the operator network, leaving developers free to worry instead about the actual application.

Evolved IP Multimedia Architecture and Services

13.1 INTRODUCTION

This chapter aims to explain the main new concepts and evolution of IMS supporting the mobile telephony evolution. The intent is that it will provide background and create awareness for how value-added service developers need to understand this evolution.

13.2 OVERVIEW OF THE EVOLVED IMS ARCHITECTURE

The main driver of the IMS and IMS service standard since 3GPP Release 7 has been to enable and support the evolution and migration of mobile telephony. The IMS architecture and generic service concepts were finalized in 3GPP Release 7 and it coincided with the handover of the MMTel service developed in ETSI TISPAN into 3GPP. These two have been the cornerstones for evolution of mobile telephony developed in 3GPP since then. The prime driver for that a new operator-provided packet-based telephony service is required is because the LTE access is packet only. The mobile industry has chosen 3GPP MMTel to be the packet-based telephony service for LTE. Figure 13.1 illustrates the evolution and objectives of bringing MMTel into 3GPP and its role as a foundation for GSMA VoLTE.

In order to facilitate the introduction, a minimum profile targeting IMS telephony service introduction for early LTE deployments has been developed. It was developed within the One Voice consortia, a group of leading mobile network operators, phone, and network equipment vendors, and was handed over to the GSM Association (GSMA) upon completion. GSMA is now maintaining and evolving the profile.

In most mobile networks, the LTE coverage will be introduced as a partial overlay on existing 2G/3G cellular radio coverage. In order to handle this situation, the migration concepts of IMS Centralized Services and Single Radio Voice Call Continuity are introduced. Figure 13.2 provides an overview of these key new mechanisms introduced to support an IMS telephony service in a network without "full" LTE coverage with an underlying GSM/UMTS network.

IMS Centralized Services (ICS) allow the MMTel telephony service in the IMS service domain to be used by IMS users when the access is provided via a circuit-switched network. Single Radio Voice Call Continuity (SRVCC) allows an ongoing MMTel voice call over LTE to be handed over to the circuit-switched domain in GSM or WCDMA when losing LTE coverage, in order to reduce the number of dropped calls.

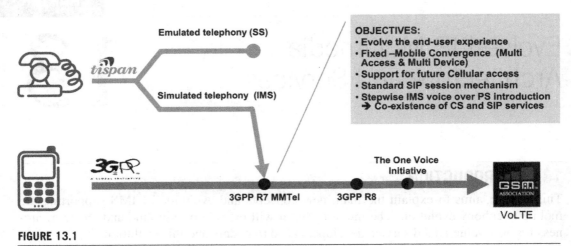

FIGURE 13.1

The evolution of Multimedia Telephony (MMTel from TISPAN to GSMA VoLTE).

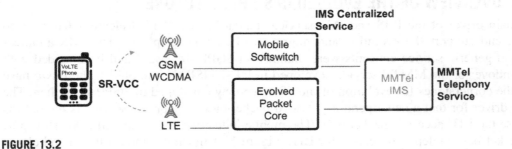

FIGURE 13.2

Single Radio Voice Call Continuity (SRVCC) is a handover mechanism and IMS Centralized Services is a mechanism to provide MMTel telephony over the CS access.

13.3 GSMA VoLTE – IMS PROFILE FOR VOICE AND SMS

In 2009 several vendors and operators in the telecom industry reacted to the threats to fragment the evolution of the mobile telephony service upon the introduction of LTE. To align the industry in support of 3GPP standards, several vendors and network service providers jointly formed the One Voice initiative with the mission to create a user equipment to network interface (UNI) profile of existing 3GPP specifications suitable for telephony over LTE in an introduction phase. This profile specifies a minimum set of features that terminal and network vendors should comply with to guarantee interoperability. The initiative has been much appreciated in the industry and GSMA now owns and maintains the specification as GSMA PRD IR.92 "IMS Profile for Voice and SMS", often referred to as the VoLTE profile.

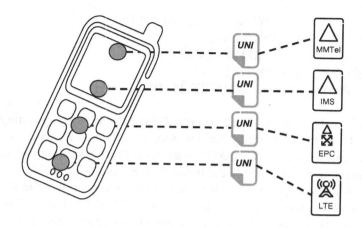

FIGURE 13.3

The GSMA PRD IR.92 IMS profile for voice and SMS, also known as the "VoLTE profile", covers all layers of the UNI.

As shown in Figure 13.3, this UNI profiling specification for voice telephony and SMS spans every layer of the network and therefore includes profiling of telephony service features, IMS features, media features, bearer management features, LTE radio features, and common functions such as the IP version. The profile is primarily based on 3GPP Release 8 specifications with some specific additions that are based on 3GPP Release 9.

The telephony service features often referred to as supplementary services that are included in this profile are:

- Originating Identification Presentation 3GPP TS 24.607
- Terminating Identification Presentation 3GPP TS 24.608
- Originating Identification Restriction 3GPP TS 24.607[1]
- Terminating Identification Restriction 3GPP TS 24.608[1]
- Communication Forwarding Unconditional 3GPP TS 24.604[1]
- Communication Forwarding on Not Logged in 3GPP TS 24.604[1]
- Communication Forwarding on Busy 3GPP TS 24.604[1]
- Communication Forwarding on Not Reachable 3GPP TS 24.604[1]
- Communication Forwarding on No Reply 3GPP TS 24.604[1]
- Barring of All Incoming Calls 3GPP TS 24.611[1]
- Barring of All Outgoing Calls 3GPP TS 24.611[1]
- Barring of Outgoing International Calls 3GPP TS 24.611
- Barring of Outgoing International Calls – ex Home Country 3GPP TS 24.611

---

[1] Specific implementation options profiled within the 3GPP supplementary service specifications.

- Barring of Incoming Calls – When Roaming 3GPP TS 24.611[1]
- Communication Hold 3GPP TS 24.610
- Message Waiting Indication 3GPP TS 24.606[1]
- Communication Waiting 3GPP TS 24.615[1]
- Ad-Hoc Multi Party Conference 3GPP TS 24.605.[1]

The list above is a subset of the 3GPP standardized MMTel supplementary services. It is selected to provide an IMS telephony service with a similar user experience as the circuit-switched service used today in 2G and 3G networks. Over time more of the standardized services are expected to be included in the defined interoperability profiles.

The profile mandates supplementary service self-management over the Ut reference point using XCAP procedures based on XML according to 3GPP TS 24.623. The use of Ut and XCAP enables the use of more complex and powerful expressions in the future than the legacy equivalent 3GPP TS 24.010 protocol, which is used today. Examples are the rules expressed in XML for black- or white-list based barring and time-dependent diversions, which are part of the full MMTel barring and diversion supplementary services. The minimum profile, however, uses only basic GSM-like, less complex conditions and actions.

The IMS features specifically mandated in the profile are:

- Support for IMS authentication and key agreement (IMS-AKA) and IMS subscriber identity module (ISIM). A universal SIM (USIM) may be used if the network service provider has not deployed ISIM.
- IPSec protection of signaling over the UNI.
- Both Tel URI and SIP URI addressing schemes shall be supported.
- SIM-based authentication of supplementary service self-management operations using XCAP over Ut according to the GBA architecture is recommended. However, the profile also allows the use of HTTP digest authentication.
- Support of "early SIP dialogs".
- IMS emergency procedures.
- Support of AMR narrowband and AMR wideband voice codecs.

The telephony service requirements put on bearer management and charging mandate that an LTE/Evolved Packet Core (EPC) system used for MMTel according to the VoLTE profile must be equipped with flow-based QoS and charging control functions. These enable setup and use of suitable 3GPP Evolved Packet System (EPS) bearers for the telephony service. The characteristics of these bearers are signaled via a QoS Class Identifier (QCI). Two standardized classes are used for VoLTE flows; QCI = 1 – a guaranteed bit-rate bearer for VoIP media; and QCI = 5 – a high-priority non-guaranteed bit-rate bearer for IMS SIP signaling and XCAP.

In conjunction with the IR.92 work in GSMA, the main industry players have concluded that the IMS roaming architecture with P-CSCF in the visited network shall be used for

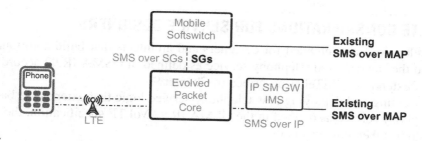

FIGURE 13.4

The two alternatives for transporting SMS over LTE.

VoLTE traffic. The use of this architecture enables the visited network to be service aware, which is required for lawful intercept and service-based revenue sharing, and gives the possibility to later provide optimized media routing. Furthermore, GSMA have agreed that an APN with a well-known name, i.e. "IMS", shall be used for IMS signaling and media by VoLTE UEs.

The IR.92 profile also includes an annex targeting today's 3GPP network operators, which operates a CS telephony service on GSM or UTRAN. The annex profiles 3GPP Single Radio Voice Call Continuity (SRVCC) for use in a VoLTE context. SRVCC is a handover mechanism between IMS telephony over Evolved Packet Core access to CS access in networks lacking sufficient voice-capable LTE coverage. It is described in more detail later in this chapter.

The existing SMS service will also be used over LTE. It supports the same existing SMS features as in 2G and 3G: concatenated SMS, EMS (small pictures, sounds, animations, and formatted text), notification SMS, configuration SMS, etc. The message content and size is limited to what the SMS service provides. Existing SMS in 2G/3G uses DTAP signaling over the CS access to transport the SMS to and from the device. The 3GPP standardization has devised two new mechanisms to transport the SMS over the LTE access instead of using signaling according to 3GPP TS 24.008 over the CS access: SMS over SGs and SMS over IP (see Figure 13.4).

In SMS over IP, the SMS is transported in a SIP MESSAGE over the PS access via a new logical entity called IP-SM-GW. The IP-SM-GW is located in IMS on the ISC interface as an application server and communicates with the existing SMS-C and HSS to perform domain selection, i.e. routing the message through IMS or MSC depending on where the targeted user is camping.

In SMS over SGs, the SMS is then sent via a new interface (SGs) to the MME in the evolved packet core and from there is transported to the terminal using existing Evolved Packet Core – Network Attach signaling. The SMS over SGs method allows the SMS to be sent via the MSC in the visited domain as today according to existing CS roaming relations.

The VoLTE profile includes support for both these methods.

13.4 VoLTE CONSIDERATIONS FOR SERVICE DESIGNERS

These considerations are intended for designers and architects that build additional service logic around the standardized telephony service as defined in GSMA IR.92 according to the principles of extending MMTel as described in Chapter 9.

When designing and deploying new telephony service logic that extends the basic standardized telephony service as defined in the GSMA IR.92 VoLTE specification, the following points should be taken into account:

- The supplementary service set defined in the GSMA IR.92 specification is a minimum set agreed as good enough when introducing IMS telephony in the mobile networks. This set is very likely to be extended in future industry agreements, at least to the ones already standardized in 3GPP.
- Special care should be taken when considering adding new service logic in the ISC chain before or after the MMTel AS, since the set of standardized supplementary services it provides is likely to grow and be modified with future releases. The reason is that when extending MMTel by ISC chaining, you build around a defined logic. When this defined logic is upgraded and modified to include more supplementary services, your extension logic may need to be updated as well.

13.5 SINGLE RADIO VOICE CALL CONTINUITY (SRVCC)

It is expected that most deployments of LTE, which has PS-only access, is to be done by operators that already operate a network with CS voice services. Many of these will initially introduce LTE in parts of the network and rely on underlying CS coverage when leaving PS voice-capable LTE coverage. Targeting such a deployment model SRVCC is defined as the procedures to hand over an ongoing IMS telephony call in LTE to CS access over GSM/UMTS to avoid dropping the calls, according to the scenario shown in Figure 13.5.

A single radio UE uses one radio transmitter/receiver for one radio access technology at a time, e.g. GSM, UMTS, or LTE. A Single Radio Voice Call Continuity (SRVCC) handover is performed when the UE is moving out of IMS voice-capable coverage in order to avoid dropping the call.

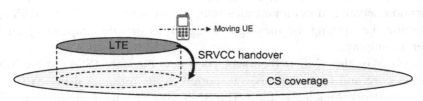

FIGURE 13.5

SRVCC handover scenario.

SRVCC covers the actual handover procedures, remaining mid-call procedures, and call termination procedures of an ongoing call. The handover case that is standardized is when moving outside of "IMS voice-capable PS coverage" in LTE or 3G with HSPA access during an ongoing call and handing over to CS in GSM or UMTS. In SRVCC procedures it is assumed that the UE has a single radio capable of using multiple radio access technologies but only one at a time. It is anticipated that SRVCC will pave the way for introducing IMS Centralized Services in the networks in order to provide a consistent service behavior both in LTE coverage and when making calls over the CS access.

13.5.1 SRVCC Architecture in 3GPP Release 9

The Single Radio Voice Call Continuity (SRVCC) architecture has been developed over a number of 3GPP releases. The first version of SRVCC was completed in 3GPP R9; its architecture is shown in Figure 13.6. The figure shows the nodes and functions used to support handover and re-establishment of voice telephony. Handling of non-telephony packet data is not shown.

The SRVCC solution impacts the UE, eNodeB, MME, HSS, and MSC server, and requires service continuity support in the IMS domain; the SCC AS is where the call is anchored at IMS level.

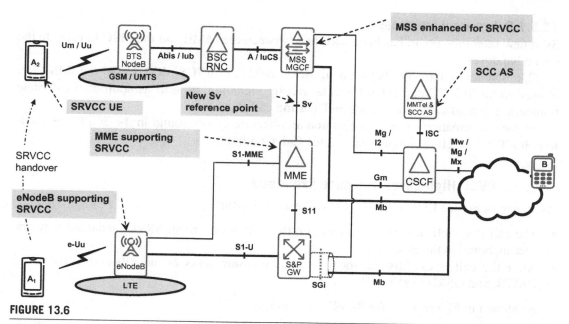

FIGURE 13.6

Functions and nodes is the network supporting SRVCC according to 3GPP R9.

In addition to the handover between MME and MSC, SRVCC is a transfer on the IMS level between access legs. This is handled by the Service Centralization and Continuity Application Server (SCC AS). A special number, the Session Transfer Number for SRVCC (STN-SR), is used to set up the call leg from the MSC executing SRVCC to the SCC AS. In order to determine that access legs are correlated, i.e. different call legs go to the same UE, a specific associated identity is used called Correlation MSISDN (C-MSISDN).

13.5.1.1 Main Nodes with SRVCC Functionality

The Service Centralization and Continuity application server (SCC AS) is a new logical IMS application server that for SRVCC is used primarily as a call signaling anchor when switching between the access legs. It acts as a SIP back-to-back user agent (B2BUA), maintaining service continuity by sitting in the signaling paths and executing the transfer between the access legs. The SCC AS is located on the ISC reference point.

The MSC needs to support SRVCC and the Sv reference point. SRVCC handover of a single active call can be done without support of the new SIP-based MSC interface for SRVCC and ICS I2, i.e. by using an MGCF for CS/IMS interworking. However, the I2 interface is required to support SRVCC handover of calls in alerting state, held calls, and calls in conference state.

The Mobility Management Entity (MME) in the Evolved Packet Core architecture needs to support SRVCC by indication of SRVCC capability, receiving SRVCC indication from eNodeB, supporting the handover via the new reference point Sv, and should include a bearer splitting function.

13.5.1.2 Important Interfaces

Sv, a new reference point, has been specified between the MME and the MSC to execute the actual handover.

I2, a new reference point between an MSC enhanced for ICS/SRVCC and the IMS core, is used in an SRVCC context for access transfer signaling, as well as additional call state transfers, e.g. held calls, conference and alerting call states.

Further information on the concept and architecture can be found in the 3GPP specification 3GPP TS 23.216 – Single Radio Voice Call Continuity.

13.5.2 SRVCC High-Level Use-case Explained

As shown in Figure 13.7, with SRVCC call establishment:

- The call is established while camping on the LTE access in an SRVCC-enabled network and anchored in the SCC AS.
- After the call is established, the UE sends measurements on neighboring cells (LTE, UMTS, and GSM) to eNodeB.

As shown in Figure 13.8, for SRVCC execution:

- When moving out of LTE coverage, the eNodeB decides to initiate SRVCC handover to GSM.

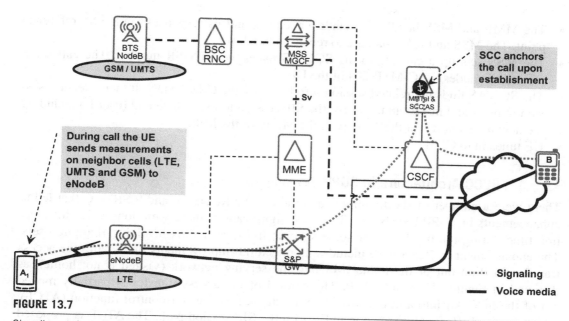

FIGURE 13.7

Signaling and media flows for SRVCC call establishment according to 3GPP R9.

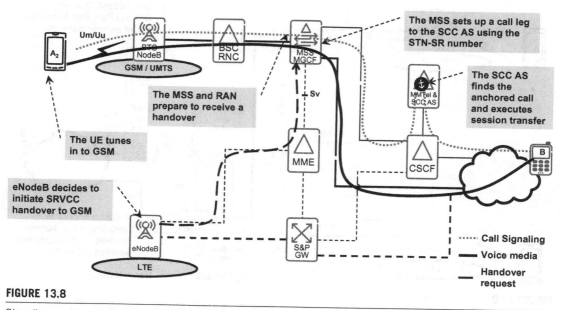

FIGURE 13.8

Signaling and media flows for SRVCC execution according to 3GPP R9.

- The MME and MSS handle the handover request and its responses via the Sv reference point. The MSS and RAN prepare to receive a handover.
- The MSS sets up a call leg to the SCC AS using the STN-SR number. The call setup message includes the C-MSISDN of the UE.
- The SCC AS finds the related session info by using the C-MSISDN. It then executes session transfer and makes an update of the SIP session to the remote end (user B) including the new media ports for the A-party in the MGW of the MSS.
- UE tunes in to GSM.

13.5.3 SRVCC Architecture in 3GPP Release 10

There were a number of shortcomings identified in the Release 8 and 9 SRVCC that led to enhancements in 3GPP Release 10. The most significant of these were long voice interruption time during the handover for some specific traffic scenarios involving roaming users. The enhancement in Release 10 includes architectural improvements by moving the access transfer function out of the home IMS into the serving network (visited if not home), as depicted in Figures 13.9 and 13.10. The control of the access transfer is partially moved out of the SCC AS into a new fictional entity, the access transfer control function (ATCF), which acts as SIP signaling anchor and sits on the SIP session path. The ATCF is proposed to be co-located with the P-CSCF.

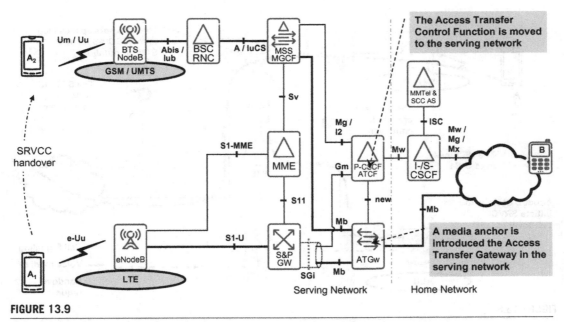

FIGURE 13.9

Functions and nodes in the network supporting SRVCC according to 3GPP R10.

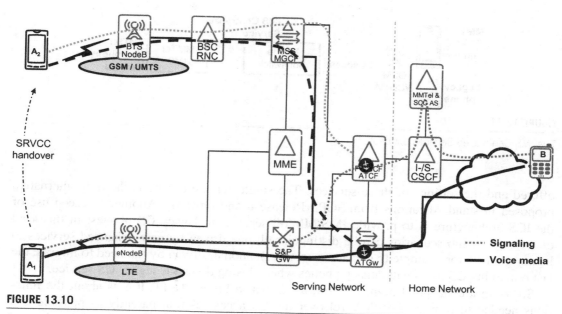

FIGURE 13.10

Signaling and media flows for SRVCC according to 3GPP R10.

The real performance gains were achieved by introducing a media plane function in the served network, the access transfer gateway (ATGW), controlled by the ATCF. Both ATCF and ATGW are needed to ensure that a session transfer call leg sent by an SRVCC-enhanced MSC server in the serving network does not need to be routed to the home network and that a potentially slow remote end update due to long distance is not needed. The access transfer functions ATCF and ATGw are located in the serving network.

Further information on the concept and architecture can be found in the 3GPP specification 3GPP TS 23.237 – IMS Service Continuity Release 10.

Two other major improvements of SRVCC are also included in 3GPP Releases 9 and 10: support of handover of calls in a mid-call state and support of handover of calls in alerting state. Both of these require use of the I2 interface from the anchor MSC.

13.6 IMS CENTRALIZED SERVICES (ICS)

The prime architecture and functional driver for IMS Centralized Services (ICS) is to provide the means to use the IMS/MMTel service over the GSM/UTRAN CS access in conjunction with SRVCC to obtain consistent service behavior during the rest of the call and for subsequent calls when camping on CS. It was concluded in 3GPP that a single service engine is a better solution for providing consistent service behavior compared with two service engines, one for CS and one for IMS, which need to be synchronized for the services

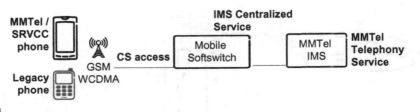

FIGURE 13.11

The IMS Centralized Services concept.

offered and the settings of these services. The single service engine is the only alternative proposed in standardizations from 3GPP Release 8 and onwards. Another foreseen use of the ICS architecture is to provide IMS/MMTel services to legacy CS phones. In this kind of usage, two sub-scenarios can be distinguished, one driven by new enhanced services to legacy phones, for example (e.g. centrex services), another driven by the need to migrate the last remaining CS users with legacy phones when closing down the legacy CS service.

So, from a conceptual point of view, as shown in Figure 13.11, ICS is about the functions needed to provide IMS/MMTel over the CS access as transparently as possible and provide functionality in IMS to select the right access path (PS or CS) to the terminal when receiving calls. The functionality in IMS for selection of access path to the UE for terminating calls is called terminating access domain selection (T-ADS) and is located in the logical node SCC AS.

The 3GPP standardization as of Release 9 has not come to a conclusion, but left four alternatives. These are:

- Network-based solution with enhanced MSC.
- Network-based solution using existing ISUP and CAMEL.
- UE-based solution based on SIP over Gm, relying on simultaneous CS and PS connectivity always being present.
- UE-based solution based on compressed SIP signaling being sent over USSD.

Coinciding in time with the GSMA VoLTE profiling work during the autumn of 2010, the industry has finally concluded that the network-based solution with enhanced MSC is the preferred one and is considered to be the target in networks providing an MMTel/ICS service to its own users. The main reason for this is that it is the only solution that supports service centralization before, during, and after SRVCC across all required radio access technologies (LTE, UTRAN, and GSM). It should be noted that the enhanced MSC solution still enables the use of Gm over the PS access for other services, e.g. for RCS while camping on 3G and 2G. If concurrent use of theses services with MMTel is required, UMTS or GSM DTM is needed.

When MMTel/ICS users are roaming in networks not enhanced for ICS, the network solution based on ISUP and CAMEL is anticipated to be used.

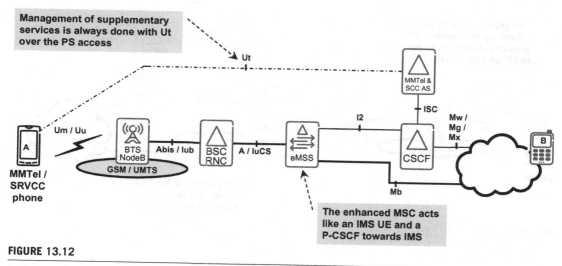

FIGURE 13.12

Anticipated main ICS solution for networks with ICS and SRVCC.

In scenarios for IMS Centralized Services, two classes of UE are discussed. The first is the MMTel UE that also supports MMTel over the CS access enhanced for ICS and the second is legacy CS telephony only with UE that is not MMTel aware. The MMTel UE is expected to use Ut, the XCAP-based interface for subscriber service self-management defined in TS 24.623 (e.g. changing the forwarding number in the communication forwarding supplementary service) also when camping on the CS access.

Further information on these concepts and architectures can be found in the 3GPP specification 3GPP TS 23.292 – IMS Centralized Services.

13.6.1 ICS Solution with Evolved MSC

In this solution, illustrated in Figure 13.12, the MSC is evolved so that it acts like a multiple SIP terminal gateway for the ICS-enabled user devices that are served by a particular MSC. The evolved MSC server maps between 3GPP TS 24.008 signaling that is sent over the UNI to 3GPP TS 24.229 signaling (SIP) to the IMS and register in IMS on behalf of the UE so that a CS attached terminal can use IMS service procedures for call control according to IMS/MMTel principles used by a SIP device over the Gm reference point. This allows a very high degree of service behavior alignment between MMTel over CS and MMTel over PS. Registration, alerting, call establishment, mid-call and call termination procedures and behaviors can all be supported over the CS access with this architecture.

The MSC is enhanced and supports ICS function and I2.

The network solution with evolved MSC, as shown in Figure 13.13, is the only solution that gives appropriate support for legacy CS phones, including a possible translation of CS service management using 3GPP TS 24.010 procedures to XCAP over the new I3 interface.

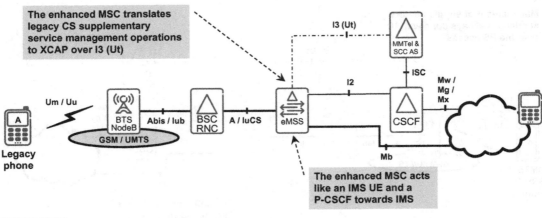

FIGURE 13.13

Supporting legacy phones using enhanced MCS solution for ICS.

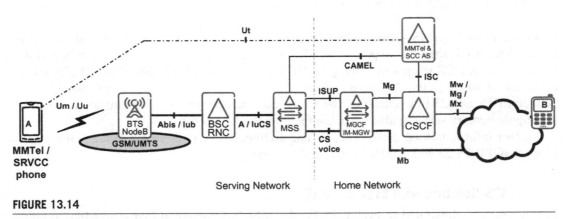

FIGURE 13.14

ICS implemented using a non-enhanced MSC with CAMEL and ISUP/Mg.

In this solution alternative, the UE is not required to have any specific ICS functions. Hence this architecture can also be used in scenarios where legacy CS telephony UE is used with an IMS service engine.

13.6.2 ICS Solution Using Existing ISUP/Mg and CAMEL

This solution is anticipated to be the typical alternative when serving inbound roamers in networks that are not yet enhanced for their own users and in visiting networks where no IMS-based roaming agreement to the home network exists. Figure 13.14 depicts the architecture principle for ICS implemented using CAMEL and ISUP/Mg.

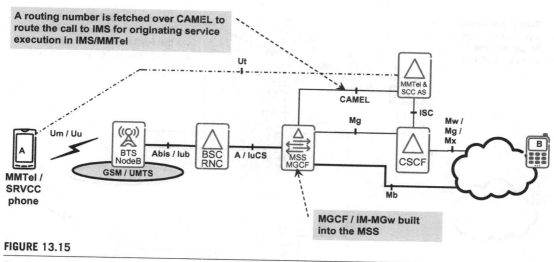

FIGURE 13.15

Routing of MSC originated calls to the IMS service domain using CAMEL.

This solution is based on existing inter-operator signaling mechanisms CAMEL and ISUP call control. CAP, the mobile inter-operator IN protocol, is used to force calls originated over CS access to be routed via the home IMS service engine in order to apply originating supplementary services, as illustrated in Figure 13.15.

13.6.3 Terminating Access Domain Selection (T-ADS)

In ICS, all terminating calls are routed to IMS and the IMS includes the terminating access domain selection (T-ADS) in the SCC AS so that calls can be delivered to the UE over the access best suited for the coverage situation of the user location. If the UE is IMS registered the SCC AS needs to query the serving MME or SGSN to determine whether the UE is in PS coverage that supports "IMS voice" or not, as illustrated in Figure 13.16. The scenario in Figure 13.16 is that UE is IMS registered but not in IMS voice-capable PS coverage, since it is attached over UMTS access (in this example not supporting IMS voice over PS).

According to the 3GPP specifications the SCC AS that includes the T-ADS should be the last logical application function on the terminating side. If the SCC AS is invoked over ISC it must be the last in the ISC chain.

13.7 SRVCC AND ICS CONSIDERATIONS FOR SERVICE DESIGNERS

The migration mechanisms for telephony, SRVCC, and IMS Centralized Services are anticipated to be used primarily in a voice telephony context. Therefore service designers need to

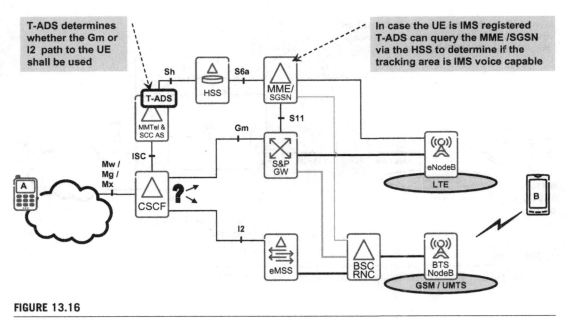

FIGURE 13.16

T-ADS functions in the network.

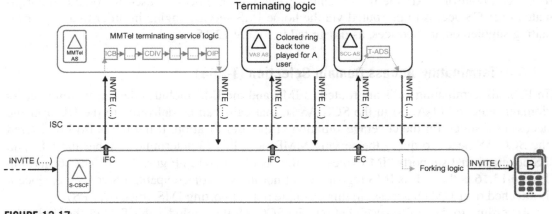

FIGURE 13.17

An AS providing "colored ring back tone" linked in between MMTel Service logic and the T-ADS function.

understand these mechanisms and how they use the IMS architecture to work with services extending standard telephony, also known as value-added service on telephony.

As explained above, the SCC AS provides the major part of the IMS functionality relating to SRVCC and ICS. Since it acts like an anchor for SRVCC in Release 9 and as a terminal path selector in the T-ADS function for ICS, it must be located closest to the terminal if there are several AS entities in an ISC chain. This means that it has to be first on the originating side and last on the terminating side.

Figure 13.17 shows an example where a value-added service "Colored ringback tone" is added in the ISC chain between MMTel terminating service logic and T-ADS logic in the SCC AS.

When designing and deploying new telephony service logic that extends the basic standardized telephony service in networks where IMS Centralized Services are in use, the following points should be taken into account:

- In the case of network-based ICS (expected to be predominant in the networks), the service logic in the network and device is constrained by the CS signaling protocol 3GPP TS 24.008.
- No service logic may be placed after the SCC AS in the ISC chain on the terminating side.
- No service logic may be placed before the SCC AS in the ISC chain on the originating side if SRVCC Release 9 is used.
- Special care should be taken in deployments where there is forking to multiple terminals in combination with T-ADS, and where the new service intends to influence the terminal selection.

Future Outlook: Market and Technology

14

14.1 WHAT IS NEXT IN STORE FOR IMS?

As we have discussed before, IMS is essentially the way the telecommunications industry has agreed to embrace IP technology, choosing SIP as the basic control machinery, wrapping it in an architecture that delivers the multi-operator and multi-vendor properties expected, and ensuring that it allows efficient interoperability with the existing networks supporting billions of legacy devices. It is in this light we can see IMS as being the basis for the next generation of fixed telephony networks (replacing aging switching and access technology), as well as being the natural choice for voice services over LTE (known as VoLTE; fundamentally based on IP paradigms). Thus, IMS is very likely to sneak up on us, all of a sudden being there as a natural part of the technology background that we will take for granted, just as circuit-switched telephony has been for over a century. Having said that, what are the applications that we can imagine will be supported by IMS in the near to somewhat more distant future? Here follow a few suggestions. We certainly hope – indeed, that is really one of the major reasons for writing this book – that you will create more!

14.2 TV

The Open IPTV Forum has created a number of IPTV profiles, one of them being based on IMS. Currently, the approach is more device (TV, set-top box) oriented, but as the TV becomes more interactive and personal, the intrinsic capabilities of IMS (security, identification, multi-device features) can also be brought to bear on this domain. In particular, this will apply to three-screen TV (TV set, PC, mobile device).

14.3 SMART PIPES

Bandwidth requirements keep going up (witness higher and higher definition displays, higher frame rates, 3D, etc.); also, some applications may not be as bandwidth hungry but require guaranteed low latency on request (remote surgery, for instance). Thus, the ability to declare QoS requirements and to ensure that the requested performance is delivered to the right end-points is crucial. The positive identification and session negotiation mechanisms built into IMS form an excellent foundation for these kinds of applications.

14.4 HOME NETWORKS

The home network of today is a rather complicated affair, showing no tendencies to slow its rate of development. An IMS gateway on the edge of the home network can be the ideal demarcation point between the outside network controlled by the operator and the inside network managed by autoconfiguring architectures like DLNA. Thus, the contract with the upstream operator(s) will be in terms of delivered services; the inside of the home network need not be known to the outside world.

14.5 WEB CLIENTS

Various initiatives are under way to reduce the need for middleware support. This will make it easier for the developer to incorporate IMS technology; as we discussed before, HTML5 and similar developments will also help bring media-related functionality closer to the web-savvy developer. Additionally, the advent of the RESTful IMS enabler APIs currently being designed by OMA – as requested by the RCS industry committee – will also help developers making creative applications on the basis of the RCS components: IMS chat, file transfer, presence, and streaming live or pre-stored video and/or images during calls.

14.6 MACHINE TO MACHINE (M2M)

It has been estimated that the next order of magnitude growth in communications usage will be when more devices are connected, possibly going from today's close to 5 billion up to 50 billion connected devices in just a few years. Finding and connecting to all these devices securely will be a challenge; IMS can in our view play an important part here.

14.7 VEHICLE AUTOMATION

A very interesting combination of the home network and the M2M scenario is when a vehicle (e.g. the family car) becomes connected. Secure connections for the car to talk to the maintenance system will be vital (one frequently recurring idea is to rent and download power upgrades on demand), and access to the home network for video and music playout will be very convenient. In all such cases, IMS properties like positive identification and the ability to secure resources come in very handy.

14.8 WAC AND OTHER APP STORES

With the advent of app stores, fuelled by initiatives like the Wholesale Applications Community (WAC), some of these applications will depend on communications capabilities.

The operator(s) providing them will provide the APIs to achieve this; the application developers will demand commonality in these APIs to avoid having to build special variants for every network. As usual, profitability depends on volumes, in particular in markets where the price for an app is expected to be low (or zero, in the case of ad-financed apps). To achieve volumes, specialization must be avoided.

14.9 SECURE, NON-ANONYMOUS COMMS: THE ALTERNATIVE NETWORK

Finally, perhaps IMS is the way to provide what the world really needs: a dependable network where spammers and hackers cannot be anonymous, where resource consumption can always be tracked and accounted for (and therefore misuse and abuse is discouraged), where revenue share can be managed (and therefore working business models can be implemented), and where privacy can be upheld and controlled by the end-user. The open Internet will always be there, being the creative and open-ended environment it has proven to be, but maybe we would all benefit from having a part of it dedicated to communications we can trust?

14.10 CONCLUSION

We have taken you on a journey from the basic premises of IMS – technical as well as business oriented – through some details of how services work on IMS, via how you can actually get money out of working with IMS applications, further through the evolution of IMS, and then to some thoughts on where it might go from here. We hope you have enjoyed the ride, we wish you the best of luck with your IMS applications, and we sincerely hope to be able to use them in the future!

References

For an overview of specifications and standards related to SIP and IMS, please see the companion website: www.elsevierdirect.com/companions/9780123821928, Appendix C.

Article 19, Human Rights Convention.

Boland, R., & Collopy, F. (2004). *Managing as Designing*. Stanford, CA: Stanford University Press.

Bond, G., Cheung, E., Fikouras, I., & Levenshteyn, R. (2009). Unified telecom and web services composition: problem definition and future directions. *Proceedings of the 3rd International Conference on Principles, Systems and Applications of IP Telecommunications*. New York, ACM (ISBN 978-1605587677).

Campbell-Kelly, M. (2003). *From Airline Reservations to Sonic the Hedgehog: A History of the Software Industry*. Cambridge, MA: MIT Press.

Chesborough, H. (2010). Business Model Innovation: Opportunities and Barriers. *Long Range Planning, 43*, (2).

Chesborough, H., & Rosenbloom, R. (2002). The role of the business model in capturing value from innovation: evidence from xerox corporation's technology spin-off companies. *Industrial and Corporate Change, 11*(3), 529–555.

Christensen, C. M. (2001). *The Innovator's Dilemma*. New York: Harper Business (ISBN 978-0060521998).

Churchman (1967). [Details?]

Coase, R. (1937). The nature of the firm. *Economica, 4*(16), 386–405.

Dahlman, E., Parkvall, S., & Skold, J. (2008). *4G: LTE/LTE-Advanced for Mobile Broadband*. Oxford: Elsevier (ISBN 978-0123854896).

Dunne, D., & Martin, R. (2006). Design thinking and how it will change management education: an interview and discussion. *Academy of Management Learning and Education, 5*(4), 512–523.

Economides, N. & Katsamakas E. (2006). Two-sided competition of proprietary vs. open source technology platforms and implications for the software industry.

Eriksson, H. (2010). Ericsson CTO. CTIA presentation.

Galbraith, J. K. (1958). *The Affluent Society*. New York: Mariner Books (ISBN 978-0395925003).

Galbraith, J. K. (Year?). *A History of Economics: The Past as the Present*, Penguin, UK (ISBN 978-0140153958).

Garlan, D., & Shaw, M. (1993). An Introduction to software architecture. In V. Ambriola & G. Tortora (Eds.), *Advances in Software Engineering and Knowledge Engineering*, Vol 2. Singapore: World Scientific, pp. 1–39.

Jackson, M., & Zave, P. (October 1998). Distributed feature composition: a virtual architecture for telecommunications services. *IEEE Transactions on Software Engineering, 24*(10), 831–847.

Johnson, M., Christensen, C., & Kagermann, H. (2008). Reinventing your business model. *Harvard Business Review*, December 51–59.

Ljungberg (2010). [Details?]

Martin, R. (2010). *The Design of Business: Why Design Thinking is the Next Competitive Advantage*. Cambridge, MA: Harvard Business School.

McKinsey, Malik, Y., Niemeyer, A., & Ruwadi, B. (2011). Building the Supply Chain of the future, McKinsey Quarterly.

Mulligan, C. (2011a). The Emerging Economics of Open Mobile APIs; http://blog.programmableweb.com/2011/01/27/the-emerging-economics-of-open-mobile-apis

Mulligan, C. (2011b). *The Communications Industries in the Era of Convergence*. Oxford: Routledge (ISBN 978-0415584845).

Naughton, J. (2010). The internet: everything you ever need to know. *The Observer, 20 June*.

Niemöller, J., Fikouras, I., de Rooij, F., Klostermann, L., Stringer, U., & Olsson, U. (2009). Ericsson Composition Engine. Available from: http://www.ericsson.com/ericsson/corpinfo/publications/review/2009_02/files/NGIN.pdf

Nolan, P. (2009). *Crossroads: The End of Wild Capitalism*. UK: Marshall Cavendish (ISBN 978-0462099682).

Noldus, R. (2006). *CAMEL, Intelligent Networks for the GSM, GPRS and UMTS Network*. USA: Wiley (ISBN 978-0470016947).

Olsson, M., Sultana, S., Rommer, S., Frid, L., & Mulligan, C. (2009). *SAE and the Evolved Packet Core: Driving the Mobile Broadband Revolution*. UK: Elsevier (ISBN 978-0123748263).

Oracle Archives (1998). Available from: http://www.oracle.com/us/corporate/history/index.html

Pandza, K., & Thorpe, R. (2010). Management as design, but what kind of design? An appraisal of the design science analogy for management. *British Journal of Management*, 21(1), 171–186.

Penrose, E. (1995). *The Theory of the Growth of the Firm*. Oxford: Oxford University Press.

Rasmussen, J. (1990). *A Taxonomy for Cognitive Work Analysis*. Norway: Riso National Laboratory. Report.

Smith, T. and Bond, G. (2007). Echarts for SIP Servlets: A State-Machine Programming Environment for VoIP Applications. *Proceedings of the 1st International Conference on Principles, Systems and Applications of IP Telecommunications*, pp. 89–98. New York, ACM (ISBN 978-1605580067).

Tuffin, S., & Jestin, J. (2008). *Feedbacks from a Large-Scale Pre-IMS Deployment*. ICIN.

Vargo, S., & Lusch, R. (2008). *From Goods to Services: Divergences and Convergences of Logics*. Industrial Marketing Management, Vol 37(3), pp. 254–259.

Verwaayen, B. (2010). Alcatel-Lucent, Q3 2010 results.

Vestberg, H. (2010). Ericsson CEO, Q3 2010 results.

Appendix A

A.1 VIRTUAL CALL CENTER USE-CASE IMPLEMENTATION

This section provides a complete source code for the classes according to the JSR289 API as sketched and described in chapter 5.

A.1.1 Implementation of the Announcement Component

A.1.1.1 Class AnnouncementB2B

```java
package com.ericsson.services;

import java.io.IOException;
import java.util.logging.Logger;

import javax.annotation.Resource;
import javax.servlet.ServletException;
import javax.servlet.sip.Address;
import javax.servlet.sip.SipFactory;
import javax.servlet.sip.SipServlet;
import javax.servlet.sip.SipServletRequest;
import javax.servlet.sip.SipServletResponse;
import javax.servlet.sip.SipSession;
import javax.servlet.sip.SipURI;
import javax.servlet.sip.SipSession.State;
import javax.servlet.sip.ar.SipApplicationRoutingDirective;

@javax.servlet.sip.annotation.SipServlet
        (applicationName = "AnnouncementB2B",
            name = "AnnouncementB2B",
            loadOnStartup = 1)
public class AnnouncementB2B extends SipServlet {

    private static final long serialVersionUID = 1L;
    private static final String SERVER_RESPONSE = "server_response";
    // URI to be used for routing SIP packets via CSCF
    private static final String CSCFADDRESS = "127.0.0.1:5081;lr;call=term:";
    private static final Logger LOGGER = Logger
                .getLogger("com.ericsson.msa.services.AnnouncementB2B");

    // Use resource injection to obtain a SipFactory object
    @Resource
    private SipFactory sipFactory;

    /***
     * Receive SIP INVITE packets and forward them
     */
    @Override
    protected void doInvite(SipServletRequest aClientRequest)
                throws ServletException, IOException {
```

```
        LOGGER.info("Recieved INVITE");
        doRequestForwarding(aClientRequest);
}

/**
 * Helper method for request forwarding
 */
protected void doRequestForwarding(SipServletRequest req)
            throws ServletException, IOException {
    SipServletRequest serverRequest;
    if (req.isInitial()) {
req.getSession().setAttribute("role", "incoming");
req.getSession().setAttribute("status", "active");

// create outgoing leg
serverRequest = sipFactory.createRequest(
            req.getApplicationSession(), req.getMethod(),
            req.getFrom(), req.getTo());

if (req.getContentLength() > 0)
        serverRequest
                    .setContent(req.getContent(), req.getContentType());

serverRequest.setRoutingDirective(
            SipApplicationRoutingDirective.CONTINUE, req);
SipURI routeURI = sipFactory.createSipURI(null, CSCFADDRESS);
routeURI.setLrParam(true);
serverRequest.pushRoute(routeURI);

serverRequest.getSession().setAttribute("role", "outgoing");
// Associate outgoing request with the incoming request
serverRequest.getSession().setAttribute("linkedrequest", req);
serverRequest.getSession().setAttribute("status", "inactive");
req.getSession().setAttribute("outgoing", serverRequest);

// create announcement leg
String dest_addr = req.getHeader("new_dest");
Address new_addr;
if ((dest_addr != null) && (dest_addr.length() > 0)) {
    if (!dest_addr.startsWith("sip"))
            dest_addr = "sip:" + dest_addr;
    new_addr = sipFactory.createAddress(dest_addr);
    // Create annoucement request
    SipServletRequest new_request = sipFactory.createRequest(req
                .getApplicationSession(), req.getMethod(), req
                .getFrom().getURI(), new_addr.getURI());
    if (req.getContentLength() > 0)
            new_request.setContent(req.getContent(),
                            req.getContentType());
    new_request.addHeader("x-ac-composition-directive",
                    "no-composition");
```

```
                new_request.addHeader("x-ac-leg", "announcement-leg");
                new_request.setRequestURI(sipFactory.createURI(dest_addr));
                // Route via the CSCF
                new_request.pushRoute(routeURI);
                new_request.getSession().setAttribute("role", "announcement");
                new_request.getSession().setAttribute("status", "active");
                // Associate announcement request with incoming request
                new_request.getSession().setAttribute("linkedrequest", req);
                req.getSession().setAttribute("announcement", new_request);
                // Associate incoming request with announcement request
                req.getSession().setAttribute("linkedrequest", new_request);
                new_request.send();
        }
    } else {
        LOGGER.info("Non-initial request:" + req.getMethod());
        SipServletRequest serverSession = (SipServletRequest) req
                    .getSession().getAttribute("linkedrequest");
        if (serverSession == null) {
            LOGGER.warning("No linked session found for request with callId: "
                            + req.getCallId());
            return;
        }
        serverRequest = serverSession.getSession().createRequest(
                    req.getMethod());
        LOGGER.info("create request:\n" + serverRequest);
}

            serverRequest.send();
            LOGGER.info("Created request:\n" + serverRequest
                            + "\n out of incoming request:\n" + req);
        }

        @Override
        protected void doProvisionalResponse(SipServletResponse aServerResponse)
                    throws ServletException, IOException {
            LOGGER.info("Recieved Provisional Response");
            doResponseForwarding(aServerResponse);
        }

        /**
         * Process call leg establishment confirmation (e.g. SIP 200 OK)
         */
        @Override
        protected void doSuccessResponse(SipServletResponse aServerResponse)
                    throws ServletException, IOException {
            LOGGER.info("Recieved Success Response: " + aServerResponse.getMethod()
                            + "/" + aServerResponse.getStatus());
            if (aServerResponse.getMethod().equalsIgnoreCase("INVITE")) {
                    String invitePurpose = (String) aServerResponse.getSession()
                        .getAttribute("role");
```

```
                    String status = (String) aServerResponse.getSession().getAttribute(
                        "status");
                    LOGGER.info("role " + invitePurpose + " status " + status);
                    if (invitePurpose != null
                            && invitePurpose.equalsIgnoreCase("announcement")) {
                        // Received OK for announcment leg
                        SipServletRequest ackRequest = aServerResponse.createAck();
                        ackRequest.send();
                        SipServletRequest linkedRequest = (SipServletRequest)
aServerResponse
                            .getSession().getAttribute("linkedrequest");
                    if (linkedRequest != null) {
                        SipServletResponse tSessionProgress = linkedRequest
.createResponse(SipServletResponse.SC_SESSION_PROGRESS);
                        tSessionProgress.setContent(aServerResponse.
                    getContent(),
                                    aServerResponse.getContentType());
                        tSessionProgress.send();
                    }

                    return;
        }
        if (invitePurpose != null
                    && invitePurpose.equalsIgnoreCase("outgoing")
                    && (status.equalsIgnoreCase("inactive"))) {
            // Received OK to outgoing request. user picked up. switch
        active session
            SipServletRequest initrequest = (SipServletRequest)
        aServerResponse
                        .getSession().getAttribute("linkedrequest");
            // kill announcement leg
            SipServletRequest announcement_request = (SipServletRequest)
        initrequest
                        .getSession().getAttribute("announcement");
            if (announcement_request.getSession().getState() == State.
        CONFIRMED) {
                    // announcement is already playing
                    SipServletRequest bye = announcement_request.getSession()
                            .createRequest("BYE");
                    bye.send();
                    LOGGER.info("send BYE\n" 1 bye);
            } else { // announcement has not started yet
                    announcement_request.getSession().createRequest("CANCEL")
                            .send();
                    LOGGER.info("send CANCEL\n");
            }
            if (initrequest.getSession().getState() == State.CONFIRMED) {
                    // create reInvite for incoming leg
```

```
                    SipServletRequest reInvite = initrequest.getSession()
                                .createRequest("INVITE");
                    if (aServerResponse.getContentLength() > 0)
                            // copy SDP from OK to reINVITE
                            reInvite.setContent(aServerResponse.getContent(),
                                        aServerResponse.getContentType());
                    LOGGER.info("final ReINVITE\n" + reInvite);
                    reInvite.getSession().setAttribute("role", "reinvite");
                    reInvite.send();

                    // send ack for OK
                    aServerResponse.createAck().send();
                } else {
                    doResponseForwarding(aServerResponse);
                    aServerResponse.createAck().send();

                }
                // clean up & change status
                aServerResponse.getSession().setAttribute("status", "active");
                initrequest.getSession().setAttribute("linkedrequest",
                            aServerResponse.getRequest());
                initrequest.getSession().removeAttribute("announcement");
                announcement_request.getSession().invalidate();
                return;
            }

            if (invitePurpose != null
                        && invitePurpose.equalsIgnoreCase("reinvite")) {
                // Received 200 OK from reinviteLeg
                SipServletRequest ackRequest = aServerResponse.createAck();
                ackRequest.send();
                aServerResponse.getSession().setAttribute("role", "incoming");
                return;
            }
            doResponseForwarding(aServerResponse);
            // Store the incoming response, so we can ACK it later.
            aServerResponse.getSession().setAttribute(SERVER_RESPONSE,
                        aServerResponse);
        } else if (aServerResponse.getMethod().equalsIgnoreCase("BYE")) {
            LOGGER.finest("Received 200 OK to BYE. ");
            // 200 OK to BYE doesn't need to be forwarded.
        } else if (aServerResponse.getMethod().equalsIgnoreCase("CANCEL")) {
            LOGGER.finest("Received 200 OK to CANCEL");
            // 200 OK to CANCEL doesn't need to be forwarded.
        }

    }

    /**
     * Helper method for response forwarding
```

```
        */
    private void doResponseForwarding(SipServletResponse aServerResponse)
            throws ServletException, IOException {
    LOGGER.info("doResponseForwarding for "
                    + aServerResponse.getSession().getAttribute("role"));
    // Get the linked request
    SipServletRequest linkedRequest = (SipServletRequest) aServerResponse
                    .getSession().getAttribute("linkedrequest");
    if (linkedRequest != null) {
            if (((String) linkedRequest.getSession().getAttribute("status"))
                            .equalsIgnoreCase("active")) {
                    // If linked session is still active and not confirmed yet,
                    // propagate a response
                    if (linkedRequest.getSession().getState() != State.CONFIRMED) {
                            SipServletResponse clientResponse = linkedRequest
                                            .createResponse(aServerResponse.getStatus(),
        aServerResponse.getReasonPhrase());
                                    if (aServerResponse.getContent() != null) {

    clientResponse.setContent(aServerResponse.getContent(),
                                                    aServerResponse.getContentType());
                                    }
                                    LOGGER.info("Created response:\n" + clientResponse
                                                    + "\n out of incoming response:\n"
                                                    + aServerResponse);
                                    clientResponse.send();
                    } else {
                            LOGGER.info("session is aleady confirmed");

                    }
            } else {
                    LOGGER.info("session is not active. dont forward OK");
            }
    } else {
            LOGGER.warning("No linked client request found for response with
        callId: "
                            + aServerResponse.getCallId());

    }

    }

    /**
     * Process ACKs
     */
    @Override
    protected void doAck(SipServletRequest aClientRequest)
            throws ServletException, IOException {
        LOGGER.finest("Recieved ACK:\n" + aClientRequest);
        // Get linked request
        SipServletRequest linkedRequest = (SipServletRequest) aClientRequest
```

AQ1

```
                              .getSession().getAttribute("linkedrequest");
        if (linkedRequest != null) {
            SipServletResponse aServerResponse = (SipServletResponse)
          linkedRequest
                             .getSession().getAttribute(SERVER_RESPONSE);
            SipServletRequest serverRequest = aServerResponse.createAck();
            LOGGER.finest("Created ACK message: \n" + serverRequest
                            + "\n to response: " + aServerResponse);
            serverRequest.send();
        } else {
            LOGGER.warning("No server session found for request with callId: "
                            + aClientRequest.getCallId());
        }
    }

    /**
     * Termiate a SIP session correspondig to a given SIP request
     */
    private void terminateSession(SipServletRequest inRequest) {
        if (inRequest == null)
            return;
        SipSession tSession = inRequest.getSession();
        SipServletRequest terminateRequest = null;
        if (tSession.getState() == State.CONFIRMED)
            terminateRequest = tSession.createRequest("BYE");
        else if (tSession.getState() != State.TERMINATED)
            terminateRequest = inRequest.createCancel();
        if (terminateRequest != null)
            try {
                terminateRequest.send();
                tSession.setAttribute("terminateRequest", terminateRequest);
            } catch (IOException e) {
                terminateRequest = null;
                e.printStackTrace();
            }
    }
    private void terminateAllSessions(SipServletRequest aClientRequest) {
        String role = (String) aClientRequest.getSession().getAttribute("role");
        if (role.equalsIgnoreCase("incoming")) {
            terminateSession((SipServletRequest) aClientRequest.getSession()
                        .getAttribute("outgoing"));
            terminateSession((SipServletRequest) aClientRequest.getSession()
                        .getAttribute("announcement"));
            // } else if (role.equalsIgnoreCase("announcement")) { // actually
            // should not happen
            // SipServletRequest linkedrequest = (SipServletRequest)
            // aClientRequest.getSession().getAttribute("linkedrequest");
            // terminateSession(linkedrequest);
            // terminateSession((SipServletRequest)
```

```
                    // linkedrequest.getSession().getAttribute("outgoing"));
          } else { // bye from outgoing leg - end of session
                  SipServletRequest linkedrequest = (SipServletRequest) aClientRequest
                              .getSession().getAttribute("linkedrequest");
                  terminateSession(linkedrequest);
                  terminateSession((SipServletRequest) linkedrequest.getSession()
                              .getAttribute("announcement"));
          }
    }

    /**
     * Process BYE events
     */
    @Override
    protected void doBye(SipServletRequest aClientRequest)
               throws ServletException, IOException {
        aClientRequest.createResponse(SipServletResponse.SC_OK).send();
        terminateAllSessions(aClientRequest);
    }

    /**
     * Process CANCEL events
     */
    @Override
    protected void doCancel(SipServletRequest aClientRequest)
               throws ServletException, IOException {
        terminateAllSessions(aClientRequest);
    }

    /**
     * Process error responses
     */
    @Override
    protected void doErrorResponse(SipServletResponse aServerResponse)
               throws ServletException, IOException {
        LOGGER.info("Recieved Error Response");
        doResponseForwarding(aServerResponse);
    }
}
```

A.1.2 Implementation of the Call Transfer Component
A.1.2.1 Class CallTransferB2BFrontend
```
package com.ericsson.services;

import java.io.IOException;
import java.io.PrintWriter;
import java.util.Iterator;
import java.util.logging.Logger;
```

```java
import javax.annotation.Resource;
import javax.servlet.ServletConfig;
import javax.servlet.ServletException;
import javax.servlet.http.HttpServlet;
import javax.servlet.http.HttpServletRequest;
import javax.servlet.http.HttpServletResponse;
import javax.servlet.sip.B2buaHelper;
import javax.servlet.sip.SipApplicationSession;
import javax.servlet.sip.SipFactory;
import javax.servlet.sip.SipServletRequest;
import javax.servlet.sip.SipSession;
import javax.servlet.sip.SipSessionsUtil;
import javax.servlet.sip.SipURI;
import javax.servlet.sip.SipSession.State;
import javax.servlet.sip.ar.SipApplicationRoutingDirective;

/**
 * Servlet implementation class GenericB2Bfrontend
 */
public class CallTransferB2Bfrontend extends HttpServlet {
        private static final long serialVersionUID = 1L;
        private static Logger LOGGER = Logger
                        .getLogger(CallTransferB2Bfrontend.class.getName());

        // The Address for the CSCF, needed to Route Requests to.
        private static final String CSCFADDRESS = "10.42.161.249:5060";

        // http parameter names
        private static final String PARAM_CALL_ID = "callid";
        private static final String PARAM_NEW_DEST = "new_dest";

        @Resource
        private SipFactory sipFactory;

        @Resource
        SipSessionsUtil sipSessionsUtil;

        /**
         * Default constructor.
         */
        public CallTransferB2Bfrontend() {
        }

        @Override
        public void init(ServletConfig config) throws ServletException {
                super.init(config);
        }

        /**
         * @see HttpServlet#doGet(HttpServletRequest request, HttpServletResponse
         * response)
```

```
             */
         protected void doGet(HttpServletRequest httpReq, HttpServletResponse httpRes)
                   throws ServletException, IOException {
             PrintWriter out = httpRes.getWriter();

             SipApplicationSession tAppSession = sipSessionsUtil
                       .getApplicationSessionByKey("_MacLeod_", false);
             if (tAppSession == null || !tAppSession.isValid()) {
                   LOGGER.severe("Application session not found or invalid");
                   out.close();
                   return;
             }
             // check parameters
             String callid = httpReq.getParameter(PARAM_CALL_ID);
             if ((callid == null) || (callid.isEmpty())) {
                   out.println("error: callid parameters not found");
                   out.close();
                   return;
             }
             callid = callid.trim();

             String tNewDest = httpReq.getParameter(PARAM_NEW_DEST);
             if ((tNewDest == null) || (tNewDest.isEmpty())) {
                   out.println("error: new destionation for the transfer not found");
                   out.close();
                   return;
             }
             if (!tNewDest.startsWith("sip:"))
                   tNewDest = "sip:" + tNewDest;

             // get all sip session and search for session with specific call id
             Iterator<?> sipSessionsIter = (Iterator<?>) tAppSession
                       .getSessions("SIP");
             SipServletRequest initialRequest = null;
             SipSession tSipSession = null;
             while (sipSessionsIter.hasNext() && (initialRequest == null)) {
                   tSipSession = (SipSession) sipSessionsIter.next();
                   String tCallId = (String) tSipSession
                       .getAttribute(CallTransferB2B.LEG_CALL_ID);
                   if ((tCallId != null) && (tCallId.trim().equalsIgnoreCase
             (callid))) {
                         initialRequest = (SipServletRequest) tSipSession
                              .getAttribute(CallTransferB2B.INITIAL_REQUEST);
                         break;
                   }
             }
             B2buaHelper b2buahelper = initialRequest.getB2buaHelper();
             SipSession tLinkedSession = b2buahelper.getLinkedSession(tSipSession);
             SipServletRequest serverRequest;
             serverRequest = sipFactory.createRequest(tAppSession,
```

```
                                    initialRequest.getMethod(), initialRequest.getFrom(),
                                sipFactory.createAddress(tNewDest));
                    if (initialRequest.getContentLength() > 0)
                            serverRequest.setContent(initialRequest.getContent(),
                                    initialRequest.getContentType());
                    // new sip invite belongs to the same session
                    serverRequest.setRoutingDirective(
                                SipApplicationRoutingDirective.CONTINUE,
                            initialRequest);
                    serverRequest.setRequestURI(serverRequest.getTo().getURI());

                    b2buahelper.unlinkSipSessions(tLinkedSession);
                    b2buahelper.linkSipSessions(tSipSession, serverRequest.getSession());
                    serverRequest.getSession().setAttribute(CallTransferB2B.INVITE_PURPOSE,
                                CallTransferB2B.INVITE_PURPOSE_TRANSFER);

                    // Route via the CSCF
                    SipURI routeURI = sipFactory.createSipURI(null, CSCFADDRESS);
                    routeURI.setLrParam(true);
                    serverRequest.pushRoute(routeURI);
                    serverRequest.send();

                    // kill old leg
                    if (tLinkedSession != null && tLinkedSession.isValid()) {
                            if (tLinkedSession.getState() == State.CONFIRMED)
                                    tLinkedSession.createRequest("BYE").send();
                            else
                                    tLinkedSession.createRequest("CANCEL").send();
                    }
                    tLinkedSession.invalidate();
                    out.close();
            }
            /**
             * @see HttpServlet#doPost(HttpServletRequest request, HttpServletResponse
             * response)
             */
        protected void doPost(HttpServletRequest request,
                HttpServletResponse response) throws ServletException, IOException {
                    // TODO Auto-generated method stub
        }
    }
```

A.1.2.2 Class CallTransferB2B

```
package com.ericsson.services;

import java.io.IOException;
import java.util.HashMap;
import java.util.List;
import java.util.Map;
import java.util.logging.Logger;
```

```
import javax.servlet.ServletException;
import javax.servlet.sip.B2buaHelper;
import javax.servlet.sip.SipServlet;
import javax.servlet.sip.SipServletRequest;
import javax.servlet.sip.SipServletResponse;
import javax.servlet.sip.SipSession;
import javax.servlet.sip.annotation.SipApplicationKey;
import javax.servlet.sip.ar.SipApplicationRoutingDirective;

@javax.servlet.sip.annotation.SipServlet(applicationName = "CallTransferB2B", name =
"CallTransferB2BServlet", loadOnStartup = 1)
public class CallTransferB2B extends SipServlet {
        private static final long serialVersionUID = 1L;
        private static final Logger LOGGER = Logger.getLogger(CallTransferB2B.class
                        .getName());

        // some session's attribute names
        public static final String INITIAL_REQUEST = "InitialRequest";
        public static final String LEG_CALL_ID = "CallId";
        public static final String INVITE_PURPOSE = "invitePurpose";
        public static final String INVITE_PURPOSE_INCOMING = "incoming";
        public static final String INVITE_PURPOSE_OUTGOING = "outgoing";
        public static final String INVITE_PURPOSE_TRANSFER = "transfer";
        public static final String INVITE_PURPOSE_REINIVTE = "reinvite";

        private static final String LAST_SERVER_RESPONSE = "server_response";

        @Override
        protected void doInvite(SipServletRequest aClientRequest)
                        throws ServletException, IOException {
                doRequestForwarding(aClientRequest);
        }

        @SipApplicationKey
        public static String applicationSessionKey(final SipServletRequest request) {
                return "_MacLeod_"; // Highlander principle ... there can be only one
        }

        /**
         * @inheritDoc
         */
        protected void doRequestForwarding(SipServletRequest req)
                        throws ServletException, IOException {
                B2buaHelper b2buaHelper = req.getB2buaHelper();
        SipServletRequest serverRequest;
        if (req.isInitial()) {
                LOGGER.fine("Initial request:" + req.getMethod());

                // save initial request
                req.getSession().setAttribute(INITIAL_REQUEST, req); // make all

                        // legs
```

```
                          // symmetric
                  req.getSession().setAttribute(LEG_CALL_ID, req.getCallId());
                  req.getSession().setAttribute(INVITE_PURPOSE,
                               INVITE_PURPOSE_INCOMING);

                  Map<String, List<String>> newHeaders = new HashMap<String,
              List<String>>();
                  serverRequest = b2buaHelper.createRequest(req, true, newHeaders);

                  // new sip request belongs to the same composition session
                  serverRequest.setRoutingDirective(
                               SipApplicationRoutingDirective.CONTINUE, req);

                  // save new sip request
                  serverRequest.getSession().setAttribute(LEG_CALL_ID,
                               serverRequest.getCallId());
                  serverRequest.getSession().setAttribute(INITIAL_REQUEST,
                               serverRequest);
                  serverRequest.getSession().setAttribute(INVITE_PURPOSE,
                               INVITE_PURPOSE_OUTGOING);
          } else {
                  LOGGER.fine("Non-initial request:" + req.getMethod());
                  javax.servlet.sip.SipSession serverSession = b2buaHelper
                               .getLinkedSession(req.getSession());
                  if (serverSession == null) {
                          LOGGER.warning("No linked session found for request with callId: "
                                       + req.getCallId());
                          return;
                  }
                  serverRequest = b2buaHelper.createRequest(serverSession, req, null);
          }

          if (req.getContent() != null) { // copy SDP
                  serverRequest.setContent(req.getContent(), req.getContentType());
          }

          serverRequest.send();
          LOGGER.fine("Created request:\n" + serverRequest
                       + "\n out of incoming request:\n" + req);
  }

  @Override
  protected void doProvisionalResponse(SipServletResponse aServerResponse)
              throws ServletException, IOException {
      LOGGER.info("Recieved Provisional Response\n" + aServerResponse);
      doResponseForwarding(aServerResponse);
  }

  /**
   * {@inheritDoc}
   */
  @Override
```

```
protected void doSuccessResponse(SipServletResponse aServerResponse)
            throws ServletException, IOException {
    LOGGER.fine("Recieved Success Response: " + aServerResponse.getMethod()
                + "/" + aServerResponse.getStatus());
    if (aServerResponse.getMethod().equalsIgnoreCase("INVITE")) {
            String invitePurpose = (String) aServerResponse.getSession()
                        .getAttribute(INVITE_PURPOSE);
    if (invitePurpose != null
                && invitePurpose.equalsIgnoreCase(INVITE_PURPOSE_TRANSFER)) {

            // This is a 200 OK from the transfer leg initiate the REINVITE
            // for the other side

            // first send ACK back
            SipServletRequest ackRequest = aServerResponse.createAck();
            ackRequest.send();

            B2buaHelper b2buaHelper = aServerResponse.getRequest()
                        .getB2buaHelper();
            SipSession clientSession = b2buaHelper
                        .getLinkedSession(aServerResponse.getRequest()
                                    .getSession(false));

            // create reinvite
            SipServletRequest reInvite = clientSession
                        .createRequest("INVITE");
            reInvite.setContent(aServerResponse.getContent(),
                        aServerResponse.getContentType());
            reInvite.send();
            // mark session as reinvite session
            reInvite.getSession().setAttribute(INVITE_PURPOSE,
                        INVITE_PURPOSE_REINVITE);
            return;
    }

    if (invitePurpose != null
                && invitePurpose.equalsIgnoreCase(INVITE_PURPOSE_REINIVTE))
    {
            // This is 200 OK from reinviteLeg
            // send ack
            SipServletRequest ackRequest = aServerResponse.createAck();
            ackRequest.send();
            // clean up
            aServerResponse.getSession().removeAttribute("invitePurpose");
            return;
    }

    doResponseForwarding(aServerResponse);
    // Store the incoming response, so we can ACK it later.
    aServerResponse.getSession().setAttribute(LAST_SERVER_RESPONSE,
                aServerResponse);
```

```
    } else if (aServerResponse.getMethod().equalsIgnoreCase("BYE")) {
        LOGGER.fine("Received 200 OK to BYE. ");
        doOkafterByeForwarding(aServerResponse);
    } else if (aServerResponse.getMethod().equalsIgnoreCase("CANCEL")) {
        LOGGER.fine("Received 200 OK to CANCEL");
        // 200 OK to CANCEL doesn't need to be forwarded.
    }
}
/**
 * {@inheritDoc}
 */
@Override
protected void doAck(SipServletRequest aClientRequest)
            throws ServletException, IOException {
    LOGGER.fine("Recieved ACK" + aClientRequest);
    B2buaHelper b2buaHelper = aClientRequest.getB2buaHelper();
    SipSession serverSession = b2buaHelper.getLinkedSession(aClientRequest
                    .getSession(false));
    if (serverSession != null) {
        // get last response for another side
        SipServletResponse aServerResponse = (SipServletResponse) serverSession
                    .getAttribute(LAST_SERVER_RESPONSE);
        SipServletRequest serverRequest = aServerResponse.createAck();
        LOGGER.fine("Created ACK message: " + serverRequest
                    + "\n to response: " + aServerResponse);
        serverRequest.send();
    } else {
        LOGGER.warning("No server session found for request with callId: "
                    + aClientRequest.getCallId());
    }
}
/**
 * {@inheritDoc}
 */
@Override
protected void doBye(SipServletRequest aClientRequest)
            throws ServletException, IOException {
    String tPurpose = (String) aClientRequest.getSession().getAttribute(
                    INVITE_PURPOSE);
    if ((tPurpose == null)
                    || (!tPurpose.equalsIgnoreCase(INVITE_PURPOSE_OUTGOING))) {
            doRequestForwarding(aClientRequest);
            aClientRequest.getSession().setAttribute("BYE",
                    aClientRequest.createResponse(SipServletResponse.SC_OK));
    }
}
/**
 * {@inheritDoc}
```

```
    */
  @Override
  protected void doCancel(SipServletRequest aClientRequest)
              throws ServletException, IOException {
      LOGGER.fine("Recieved CANCEL");
      B2buaHelper b2buaHelper = aClientRequest.getB2buaHelper();
      SipSession serverSession = b2buaHelper.getLinkedSession(aClientRequest
              .getSession(false));
      if (serverSession != null) {
          SipServletRequest serverRequest = b2buaHelper
                      .createCancel(serverSession);
          serverRequest.send();
      } else {
          LOGGER.warning("No server session found for request with callId: "
                      + aClientRequest.getCallId());
      }
  }

  private void doResponseForwarding(SipServletResponse aServerResponse)
              throws ServletException, IOException {
      SipServletRequest serverRequest = aServerResponse.getRequest();
      B2buaHelper b2buaHelper = serverRequest.getB2buaHelper();
      SipServletRequest aClientRequest = b2buaHelper
                  .getLinkedSipServletRequest(serverRequest);
      if (aClientRequest != null) {
          SipServletResponse clientResponse = aClientRequest.createResponse(
                      aServerResponse.getStatus(),
                      aServerResponse.getReasonPhrase());
          if (aServerResponse.getContent() != null) {
              clientResponse.setContent(aServerResponse.getContent(),
                          aServerResponse.getContentType());
          }
          LOGGER.fine("Created response:\n" + clientResponse
                      + "\n out of incoming response:\n" +
                  aServerResponse);
          clientResponse.send();
      } else {
          LOGGER.warning("No linked client request found for response with
          callId: "
                      + aServerResponse.getCallId());
      }
  }

  private void doOkafterByeForwarding(SipServletResponse aServerResponse)
              throws ServletException, IOException {
      B2buaHelper b2buaHelper = aServerResponse.getRequest().getB2buaHelper();
      SipSession clientSession = b2buaHelper.getLinkedSession(aServerResponse
                  .getSession());
      if (clientSession != null) {
          // get last bye
```

```
            SipServletResponse clientResponse = (SipServletResponse)
    clientSession
                        .getAttribute("BYE");
        if (clientResponse != null) {
            clientSession.removeAttribute("BYE");
            clientResponse.setStatus(aServerResponse.getStatus(),
                    aServerResponse.getReasonPhrase());
            clientResponse.send();
        }
    } else {
        LOGGER.warning("No linked session found for response with callId: "
                + aServerResponse.getCallId());
        return;
    }
  }
}
```

A.2 WEB-BASED DO-NOT-DISTURB USE-CASE IMPLEMENTATION

This section provides a complete source code for the classes sketched and described in the use-case implementation section. The source code for the components that deal with calendar functions is intentionally left-out since it is extremely vendor-specific.

A.2.1 Implementation of the Send_Message Component

```java
import java.io.IOException;

import javax.servlet.ServletConfig;
import javax.servlet.ServletException;
import javax.servlet.sip.Address;
import javax.servlet.sip.Proxy;
import javax.servlet.sip.SipApplicationSession;
import javax.servlet.sip.SipFactory;
import javax.servlet.sip.SipServlet;
import javax.servlet.sip.SipServletRequest;
import javax.servlet.sip.SipURI;

public class SendSIPMessage extends SipServlet {
        /**
         * The SIP Factory. Can be used to create URI and requests.
         */
        private SipFactory sipFactory;
        private static final long serialVersionUID = -17318953482492896627L;
        private static final String PARAM_TEXT = "text";
        private static final String PARAM_FROM = "msgfrom";
        private static final String PARAM_TO = "msgto";
        private static final String PARAM_CONTENT_TYPE = "contenttype";
        public void init(ServletConfig config) throws ServletException {
                super.init(config);
```

```
            sipFactory = (SipFactory)
config.getServletContext().getAttribute("javax.servlet.sip.SipFactory");
    }
    @Override
    protected void doInvite(SipServletRequest req) throws ServletException,
              IOException {
        Address fromAddr=null;
        Address toAddr=null;
        String imFrom=null;
        imFrom=req.getHeader(PARAM_FROM);
        if ((imFrom!=null) && (imFrom.length()>0)) {
                if (imFrom.indexOf("sip:")==-1)
        imFrom="sip:"+imFrom;
                fromAddr= sipFactory.createAddress(imFrom);
        } else {
                fromAddr=req.getFrom();
        }
        String imTo=null;
        imTo=req.getHeader(PARAM_TO);
        if ((imTo!=null) && (imTo.length()>0)) {
                if (imTo.indexOf("sip:")==-1) imTo="sip:"+imTo;
                toAddr= sipFactory.createAddress(imTo);
        } else {
                toAddr=req.getTo();
        }
        String imMsg=null;
        imMsg=req.getHeader(PARAM_TEXT);

        String content_type=req.getHeader(PARAM_CONTENT_TYPE);
        if ((content_type==null) ||
    (content_type.length()==0)) content_type="text/plain";
        SipApplicationSession applicationSession = sipFactory.
    createApplicationSession();
        SipServletRequest reqInfo = sipFactory.createRequest(application
    Session, "MESSAGE", fromAddr, toAddr);
        reqInfo.setContent(imMsg, content_type);
        reqInfo.pushRoute((SipURI) sipFactory.createAddress("sip:127.0.0
    .1:5081;lr").getURI());
        reqInfo.send();
        if (req.isInitial())
          {
                Proxy proxy = req.getProxy();
                proxy.proxyTo(req.getRequestURI());
          }
    }
  }
```

A.2.2 Implementation of the Send_response Component

```java
import java.io.IOException;
import javax.servlet.ServletException;
import javax.servlet.sip.Proxy;
import javax.servlet.sip.SipServlet;
import javax.servlet.sip.SipServletRequest;
import javax.servlet.sip.SipServletResponse;
import javax.servlet.sip.SipURI;

@javax.servlet.sip.annotation.SipServlet
public class SendSipResponse extends SipServlet {
    private static final long serialVersionUID = 1L;

    protected void doRequest(SipServletRequest req)
                throws ServletException, IOException {
        String response_code=req.getHeader("response_code");
        // Optional parameters
        String content=req.getHeader("content");
        String content_type=req.getHeader("content_type");
        int code=0;
        if ((response_code!=null) && (response_code.length()>0))
            code = Integer.parseInt(response_code);
        if ((code<200) && (req.isInitial())) { // continue sip chain building
    if provisional response
            Proxy proxy = req.getProxy();
            proxy.setRecordRoute(true);
            proxy.setSupervised(true);
            proxy.proxyTo(req.getRequestURI());
        }

        if (code!=0) {
            String
    response_phrase=req.getHeader("response_phrase");
            SipServletResponse response;
            if ((response_phrase!=null) &&
    (response_phrase.length()>0)) {
                response =
    req.createResponse(code,response_phrase);
            } else {
                response = req.createResponse(code);
            }
            if ((content!=null) && (content.length()>0)) {
                if ((content_type==null) || (content_type.length()==0))
                 content_type="text/plain";
                response.setContent(content, content_type);
            }
            response.send();
        }
    }
}
```

Appendix B

The Ericsson service development environment is a freely available development and testing environment for IMS applications, http://devtools.ericsson.com. It comes with an emulated IMS core running inside the IDE, which may be used to test IMS applications at development time. The following sections give an overview of the configuration steps required to set up the IMS core inside the IDE.

B.1 CSCF AND DNS CONFIGURATION

By default the IDE is configured to use the domain name "ericsson.com". In order to avoid conflicts with the real "ericsson.com" domain, we recommend using a domain name that does not exist on the Internet. In the following text we describe how to configure the IDE for the domain "scima.msa"; of course, you are free to use any domain name you prefer as long as it does not conflict with an existing one.

To configure the CSCF in the IDE (see Figure B.1), carry out the following steps:

1. Start the IDE.
2. Go to "Window > Preferences".
3. Expand the "Ericsson SDS" and then "Server" properties tree.
4. Select CSCF in the properties tree and replace ericsson.com in the domain field with your domain.

To configure the CSCF in the IDE (see Figures B.2 and B.3), take the following steps:

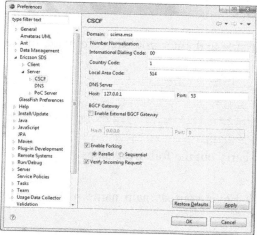

FIGURE B.1

CSCF configuration.

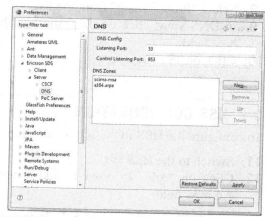

FIGURE B.2

DNS configuration menu selection.

e21

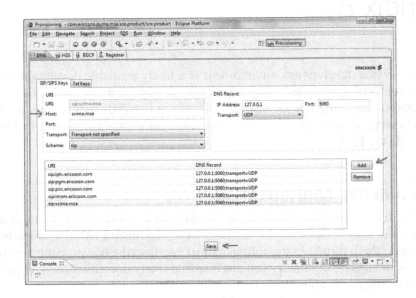

FIGURE B.3

DNS configuration.

5. Select DNS in properties tree, remove "ericsson.com" from the domain list, and add your domain.
6. Click on "OK".

Then proceed as shown in Figure B.3 to execute the following steps:

7. Go to menu Window > Open Perspective > Other.
8. Select "Provisioning" from the list.
9. In DNS view, click on the "Add" on the left side and enter your domain name in the host field.
10. Click on "Save".

B.2 HSS CONFIGURATION

To configure the HSS in the IDE (see Figure B.4), carry out the following steps:

11. Switch to the HSS tab.
12. Click on "Add" in order to add a new IFC. Enter your domain name in the Server Address field.

To create SPTs (see Figures B.5–B.7), take the following steps:

13. Switch to the "Service Point Trigger" tab. We need to add here 3 Service Point Trigger.

FIGURE B.4

HSS configuration.

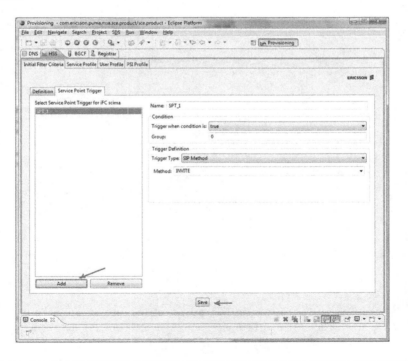

FIGURE B.5

SPT configuration.

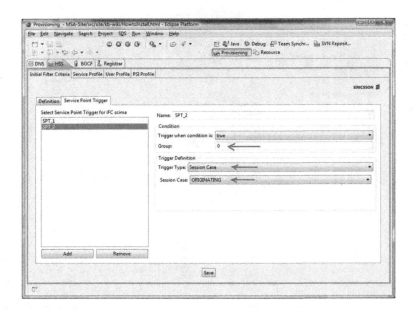

FIGURE B.6

SPT configuration.

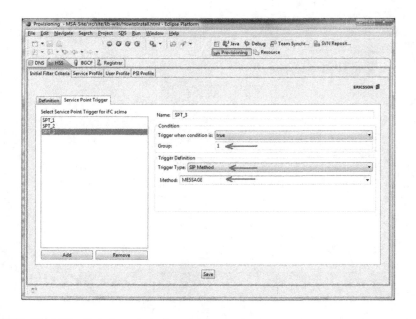

FIGURE B.7

SPT configuration.

14. Click on "Add" to add the first "Service Point Trigger" (forward INVITE to the AS).
15. Click on "Add" to add a second "Service Point Trigger" (forward INVITE only in originating session case to the AS).
16. Click on "Add" to add a third "Service Point Trigger" to forward MESSAGE to the AS. Don't forget to change the "Group Number" of the SPT to 1.

To create IFC (see Figures B.8–B.10), execute the following steps:

17. Switch to "Service Profile Tab".
18. Click on "Add" to add new service profile.
19. Select your IFC in the list of IFCs.
20. Switch to the "User Profile" tab.
21. Select one user after another and replace "ericsson.com" with your domain name in the user settings and switch default profile "Profile" to your profile created in the previous step.
22. Switch to "PSI Profile".
23. Replace "ericsson.com" with your domain name in all PSI profiles.
24. Click on "Save".

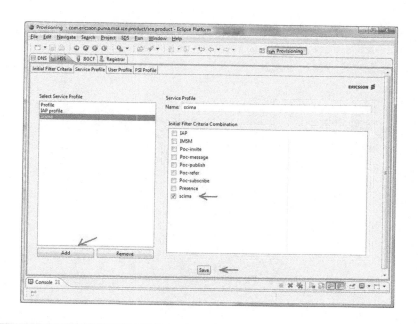

FIGURE B.8

IFC configuration.

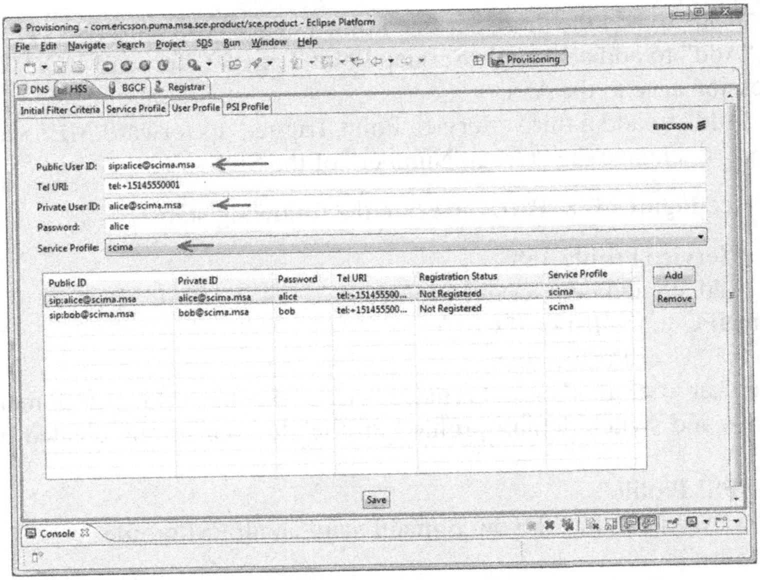

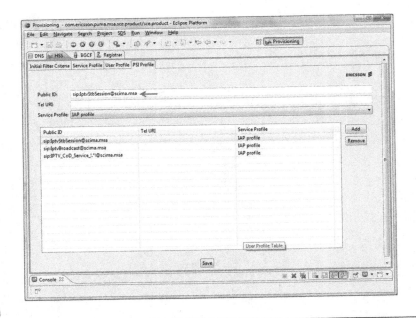

FIGURE B.9

IFC configuration.

FIGURE B.10

IFC configuration.

Appendix C

This appendix contains, among other things, an overview of SIP methods, SIP headers, and SIP response codes. New methods, headers, and response codes are regularly defined by IETF. These new methods, headers, and response codes are administered by the Internet Assigned Numbers Authority (IANA), www.iana.org. References to IETF RFCs are provided by IANA as well. The reader is invited to familiarize him/herself with the information provided by IANA and to be informed of new SIP methods, headers, and response codes that are introduced.

C.1 OVERVIEW OF SIP METHODS

Table C.1 provides an overview of the various SIP methods, their usage, and the applicable IETF standard(s).

C.2 OVERVIEW OF SIP HEADERS

Table C.2 lists the standard track (IETF RFC 3261) SIP headers.

Table C.3 lists the SIP headers that are defined for SIP extensions. For these SIP headers, the reader is referred to the RFC where this header is defined. SIP headers that are defined for a specific application, such as IMS, are generally identified with a "P" prefix (*Private header*).

Vendors may use additional, proprietary headers within an IMS network deployment. Proprietary headers should not be sent across the IMS network boundary.

C.3 OVERVIEW OF SIP RESPONSE CODES

Table C.4 lists the standard track (IETF RFC 3261) SIP response codes. Chapter 7 provides explanation of the different response code classes.

Table C.5 lists the SIP response codes that are defined for SIP extensions.

C.4 GSM/UMTS SUPPLEMENTARY SERVICES AND CORRESPONDING IMS SUPPLEMENTARY SERVICES

These are given in Table C.6.

Table C.1 SIP Methods

Method	Description	IETF RFC
ACK	ACK (acknowledgement) is used in combination with Invite transaction. For unsuccessful Invite transaction (3xx–6xx final response), the ACK method forms part of the Invite transaction. For successful Invite transaction (2xx final response), the ACK method is a separate transaction.	3261
BYE	Used to tear down a SIP session and the associated media session(s).	3261
CANCEL	Used to cancel a pending Invite transaction.	3261
INFO	INFO is used within a SIP session. It may carry a variety of information that augments a SIP service, such as DTMF tones or graphic images.	6086
INVITE	INVITE is used to initiate a SIP session and corresponding media stream(s). It may also be used within an active SIP session to change certain parameters of the SIP session or to change certain parameters of the media session.	3261
MESSAGE	MESSAGE is used for the sending of a message to an IMS subscriber (Instant message). The "message" typically has the form of a text message, but other forms of media are also supported.	3428
NOTIFY	NOTIFY is used in combination with SUBSCRIBE. SUBSCRIBE and NOTIFY enable one entity in the IMS network to obtain event notifications from another entity in the IMS network. The notifications may, for example, relate to registration state of a subscriber.	2848, 3265
OPTIONS	The OPTIONS method is used to query another IMS user about that IMS user's terminal capabilities.	3261
PRACK	PRACK (provisional acknowledgement) is used for reliable provisional response.	3262
PUBLISH	The PUBLISH method is used by a SIP user agent to publish that UA's presence information. The presence information is maintained at a presence server, from where it may be disseminated to *watchers*, other users who have subscribed to this UA's presence information.	3903
REFER	REFER is used by one SIP entity to provoke another SIP entity to establish a SIP session to a designated destination. This method may be used for, among other things, services such as call transfer.	3515
REGISTER	REGISTER is used to create a binding in registrar, to refresh an existing binding or to remove a binding.	3261
SUBSCRIBE	SUBSCRIBE is used in combination with NOTIFY. See description of NOTIFY.	2848, 3265
UNSUBSCRIBE	UNSUBSCRIBE is used in combination with SUBSCRIBE and NOTIFY. A SUBSCRIBE request includes an indication of the *duration* of the subscription, leading to implicit expiry of the subscription. To explicitly terminate a subscription, UNSUBSCRIBE may be used.	2848
UPDATE	UPDATE is used to update session-related parameters during SIP session establishment or during an active SIP session.	3311

Table C.2 Standard Track SIP Headers

SIP Header	Explanation
Accept	This header indicates which media types (described in SDP and then used on the user plane) are acceptable to the sender of the message. It may, for example, be used to indicate the support of application/sdp and text/html.
Accept-Encoding	This header is used by a user agent to signal the support of specific encoding standard(s) for the content carried in the body of a message sent to this user agent.
Accept-Language	This header may be used in a request message to indicate which languages are supported for the reason phrases, session descriptions, and status responses.
Alert-Info	Alert-Info may be included in an INVITE request or in a 180 Ringing response. When included in an INVITE request, it provides a ringtone towards the UAS (called party); when included in 180 Ringing, it provides a ringback tone. It typically has the form of an http address (wav file). Alert-Info may be included by a designated application server, acting on behalf of calling or called party.
Allow	The Allow header in a SIP message indicates the supported methods by the sender of that SIP message.
Authentication-Info	This header may be used by a UAS in a 200 OK (or other 2xx final response), to provide authentication info to be used for a subsequent authentication procedure.
Authorization	The Authorization header is used during the authentication procedure, to provide by a UA information such as authentication credentials (e.g. a response to a challenge).
Call-ID	Call-ID identifies a SIP dialog, together with the From tag and the To tag.
Call-Info	Call-Info may be used to convey additional, call-related information to the other party (calling party or called party). A calling party may, for example, include a business card or a (web-based) picture of the calling/called party.
Contact	The Contact header is used primarily to provide, to a remote party, an address where the user is contactable once a SIP session is established.
Content-Disposition	This header defines how a message body or message body part will be interpreted. It may, for example, indicate that a message body represents an icon or alerting information, to be used during call establishment.
Content-Encoding	The Content-type header indicates that the body of the message is encoded in a particular manner.
Content-Language	This header indicates that a particular message body is intended for receipt of specific language(s). A text contained in the message body could, for example, be intended for German speaking receivers.
Content-Length	The Content-Length indicates the length of the message body, in octets. It helps the receiver to parse the message.
Content-Type	This header identifies the type of content in the message body – for example, an SDP offer/answer.
CSeq	The Command sequence (CSeq) header identifies the numerical order of a transaction within a dialog. It is used to process requests in the right order. It is also possible to recognize a retransmitted message from the CSeq header.
Date	Identifies the date and time of the first transmission of a message (i.e. not retransmitted). The date will be expressed in GMT.
Error-Info	This header may be used to augment an unsuccessful final response. The reason phrase of the final response gives information about the failure of the transaction. An application server in the network may provide additional information in the form of, for

(Continued)

Table C.2 (Continued)

SIP Header	Explanation
	example, an http address pointing to a wav file. The client receiving this information may render that file to the user.
Expires	Provides a validity period for the content contained in the SIP message. An example is the validity period of a binding that is requested during registration.
From	The From header identifies the initiator of a transaction. It also carries a "tag". The tag in the From header is allocated by the initiator of the dialog. It is used together with the tag in the To header and the Call-ID to identify the dialog. The From header may be used for display purposes.
In-Reply-To	An entity that returns a call may include this header to indicate towards the callee that this call is in return to a specific previous call from that callee.
Max-Forwards	The Max-forwards header protects the network from message looping and prevents infinite message transmission. When a request message is generated, the Max-forwards header is set to an initial value (normally 70) and is decremented by 1 by every proxy or server that forwards the message. When it reaches 0, the message will not be transferred further.
Min-Expires	This header is used by a registrar to signal, in a *423 Interval Too Brief Final* response, to a client that the offered registration period is too small; the header indicates the minimum registration interval that is allowed.
MIME-Version	A message body may be encoded in accordance with the Mime (multipurpose internet mail extension) standard. The Mime-version header provides information about the Mime version used for this header.
Organization	The sender of a SIP request or response may include the name of an organization that the sender belongs to. The inclusion of such a header is purely informational and application dependent.
Priority	This header, which is also informational, may be included by the sender of an Invite request message or by an application server, to classify the request in terms of its priority. The receiver may use it to decide on the handling of it, e.g. whether or not to answer an incoming call.
Proxy-Authenticate	A client may, when sending a request, receive an authentication challenge from a proxy node. The client should re-send the request including a response to the challenge.
Proxy-Authorization	The Proxy-Authorization header contains the response requested through Proxy-Authenticate.
Proxy-Require	A client may indicate in a request that a particular capability is required from the proxy node. It may, functionally, be compared with the Require header.
Record-Route	The Record-Route header is used by proxies or application servers to indicate that they want to remain in the SIP session once the SIP session has become active.
Reply-To	A client may include the Reply-To header in an INVITE request. It includes a URI that the receiver may use for returning a call. Without this header, the receiver of the call would establish a return call to the address contained in the From header.
Require	A client (UAC) may use this header in an Invite request to indicate that specific (standard) capability is required from the receiver of the request message (UAS).
Retry-After	Various responses may contain a Retry-After header, indicating to the sender of the request that the request may be retried at a later time. The header may be generated by a network entity (proxy, application) or by the end-point (called user).

(Continued)

Table C.2 (Continued)

SIP Header	Explanation
Route	The Route header is used for routing the SIP request message.
Server	A receiver of a request (UAS) can provide information about its software version. This information may be used by the client (UAC) to adapt its further processing of the SIP session.
Subject	The sender of an INVITE request may provide an informative subject. The Subject header may be used in combination with the Priority header.
Supported	A UAC or UAS may, during session establishment, inform the other party about supported optional capability.
Timestamp	A Timestamp header may be transferred in a SIP request and corresponding response(s). It enables a UAC, as implementation option, to determine the round-trip time (RTT) of a request–response combination.
To	The To header contains an identification of the intended receiver of a request. It is for information purposes and is not used for routing. The To header contains a tag value, once the SIP dialog is established.
Unsupported	This header is included in the *420 Bad Extension* final response to inform the UAC that one or more capabilities identified in the Require header or Proxy-require header are not supported. The UAC may re-issue the request without including these required capabilities.
User-Agent	The User-Agent header informs the receiver of a SIP message about the type of phone used. An IMS network may, for example, apply designated header manipulation for specific SIP phones (when it would be known that certain SIP phones have special requirements).
Via	The Via header is used for routing of response message(s). It includes a branch identifier, which may be used by a proxy or UA to associate the response message with an internal process (client transaction).
Warning	The Warning header contains information that augments an unsuccessful final response code. Specifically, it may inform the sender of the request of the reason why a request message could not be processed.
WWW-Authenticate	A user may be requested by a proxy (in *407 Proxy Authentication)* or by the UAS (in *401 Unauthorized)* to provide authentication. The WWW-Authenticate carries the authentication details for the user.

Table C.3 SIP Headers for SIP Extensions

SIP Header	Explanation	IETF RFC
Private Headers		
P-access-network-info	The P-access-network-info (PANI) provides information about the access network and location of a UA. The PANI may be provided by the UA or may be determined by the IMS network (by a node that has a functional connection to the access network).	3455
P-asserted-identity	The P-asserted-identity (PAI) is a network-asserted identification of a subscriber issuing a request or responding to a request.	3325
P-associated-URI	This header contains a set of URIs that are associated in an implicit registration set. It is returned to P-CSCF and to UA during registration, informing the P-CSCF and UA which public user identities are registered through this registration action.	3455
P-called-party-ID	The P-called-party-ID identifies to the called party how that party was identified for the call establishment.	3455
P-charging-function-addresses	This header is received in P-CSCF and S-CSCF during registration. It is included in SIP session establishment or stand-alone transaction. It informs the various nodes about the address to which charging records should be sent for this session/transaction.	3455
P-charging-vector	The P-charging-vector is included in SIP session establishment or stand-alone transaction. It contains, among other things, an IMS charging identifier, allowing for charging record correlation.	3455
P-early-media-header	This header may be used by an application server to instruct the calling UA or border gateway to authorize media transfer during session establishment.	5009
P-media-authorization	The P-media-authorization is used between proxy and UA during session establishment. It contains *authorization tokens* that the UA may use for requesting QoS (on user plane) from the network, if needed.	3313
P-preferred-identity	This header may be used by a UA to indicate to the network how (s)he would like to be identified in the SIP signaling. Specifically, which one of his/her public identities should be used to set the P-asserted-identity.	3325
P-profile-key	The P-profile-key (PPK) is used during wildcard public service identity (PSI) service invocation. The PPK is passed from HSS to I-CSCF and form I-CSCF to S-CSCF and identifies the public service to which this service request relates.	5002
P-refused-URI-list	This header may be used in the push-to-talk over a cellular (PoC) service, to filter out certain URIs from a requested PoC service request.	5318
P-served-user	The P-served-user identifies the served party during session handling. It is used, among other things, for a diverted call, where the *served party* is typically the party that diverted the call.	5502
P-user-database	This header carries the address of the HSS where the subscriber data is located. It is used in procedures where both I-CSCF and S-CSCF have to query an SLF to find the HSS address. The I-CSCF informs the S-CSCF about the HSS address, preventing the need for a second SLF query.	4457

(Continued)

Table C.3 (Continued)

SIP Header	Explanation	IETF RFC
P-visited-network-ID	A P-CSCF handling a request from a subscriber belonging to another IMS network may include the P-visited-network-ID in the request. The Home IMS network can determine from this header whether the subscriber is allowed to receive services through that IMS network.	3455
Non-Private Headers		
Accept-Contact	The Accept-Contact header is used in combination with the Reject-Contact header and the Request-Disposition header. It allows a calling party to provide a preference to the registrar of the called party about the contacts of the called party to which the request shall be sent or shall not be sent.	3841
Allow-Events	The Allow-Events header is used for the *Event Notification* functionality. This header indicates which events notifications are supported.	3265
Event	This header is, like the Allow-Events header, used for the Event Notification functionality. It is used, among other things, to indicate the occurrence of a specific event.	3265
History-Info	The History-Info header is used to reflect the *history* of a request. For example, when an Invite request message is diverted to another destination, this header identifies the diverting party.	4244
Min-SE	This header is used for the SIP session monitoring functionality ("heartbeat"). This header defines the minimum interval between two successive session refresh (heartbeat) messages.	4028
Privacy	The Privacy header may be used in a request or response to hide the user identity towards the remote party.	3323
RAck	The RAck header is used in combination with the RSeq header. These headers are used for reliable provisional response.	3262
Reason	The Reason header carries additional information, reason, about an unsuccessful final response, about cancellation of an Invite transaction or release of a session. The Reason header may be used to set the cause value in an ISUP Release message, in the case of CS–IMS interworking.	3326
Reject-Contact	See Accept-Contact header.	3841
Request-Disposition	The Request-Disposition header enables a calling party to provide preferences with respect to how a request shall be processed. For example, to indicate that a request should not be "forked" to multiple devices.	3841
RSeq	See RAck header.	3262
Session-Expires	This header is used for the SIP session monitoring functionality (see Min-SE). This contains the value of the heartbeat message interval.	4028
Subscription-State	This header is used for the Event Notification functionality. It is used to manage the subscription for a specific event.	3265

Table C.4 Standard Track SIP Response Codes

Response Code	Explanation
1xx-class: Provisional	
100 Trying	Generated by server transaction (ST) of "next hop" to confirm receipt of an INVITE request.
180 Ringing	The Invite request has reached the destination UA, acting as UAS, and the destination UA has entered the alerting phase.
181 Call is Being Forwarded	Indicates that a server is retargeting an INVITE request to an alternative destination.
182 Queued	This response may be used by an application acting as *call distribution server*. The 182 Queued indicates that the Invite request is pending (queued).
183 Session Progress	The 183 Session Progress is a generic response message to inform the calling party about the progress of the INVITE request. It may be used for early media playing, early media authorization, preconditions, etc.
2xx-class: Successful	
200 OK	Generic indication that a request has been successfully executed.
3xx-class: Redirection	
300 Multiple Choices	This response code is used to convey a list of alternative destinations that the UA may use to contact the destination party.
301 Moved Permanently	This response indicates that the destination party shall, from now onwards, be contacted at the provided alternative address.
302 Moved Temporarily	Unlike the 301 Moved Permanently, the 302 Moved Temporarily response indicates that the destination party is *temporarily* reachable at another address. The *expires* header or *expires* parameter indicates the validity duration of this alternative address.
305 Use Proxy	This response code indicates that the addressed service shall be reached through an (additional) proxy.
380 Alternative Service	The UAC gets an indication that the addressed service or user may be reachable through other communication means, as defined in the body of this response.
4xx-class: Request Failure	
400 Bad Request	A syntax problem is detected in the request message, preventing the processing of the request message.
401 Unauthorized	A registrar or UAS may generate this response to prompt the initiator of the request to provide authentication.
402 Payment Required	Reserved.
403 Forbidden	The request could not be handled, for example because the subscriber sending an INVITE is not entitled to this service.
404 Not Found	The subscriber for whom the service is requested does not exist in the network. For example, a call established towards a non-existing subscriber in the operator domain.
405 Method Not Allowed	The receiver of the request message indicates that the method of the request is not allowed. For example, an MGCF could provide this response when receiving a REFER request, assuming that the MGCF does not support REFER.

(Continued)

Table C.4 (Continued)

Response Code	Explanation
406 Not Acceptable	The response that the UAS can provide, e.g. application/sdp, does not appear in the Accept header of the INVITE request, e.g. text/html. Hence, the UAS cannot provide a response that would be acceptable for the UAC.
407 Proxy Authentication Required	This response indicates that the proxy handling the request requires authentication from the UAC.
408 Request Timeout	The UAS or registrar determines that it has not been able to provide a response to the UAC within configurable time. This could be due to internal congestion. The UAC may reattempt the request.
410 Gone	This response indicates that the address resource is no longer reachable at this domain.
413 Request Entity Too Large	The body in the request message is larger than the proxy, registrar, application server, or UAS can process.
414 Request-URI Too Long	Compare with 413 Request Entity Too Large. The 414 Request-URI Too Long indicates that the R-URI is too long to be processed by the server.
415 Unsupported Media Type	The initiator of the request message has requested a media type that is not supported by the server.
416 Unsupported URI Scheme	The initiator of the request message has used a URI scheme in the request line that is not supported by the server, for example tel:. The client may re-initiate the request, with the R-URI converted from tel: format to sip: format.
420 Bad Extension	The client had requested a SIP extension, in the Require header or Proxy-Require header, that is not supported by the server.
421 Extension Required	The server indicates that it cannot process a request since a required extension is not supported by the client (does not appear in the Supported header).
423 Interval Too Brief	This response may be returned by a registrar to indicate that the proposed registration validity period is too small.
480 Temporarily Unavailable	A server or registrar has determined that the destination subscriber is currently not reachable.
481 Call/Transaction Does Not Exist	The UAS is not capable of processing the request since there is no corresponding server transaction (as indicated in the Branch value) or no corresponding dialog (as indicated in the From tag, To tag, and Call-ID).
482 Loop Detected	The UAS has detected the occurrence of a "SIP request loop". The SIP request was forwarded by this entity already. Forwarding the request message again may lead to continuation of the loop (and eventually depletion of the Max-forwards header).
483 Too Many Hops	The Max-forwards header has reached 0, so the request cannot be forwarded any further.
484 Address Incomplete	The server determines that the information carried in the R-URI is incomplete, e.g. an incomplete directory number. The client may re-initiate the request, with a R-URI containing additional address digits.
485 Ambiguous	The initiator of the request had provided a R-URI that (partially) matches several subscriber records. The UAS may return a list of these partially matched subscriber records, enabling the client to re-initiate the request to one of these subscribers.

(Continued)

Table C.4 (Continued)

Response Code	Explanation
486 Busy Here	The destination party is currently not able to accept the session establishment, e.g. does not wish to answer the call or is busy in another communication session.
487 Request Terminated	This response is returned by a UAS when the execution of an INVITE transaction is terminated before the session became established, e.g. due to a CANCEL request or a BYE request. Although not formally standardized, 487 Request Terminated may also be returned by an MRFC when the option "early media" is used.
488 Not Acceptable Here	A particular UAS indicates that the request is not acceptable on this device.
491 Request Pending	This response code indicates the detection of a "glare condition" (or "race condition"). The receiver of the request message could not process the request message since it was busy executing another request on this dialog.
493 Undecipherable	The receiver of the request could not decipher a Mime-encoded body.
5xx-class: Server Failure	
500 Server Internal Error	A temporary or permanent failure condition has occurred in a server (or proxy).
501 Not Implemented	A server receives a request for a method that it does not support.
502 Bad Gateway	A server has forwarded a request to a downstream server. The response from the downstream server indicates that the request cannot be processed. Hence, the server returns the 502 Bad Gateway towards the client.
503 Service Unavailable	A server indicates that it is (temporarily) not capable of handling the request. The receiver of this response may put the server address in quarantine for a certain period, e.g. as indicated in the Retry-After header in the 503 response. The receiver of this response may forward the request to another server, e.g. based on received SRV records from DNS.
504 Server Timeout	The server had forwarded the request to a further external server, but has not received a response from that external server within a configurable time.
505 Version Not Supported	This response message indicates that the server does not support the SIP version (SIP/2.0) indicated in the request.
513 Message Too Large	This response message can be compared with response 413 (Request Entity Too Large) and response 414 (Request-URI Too Long). The request message exceeded the length that could be processed by this server.
6xx-class: Global Failures	
600 Busy Everywhere	This response may be generated by the destination party when receiving a request. It indicates that the client is not able to respond to the request (i.e. busy) on any device, including voicemail system or other alternative destination.
603 Decline	The destination party explicitly rejects the incoming request, indicating that (s)he is currently not reachable on any device.
604 Does Not Exist Anywhere	A server may return this response when it can state definitively that the destination of the request (R-URI) does not exist (in any network or domain).
606 Not Acceptable	An end-user terminal may return this response code when some aspects of the request, for example SDP offer characteristics, are not acceptable on any of the person's devices (or alternative destination).

Table C.5 SIP Response Codes for SIP Extensions

Response Code	Explanation	IETF RFC
202 Accepted	This response code is part of the Event Notification feature. The receiver of a Subscriber request has accepted the subscription, but cannot immediately send a notification.	3265
422 Session Interval Too Small	This response code is used to indicate that the initiator of a SIP session offered a too low session timer duration.	4028
489 Bad Event	This response is used for the Event Notification feature, to indicate that a requested event (in SUBSCRIBE request message) or provided event (in NOTIFY request message) is not supported.	3265

Table C.6

IMS Supplementary Service		GSM/UMTS Supplementary Service	
Service	3GPP TS	Service	3GPP TS
Originating identity presentation (OIP)	24.607	Calling line presentation (CLIP)	23.081
Originating identity presentation restriction (OIR)	24.607	Calling line presentation restriction (CLIR)	23.081
Terminating identity presentation (TIP)	24.608	Connected line presentation (COLP)	23.081
Terminating identity presentation restriction (TIR)	24.608	Connected line presentation restriction (COLR)	23.081
Communication diversion (CDIV)	24.604	Call forwarding (CF)	23.082
Communication hold (HOLD)	24.610	Call hold (CH)	23.083
Communication barring (CB)	24.611	Call barring (CB)	23.088
Explicit communication transfer (ECT)	24.629	Explicit call transfer (ECT)	23.091
Conference	24.605	Multiparty call	23.084
Message waiting indication	24.606	–	
Anonymous communication rejection	24.611	Note[a]	
Malicious communication identification	24.616	–	
Advice of charge	24.647	Advice of charge (AoC)	23.086
Closed user groups	24.654	Closed User Groups (CUG)	23.085
Flexible alerting	24.239	Note[b]	
Communication waiting	24.615	Call waiting (CW)	23.083
Completion of communications to busy subscriber	24.642	Completion of Calls to Busy Subscriber (CCBS)	23.093
Completion of communications by no reply	24.642	Note[c]	
Customized alerting tones	24.182	Note[d]	

[a]A service like anonymous communication rejection may be offered through Intelligent Network (IN) service.
[b]Flexible alerting may be offered in PLMN through switch-based service or through IN service.
[c]Completion of calls on no reply (CCNR) is conceptually defined for the GSM network but has never been specified in a technical standard (TS).
[d]Customized alerting tones (CAT) may be offered in the CS mobile network through switch-based service or through IN service.

Abbreviations

3GPP	Third-Generation Partnership Project
A	Address (DNS address record for IPv4 address)
AAAA	Address (DNS address record for IPv6 address)
ACA	Accounting Answer
ACELP	Algebraic-code-excited linear prediction
ACR	Accounting Request
AD	Analog-to-digital
ADPCM	Adaptive differential PCM
AJAX	Asynchronous JavaScript and XML
ALG	Application layer gateway
AMR	Adaptive multi-rate
API	Application Programming Interface
APN	Access Point Name
AR	Application Router
ARP	Address Resolution Protocol
ARPA	Advanced Research Projects Agency
ARPU	Average Revenue Per User
AS	Application server
A-SBG	Access session border gateway
ASN.1	Abstract syntax notation version 1
ATCF	Access Transfer Control Function
ATGw	Access Transfer Gateway
ATM	Asynchronous transfer mode
AVP	Attribute value pair
B2BUA	Back-to-back user agent
BC	Bearer capability
BER	Basic encoding rules
BG	Border gateway
BGCF	Border gateway control function
BICC	Bearer independent call control
CAMEL	Customized applications for mobile network enhanced logic
CAN	Carrier access network
CAP	CAMEL application part
CCA	Credit Control Answer

CCE	Connected Consumer Electronics
CCR	Credit Control Request
CCS	Common channel signaling
CDF	Charging data function
CDR	Charging Data Record
CGF	Charging Gateway Function
CIPID	Contact Information for the Presence Information Data format
CS	Circuit switched
CS-ACELP	Conjugate-structure algebraic-code-excited linear prediction
CSCF	Call session control function
CSFB	Circuit Switched FallBack
CSRC	Contributing sources
CT	Client transaction
DAR	Default Application Router
DARPA	Defense Advanced Research Projects Agency
DFC	Distributed Feature Composition
DHCP	Dynamic Host Configuration Protocol
DLNA	Digital Living Network Alliance
DND	Do not Disturb
DNS	Domain Name System
DPCM	Differential PCM
DSL	Digital subscriber line
DTAP	Direct transfer application part
DTMF	Dual-tone multi-frequency
E-CSCF	Emergency CSCF
EE	Enterprise Edition
eDNS	External DNS
EFR	Enhanced full rate
ENUM	E.164 number database
EPC	Evolved Packet Core
EPS	Enhanced Packet System
ETSI	European Telecommunications Standards Institute
FQDN	Fully qualified domain name
FTTH	Fiber to the home
GBA	General Bootstrapping Architecture
GGSN	Gateway GPRS support node
GMSC	Gateway MSC
GPRS	General packet radio service
GSM	Global system for mobile communication
GSMA	GSM Association
gsmSCF	GSM service control function

gsmSSF	GSM service switching function
GUI	Graphical User Interface
HD	High Definition
HLC	High layer compatibility
HLR	Home location register
HSPA	High-speed packet access
HSS	Home Subscriber Server
HTML	Hypertext markup language
HTTP	Hypertext Transport Protocol
IANA	Internet Assigned Numbers Authority
IARI	IMS Application Reference Identifier
IBCF	Interconnect border control function
I-CSCF	Interrogating CSCF
ICS	IMS Centralized Services
ICSI	IMS Communication Service Identifier
iDNS	Internal DNS
IEC	International Electrotechnical Commission
IEEE	Institute of Electrical and Electronics Engineers
IETF	Internet Engineering Task Force
iFC	Initial filter criteria
iLBC	Internet low bit rate codec
IM	IP Multimedia Instant messaging
IM-MGW	IP multimedia MGW
IMPI	IMS private user identity
IMPU	IMS public user identity
IMS	IP Multimedia Subsystem
IMS-AKA	IMS Authentication and Key Agreement
IMSI	International mobile subscriber identity
IM-SSF	IP multimedia service switching function
IN	Intelligent Networks
INAP	Intelligent Networks application part
IP	Internet protocol
IRS	Implicit registration set
ISC	IMS service control interface
ISDN	Integrated Services Digital Network
ISIM	IMS subscriber identity module
ISO	International Standards Organization
ISP	Internet Service Provider
ISUP	ISDN user part
ITU	International Telecommunications Union
IVR	Interactive voice response system

JAX-RS	Java API for RESTful Web Services
JPEG	Joint Photographic Experts Group
JSON	Java Simple Object Notation
JSR	Java Specification Request
LAN	Local Area Network
LPC	Linear predictive coding
lr	Loose routing
LTE	Long-term evolution
M2M	Machine to Machine
M3UA	MTP3 user adaptation
MAC	Medium access control
MAN	Metropolitan Area Network
MAP	Mobile application part
MGC	Media gateway controller
MGCF	Media gateway control function
MGW	Media gateway
MIME	Multipurpose Internet mail extension
MME	Mobility Management Entity
M-MGW	Mobile MGW
MMTel	Multimedia Telephony
MPEG	Moving Picture Experts Group
MR-ACELP	Multi-rate ACELP
MRF	Media resource function
MRFC	Media resource function controller
MRFP	Media resource function processor
MS	Mobile station
MSC	Mobile services switching center
MSISDN	Mobile station international ISDN number
MSRP	Message Session Relay Protocol
MSS	Mobile softswitch
MTAS	Multimedia Telephony Application Server
MTP	Message transfer part
MTU	Maximum transmission unit
NAI	Network address identifier
NAPTR	Naming authority pointer
NAT	Network Address Translation
N-SBG	Network session border gateway
NBI	Northbound Interface
O&M	Operations and Maintenance
OAuth	Open Authorization
OCS	Online Charging System

OMA	Open Mobile Alliance
OSGi	Open Services Gateway initiative
OSI	Open system interconnection
PAI	P-asserted-identity
PBX	Private branch exchange
PCM	Pulse code modulation
PCMA	PCM A-law
PCMU	PCM μ-law
P-CSCF	Proxy CSCF
PDG	Packet data gateway
PDPc	Packet data protocol context
PIDF	Presence Information Data Format
PLMN	Public Land Mobile Network
PoC	Push-to-talk over cellular
PS	Packet switched
PSI	Public service identity
PSTN	Public Switched Telephony Network
PTT	Postal Telephone and Telegraph
PTT	Push to Talk
PUI	Public user identity
QCELP	Qualcomm-code-excited linear prediction
QCI	QoS Class Identifier
RAN	Radio Access Network
RCS	Rich Communication Suite
REST	Representational State Transfer
RFC	Request for comments
RGW	Residential Gateway
ROS	Remote operations
RPE-LTP	Regular pulse excitation–long-term prediction – linear predictive coder
RPID	Rich Presence Information Data
RR	Resource record
RTCP	Real-time Transport Control Protocol
RTP	Real-time Transport Protocol
RTT	Round-trip time
R-URI	Request URI
SBC	Session border controller
SBG	Session border gateway
SCC AS	Service Centralization and Continuity Application Server
SCF	Service control function
SCIM	Service capability interaction manager
SCP	Service control point

S-CSCF	Serving CSCF
SCTP	Streaming Control Transmission Protocol
SCUDIF	Service change and UDI fallback
SDH	Synchronous digital hierarchy
SDP	Session Description Protocol
SG	Session gateway
SIM	Subscriber Identity Module
SIP	Session Initiation Protocol
SIP-AS	SIP application server
SIPS	Secure SIP
SLF	Subscriber locator function
SMS	Short message service
SOA	Service Oriented Architecture
SPT	Service point trigger
SRTP	Secure RTP
SRV	Service record
SRV-CC	Single Radio Voice Call Continuity
SS7	Signaling system #7
SSF	Service switching function
SSRC	Synchronization source
ST	Server transaction
STM	Synchronous transport module
T-ADS	Terminating Access Domain Selection
TA	Terminal Adapter
TCP	Transmission Control Protocol
TDM	Time division multiplex
TISPAN	Telecoms & Internet converged Services & Protocols for Advanced Networks
TLS	Transport layer security
TMR	Transmission medium requirements
TrGW	Transition gateway
TS	Technical specification
TTL	Time-to-live
TU	Transaction user
UA	User agent
UAC	User agent client
UAS	User agent server
UDI	Unrestricted digital information
UDP	User Datagram Protocol
UE	User equipment
UMTS	Universal Mobile Telecommunications System

UNI	User equipment to Network Interface
UNI	User-Network Interface
URI	Universal resource identifier
URL	Uniform resource locator
USI	User service information
USIM	Universal Subscriber Identity Module
UTRAN	UMTS Terrestial Radio Access Network
VAS	Value-added service
VCC	Virtual Call Center
ViG	Video gateway
VLR	Visitor location register
VMSC	Visited MSC
VoIP	Voice over IP
VoLTE	Voice over LTE
VSELP	Vector-sum-excited linear prediction
WAC	Wholesale Applications Community
WAN	Wide Area Network
WCDMA	Wideband Code Division Multiple Access
WLAN	Wireless LAN
XCAP	XML Configuration Access Protocol
XDSL	(any) Digital Subscriber Line
XML	eXtensible markup language

Index

2G, 3, 107, 431, 435
3G, 3, 21–2, 26, 62, 66, 107, 223, 230, 251–69, 357–8, 406,
    410–11, 431, 435
    see also CS
3GPP, 12–13, 14, 15–17, 18–19, 21–2, 23–4, 26, 39, 49–58,
    66–75, 107, 125–9, 148, 223–32, 244–5, 250–6, 278–9,
    330–6, 337–46, 357–9, 380–1, 405–30, 431–47, 491
    see also IMS…; LTE; REL…; SCIM; TS…; UNI
    list of TSs, 223–6
3GPP2, 337
3xx response, 374–6, 392–3
4G, 62, 423
    see also LTE
100 Trying, 186–94, 244–9, 270–85, 325, 361–2, 368–71,
    380–400, 410–14
180 Ringing, 188–97, 205–9, 211–22, 244–9, 270–85, 315–16,
    361–2, 368–78, 380–400, 410–14
183 Session Progress, 191–4, 208–9, 211–22, 319–20, 367–78,
    414
199 Early Dialog Terminated, 209
200 Ok, 189–97, 205–9, 211–22, 244–9, 251–69, 270–85, 315–
    16, 318–20, 361–2, 368–78, 380–400, 409–14, 424–5
401 Unauthorized, 251–69
432 Interval Too Brief, 261
480 Temporarily Unavailable, 371
486 Busy Here, 304, 370–1

A

AAAA record, 184, 240
AAAA record queries, 184–94, 240–2, 272–85
    see also IP…
A record, 143, 184, 240, 241
A-SBG (access session border gateway), 231–2, 237–41,
    243–9, 270–85, 287–92
    see also P-CSCF; SBC; SBG; SG
abstract syntax notation, 162
abstraction hierarchy, 44
abstraction levels, 'spaces for value capture', 43–8, 56
abstraction stack, 15–17, 20–1, 43–4, 331–6
    see also applications; Communication Service layer
ACA (appropriate accounting answer), 359–62
Accept-Contact SIP header, 18–19, 347, 408–30
access forms, 2–4, 13–15, 23, 39–42, 62–4, 145, 223, 226,
    229–32, 262–6, 330–6, 402–4, 449–51
    see also 3G…; 4G; ADSL; connectivity…; Ethernet;
    fixed…; WLANs
Access Point Name (APN), 435
access session border gateway, 231, 272
Access transfer control function (ATCF), 440–1

ACK, 85–93, 147–57, 162–83, 188–97, 204–9, 211–22,
    252–69, 270–85, 315–16, 319–20, 325, 361–2, 368–78,
    380–400, 410–14, 482
    see also final responses; INVITE; PRACK
    definition, 482
ACM (address complete message), 368–78, 380–400
active address books, 14–15, 26
AD (analog-to-digital) converters, 105–9
ad-financing, 7, 31–2, 44–7, 352, 354–62, 401, 451
adaptive differential PCM, 106
adaptive multi rate, 108
add/drop media features, MMTel, 332–6
additional calling party number, 367, 382
Address Resolution Protocol, 111
AdMob, 354–5
ADPCM (adaptive differential PCM), 106–9, 116–17
AdSense, 352, 354
ADSL, 2, 3–4, 230–2, 332–6
    see also Ethernet; WLANs
agendas, 100–4
aggregation proxy (AP), 338–46
AJAX (Asynchronous JavaScript XML), 93–4, 102–4
AJAX communications monitor, 93
Alcatel-Lucent, 28, 34, 418
'allowed' response, presence, 341–6
AMR (Adaptive Multi Rate), 108, 332–6, 377, 434–5
analog technology, 7–8, 105–43
anchored users, 11–13, 242–9, 438–47
Android, 18, 33
    community, 18, 30, 31, 353
    Market, 353
Android's Market, 353
ANM (answer message), 368–78, 380–400
AnnouncementB2B, 89–93, 455–62
announcements, 82–93, 302–3, 316–20, 374–6, 455–62
    see also MRFP; user interactions
anonymous URI, 283
APIs, 9–13, 15–26, 28–32, 49–58, 329–36, 357–62, 402–4,
    422, 450–1, 455–71
app stores, 8–9, 352–4, 401–2, 450–1
appendices, 84, 88, 89, 92, 95, 455–91
Apple, 33, 34, 38, 47, 352–5, 356
    AppStore, 353, 354
    iAds, 352, 354–5
    iPhone, 33, 34, 38, 47
application layer of OSI reference model, 110–11,
    231–2
application level gateway, 231
Application Router, 72–4, 86

application servers (ASs), 10–13, 17–21, 57–80, 83–104,
226–328, 334–6, 338–46, 358–62, 363–400, 431–47
see also SCIM; SIP-ASs
concepts, 69–80, 266–9, 292–328, 334–6, 338–46, 358–62,
371–400
definition, 292–4
application usage (AU), 339–46
application-initiated registration, 268–9
see also third-party…
applicationDeployed, 73–5
applications, 2–26, 28–32, 39–42, 49–58, 59–80, 81–104, 223,
226, 230–328, 330–6, 351–62, 401–30, 449–51, 455–71
see also capital goods software; charging; consumer goods
software; RCS; service chaining; service composition;
SIP-ASs
creation methods, 59–80, 81–104, 401–30, 449–51
future prospects, 35, 38, 449–51
applicationUdeployed, 73–5
AppStore, 353, 354
AR (application router), 65–9, 70–5, 77–80, 81–104, 148–57
ARP (Address Resolution Protocol), 111, 306
ARPA (Advanced Research Projects Agency), 289–90, 306
ARPANET, 289
see also Internet
ARPU (average revenue per user), 15
ASN.1 (abstract syntax notation 1), 162–83
associated IMPU, 258
asynchronous communications, 16–21, 70–5, 93–4, 102–4,
365–400
asynchronous transfer mode, 365
ATCF (access transfer control function), 440–7
ATGW (access transfer gateway), 441–7
ATM (asynchronous transfer mode), 365–400
AU, 339–40
audio mixers, multi-party communication sessions, 129–43
audiovisual customer service portals, 18
authentication, 250
authentication functions, 4, 7, 10–13, 233–4, 237–8, 249–69,
434–47, 451
see also registration
authorities relationship circles, 5
authorization, 250
authorization functions, 10, 250–69, 341–6, 374–6
see also registration; UAA; UAR
AVP, 218–22

B

B-number analysis, 287, 364–5
B2buaHelper, 77–80
B2BUAs (back-to-back user agents), 75–80, 86–93, 158, 160–
1, 166–83, 242–9, 274–85, 294–5, 298, 300–3, 318–20,
327–8, 393, 438–47
see also proxies
definition, 158, 160, 302–3

back-end systems, legacy SCP, 395–400
backward media, 206–9, 369–71, 374–6
basic call state machine, 290
basic CS network services, 390–2
Basic Encoding Rule, 162
basic telephony services, 294
batteries, 8
Bayeux, 335
BCSM (basic call state machine), 390–2
bearer independent call control, 121, 125
bearers, 16–22, 408–30
BER (Basic Encoding Rule), 162–83
BG (border gateway), 10–13, 138–43, 226–328, 358–62
BGCF (Breakout Gateway Control Function), 226–328, 358–
62, 371–2, 380–400
egress routing, 247–8
Bharti Airtel, 423
Bi reference point, 358–62
BICC (bearer independent call control), 121–2, 125–9, 244–5,
364–400
billing, 351, 357–62, 395–400
see also charging
binding, 232
binding information, 141–3, 232–4, 249–69, 273–85
see also contact addresses; PUI
black box view of IMS, 15–17, 20–1, 49–51, 331–6
black-list, 333, 334
black-lists/white-lists, 55–6, 331–6, 434–5
'blocked' response, presence, 341–6
Bo reference point, 359–62
Boolean operators, 65–9
bothway media, 206–9
BPEL, 428
branch, 152
branch identifier, 168, 169, 188
branch value, 169
branches, concepts, 76–80, 169–83, 184–97, 242–9
break-out gateway control function, 247–8, 290
break-out initiation, 247–9
see also gateways
Brenner, Michael, 404
broadband, 3–4, 6–8, 25–6, 35–8, 332–6, 402–4, 405, 418–22,
427, 449
broadband access, 418–19
broadcasting scenario, RCS, 429–30
brokers, 354–7, 362
BSC RNC, 439–40
business analysts, 428
business case for IMS, 38–42
business contacts aspect of the relationship circle, 5
business logic, 61, 81, 83–7, 98, 100
business models, 1–2, 7–8, 15, 26, 27–48, 59–60, 351–62,
401–2, 425–30, 450–1
see also charging; economics

concepts, 27–48, 59–60, 351–62, 401–2, 425–30, 450–1
definition, 34, 42, 43–4
IMS examples, 47–8
the individual, 35–7, 403–4, 426–30
reachability factors, 47–8, 426–30
'spaces for value capture', 43–8, 56, 450–1
UPS example, 47–8
business-to-business interfaces, 10–13, 49–58, 455–62
BUSY, 98–104, 205, 304, 433
Busy Here, 304, 370–1
BYE, 84–93, 151–7, 162–83, 199–209, 212–22, 252–69,
    270–85, 319–20, 325–8, 368–78, 408–14, 482
    see also termination of SIP sessions
    definition, 482

C

C-MSISDN (Correlation MSISDN), 438–47
cable connectivity, 6, 26, 136–43, 230–2, 332–6
calendar functions, 55–6, 93–104, 336, 471
calendar_login, 97–104
call assist services, 393–4
call barring, 293–4, 333–6, 381–9, 433–4
call centers, virtual call center use-case implementation, 81–93,
    455–71
call diversion, 55–6, 295–8, 331–6, 392–3, 434–47
call forwarding, 293–4, 333–6, 381, 388–9, 433–47
call hold/resume, interworking with legacy networks, 381,
    386–8
Call-id, 198, 242
Call-ID system header, 163, 168–83, 198–209, 242–9, 252–69,
    272–85, 324–8
call pickups, 392–3
call screening, 390–2, 394–400
call transfer component, 82–93, 293–4, 462–71
calling line presentation, interworking with legacy networks,
    381–3
call out of the blue, 312
call_transfer/call_transfer_frontend, 89–93
called party number, 292, 367
calling line presentation, 199, 382
calling line presentation restriction, 383
calling party number, 367
CallTransferB2B, 77–80, 92, 465–71
CallTransferB2BFrontend, 77–80, 92–3, 462–5
CAMEL, 24, 298
CAMEL application part, 298, 391, 392, 445
CAMEL protocol versions, 12, 24–5, 66–7, 148, 290, 298–300,
    390–400, 442–7
cameras, 5, 8–9, 13, 22–3
CANCEL, 76–7, 90–3, 152–7, 162–83, 194–7, 205–9, 252–69,
    315–16, 325–8, 372–6, 482, 490
CAP (CAMEL Application Part), 148, 293–4, 298–300, 391–400
capabilities, 8, 38, 59–60, 65–9, 71–5, 257–69, 279–85, 302–3,
    317–20, 321–3, 331–6, 348–9, 371–6, 390–400, 405, 423–4

capability discovery, 424
capability discovery, RCS-e, 423–5
CAPEX spend, 14, 393–9
capital goods, 32
capital goods software, 27, 32–48, 81–104, 401–2
capitalism, 36
CAPv2, 394–400
CAPv3, 396–400
Carrier (Infrastructure) ENUM, 290
cashless calling, 394–400
cause value, 371
CCA (Credit Control Answer), 359–62
CCE (Connected Consumer Electronics), 8–9
CCR (Credit Control Request), 359–62
CCS (common channel signaling), 121–2
CDF (charging data function), 258–69, 357–62
    see also charging
CDMA, 117, 402–4
CDRs (Charging Data Records), 357–8
CelB, 117
Centrex services, 387–8, 394–5
CGF (Charging Gateway Function), 358–62
CgPN (calling party number), 368–78, 382–9
challenge-response mechanisms, authentication functions, 255–69
charging, 4, 7, 10–15, 18, 31–2, 68–9, 233–4, 258–69, 274–85,
    299–300, 330–6, 351–62, 365, 394–402, 416, 451
    see also ad-financing; app stores; business models; CDF;
        donations; pay-per-use…; revenue…; sponsorship…;
        subscription fees
    billing contrasts, 351
    community-based workflows, 37
    concepts, 7, 31–2, 233–4, 258–9, 351–62, 365, 394–400,
        401–2, 416, 451
    definition, 351–2
    legacy VAS, 394–400
    mechanics, 357–62
    methods, 7, 31–2, 233–4, 258–9, 351–62, 401–2, 451
    online/offline charging models, 357–62
    sources of revenue, 352, 451
    tables, 395–400
charging data function, 258, 359
charities, 352, 356–62
chat, 22–3, 49–58, 64, 405, 414–16, 418–20, 450
cheapness benefits, 9
chess applications, 346–9
churn rates, definition, 15
CIPID (Contact Information in Presence Information Data
    format), 344–6
circuit switched network, 105, 378
circuit-switched technologies, 14–15, 21–4, 25–6, 68–9, 105–9
client-server architecture, 17–21, 49–58, 148–222, 229, 273–85
client-side applications, 49–52, 53–5, 56–8, 270–85, 336, 450
    see also web…
    definition, 49–50

client-side architecture, 9–10, 16–26, 49–58, 336
client transaction, 156, 203
ClientLogin, 97–104
cloud computing capabilities, 38
CNP, 339
co-location concepts, 227–32
co-workers relationship circles, 5
coax concepts, 110–11
codecs (coder-decoder), 9–10, 106–43, 210–22, 332–6,
    366–400
    see also decoding; encoding; G.711; PCM
    definition, 106–8
codes, 105–43
    see also DPCM; PCM; sampling
Combinational Services, 14, 230–2, 406–7
Comet, 104, 335
command sequence, 152–3, 196
common channel signaling, 121
communication end-points, SIP/media sessions, 146–7
communication infrastructures, requirements, 2, 8–9, 14
Communication Service layer, 16–20
communications industries
    future prospects, 35, 38, 449–51
    privacy regulations, 41–2, 333–6
    value creation/capture, 33–48, 425–30, 431–47, 449–51
community-based workflows, 37, 427–30
complementary good, 48
complexity, 42–4, 400, 406–16
conference bridging, 25–6
conference calls, 12, 15, 23, 25–6, 52, 129–43, 302–3, 333–6,
    427, 434
conferencing, 333
confirmed dialogs, 198–209, 376
Connected Consumer Electronics, 8
connected line presentation, 383–6
connected line presentation restriction, 383–6
connected line presentation, interworking with legacy
    networks, 381, 383–6
connectivity mechanisms, 3–9, 25–6, 35–8, 39–42, 52–8, 62–4,
    136–43, 229–32, 251–69, 293–328, 332–6, 352–62, 392–3
    see also access...
consumer goods software, 27, 33, 401–30
    see also applications; RCS
consumer products and services, 27
consumers, 27, 33, 36–7
Contact, 162
contact addresses, 133–6, 165–6, 179–83, 202–9, 232–4,
    249–85, 288–92, 408–14
    see also binding information; target sets
contacts, relationship circles, 5, 162–3
content, 4, 203–9, 215–22, 325–6
    see also service layer
Content-Length headers, 215–22, 325–8

Content-Type headers, 215–22, 325–6
contracts, 28
control plane, 118–29, 139–43, 145–222, 223, 226–32, 244–9,
    278–85, 293–4, 330–6, 365–400
    see also CSCF; ISUP; MGCF; SBC; signaling; SIP
converged fixed and multimedia services, 22–3, 25–6, 35, 39–
    42, 59, 60–80, 93–104, 271–85, 329–36, 402–4, 449–51
coordination of workflows, 427
copper wire connections, 110–11
core IMS network, 4, 18–19, 20–1, 23–4, 39, 50–8, 62–4, 68–9,
    223–328, 337–46, 358–62, 389–400, 403–4, 414–30,
    437–47, 475–80
    see also control plane; DNS; HSS; I-CSCF; IMS network;
    media plane; P-CSCF; S-CSCF; services; SLF; user plane
    concepts, 63–4, 226–9, 232–42, 293–4, 394–400, 403–4,
    414–30
    definition, 63, 226–9, 232–4
costs, 1–2, 13, 24, 27–33, 34–5, 38, 43–4, 329–36, 423–4,
    427–8
coupling concepts, service composition, 61–4
courier companies, 47–8
CPG, 370–1, 388–9
CPM, 416, 421
createProxyBranches, 76–80
creative destruction, 30
creativity, 42–4
credit cards, 5, 36–7
credit notifications, 361–2
critical technical systems, 32–3
cross-network proxy (CNS), 339–46
crowd sourcing, 36, 426–7
CS (circuit-switched), 14–15, 21–4, 25–6, 105–9, 116–29,
    207–9, 226–328, 358–62, 363–400, 406–30, 431–47, 449
    see also 3G; ICS; IN (Intelligent Network); ISDN; ISUP;
    PLMN; PSTN; SCP; SRVCC; TDM
    application servers, 292–3, 358–62
    basic services, 390–2
    interworking with legacy networks, 363–400, 431–47
    legacy VAS, 389–400
    network services, 390–2, 431–47
    protocol mapping, 364–71
    supplementary services, 380–9, 390–2, 433–47
    video communications, 378–80
CS-ACELP, 116
CS-MGW, 246–9
CS/GSM-specific legacy VAS services, 393–400
CSCF (call state control function), 10–13, 21–2, 25–6, 57–8,
    62–9, 83–93, 94–104, 224, 226–328, 330–6, 339–46,
    392–400, 437–47, 475–6
    see also Cx...; Dx...; Gm...; I-...; Mw...; P-...; S-...
CS networks, 105
CSeq (command sequence), 151, 152–4, 163–83, 192–4,
    196–7, 252–69, 325–8

CSI (Combinational Services), 14, 230–2, 406–7
CSRC (contributing source identifiers), 114–17
CT (client transaction), 157–61, 168–83, 197–209, 273–85
Cx, 228
Cx reference point, 63, 226–32, 238–40, 254–69
   see also CSCF; HSS

D
DAR, 73
data connection, 185–94
data link layer of OSI reference model, 109–11
   see also Ethernet
data model, presence, 338, 343–6
data packet loss, digital speech transmission, 108–9, 113–17, 333–6
data transmission using real-time transport protocol, 111–43
de-registration, 136–7, 260–9, 310–12
decapsulation, 111–43
decoding, 105–43, 366–78
   see also codecs
default application router (DAR), 73–5
default handling, 65–9, 72–5
default listeners, VoIP, 123–4
deferred content delivery, RCS, 405, 421–2
DELETE, 104, 340–6
deployment concepts, 49–58, 73–80, 227–328, 331–6
de-registration, 136
design-thinking methodologies, 42–4, 436, 445–7
destination parties, 239–40, 247–9, 252–69, 270–92, 295–8
Deutsche Telecom, 423
developers, 1–26, 28–48, 49–58, 59–80, 81–104, 223, 293–328, 329–36, 351–62, 401–30, 431–47, 449–51
   see also RCS
   application-creation methods, 59–80, 81–104, 401–30, 449–51
   black box views, 15–17, 20–1, 49–51, 331–6
   business models, 27–48, 351–62, 401–2, 425–30
   capital goods software, 27, 32–48, 81–104, 401–2
   consumer goods software, 27, 33, 401–30
   evolution of IMS architecture and services, 10–14, 431–47, 449–51
   global interoperability, 39–42, 52–8, 329–36, 402–4, 422–3, 426–30, 432–47, 449–51
   Open APIs, 28–32, 402–4
   personal data, 31–5
   service deployment patterns, 49–58, 331–6
   the service story, 15–26
device-related information, 'Presentity' concepts, 343–6
devices, 3–26, 39–42, 50–8, 59–60, 68–9, 132–43, 226–328, 329–36, 343–9, 401–30, 449–51
DFC (Distributed Feature Composition), 61–4, 72–5
   see also service chaining
Dh reference point, 236

   see also SIP-ASs; SLF
DHCP, 131, 134, 136, 139–43, 252–69
   see also IP
diagramming techniques, 44–8, 81–104
dialogs, 168, 197–222, 242–9, 252–69, 270–85, 296–8, 318–19, 324–8, 372–89, 434
   see also Call-ID...; From tag; To tag
Diameter, 254, 255, 364–5
Diameter base protocol, 254
Diameter protocol, 224–32, 236, 254–69, 275–85, 292, 309–16, 321–3, 358–62, 364–5
   see also LIA; LIR
differential PCM, 106
digital circuits, definition, 105
digital economy, 21, 29, 30, 35
digital speech transmission, 105–43, 333–6
   see also VoIP
direct PSI triggering, 305, 307–12
direct transfer application part, 391
DISCONNECT, 408–14
dispatcher, 19–20
Display names, 199–209
displays, 8
distinct PSIs, 305–12
Distributed Feature Composition, 61, 72
diversity issues, devices, 5–6, 262–6, 330–6
DLNA, 13–14, 450
DND *see* web-based do-not-disturb use-case implementation
DNS, 57–8, 133–4, 138–43, 166–94, 196–7, 227–328, 475–6
   see also domain names; ENUM; NAPTR
do-not-disturb web-based use-case implementation, 81–2, 93–104, 471–3
Document Object Model, 104
doGet, 70
doInvite, 76–80, 95–104
domain name server, 134, 143
domain names, 18–19, 142–3, 166–94, 248–9, 251–69, 277–85, 288–92, 306–12, 369–71, 386–7, 435
   see also DNS
domains, 18–19, 132–43, 166–94, 248–69, 277–85, 288–92, 306–12, 369–71, 386–7, 435
   see also ENUM; realms
donations, 352, 356–62
doPost, 70
doRequest, 96–104
doRequestForwarding, 91–3
doSipEvent, 70–5
doSuccessResponse, 70, 89–93
DPCM (differential PCM), 106–9
DTAP (direct transfer application part), 383–9, 391–2, 435
DTMF (dual-tone multifrequency), 218, 317–20, 367–78
DVI4, 116–17
   see also VDVI

Dx, 228
Dx reference point, 63, 226–32, 236, 238–40
   see also CSCF; SLF
dynamic subscriber data, 141–3

E

E-CSCF (emergency CSCF), 233–4
E1 communication links, 107, 124–9, 245–9, 366–78
E.164, 132–3, 233–4, 288–92, 329
   see also ENUM
early media, 191, 206–9, 372–6
   see also backward…; bothway…; forward…
early media authorization, 374–5
eBay, 352
ECE, 104
economies of scale, 27–8, 31
eCharts, 70, 81–2
echo cancellation, 333–6
EchoServlet, 78–80
economics, 1–2, 13, 27–48, 351–62, 393–9, 401–2, 426–30
   see also business models; charging
   capital goods software, 27, 32–48, 81–104, 401–2
   economies of scale, 13, 27–8, 31–2, 34–5
   factors of production, 32–3, 36, 401–2, 426–30
   nexus of contracts, 34–5, 37–8
   Open APIs, 28–32, 402–4
economies of scale, 13, 27–8, 31–2, 34–5
egress routing, BG, 242–9
ELSE, 90–3
emergencies, 233–4, 330, 434–5
EMS, 435
enablement value proposition of the network vision, 3–4,
   38–42, 56, 329–49, 404–5
encapsulation, 108–9, 111–43, 245–9
encoding, 105–43, 366–78
   see also codecs
end-point APIs, 9–13, 49–58, 146–7, 159–61, 331–6
end-users, 1–26, 29–32, 33–48, 49–58, 120–9, 201–9, 281–5,
   300–4, 351–62, 372–6, 392–3, 401–30, 446–7
Enhanced Address Book (EAB), RCS, 405
Enhanced Call, 402
Enhanced Chat, RCS, 405, 414–16
Enhanced Messaging, 414
   see also MMS; SMS
   RCS, 405, 414–16
Enhanced Presence, RCS, 405
eNodeB, 437–47
Enriched Call, 406, 419–20
Enriched Call Image, 418
Enriched Call Video, 418
Enriched Call through image/video sharing, RCS, 402, 405–14,
   419–20
enriched messaging – chat, 415

Enriched Phone Book, RCS, 402, 405, 406, 416, 417, 420
enterprise opportunities, 13
enterprise platforms, 32, 37–8
   see also Oracle; SAP
enterprise subscriber service data/logic, legacy SCP, 395–400
enterprise systems, 427–30
ENUM (E.164 numbering plan), 227–328, 369–71
   see also E.164
   concepts, 288–92, 298–300, 311–12, 313, 369
   definition, 288–90
   phone numbers in IMS, 286–92, 297–8
EPC, 34–5, 433, 434–5
EPS (Evolved Packet System), 434–47
Ericsson, 28, 31, 34, 40, 70, 81–2, 93–4, 400
   Composition Engine, 70, 81–2
   economies of scale, 28, 31
   'Next-generation intelligent networks: migrating to IMS'
     paper, 400
error codes *see* response codes
establishment of SIP sessions, 161–83, 194–7, 209–22, 232–4,
   237–49, 252–69, 270–85, 286–92, 293–328, 363–400,
   407–30, 438–47
   see also INVITE
Etag mechanism, presence, 342–6
Ethernet, 3, 6, 110–11, 132–43, 186–94, 230–2
evolution of IMS architecture and services, 10–14, 431–47,
   449–51
   see also ICS; LTE; SRVCC; VoLTE
existing services, SOA, 59–80, 81–104
expired bindings, 252, 260–1
explicit user relationships, 122–9, 347
extensions to existing behaviours, 6, 334–6, 375–6, 436
external DNS (eDNS) contrasts, internal DNS, 185, 229,
   240–2, 270–85
external DNS, 185

F

Facebook, 6, 403, 422–3
factors of production, 32–3, 36, 401–2, 426–30
   see also economics
family relationship circles, 5
favorite links, 405
feature composition, 61–4, 433–47
   see also SOA
feature tags, CSI, 406–16, 424–5
fibre connectivity, 6, 26, 230–2, 332–6
file/image/video sharing, 14, 22–3, 49–58, 333–6, 349, 405–22,
   425–30, 450
File Transfer, 414–15
File Transfer sessions, RCS, 405, 414–16, 418–20, 425, 450
final responses, 148–57, 174–83, 185–97, 204–9, 257–69,
   293–328, 359–62, 369–89, 424–5
   see also ACK

find-and-connect services, 4, 21
  see also IMS
finding functions, 4
fixed networks, IMS background, 1–26, 39–42, 62–3, 223, 271–85, 332–6, 449–51
follow-on calls, 302–3
fonts, 325–8
forking, 202
forking concepts, 4, 201–9, 281–5, 300–4, 372–6, 392–3, 446–7
forking proxy, 201, 203
forward media, 206–9, 330–6, 374–6
forwarding proxy, 201
forwarding to extension logic, 334, 335–6
FQDN (fully qualified domain name), 174–83
framing, 365
freedom of speech, 41–2
freephone, 394–5
freeware, 351, 356–62
fremium models, 353
friends relationship circles, 5–6
  see also social networks
From tag, 168–9, 198–209, 242–9, 252–69, 272–85, 286–92, 324–8, 382–9
  see also dialogs; INVITE
front-end service logic, legacy SCP, 394–400
FTTH (fibre to the home), 230–2
  see also fibre…
fully conversational voice, 331–6
fully qualified domain name, 174
future prospects, communications industries, 35, 38, 449–51

G

G.711, 106–9, 111–17, 213, 366–78
  see also codecs
G.722, 108, 111–13, 115–17, 213, 375
G.723.1, 108, 116, 119, 127–9, 131, 367, 379–80
G.726, 108, 116, 367
G.728, 116
G.729, 108, 116, 367
G.992.1, 230
Ga reference point, 358–62
gaming, 5, 16–17, 21, 51–8, 346–7, 353
Gateway MSC, 390–2
gateways, 10–13, 24–5, 125–9, 138–43, 226–328, 358–62, 363–400
geo-location, 405, 421–2
Geopriv, 344–6
GET, 92–3, 96–104, 318–20, 341–6, 417
getHeader, 95–104
getNextApplication, 73–5
getProxy, 75–80
getSession, 91–3

GGSN, 252–69, 337–46
'ghost ringing', 346–7
global failure responses, 150–7
global interoperability, 1–4, 6, 36–48, 52–8, 329–36, 402–4, 422–3, 426–30, 432–47, 449–51
  see also multi…
global number strings ('+' character), 286
Gm, 228
Gm reference point, 63, 226–32, 237–40, 324–8, 437–47
  see also P-CSCF; SIP; UE
GMLC, 337–46
GMSC (gateway MSC), 390–2, 397–400
Google, 18, 30, 31, 47, 55–6, 96, 354–5
  AdMob, 354–5
  Android community, 18, 30, 31, 47
  Calendar, 55–6, 93, 96
  Maps, 34
Google AdSense, 352, 354–5
Google Android, 30, 47
Google Calendar, 55, 93, 96
GPRS, 139–43, 250–1
Grandstream GXV3000, 219–22
granularity concepts, 60–4, 402–4
  see also SOA
group calls, 392–3, 405
GSM (Global System for Mobile Communications), 3–4, 12–13, 21–2, 26, 34, 39–40, 56, 68–9, 105, 107–9, 116–17, 121–9, 234, 260, 293–4, 324, 332–6, 345–6, 357–8, 364–400, 421, 431–47
  see also 2G; 3G…; PLMN; RAN
  IMS comparisons, 293–4, 324
  migration to IMS, 68–9, 431–2, 445–7
  protocol mapping, 364–71
gsm Service control function, 390
gsm Service switching function, 390
GSM/UMTS supplementary services and corresponding IMS supplementary services, 128–9, 380–9, 482, 491
GSMA (GSM Association), 20–1, 34, 49–58, 335–6, 337, 390–2, 398–400, 401–6, 408–30, 431–47
  see also RCS; VoLTE
  OneApi project, 20–1, 34, 58, 335–6
  PRD IR.92 (VoLTE), 433–47
GSMA OneAPI, 20, 335–6
GSMA RCS, 420, 422
gsmSCF, 390–2
gsmSSF, 390–2, 398–400
guaranteed signaling, 185–94

H

H.223, 379–80
H.225, 145
H.245, 145, 379–80
H.248, 125–9, 293, 320, 379–80, 391

H.261, 117
H.263, 117, 119, 222, 332, 379–80, 413
H.264, 114–17, 332, 423
H.323, 145
hackers, future prospects, 451
half-duplex voice communications, 331–6
    see also push-to-talk
handover procedures, SRVCC, 431–47
hardware sales, 352, 356–62
HD video, 2, 13
headers, SIP headers, 18–19, 63–80, 95–104, 148–57, 162–83,
    193–4, 196–7, 203–9, 215–22, 234, 242–9, 252–69,
    272–85, 481, 483–7
header parameter, 163
helpdesks, 310–11
high layer compatibility, 369
History-Info SIP header, 389
HLR, 10, 24
HLR (home location register), 10–13, 24, 378–89, 390–2
home concepts, 12–13, 24–6, 234, 237, 332–6
    see also residential opportunities
Home IMS network, 234, 237–8, 243–4, 251–69, 445–7
home IMS realm, 251
home location register, 391
home networks, 12–13, 24–6, 234, 237–8, 450
home realm, 134
Home Subscriber Server, 228, 234–5
hop counters, 167–97, 282–5, 298–300, 368–78
Houghton, Dr Robert, 27–48
HSPA (High Speed Packet Access), 14, 23–4, 136–43, 230–2,
    332–6, 437
    see also Turbo3G
HSS (Home Subscriber System), 10–13, 25–6, 57–8, 62–9,
    83–93, 94–104, 224, 226–328, 330–6, 337–46, 435,
    437–8, 446–7, 476–80
    see also Cx…; I-CSCF; IFC; IMPI; IMPU; Sh…; SLF;
    subscriber profiles
    concepts, 227–8, 233–69, 275–85, 292, 293, 295–8, 308–12,
    314–16, 321–3, 330–6, 337–46, 435, 437–8, 446–7
    definition, 233, 234–5, 238–9
HTML-5, 41–2, 102–4, 326, 336, 450
HTTP (Hyper Text Transfer Protocol), 16, 50, 53–8, 60, 70–5,
    87, 89–104, 143, 145, 317–20, 331–6, 345–6, 434–5
    see also REST style APIs; SIP
HTTP/1.1, 149
HTTP GET, 92, 96, 99, 100, 103, 318
HTTP interface, 87, 89
HTTP listener, 102, 104
HTTP PUT, 97, 100
HTTP servlet, 70, 91, 92, 93
HTTP URL, 93
HTTPServlet, 70
Huawei, 28

human rights, 41–2
human-centric concepts, 4–6, 17–21
    see also personalization…
'hybrid' radio, 406, 410–11
Hyper Text Transfer Protocol, 145

I

I-call-you models, 6
I-CSCF (interrogating CSCF), 63, 159–61, 226–328, 358–62,
    376
    see also HSS; SLF
    concepts, 226–34, 236, 237–42, 245–9, 251, 358–62
    definition, 233, 236, 238–9, 279
I2 reference point, 437–47
I3 reference point, 443–7
IAD, 8
iAds, Apple, 352, 354–5
IAM (initial address message), 367–78, 380–400
IANA (Internet Assigned Numbers Authority), 481
    see also IETF
IARI (IMS Application Reference Identifier), 16–17, 18–20
IBCF (interconnect border control function), 226–328
    see also N-SBG
    concepts, 238, 242–9, 253–69, 270–85, 306–12, 313
    definition, 242
IBM, 31, 32, 35, 39, 43
ICMP error message, 262
icons, 50–8
ICS (IMS Centralized Services), 225, 431–47
    see also MMTel
    definition, 431–2, 441–3
ICSI (IMS Communication Service Identifier), 17–21
IDP (Initial DP), 396–400
IEEE, 62, 230
IETF (Internet Engineering Task Force), 14–15, 16–17, 18–19,
    26, 74–80, 107, 114–18, 132–3, 145–222, 223–6, 337–46,
    481–91
    see also IANA; PIDF; RFC…; SIP
    list of RFCs, 223–6
IFC chaining, 62–4, 298–300, 334–6
    see also service chaining
IFC (initial filter criteria), 10–13, 19–20, 21–2, 57–8, 62–9,
    235, 258–69, 295–300, 327–8, 330–6, 393–400, 446–7
    see also HSS
    definition, 62–3
iLBC (Internet Low Bitrate Codec), 107–8, 220–2
ILCM, 294–8
ILSM, 294–8
    see also S-CSCF
IM application server, 416
IM-MGW (IP multimedia–media gateway), 243–9, 363–400,
    445–7
    definition, 244, 363, 365–7

protocol mapping, 364–71

IM-SSF (IP multimedia service switching function), 59–60, 66–9, 396–400

Image Share, RCS, 407–10, 412–14, 415, 418–19, 421–2, 425–30, 450

images, 14, 22–3, 49–58, 333–6, 402, 405–14, 415, 418–19, 421–2, 425–30, 450

immune, 348

IMPI (IMS private user identity), 249–69
   see also NAI; PUI

implicit registration, 263

implicit user relationships, 122–9

IMPU (IMS public user identity), 235–8, 249–69, 270–85, 291–2, 302–4
   see also IRSs; PUI; URIs

IMS application servers, 60, 62, 71, 75, 93
   see also application…; SIP-ASs
   concepts, 69–80

IMS architecture, 9–14, 20–1, 39–42, 59–80, 83–104, 223–328, 330–6, 408–10, 412–14, 416–17, 431–47, 449–51
   see also core IMS network; SIP; VoLTE
   closer look, 232–49, 331–6
   evolution of IMS architecture and services, 10–14, 431–47, 449–51
   global view, 224–32
   overview, 9–13, 25, 62–4, 224–32

IMS call control, 93

IMS core network *see* core IMS network

IMS (IP Multimedia Subsystem), 1–26, 35, 37–48, 59–80, 207–9, 223–328, 351–62, 363–400, 401–4, 431–47, 449–51
   see also ISC; MMTel; presence
   application-creation methods, 59–80, 449–51
   benefits, 1–4, 6, 7, 9, 10–13, 38–42, 47–8, 392–3, 425–30, 449–51
   black box views, 15–17, 20–1, 49–51, 331–6
   business case, 38–48, 351–62
   business model examples, 47–8
   concepts, 1–26, 35, 37–42, 59–80, 207–9, 223–328, 363–400, 401–4, 449–51
   definition, 1–2, 16, 21, 37, 293–4, 449
   evolutionary scenarios, 10–14, 431–47
   future prospects, 35, 38, 449–51
   GSM comparisons, 293–4, 324
   interworking with legacy networks, 363–400, 431–47, 449–51
   legacy VAS, 363–4, 389–400
   non-IMS networks, 363–400
   purposes, 1–26, 37–42, 223–7
   reachability factors, 47–8, 426–30
   releases, 223–6, 230
   services, 12–26
   standards, 18–19, 127–9, 148, 223–6, 491

supplementary services, 55–8, 128–9, 331–6, 380–9, 433–47, 482, 491

universal service access, 1–4, 229–32

value propositions, 3–4, 38–42, 56, 329–49, 389–400

Web 2.0, 20–1, 68

IMS network, 4, 9–13, 18–19, 20–1, 23–4, 39, 50–8, 59, 61–80, 86–93, 138–43, 157–61, 207–9, 223–328, 334–6, 363–400, 436, 445–7, 475–80
   see also application servers; core IMS network; ENUM; ISC; registration
   concepts, 63, 207–9, 223–328, 363–400
   early media, 207–9
   establishment of sessions, 161–83, 194–7, 209–22, 232–4, 237–49, 252–69, 270–85, 286–92, 293–328, 392–400, 438–47
   interworking with legacy networks, 363–400, 431–47, 449–51
   introduction, 63, 223–32
   legacy VAS, 389–400
   logical separation, 226–32
   media plane, 63–4, 207–9, 223, 227–32
   messaging communications, 324–8, 330
   network border gateways, 10–13, 138–43, 229, 242–9, 313, 363–400
   phone-number usage, 285–92
   protocol mapping, 364–71
   public services, 304–12
   service chaining, 59, 61–80, 86–93, 157–61, 298–300, 334–6, 372–6, 436, 445–7
   supplementary services, 55–8, 128–9, 331–6, 380–9, 392–3, 433–47, 482, 491
   termination of sessions, 162–83, 212–22, 228, 234, 242–3, 270–85, 295–300, 302–3, 321–3, 341–6, 368–78, 409–14, 433–47
   unregistered service invocation, 320–3
   user interactions, 316–20

IMS roaming, 234

IMS service composition, 64, 82, 93

IMS service creation, 60–9
   see also applications; service chaining; service composition

IMS services, 22, 62, 93, 223, 280

IMS Services API, 16–17, 50–8
   see also JSR281 API

IMS stack, 17, 18–20, 87–93

IMS supplementary services and corresponding GSM/UMTS supplementary services, 128–9, 380–9, 482, 491

IMS-AKA (IMS authentication and key management), 434–5

IMS-ALG (application level gateway), 230–2

IMSI (International Mobile Subscriber Identity), 250–69, 287

IN (Intelligent Network), 55–6, 298, 357–8, 363, 389–400, 445–7
   see also CS; legacy VAS

in-session transactions, 156–7, 167–8, 173–83

INAP (Intelligent Network application part), 298
inbound SIP servers, 143, 167–83, 279
indirect PSI triggering, 307–12
the individual, 35–7, 403–4, 426–30
    see also personalization…
industrial change, 33
industrial subsystem, 43
industry verticles, 37–8
    see also enterprise platforms
INFO, 252–69, 482
infrastructure requirements, 2, 4, 8–9, 14
ingress routing, BG, 242–9
initial registration, 133–6, 143, 232–6, 237–40, 246–9, 250–69, 270–85, 295–8, 424–5
initial request message routing, 163–83, 185–94, 234–49, 270–85, 295–8, 367–78, 424–5
innovative/differentiating services, market needs, 1–2, 21, 30–2
instant messaging, 6, 325–8, 425
intellectual property, 36
internal DNS (iDNS), external DNS contrasts, 185, 229, 240–2, 270–85
Internet, 33, 36–7, 59–60, 67, 70, 93–104, 107–43, 289–90, 329–49, 402–4, 426, 451
    see also PS
interrogating proxies, 139–43
Interval Too Brief, 261
investors, 351–62
'invisible change' requirements, 7–8
INVITE, 17–19, 21–2, 64–9, 70–5, 78–93, 98–104, 148–57, 159–61, 162–222, 244–9, 252–69, 270–92, 293–328, 333–6, 346–9, 360–2, 367–78, 382–9, 400, 406–16, 421, 446–7, 482, 484–5, 488–90
    see also ACK; establishment of SIP sessions; From tag; To tag
    concepts, 148–57, 159–83, 186–97, 211–22, 244–9, 252–69, 270–92, 293–304, 306–23, 333–6, 360–2, 367–80, 382–9, 400, 406–16, 421, 446–7, 482
    definition, 148, 194, 482
invitePurpose, 91–3
IP, 2, 4, 14, 16–22, 24–6, 62–4, 105–43, 162–83, 184–94, 203–9, 210–22, 223–32, 245–9, 251–69, 272–85, 288–92, 329–36, 363–400, 410–14
    see also DHCP; Mb…; P-CSCF; RTP
    addresses, 4, 251–69, 288–92
    basic properties, 4
    bearers, 16–22
    encapsulation, 112–13, 245–9
    infrastructure requirements, 2, 4, 14
    interworking with legacy networks, 363–400, 431–47
    VoIP, 14, 105–43, 405
IP-based real-time communications, 105–43
    see also VoIP
IP-CAN (carrier access network), 229–32
IP centrex, 13

IP-SM-GW, 435–47
iPhone, 33, 34, 38, 47
IPSec, 7, 434–5
IPTV, 349, 449
IPv4, 110, 112–13, 119, 184–94, 217, 220–2, 242–9, 272–85
IPv6, 110, 132, 184, 217, 230–2, 242–9, 272–85
IPX (IP transmission network), 278–85
IR.74, 412
IR.79, 408, 409
IR.92, 418
IRSs (implicit registration sets), 250–69, 275–85, 302–3
    see also IMPU; registration
ISC (IMS service control interface), 21, 62–4, 66–9, 78–80, 83–93, 94–104, 226–328, 330–6, 392–400, 435–47
    see also MMTel; S-CSCF; SIP
    definition, 294–5
ISC chaining, 334
ISC reference point, 294
ISDN, 21, 68–9, 105, 107, 121–9, 168, 244–9, 298, 332–6, 363–400, 442–7
    see also CS; ISUP
ISIM (IMS subscriber identification module), 250–69, 434–5
ISO (International Organization for Standardization), 109, 116
ISPs, 13, 43–4, 52–3
ISUP (ISDN User Part), 121–9, 168, 244–9, 292, 293–4, 298, 306–12, 332–6, 363–400, 442–7
    see also control plane; ISDN; MTP3
    protocol mapping, 364–71
IT/computing networks, IMS background, 1–26, 39–42
IT inside, 38
ITU TISPAN, 14, 15, 26, 337–46, 381, 431
ITU-T, 14, 26, 107, 114–18, 121–2, 132–3, 145, 230, 233–4, 288–92, 370–1
    see also G…; H…
    E.164, 132–3, 233–4, 288–92
    Q.850, 370
    Q.1902, 121
    X.880, 148
IVR (interactive voice recognition), 82–93, 335–6, 377–8

J

JAIN SLEE, 69–75
Java, 16–17, 50–8, 67–8, 73–5, 81–104, 340–6
Java Enterprise Edition (Java EE), 60, 64, 67–8, 69–80, 83–104
    see also off-the-shelf software technologies
Java EE AS, 70
JavaME, 16–17, 21, 50–8
JAX-RS API, 83
    see also Java…; XML
jitter problems, 108–9, 113–17, 333–6
JPEG, 117
JSON, 96–7
JSR116 API, 50–1, 60, 69–75, 77–80

JSR180 API, 16–17
JSR224 API, 67–9
JSR281 API, 16–17, 21, 50–8
    see also IMS Services API; JavaME
JSR289 API, 18–20, 50–1, 58, 65, 67–80, 81–104, 335, 455–75
JSR311 API, 95–104
JSR315 API, 70–5

K

key management, 434–5

L

L8, 116
L16, 116, 220–2
LANs (local area networks), 111–12, 136–43, 230–2, 251–69
laptops, 3–4
latency, 107–9, 113–14, 449
lawful intercepts, 41
layered architecture, 125
legacy networks, 3, 7, 11–13, 21–2, 23, 39–42, 55–6, 59–60,
    68–9, 128–9, 226–328, 329–36, 357–8, 363–400, 405,
    417, 421–2, 431–47, 449–51, 482, 491
    see also CS; IM-MGW; MGCF
    definition, 363
    protocol mapping, 364–71
legacy SCP, 394–400
legacy services, 363
legacy VAS, 363–4, 389–400
    see also IN (Intelligent Network)
    categories of services, 393–9
    CS/GSM-specific services, 393–400
    definition, 389–90
    safeguarding investment challenges, 393–9
    starting point, 389–93
LIA (location information answer), 308–16, 321–3
licensing agreements, 352, 353–4
LinkedIn, 403–4
LIR (location information query), 308–12, 313–16, 321–3
listing considerations, service deployment patterns, 49–58
load controls, presence, 342–6
local area network, 230
location considerations, 98–104, 137–41, 223, 227–32, 236,
    251–69, 279–85, 307–16, 405, 421–2
    see also SLF
location servers, 141–3
LOGGER, 455–71
logging functions, 60
loose routing, 165–83
Lotus Notes, 33
LPC, 116
LTE (Long-Term Evolution), 3, 7, 14, 22, 23–4, 26, 28, 34–5,
    39–40, 230–2, 252, 330–6, 402–5, 418–19, 422–3, 425,
    429, 431–47, 449
    concepts, 402–5, 418–19, 422–3, 431–47, 449
    definition, 431–3

M

M-MGW, 24–5, 243–9, 391–2
M2M (machine to machine), 38, 450
M3UA (MTP3 user adaptation), 245–9, 365–78
MAC addresses, 111
Ma reference point, 239, 307–8, 327–8, 392–3
MANs (metropolitan area networks), 230–2
manufacturers, 28, 35, 39–42, 402, 450–1
    see also devices
MAP (mobile application part), 148, 391–2
mapping needs, protocols, 364–71
MAR (multimedia authentication answer), 255–69
MAR (multimedia authentication request), 255–69
Market, 353
market needs, 1–2, 20–1, 30–2
mashup applications, 93–104
mass-market services, 394–5
Max-Forwards header, 167–83, 252–69, 325–6, 385–9
Mb reference point, 226–32, 316–20, 437–47
Media Gateway nodes, 12, 284–5
media gating, concepts, 284–5
media plane, 63–4, 119–29, 206–22, 223, 227–32, 278–85
    see also core IMS network; media transmission; user plane
media players, 353
media production company example, diagramming techniques,
    44–8
Media Resource Function, 83
media server, early media, 207–9, 372–6
media sessions, SIP sessions, 145–7, 173–83, 191–222, 283–5,
    293–328, 367–400
media streams, 22–3, 206–22, 225, 229–328, 329–36, 372–89, 392
media transmission, 209–22, 223, 225, 227–32, 270–85, 312–20,
    328, 330–6, 379–89
    see also media plane
mergers and acquisitions, 38
MESSAGE, 95–104, 162–83, 233–4, 252–69, 320–3, 324–8,
    421, 435, 482
message body, 148
message structure, SIP messages, 162–83
message transfer part, 366
messaging communications, 2, 6, 8–9, 11–13, 14–15, 17–21,
    24–6, 49–58, 64, 148–222, 224–5, 324–8, 330, 357,
    392–3, 405, 414–16, 420–1, 425, 432–47
    see also MMS; SMS
    concepts, 2, 8–9, 324–8, 357, 414–16, 420–1
    definition, 324–5
messaging sessions, 325, 328
META-INF, 74
method, 148
Metro Ethernet, 3, 6

Mg reference point, 437–47
MGC (media gateway controller), 207
MGCF (Media Gateway Control Function), 12–13, 226–328, 358–62, 363–400, 439–47
  see also control plane
  concepts, 243–9, 284–5, 290–2, 306–12, 358–62, 363–78
  definition, 244, 363, 365–7
  protocol mapping, 354–71
MGW (media gateway), 12, 24–5, 125–9, 226–328, 365–400, 440–7
Mi reference point, 239
Microsoft Office, 33
mid-call announcements, 302–3, 443–4
mid-point applications, 49–50, 55–8
  see also supplementary services
middleware support, future prospects, 450
migration considerations, 68–9, 431–2, 445–7
MME (Mobility Management Entity), 437–47
MMI (man–machine interface), 129–43
MMS, 403–5, 414–16, 420–1, 426
  see also messaging…
MMTel (Multimedia Telephony), 7–8, 14–15, 18, 19–20, 21–6, 49–50, 55–6, 63–4, 66–9, 83–93, 94–104, 114, 129, 293–4, 316–20, 329–49, 359–62, 381–2, 392–400, 405, 418–19, 431–47
  see also ICS; SIP-ASs; SRVCC; VoIP
  benefits, 21–3, 26, 329–36, 392, 431–47
  calendar-based routing example, 336
  concepts, 7–8, 14–15, 21–6, 55–6, 66–9, 83–93, 94–104, 114, 129, 293–4, 316–20, 329–49, 359–62, 381–2, 392–400, 405, 418–19, 431–47
  costs, 24, 329–36
  definition, 7–8, 21–3, 55, 66–7, 129, 329–41, 392
  driving requirements, 23–6, 329–36
  evolution of IMS architecture and services, 431–47
  extensions, 334–6, 436
  forwarding to extension logic, 334, 335–6
  northbound reference point, 294, 330–6
  presence, 330, 331, 333–6
MO (mobile oriented), 65–9
mobile broadband connectivity, 3–4, 6, 25–6, 35–8, 136–43, 332–6, 352–62, 402–4, 405, 418–22, 427, 449
mobile broadband platform, 33–8
mobile handsets, 1–26, 28, 31–2, 34–7, 47–8, 223, 262–6, 271–85, 298–300, 329–36, 354–62, 401–4, 449–51
  constraints, 354–5
  features, 5–6, 8–9, 47–8
  importance, 4–5, 47–8
  major players, 28, 423
Mobile IMS, 271–85, 367–400
mobile networks, 1–26, 28, 33–5, 39–42, 243–9, 252–69, 271–85, 332–6, 351–62, 423, 449–51
  see also PLMN

mobile services switching centre, 391
mobile station, 391
mobile switching services centre, 121
mood, presence, 344–6
MP2T (MPEG-2 transport streams), 117
MP3s, 44
MPA, 116
MPEG, 116
MPV (MPEG video streams), 117
MRE, 226–328
MRF (Multimedia Resource Function), 62–4, 83–93, 316–20, 335–6, 373–8
MRFC (media resource function controller), 10–13, 226–328, 358–62
  see also user interactions
  definition, 316–17
MRFP (media resource function processor), 10–13, 226–328, 332–6
  see also announcements
  definition, 316–17
MSC-S, 24, 26, 437–8
MSCs (Mobile Switching Centres), 12, 24–6, 121–9, 287, 293–4, 337–46, 363–400, 437–8, 442–7
MSISDN, 403–4, 419, 438–47
MSRP (Message Session Relay Protocol), 227–328, 346–9, 407–14
MSS (mobile softswitch), 246–9, 365–7, 419–20, 432, 435, 439–47
MTAS (MMTel AS), 67, 332–6, 359–62
MTP2, 245–9
MTP3, 245–9, 365–78
  see also ISUP; M3UA
MTU (maximum transmission unit), 186–94
Mulligan, Dr C.E.A., 27–48
multi-access, 39
multi-access aspects of the network vision, 2–4, 7, 13, 22, 23, 39–42, 229–32, 262–6, 330–6, 402–4, 449–51
  see also GSM; LTE; WCDMA
multi-device, 39–40
multi-device aspects of the network vision, 3–4, 7, 13, 22–3, 39–42, 132–43, 262–6, 330–6, 346–9, 401–30, 449–51
multi-device environment, 419
multi-operator, 39
multi-operator support, 2–9, 39–42, 329–36, 401–2, 449–51
  see also global interoperability
multi-party communication sessions, 15, 23, 52, 129–43
multi-party conferencing, 15, 23, 52, 129–43, 333–6, 427
multi-service mobiles, 8–9, 13, 17–18, 34–5, 298–300, 329–36, 401–4, 449–51
multi-talented mobile handsets, 5–6
multi-vendor, 39
multi-vendor framework, 7, 23, 39–43, 330–6, 401–4, 449–51
  see also operators

multimedia features, 6, 7, 8–9, 13, 223–328, 329–49, 365, 367–8, 378–400, 425, 433–47
see also audio; images; MMTel; video
multimodal communications, infrastructure requirements, 2
multiplayer games, 21, 51–8, 353–62
multiple early dialogs, 201–5, 372–6, 434
multiple terminals, 22, 137–8, 237–8, 262–9, 363–400, 401–4, 447, 449–51
multiple user identities, 22
multiplexing, 107–19, 121–2, 129–43, 245–9, 293–4, 363–400
see also TDM
music industry, 44–7
music players, 5
Mw reference point, 63, 226–34, 237–8, 242–9, 313–16, 324–8, 393–400, 437–47
see also CSCF; SIP
Mx reference point, 238–40, 242–9, 437–47

N

N-SBG (network session border gateway), 242–9
see also IBCF; TrGW
NAI (Network Access Identifier), 249–69
see also IMPI
namespaces, domain names, 18–19
NAPTR (Naming Authority Pointer Record), 143, 183–94, 240–2, 270–85, 289–90
see also DNS
NAT (Network Address Translation), 10–13, 390–2
NB (narrowband), 25–6, 118, 332–6, 365–400, 434–5
NDAs (non-disclosure agreements), 29–32
Network Address Book, RCS, 402, 405, 406, 417, 420
network border gateways, 10–13, 138–43, 229, 242–9, 313, 363–400
network domains, 142–3, 288–92
see also domains
network layer of OSI reference model, 109–11
see also IP...
network vision, 2–9, 13, 15–26, 39–42, 49–58
network-based multi-party calls, 130–43
network-side architecture, 9–13, 24–6, 49–58, 62–80, 223–328, 331–6, 408–10, 412–14, 416–17, 431–47
see also server-side...
Nextel, 15
nexus of contracts, 34–5, 37–8
NG-IN (Next-Generation Intelligent Network), 68–9, 357–8
NGSI (Next Generation Service Interfaces), 58
nicknamed buddy-invite, 405
NNI (network-to-network interface), 11–13
noise reduction, 333–6
Nokia, 30, 40, 47
nomadicity registration, 137
non-anonymous spammers/hackers, future prospects, 451
non-IMS networks, 363–400

see also CS
nonce, 255
non-transparent subscription data, 235
northbound interface, 335
northbound reference point, MMTel, 294, 330–6, 398–400
note-takers, 5
notepad software, 353
NOTIFY, 162–83, 252–69, 325–8, 340–6, 417, 482, 491
NP (number portability), 306–12
number normalization, 233–4, 286–92
number reformatting by I-CSCF, 291–2
NVAS (Network Value-Added Service), RCS, 422, 429–30
Nyquist theorem, 106

O

O&M, 15
O2, 28
OAuth API, 97–104
OCS (Online Charging System), 359–62
off-the-shelf software technologies, 60–80, 81–104
see also Java Enterprise Edition
offer–answer model, 209–22, 225, 270–85, 312–16, 328, 373–89
see also SDP
offline/online charging models, 357–62
OK, 85–93, 97–104, 151–7, 189–97, 205–9, 211–22, 244–9, 251–69, 270–85, 315–16, 318–20, 343–6, 359–62, 368–78, 380–400, 409–14, 421, 424–5
OLCM, 294–8
oligopolies, 28
OLSM, 294–8
see also S-CSCF
OMA (Open Mobile Alliance), 15, 20–1, 49–58, 336, 337–46, 402–5, 414–30, 450
see also presence
OMA CPM, 416
OMA IM, 415, 420
OMA Presence, 417
OneAPI, 20, 58, 335
one call half SIP-AS interactions, 299–300
One Voice initiative, 7, 431–2
see also VoLTE
online/offline charging models, 357–62
Open APIs, 28–32, 330–6, 402–4
see also APIs
Open IPTV, 349, 449
OpenSource tools, 70, 351–62
operators, 1–26, 28, 34–5, 39–43, 47, 49–58, 243–4, 292–328, 329–36, 351–62, 393–400, 401–30, 431–3, 449–51
see also charging; multi-vendor framework; roaming
OPEX spend, 14, 393–9
OPTIONS, 162–83, 233–4, 252–69, 325–8, 375–6, 408–14, 424, 482
see also RCS-e

Oracle, 32, 35, 37–8, 39
Orange, 423
orchestration service logic, 399–400
   see also SCIM
order problems, digital speech transmission, 108–9, 113–17, 333–6
original dialog identifier, 296
originating identity presentation, 384
originating identity presentation restriction, 384
originating parties, 239–40, 246–9, 252–69, 270–85, 295–300, 314–16, 322–3, 325–8, 334–6, 433–47
Originating_unregistered, 295
OSA, 66–9
OSGi, 13
OSI (Open System Interconnection) reference model, 109–43
outbound routing, 277–85, 313–16

P
'P' prefixes, 481, 486–7
   see also private headers; SIP headers
P-associated-URIs, 258–69
P-called-party-ID, 264–6, 281–5
P-charging-function-addresses, 258–69, 274–85
P-CSCF (proxy CSCF), 63–4, 226–328, 358–62, 435, 440–1
   see also A-SBG; Gm...; IP...; security...; UE; user-to-network...
   concepts, 63, 226–35, 237–49, 251–69, 270–85, 286–92, 296–8, 319–20, 324–8, 358–62, 435, 440–1
   definition, 231–2, 237–8
P-preferred-identity, 237–8, 264–6
packet switched networks, 105
packetization, 111
PAI (P-asserted-identity), 237–8, 274–316, 323, 326–8, 340–6, 381–9
parallel alerting, 22, 302–3, 392–3, 443–4
parameters, 163–83, 242–9
Parlay X, 53, 294, 335, 336, 357, 398–400
   see also SOAP web service tool
participatory value chains, 34–5
passwords, 256, 259
   see also authentication functions
patents, 36
paths, 146–7, 173–83, 191–222, 234, 254–69, 283–5, 293–328
pay-per-use concepts, 352, 354–62
payers, phone calls, 18
payload, 215, 218
payload type field in the RTP header, 114–18, 163, 215–16, 220–2
PayPal, 353, 356
PBX, 13, 374–5
PCM (pulse code modulation), 26, 106–9, 111–43, 220–2, 245–9, 366–400
   see also codecs
PCMA, 116

PCMU, 116
PDG, 252
PDP, 252–69
PEAs (Presence External Agents), 337–46
peer-to-peer communications, 10–13, 14–15, 17–20, 21–3, 37, 56, 128–43, 214–22
'pending' response, presence, 341–6, 417
periodic location update, GSM, 260
person-related information, 'Presentity' concepts, 343–6
person-to-person communications, 10–13, 14–15, 17–20, 21–3, 37, 56, 128–9, 145–7, 401–30
person-to-service/content, 10–13
personal data, 31–5, 344–6
   see also private communications
personalization concepts, 4–6, 35–7
   see also human-centric...; individual
phone calls, 18, 22–3, 49–58, 257–8, 284–92
phone numbers in IMS, 285–92, 297–8, 306–12, 329–49, 367–78
phone-context, 257, 286–92
PIDF (Presence Information Data Format), 337–46
'piggybacking', 22–3
placement considerations, service deployment patterns, 49–58
platforms, 5–9
play_announcement, 89–93
PLMN (Public Land Mobile Network), 3, 243–9, 252–69, 290, 306–12, 363–400
PNAs (Presence Network Agents), 337–46
PoC (push-to-talk over cellular), 15
poker games, 21
'politely blocked' response, presence, 341–6
political landscape, 1–2
polling applications, 17, 425
POST, 100–4
postpaid charging, 351, 357–8
POTS, 26, 332–6
PRACK (provisional acknowledgement), 191–4, 204–9, 211–22, 318–20, 367–78, 414, 482
   see also ACK; provisional responses; RAck
PRD IR.92 *see* VoLTE
preconditions, Mobile IMS, 271–2
preferred long-distance carriers, 247–8
premium rates, 394–5
prepaid charging, 351, 357–8, 390–3, 395–400
presence, 2, 5, 14–15, 17–18, 26, 49–58, 145–222, 230–2, 330, 331, 333–6, 337–46, 349, 404–5, 418–20, 426–30
   see also XCAP
   architecture, 337–8
   authorization functions, 341–6
   concepts, 337–46, 349, 404–5, 418–20, 426–30
   data model, 338, 343–6
   definition, 337–40
   infrastructure requirements, 2, 14–15, 26, 330
   load controls, 342–6

MMTel, 330, 331, 333–6
personal data, 344–6
'Presentity' concepts, 338, 340–6
publishing data, 342–3
SUBSCRIBE, 340–6
watchers, 337–46
presence server (PS), 337–46, 424
presence-style postings, 5, 17–18, 340–6
presentation layer of OSI reference model, 110–11
'Presentity' concepts, 338, 340–6
priority, 202
privacy, 278
'privacy by design', 42, 333–6
privacy issues, 42, 277–85, 333–6, 368–78, 382–9, 403
private communications
    see also personal data
    regulations, 41–2
Private ENUM, 290
private headers, 382–9, 481, 486–7
    see also SIP headers
processing capabilities, 8, 257–69, 302–3, 331–6, 371–6,
    390–400
profile_info, 95–104
promotional bundles, 355–6
proprietary headers, 335–6, 481
protocols, 7–13, 15–26, 49–58, 59–80, 104, 109–43, 145–222,
    223–328, 329–49, 363–400
    see also Diameter...; SIP
provision aspects of RCS, 420–1
provisional responses, 148–57, 183, 185–94, 197–201, 204–9,
    211–22, 318–20, 325, 367–78, 414
    see also PRACK
provisioning relationships, 142–3, 148–57, 259–69
proxies, 69–80, 88–93, 126–9, 138–43, 146–222, 227–328,
    393–400
    see also B2BUAs; ISC; redirect_sip; RFC 3261; stateful...;
    stateless...
proxy call state control function, 237
proxy–registrar relationships, 141–2, 259–69
ProxyBranches, 76–80
proxyTo, 75–80
PS (packet switching), 24–6, 105–43, 245–9, 366–78, 406–14,
    417, 423, 435–47
    see also Internet
PSI (public service identity), 304–12
PSTN (Public Switched Telephony Network), 3, 6–7, 21,
    25–6, 68–9, 128–9, 243–4, 247–9, 269, 276, 290, 306–12,
    329–36, 363–400
PTT (push-to-talk), concepts, 15, 19–20, 21–2, 331–2, 349
PUAs (Presence User Agents), 337–46
Public ENUM, 290
public services, 304–12
    see also PSI
PUBLISH, 53–4, 341–6, 424–5, 482

publish-and-subscribe, presence-style postings, 5, 17–18, 340–6
publishing data, presence, 342–3
PUI (public user identity), 132–43, 179–83, 206–9, 232–40,
    249–69, 270–85, 305
    see also IMPI; IMPU; registration; VoIP
    concepts, 132–7, 249–69, 270–85, 305
    definition, 132–3, 249–50, 305
push-to-talk, 15, 19–20, 21–2, 331–2, 349
    see also half-duplex voice communications
pushRoute, 78–80
PUT, 97–104, 341–6, 417

Q

Q.1902, 121
q parameter, 202
q-value, 348
QCELP, 117
QCI (QoS Class Identifier), 434–5
QoS Class Identifier (QCI), 434
QoS (quality of service), 2, 52–3, 330–6, 434–5, 449
querydb, 89–93

R

R-URI (Request URI), 162–94, 196–7, 199–209, 239–40,
    245–9, 264–9, 279–85, 286–92, 297–8, 301–4, 306–12,
    314–20, 321–3, 383–9, 400
RAck (response acknowledgement), 192–4
    see also PRACK
radio access network, 107, 127, 246
radio capability exchange, 406–14
radios, 5, 23–4, 34–42, 107, 125–9, 406–14, 433–47
RAN (Radio Access Network), 107, 125–9
RCS 1.0, 14–15, 404–17, 420, 424
RCS 2.0, 14–15, 405, 418–21, 424
RCS 3.0, 405, 421–3
RCS 4.0, 404–5, 416, 422–3
RCS (Rich Communication Suite), 14–15, 18–20, 26, 114,
    230–2, 337–46, 401–30, 450
    basics, 402–4
    benefits, 402–4, 425–30, 450
    broadcasting scenario, 429–30
    concepts, 14–15, 26, 230, 401–30, 450
    definition, 14–15, 401–4
    functionality overview, 404–5
    global interoperability, 402–4, 422–3, 426–30
    provision aspects, 420–1
    releases' overview, 404–5
    scenarios, 426–30
    Social Presence Information, 417, 421–2, 424
    subscription-centric social networking, 402–4, 426–30
    value creation/capture, 425–30, 450
RCS Chat, 415–16
RCS-e, 423–30
    see also OPTIONS

RCS file transfer, 416
RCS Image Share, 408–10
RCS provisioning, 420–1
re-registration, 136–7, 260–9
reachability factors, IMS business models, 47–8, 426–30
real-time person-to-person services, 21–3, 56, 105–43
   see also MMTel
real time transport control protocol, 118
real time transport protocol, 111–17
real-time voices, 21–3, 56
realms, 138–43, 369–71
   see also domains
Reason header, 371
Reason phrase, 163
record queries, 183–94
record routing, 74–80, 159–61, 162–94, 203–9, 242–9, 282–5, 297–8, 327–8
Record-Route, 74–80, 162–83, 203–9, 242–9, 282–5, 298, 327–8
redirection responses, 149–57, 374–6, 388–9
redirect_sip, 76–80, 85–93
   see also individual reference points; proxies
redundancy issues, 228–9, 236
REFER, 162–83, 252–69, 293–4, 325–8, 375–6, 392–3, 482, 488
reference architecture, concepts, 2–9
reference points, 63, 226–49, 293–8, 313–20, 324–8, 358–62, 390–400, 437–47
   see also northbound…
references, 453–4
REGISTER, 78–80, 148–57, 162–83, 238–40, 251–69, 273, 281–2, 340–6, 348, 424–5, 482
registrars, 133–43, 146–7, 223, 232–4, 249–69
   see also S-CSCF
   location considerations, 137–41, 223, 251–69
registration, 51–2, 65–9, 78–80, 130–43, 145–222, 226–36, 237–40, 249–69, 273, 279–85, 295–8, 313, 320–3, 330–6, 340–6, 348, 424–5, 430, 482
   see also authentication…; authorization…; IRSs
   call establishment, 133–6, 143, 148–57, 201–9, 232–4, 237–49, 250–69, 270–85
   current registration status of administered VoIP network subscribers, 141–2
   de-registration, 136–7, 260–9, 310–12
   definition, 130–3, 249–50
   expired bindings, 252, 260–1
   initial registration, 133–6, 143, 232–6, 237–40, 246–9, 250–69, 270–85, 295–8, 424–5
   multiple terminals, 137–8, 237–8, 262–9
   nomadicity registration, 137
   re-registration, 136–7, 260–9
   registrar-location considerations, 137–41, 223
   relationships, 141–3, 259–69
   steps, 249–69
   third-party registration, 266–9, 313

unregistered service invocation, 320–3
regular expression, 289
regulations, 31–5, 41–2, 330
re-INVITE, 90–3, 148–57, 214, 376–7
REL, 368–78
REL-5 version, 66, 223–6, 230
REL-6 version, 225, 359
REL-7 version, 225, 431–2
REL-8 version, 225, 432, 433–4, 440
REL-9 version, 225, 433–4, 437–9, 440–1, 446–7
REL-10 version, 225, 440–1
relational databases, 95–104
relationship circles, 5
relationships, registration, 141–3, 259–69
RELEASE, 408–14
RELEASE COMPLETE, 408–14
releases
   see also RCS; REL…
   IMS, 223–6, 230
   RCS overview, 404–5
reliability of SIP requests/responses, 185–94, 204–9, 318–20, 372–6
reliable provisional response, 192, 193, 372–3
Remote Operation, 148
remote source filtering, 373
remote targets, 148, 180–3, 312–28, 376–89, 435
replace_system_headers, 89–93
request failure responses, 149–57, 370–1
request lines, SIP session establishment, 162, 164–83, 270–85
request messages, 148, 162–83, 185–97, 234–49, 252–69, 270–85, 293–328, 367–78, 408–30, 438–47
Request-URIs, 65–9
requirements of communication infrastructures, 2, 8–9, 14
residential opportunities, 13–14
   see also home concepts
residential subscriber service data/logic, legacy SCP, 395–400
resource equipment, user interactions, 316–20
resource-list server (RLS), 338–46, 417
response acknowledgement, 192
response codes, 149–57, 167–8, 185–94, 293–328, 375–6, 481, 488–91
response messages, 162–83, 185–94, 199–209, 257–69, 270–85, 296–328, 341–6, 359–62, 367–78, 424–5
response sequence, 192
REST, 53
REST-based API, 335
REST style APIs, 20–1, 49–58, 67–9, 81–104, 335–6, 357
   see also HTTP
RESTful services, 81–104, 450
retargeting, 264, 294
retransmitted final responses, 188–90
retransmitted request messages, 186–94
reusable services, 59–80, 81–104, 394–400
   see also granularity…; legacy VAS; SOA

revenue-shared financing, 7, 31–2, 351–2, 355–62, 401–2, 451
Rf reference point, 358–62
RFC 1889, 115
RFC 1890, 114–18
RFC 2543, 145, 165, 200, 301–2
RFC 2782, 240–2
RFC 3261, 74–80, 145, 150, 156–7, 162–3, 165, 169–70, 200, 204–5, 225, 232–3, 294, 305, 348, 481–2
   see also proxies
RFC 3262, 191–2, 225
RFC 3263, 167–8, 184, 225
RFC 3264, 210, 225
RFC 3311, 375
RFC 3312, 272
RFC 3323, 225
RFC 3325, 225
RFC 3326, 225, 371
RFC 3404, 240–1
RFC 3428, 324
RFC 3455, 225
RFC 3515, 375
RFC 3550, 114–18
RFC 3551, 114–18, 218
RFC 3588, 225, 254–5, 359
RFC 3605, 219
RFC 3608, 225, 258
RFC 3711, 218
RFC 3761, 289
RFC 3840/41, 18–19, 346–7
RFC 3841, 225
RFC 3966, 132–3, 287, 305
RFC 3986, 132–3
RFC 3994, 326
RFC 4006, 359
RFC 4028, 225
RFC 4240, 225, 317
RFC 4244, 225, 369, 389
RFC 4282, 249–50
RFC 4457, 225
RFC 4566, 210, 216–17, 219
RFC 4733, 375–8
RFC 4825, 346
RFC 4916, 200–1, 386
RFC 4975, 328, 346–7, 409–10
RFC 5002, 225
RFC 5009, 225, 369
RFC 5246, 132–3
RFC 5407, 213, 376
RFC 5502, 225
RFC 5526, 290
RFCs
   see also IETF; RFC...
   list of RFCs, 223–6
RFID, 38

RGW, 13, 242, 243, 278
Rich Call Image, 409
Rich Call Video, 409
rich communication suite, 14–15, 230, 401
Rich Messaging, RCS, 402, 405
ring-back tones, 394–5, 446–7
Ringing, 188–97, 205–9, 211–22, 244–9, 270–85, 315–16, 361–2, 368–78, 380–400, 410–20
ringtones, 244–9, 368–78
RJ-11, 333
RLS XDMS, 339–46, 417
Ro reference point, 359–62
roaming, 12–13, 22, 24–5, 28, 52–3, 234, 238, 243–4, 359, 434–47
   see also operators; Visited IMS network
ROS (Remote Operations), 148
route headers, 164–83, 193–4, 196–7, 203–9, 252–69, 270–85, 296–8, 308–12, 322–3, 324–8, 367–78
routing, 9–14, 17–21, 25–6, 51–8, 63–80, 81–104, 112–43, 145–222, 226–328, 330–6, 359–62, 363–400, 407–30, 431–47
RPID (Rich Presence Information Data), 344–6
RSeq (response sequence), 192–4
RTCP (Real-Time Transport Control Protocol), 118–43, 219–22, 242–9, 278–85, 365–400, 411–14
   see also user plane
   definition, 118
RTCP receiver reports, 411
RTCP sender reports, 411
RTP (Real-Time Transport Protocol), 8, 24–5, 83–93, 94–104, 109–43, 209, 218–22, 227–328, 333–6, 364–400, 410–14
   see also G...; H...; IP; user plane
   definition, 110, 111–14
   fields in the header, 113–17
   header structure, 113–17
RTP session, 411

S

S-CSCF (serving CSCF), 21–2, 62–9, 71–5, 83–93, 94–104, 226–328, 334–6, 337–46, 358–62, 372–400, 446–7
   see also Cx...; Dx...; HSS; ILSM; ISC; MMTel; Mw...; OLSM; SCIM; SIP-ASs
   capabilities, 240
   definition, 232–4, 238–9
   ENUM, 289–90
   messaging, 324–8
   number normalization, 233–4, 286–92
   originating/terminating SIP sessions, 274–85, 341–6
   phone numbers in IMS, 286–92, 297–8, 306–12
   registration, 250–69, 340–6
   residence issues, 234, 251–69
   service-initiated session establishment, 312–16, 321–3
   services provided to SIP, 233–4
   subscriber addresses, 232–4, 250–69
   unregistered service invocation, 320–3

SAA (server assignment answer), 257–69, 321–3

SaaS (Software-as-a-Service), 59

SailFin, 83

sampling, 105–43

   see also codes

SAP, 32, 37–8

SAR (server assignment request), 257–69, 321–3

Save the Children, 357

SAVP, 218–22

SBC (session border controller), 231–2

   see also control plane

SBG (session border gateway), 10–13, 138–43, 229–32

scale levels, 'spaces for value capture', 43–8

SCC AS (Service Centralization and Continuity Application
   Server), 438–47

schemas, PUI, 132–3, 249–69, 286–92

SCIM (Service Capability Interaction Manager), 59–60, 65–9,
   71–5, 279–80, 399–400

   see also S-CSCF

SCPs (service control points), 292–4, 390–2, 394–400

   see also application servers; CS

SCTP, 110–11, 169–83, 242–9, 253–69, 366–78

SCUDIF (Service Change and UDI Fallback), 127–9

SDH (synchronous digital hierarchy), 124–9

SDP (Session Description Protocol), 63–4, 115, 163–83, 209–
   22, 225, 270–85, 294, 315–16, 360–2, 373–89

   see also media...; offer–answer model; SIP-ASs
   definition, 210–12, 217–19, 294

SDP answer, 210

SDP offer, 210

SDS (Service Development Studio), 475–80

security, 15, 41–2, 53–8, 231–2, 237–8, 242, 250–69, 330–6,
   434–5, 449, 451

   see also authentication functions; P-CSCF; TLS

security association, 232

Send_Message component, web-based do-not-disturb use-case
   implementation, 93–104, 471–2

send_response, 96–104

Send_Response component, web-based do-not-disturb use-case
   implementation, 93–104, 473

send_sip_message, 95–104

send_sip_response, 99–104

separation of concerns, 61–4

server assignments, registration, 257–69, 321–3

server failure responses, 149–57

server-side APIs, 49–58, 62–80, 353–4, 357–62

server-side end-point applications, 51–3, 357–62

   see also web...

server transaction, 148

service assertion, concepts, 293–4

service broker, 363

service capability interaction manager, 65–7, 399–400

Service Centralization and Continuity Application Server (SCC
   AS), 438

service chaining, 59, 61–80, 86–93, 157–61, 298–300, 334–6,
   372–6, 436, 445–7

   see also DFC; IFC chaining

Service Change and UDI Fallback, 127

service composition, 60–9, 72, 77–80, 81–104

service deployment patterns, 49–58, 331–6

service designs, 81–104, 436, 445–7, 455–73

service development, 81–104, 401–30, 449–51, 455–73,
   475–80

service layer, 4, 16–20, 62–4, 292–328, 329–49, 358–62

   see also application servers; content; MRFC; SIP-ASs

service providers, 351–62

service-initiated session establishment, 312–16, 321–3

Service-Oriented Architecture, 59, 61, 80

service-related information, 'Presentity' concepts, 343–6

Service-route, 258

Service-Route header, 258–69, 272–85

services

   see also IPTV; MMTel; presence; PTT; Video Share

   chess applications, 346–9

   concepts, 12–26, 154–7, 329–49, 390–400, 449–51

   enablers, 329–49, 404–5, 449–51

   evolution of IMS architecture and services, 10–14, 431–47,
     449–51

   finding the right devices, 346–9

   useful work, 15–21

serving call state control function, 232–4

servlets, 69

session border controller, 231

session border gateway, 231

session description information, 65–9

session description protocol, 163, 210

session gateway, 231

session indicator, 296

session initiation protocol, 118

session layer of OSI reference model, 109–222

   see also RTCP; RTP; SIP

Session Progress, 191–4, 208–9, 211–22, 319–20, 367–78, 414

session signals, 17–21, 148–222

set-top boxes, 13, 449

setRecordRoute, 74–80

SG (session gateway), 231–2, 435–47

   see also user plane

SGSN, 445–7

SGW, 366–78

Sh reference point, 235, 446–7

   see also HSS; SIP-ASs

shareware, 352, 356–62

sharing functions, 14, 22–3, 49–58, 329–36, 339–46, 349,
   405–30, 450

signaling, 9–13, 18–20, 71–80, 81–104, 110–43, 148–222,
   224–328, 330–6, 364–400, 407–14, 434–47

   see also control plane; transport...

signaling gateway, 245

SIM, 7, 10–13, 434–5

SIM card, 10

SIMPLE, 420

simplification value proposition of the network vision, 3–4, 38–42

simulation servers, 62–4

single point of control principle, 68–9

Single Radio Voice Call Continuity (SRVCC), 431, 432, 436–41

SIP, 9–10, 118

SIP addresses, 47–8, 249–69

SIP application server, 305

SIP CEA, 67–9

SIP chain, 71, 76, 86, 92

SIP connectivity, 293

SIP container, 65, 70, 69, 70, 72, 74, 75, 81, 83, 90

SIP extensions, 334–6, 375–6, 481, 486–7, 491

SIP firewalls, 242–9

SIP gateway model, 303

SIP headers, 18–19, 63–80, 95–104, 148–57, 162–83, 193–4, 196–7, 203–9, 215–22, 234, 242–9, 252–69, 272–85, 296–8, 309–12, 323, 325–6, 334–6, 375–6, 382–9, 407–30, 481, 483–7

   see also private headers

   overview, 252, 481, 483–7

SIP INVITE, 409, 410

SIP methods, 148–57, 481, 482

   see also transaction models

SIP profile, 245

SIP proxy, 74, 75–7

SIP response codes, 149–50, 167–8, 185–94, 293–328, 375–6, 481, 488–91

SIP RFC, 74

SIP servlet, 60, 67, 68, 69, 70, 71, 72, 75, 77, 78

SIP Servlet standard in JSR289, 18–20, 50–1, 58, 60, 64, 65, 67–80, 81–104, 455–73

SIP session, 145–7, 409–10

SIP (Session Initiation Protocol), 6, 8, 9–26, 28–9, 47–8, 50–8, 60–80, 83–93, 105–43, 145–222, 225, 228–34, 237–49, 252–69, 285–328, 332–6, 337–46, 363–400, 407–30, 434–47, 449

   see also control plane; Gm…; HTTP; IMS architecture; Mw…; presence; protocols; registration; RFCs

   cancelations, 76–7, 90–3, 152–7, 162–83, 194–7, 315–16, 482, 490

   command sequence, 151, 152–4, 163–83, 192–4

   concepts, 9–13, 28–9, 47–8, 62–4, 145–222, 228–32, 238–40, 252–69, 293–328, 332–6, 337–46, 363–400, 434–47, 449

   definition, 9–10, 12, 16, 62–3, 145

   dialogs, 168, 197–209, 372–6

   early media, 206–9, 372–6

   establishment of SIP sessions, 86, 161–83, 194–7, 209–22, 232–4, 237–49, 252–69, 270–85, 286–92, 293–328, 363–400, 407–30, 438–47

   final responses, 148–57, 174–83, 185–97, 257–69, 293–328, 359–62, 369–78, 424–5

   HTTP syntax, 16

   media sessions, 145–7, 173–83, 191–222, 283–5, 293–328, 367–400

   message structure, 162–83, 215–22

   MGCF signaling capabilities, 371–6

   offer–answer model, 209–22, 225, 270–85, 312–16, 328, 379–89

   phone numbers in IMS, 285–92, 297–8, 306–12, 329–49, 367–78

   protocol mapping, 364–71

   provisional responses, 148–57, 183, 185–94, 197–201, 204–9, 211–22, 318–20, 325, 367–78, 414

   proxies, 69–80, 88–93, 146–7, 157–61, 167–83, 185–97, 201–9, 233–4, 242–3, 247–8, 283–5, 294–5, 300–2, 327–8, 393–400

   reliability of SIP requests/responses, 185–94, 204–9, 318–20, 372–6

   the standard, 145

   subsequent SIP requests, 173–83, 234, 254–69

   termination of SIP sessions, 162–83, 228, 234, 242–3, 270–85, 295–300, 302–3, 321–3, 341–6, 368–78, 409–14, 433–47

   transaction models, 145, 147–57, 186–94, 324–8, 393–400

   transport considerations, 183–94

   version, 162

SIP signaling, 75, 76, 83, 90, 87, 88, 93

SIP TEL URL, 409

SIP trapezoid, 146–7

SIP-ASs (SIP application servers), 19, 21–2, 62–80, 81–104, 226, 233–6, 239–40, 266–9, 292–328, 358–62, 363–400, 409–30

   see also application…; Dh…; I-CSCF; ISC; Ma…; MMTel; S-CSCF; SDP; Sh…; SLF

   concepts, 62–3, 266–9, 292–328, 358–62, 363–400, 409–30

   definition, 62–3, 293–4

   interworking with legacy networks, 363–400

   legacy VAS, 389–400

SIP-I, 244–5, 365–400

SIP/2.0, 162–83, 184–94, 196–209, 239–40, 252–69, 272–85, 287–92, 296–8, 320–3, 325–6, 372–89

SipApplicationRouter, 73–5

SipFactory, 90–3, 455–73

SipServletRequest, 74–80, 89–93, 96–104

SipSession, 87–93, 291–2, 455–73

SipURI, 87–93, 291–2, 455–73

sip.xml, 77–80

Skype, 2, 6, 34, 40

   see also VoIP

SLAs (service level agreements), 18, 28, 29–32

SLEE (Service Logic Execution Environment), 69–75, 276–7, 392–3

SLF (subscriber locator function), 226–328

   see also Dx…; HSS

smart pipes, future prospects, 449
smartphone, 426
SMS, 8–9, 11–12, 20, 26, 324, 345–6, 354, 403–5, 414–16, 420–1, 432–47
   see also messaging…; VoLTE
SMS over LTE, 435
SMSC, 435
SOA (Service-Oriented Architecture), 20–1, 59–80, 81–104
   see also feature composition; granularity…
SOAP web service tool, 53, 67–8, 82–104
   see also Parlay X
social networks, 6, 29–32, 93–104, 402–30
   see also friends…
Social Presence Information, RCS, 417, 421–2, 424
Software-as-a-Service, 59
source code, 84, 88, 89, 95, 455–73
'spaces for value capture', 43–8, 56, 450–1
spammers, future prospects, 451
sponsorship financing, 7, 351–62
SPTs (Service Point Triggers), 65–9, 476–80
SRV, 169–94, 225, 240–2, 249, 270–85
SRV record, 183, 240
SRVCC (Single Radio Voice Call Continuity), 431–47
   see also MMTel
   definition, 431–2, 435, 436–41
SS7 (signaling system no. 7), 27, 245–9, 366–78
SSRC (synchronization source identifiers), 114–19
ST (server transaction), 157–61, 168–83, 197–209, 273–85
standard track SIP headers, overview, 481, 483–5
standard track SIP response codes, overview, 481, 488–90
standardized services, market needs, 1–3, 7–8, 22, 38–48
standards, IMS, 18–19, 127–9, 148, 223–6, 491
startProxy, 76–80
state model, 154–7
stateful SIP proxy, 158–61
stateless SIP proxy, 158
static subscriber data, 141–3
status line, 163
STM, 245–9
STN SR (Session Transfer Number for SRVCC), 438–47
streaming, 22–3, 49–58, 421–30, 450
subdomain-based PSI triggering, 307–12
subdomains, 143
SUBSCRIBE, 156–7, 162–83, 252–69, 340–6, 417, 482, 491
subscriber addresses, S-CSCF, 232–4, 250–69
subscriber locator function, 236
'subscriber pays' models, 7
subscriber profiles, 62–9, 141–3, 226–328, 351–62
   see also HSS
subscription data, HSS, 235
subscription fees, 351, 353–62
   see also charging
subscription-centric social networking, RCS, 402–4, 426–30

subscriptionHandler, 103–4
subsequent SIP requests, 173–83, 234, 254–69
subsequent transactions, 156–7, 173–83, 234, 254–69
supplementary services, 55–8, 121, 128–9, 331–6, 380–9, 390–3, 433–47, 482, 491
   see also mid-point applications
   definition, 55, 331, 333–4, 380–1
supply chain of digital information, 427
supply chains, 32–3, 38, 43–4, 427–8
   see also business models
surfing scenario, RCS, 429
Sv reference point, 437–47
Symbian, 30, 47
Symbian platform, 30, 47, 50
synchronization source, 115
synchronous multiplexing, 107–19, 378–80
system design concepts, 20–1
system integrators, 401

T

T-ADS (terminating access domain selection), 442–7
T-Mobile, 28, 423
T1 communication links, 366–78
tag, 168
'take and transfer picture' button, 22–3
target sets, 205–6, 280–5
   see also contact addresses
TAs (terminal adapters), 23
TCP, 110–43, 169–94, 219–22, 232, 242–9, 253–69, 272–85, 366–400
TDM (time division multiplexing), 107–9, 121–2, 245–9, 293–4, 364–400
   see also CS; user plane
technical specification, 224
Telecom Italia, 423
Telefonica, 423
Telenor, 423
telephony event, 377
telephony softswitch, 246
TeliaSonera, 6
   see also VoIP
terminal adapter, 23
terminal-based multi-party calls, 129–43
Terminating access domain selection (T-ADS), 445
terminating identity presentation, 384, 385
terminating identity presentation restriction, 384, 385
terminating_unregistered, 295, 322
termination of SIP sessions, 162–83, 212–22, 228, 234, 242–3, 270–85, 295–300, 302–3, 321–3, 341–6, 368–78, 409–14, 433–47
   see also BYE
testing, 57–8
text communications, MMTel, 332–6, 346–7

third-party registration, 266–9, 313
time division multiplex, 366
time-to-live, 184
time to market, 1–2
timers, 156–7, 186–94, 358–62, 387–8, 425
TISPAN, 14, 15, 26, 337–46, 381, 431
TLS (transport layer security), 231–2, 237–8
To tag, 168–9, 198–209, 242–9, 252–69, 272–85, 286–92,
    324–8, 382–9
    see also dialogs; INVITE
topology hiding, 242–9, 278–85
transaction branch, 16
transaction costs, 1–2, 28–32, 35, 43–4, 358–62
    see also economics
transaction models, 145, 147–57, 163–83, 192–7, 393–400
    see also response codes; SIP; UAs
transaction state models, 154–7, 186–94, 324–8
transaction trails, 168–83, 184–94
transcoding cases in IM-MGW, 366–7, 376–7
transit transmission networks, 363–400
Transition Gateway, 238
transmission medium requirements, 369
transparent subscription data, 235, 394–400
transport considerations for SIP, 183–94
    see also signaling
transport indicator, 169
transport layer of OSI reference model, 109–43, 163–94,
    196–7, 218–22, 232, 242–9, 252–69, 272–85, 365–78
    see also SCTP; TCP; UDP
transport layer security, 231
travellers, RCs scenarios, 426–7
TRCP, 109–11
TrGW (transition gateway), 226–328
    see also N-SBG; user plane
    concepts, 238, 242–9, 278–85
triggering services, 65–9, 233–4, 295–8, 307–12, 322–3, 476–80
trusted authentication mechanisms, 7, 52–3, 237–8, 269, 451
    see also authentication…
Trying, 186–94, 244–9, 270–85, 325, 361–2, 368–71, 380–400,
    410–14
TS 22.004, 381
TS 22.081, 381
TS 22.082, 381
TS 22.083, 381
TS 22.088, 381
TS 22.091, 381
TS 22.228, 224
TS 22.340, 224
TS 23.002, 224, 228–9
TS 23.003, 224, 250, 383
TS 23.008, 224
TS 23.167, 224
TS 23.172, 127–9

TS 23.216, 438
TS 23.218, 224
TS 23.228, 224, 228–9, 324
    see also IMS
TS 23.237, 441
TS 23.278, 396
TS 23.292, 443
TS 24.008, 407–8, 411–12, 435, 443, 447
TS 24.010, 434, 443–4
TS 24.173, 381, 418–19
TS 24.229, 18–19, 224, 443
TS 24.604, 381, 433
TS 24.605, 434
TS 24.606, 434
TS 24.607, 381, 433
TS 24.608, 381, 433
TS 24.610, 381, 434
TS 24.611, 381, 433–4
TS 24.615, 434
TS 24.623, 434, 443
TS 24.629, 381
TS 29.002, 148
TS 29.078, 148
TS 29.163, 244–5, 368–9, 382–4, 386, 389
TS 29.198, 398
TS 29.199, 398
TS 29.228, 224, 228, 254–5
TS 29.229, 224, 228, 254–5
TS 29.278, 396
TS 29.279, 406
TS 32.240, 357–8
TS 32.260, 224, 357–8, 361
TS 33.203, 256
TSs (technical specifications)
    see also 3GPP; TS…
    list of TSs, 223–6
TSS (telephony softswitch), 246–9, 419–20, 432, 435
TTL (time-to-live), 184–94
Turbo3G, 14
    see also HSPA (High Speed Packet Access)
TV, future prospects, 13, 349, 449
two-tier composition, 65–9

U

UAA (user authorization answer), 254–69
UACs (user agent clients), 77–80, 148–222, 294–5, 300, 303–4,
    312–16, 327–8, 393–400
UAR (user authorization request), 254–69
UAs (user agents), 77–80, 146–222, 252–69, 274–85, 325–8,
    360–2
    see also transaction models; UE
UASs (user agent servers), 77–80, 148–222, 282–5, 294–5,
    300, 303–4, 327–8, 393–400

UDI, 127–9
UDP, 110–17, 119–43, 163–83, 194, 196–7, 218–22, 232, 242–9, 252–69, 272–85, 366–78, 413–14
UE (user equipment), 147, 226–328, 363–400, 417, 435–47
    see also Gm…; IP…; P-CSCF; UAs; user plane
UMA (Unlicensed Mobile Access), 3–4
UML Statecharts language, 70, 81–2
UMTS, 105, 107–9, 136–43, 250–1, 390–2, 431–47
Unauthorized, 251–69
UNI (user-to-network interface), 9–13, 138–43, 179–83, 226–40, 406, 432–3
unified web services, definition, 67–8
Unlicensed Mobile Access, 3
universal service access of IMS, 1–4, 229–32
Unmehopa, Musa, 404
unregistered service invocation, 320–3
UNSUBSCRIBE, definition, 482
UPDATE, 213–14, 367–78, 380–400, 414, 482
UPS business model example, 47–8
URI parameter, 163
URIs (Universal Resource Identifiers), 65–9, 73–80, 87–93, 96–104, 132–43, 162–83, 196–7, 199–209, 233–4, 239–40, 245–69, 274–92, 296–8, 301–12, 324–8, 335, 336, 346, 367–78, 382–9, 410–30, 455–73
    see also IMPU; PSI
URNs, 18–20, 409–10
use-case implementation, 81–104, 455–73
    see also virtual call center…; web-based do-not-disturb…
User agent, 147
user agent client, 77, 148, 304
user agent server, 77, 148, 303
user equipment, 147, 229
user interactions, 316–20
    see also announcement…; MRFC
user plane, 118–29, 209–22, 223, 226–32, 238, 278–85, 293–8, 302–4, 316–20, 365–400
    see also media plane; RTCP; RTP; SG; TDM; UE
user registration see registration
user relationships, 122–43
user self-care, legacy SCP, 395–400
user–registrar relationships, 141–2, 259–69
user service information, 369
user-to-network proxy, 9–13, 138–43, 179–83, 226–40
    see also P-CSCF
user = phone parameter, 291–2, 367–71, 383–9
USIM (universal SIM), 434–5
Ut, 331
Ut reference point, 434
Ut/XCAP connection, 331–6, 434–5, 443–7
UTF, 78–80, 326
UTRAN, 441–7

V

value added services, 389

value chains, 1–2, 26, 27, 32–48, 351–62, 425–30
value creation/capture, 33–48, 56, 425–30, 431–47, 449–51
value-added services, 21–2, 59, 62–3, 65–9, 293–328, 333–6, 363–4, 389–400, 405, 421–2, 425–30, 431–47, 450–1
VAS (value-added services), 21–2, 59, 62–3, 65–9, 293–328, 333–6, 363–4, 389–400, 405, 421–2, 425–30, 431–47, 450–1
    see also legacy VAS
vCard, 417
VCC see virtual call center use-case implementation
VDVI, 117
    see also DVI4
vehicle automation prospects, 450
vendors, 7, 23, 39–43, 47, 330–6, 401–4, 431–3
very-large-scale implementation, 7
Via header, 163, 169–83, 184–94, 196–7, 203–9, 242–9, 252–69, 272–85, 324–8
video communications, 2, 17–18, 22–3, 44–7, 49–58, 64, 215–22, 246–9, 283–5, 324, 329–36, 349, 365, 367–8, 378–400, 402, 403–14, 418–30, 450
video conferences, 52–3, 333–6, 427, 434
video gateway, 246, 378
Video Share, 349, 407, 410–14, 418–19, 421–2, 425–30, 450
ViG (video gateway), 246–9
virtual call center use-case implementation, 81–93, 455–71
virtual private networks, 55–6
vision, network vision, 2–9, 13, 39–42
Visited IMS network, 237, 238, 254–69
    see also roaming
visited location register, 391
visited MSC, 391
visited networks, MMTel, 24–6
visual separators ('-' character), 286
VLR (visited location register), 391–2
VMSC (visited MSC), 391–2
Vodafone, 28, 423
Voice Call Continuity, 225
voice communications, 2, 6–8, 14–26, 41–2, 64, 215–22, 247–9, 283–5, 324, 328, 331–6, 363–400, 403–30, 431–47
    see also VoLTE
voice memo recorders, 5
voicemail, 321–3, 333–6
VoiceXML, 83–93, 316–20
VoIP (Voice-over-IP), 2, 6–8, 14, 25–6, 105–43, 166–83, 184–94, 202–9, 290, 405, 418–19, 423
    see also digital speech transmission; MMTel; Skype; TeliaSonera; Vonage
    basics, 105–30
    concepts, 6–8, 14, 105–43, 184–94, 290, 405, 418–19, 423
    current registration status of administered VoIP network subscribers, 141–2
    de-registration, 136–7
    default listeners, 123–4
    definition, 105–6

existing IP networks, 14
multi-party communication sessions, 129–43
nomadicity registration, 137
PUI, 132–43
re-registration, 136–7
registrar-location considerations, 137–41
registration, 130–43, 202–9
user/media plane separation, 126–9
VoLTE (voice over LTE), 330–1, 418–19, 423, 425, 429,
    431–47, 449
    concepts, 431–47, 449
    definition, 431–5
Vonage, 6
    see also VoIP
voucher upgrades, legacy SCP, 395–400
VPNs, 21–2, 333–6, 390–3, 428–9

W

W3C (World Wide Web Consortium), 319–20
WAC (Wholesale Application Community), 34, 353, 450–1
WalMart, 38
WANs (wide area networks), 230–2
watchers, presence, 337–46
WCDMA, 3–4, 12–13, 34, 39–40, 332–6, 403–4, 431–2, 442–7
Web 2.0, 20–1, 68
web client-side end-point applications, 53–5, 336, 450
    see also client-side…
web server-side end-point applications, 20–1, 49–58, 67–9,
    81–104, 294, 335–6, 357
    see also server-side…
web services, 19–21, 402–4, 426
web-based do-not-disturb use-case implementation, 81–2,
    93–104, 471–3
white-label products, 355–6
    see also charging
white-list, 331

Wholesale Application Community, 353
wide area network, 230
wideband AMR, 332–6, 434–5
WiFi, 3–4
Wikipedia, 36
wildcard ENUM, 305–6
wildcard PSIs, 305–12
Windows Live, 2
wireless access, 3–26, 62–3, 111, 132–43, 223
Wireshark, 219–22
WLANs (wireless LANs), 3–4, 13–14, 22, 62–3, 111, 132–43,
    223, 230–2, 251–69
World of Warcraft, 352
WS CEA, 67–9

X

X-Lite, 219–22, 252–69, 326
X.880, 148
XCAP (XML Configuration Access Protocol), 331–6, 337–46,
    417, 434–5, 443–7
XDM agents, 338–46
XDMC, 338–46
XDMSA, 417
XDMSs, 338–46, 417
xDSL, 6, 26
XHR, 102–4
XML, 77–80, 83–94, 96–104, 316–20, 326–8, 331–6, 337–46,
    417, 434–5, 443–7
    see also AJAX; JAX-RS API; VoiceXML; XCAP
XML Configuration Access Protocol, 337
XML web services, 83
XPath, 346

Y

Yahoo Calendar, 96
YDMS, 417

Printed in the United States
by Bookmasters

Printed in the United States
By Bookmasters